High-Efficient Exploration and Development of Oil & Gas from Ocean—2nd Edition

High-Efficient Exploration and Development of Oil & Gas from Ocean—2nd Edition

Guest Editors

Mianmo Meng
Wenming Ji
Guodong Cui

MDPI

Basel • Beijing • Wuhan • Barcelona • Belgrade • Novi Sad • Cluj • Manchester

Guest Editors
Mianmo Meng
Hubei Key Laboratory of Marine Geological Resources
China University of Geosciences
Wuhan
China

Wenming Ji
School of Geosciences
China University of Petroleum (East China)
Qingdao
China

Guodong Cui
Faculty of Engineering
China University of Geosciences
Wuhan
China

Editorial Office
MDPI AG
Grosspeteranlage 5
4052 Basel, Switzerland

This is a reprint of the Special Issue, published open access by the journal *Journal of Marine Science and Engineering* (ISSN 2077-1312), freely accessible at: https://www.mdpi.com/journal/jmse/special_issues/56857612K5.

For citation purposes, cite each article independently as indicated on the article page online and as indicated below:

Lastname, A.A.; Lastname, B.B. Article Title. *Journal Name* **Year**, *Volume Number*, Page Range.

ISBN 978-3-7258-3671-0 (Hbk)
ISBN 978-3-7258-3672-7 (PDF)
https://doi.org/10.3390/books978-3-7258-3672-7

Contents

Journal of
Marine Science and Engineering

MDPI

ditorial

Iigh-Efficient Exploration and Development of Oil & Gas from Ocean—2nd Edition

Iianmo Meng [1,2,*], Wenming Ji [3,*] and Guodong Cui [4,*]

1 Hubei Key Laboratory of Marine Geological Resources, China University of Geosciences, Wuhan 430074, China
2 College of Marine Science and Technology, China University of Geosciences, Wuhan 430074, China
3 School of Geosciences, China University of Petroleum (East China), Qingdao 266580, China
4 Faculty of Engineering, China University of Geosciences, Wuhan 430074, China
* Correspondence: mengmianmo@cug.edu.cn (M.M.); jiwenming@upc.edu.cn (W.J.); cuiguodong@cug.edu.cn (G.C.)

Nowadays, given the lack of resources that is threatening the globe, the exploitation and exploration of oil/gas, especially unconventional reservoirs, has become a priority [1–3]. The ocean is important to human life and research focused on related scientific issues [4,5]. For example, it has abundant unconventional resources such as shale and tight oil/gas, where subsurface formations can be the crucial solution to resource supply challenges [6,7].

Under such circumstances, this Special Issue concentrates on the exploration and development of oil/gas with high efficiency in the ocean. Some critical problems regarding the marine geometry and oil/gas exploitation exist currently. First, the evaluation and characterization of the targeted layers has lacked sufficient and comprehensive research, particularly in quantifying rock–physical property variations and clarifying reservoir formation mechanisms. Second, critical parameters such as permeability and mineral composition and diagenetic alteration patterns have not been fully analyzed due to limited well-logging data and 3D seismic interpretation methods, as well as less relevant models matching with them. Third, hydraulic fracturing has achieved enhanced oil/gas recovery with the development of engineering, but the mechanism of fracture control is as yet unclear. Last but not least, the research on the heterogeneity of lithofacies, nano-scale pore structures, and pore connectivity has shown certain inadequacies. To solve these problems, multiple studies were conducted for this Special Issue. The main contributions of each article are included below.

eceived: 20 February 2025
ccepted: 24 February 2025
ublished: 17 March 2025

itation: Meng, M.; Ji, W.; Cui, G. Iigh-Efficient Exploration and Development of Oil & Gas from Ocean—2nd Edition. *J. Mar. Sci. Eng.* 025, *13*, 588. https://doi.org/).3390/jmse13030588

Contribution 1 discovered that Early Jurassic gypsum in East African basins controlled gas reservoirs because of the delaying of Lower Jurassic source rock maturation by gypsum. Post-183 Ma graben-deposited thick gypsum enhanced sealing, enabling conventional gas accumulation and shale gas potential in non-faulted areas.

Contribution 2 applied the combined finite–discrete element method (FDEM) to simulate laboratory earthquakes, promoting earthquake precursor identification and connecting laboratory experiments to natural seismic prediction. Using Light Gradient Boosting Machine (LightGBM) models and SHapley Additive exPlanations (SHAP) analysis, friction coefficients were predicted ($R^2 = 0.94$) with optimized critical features.

Contribution 3 revealed that mass transport deposits (MTDs) in the Qiongdongnan Basin uplifted the gas hydrate stability zone (GHSZ), accelerating hydrate dissociation. The resulting gas seepage formed spiny (moderate-speed) and domed (low-speed) seamounts, establishing a model combining MTDs, hydrate dynamics, and seabed geomorphology.

Contribution 4 defined the lower limits of porosity and permeability for Qiongdongnan Basin gas reservoirs in deltas, submarine canyons, and fans. Sedimentary facies, grain size, and transport distance controlled these thresholds, with Red River-sourced sandston achieving superior porosity and permeability through CO_2 and organic acid-induce feldspar dissolution.

Contribution 5 discovered that the Huaiyuan Movement formed a regional unco formity in Jiyang Depression's lower Paleozoic, with karst breccias and vertical zonatio (vadose/underflow zones). Reservoir development in the Fengshan–Yeli–Liangjiasha Formations was controlled by sedimentary microfacies, dolomitization, and karstific tion. High-energy beach granular dolomite exhibited optimal cavity-type reservoirs alon the unconformity.

Contribution 6 investigated cement sheath integrity failure during multi-stage fra turing in offshore tight oil wells. Triaxial cyclic loading tests (TCLTs) revealed cumulativ plastic deformation, which reduced compressive strength and elastic modulus, and i creased permeability. Micro-annuli formed during cyclic loading and matched with 3 finite element simulations, explaining failure mechanisms, thus aiding cement sheat design for offshore tight oil reservoirs.

Contribution 7 simulated the shallow marine shelf sedimentation in West Siberia Jurassic strata, which was influenced by "two depressions, one uplift" tectonics and tide Tidal amplitude positively affected sand body parameters, while higher initial water leve limited tidal bars.

Contribution 8 established a hybrid-driven method combining physics-based mode and machine learning to predict fluid sampling time in offshore formation testing, where digital twin generated 6000 simulated cases with physics-enhanced feature correlation. Th machine learning achieved 92% accuracy, optimized to 95%, enabling efficient real-tim prediction for oil/gas development.

Contribution 9 focused on Ek2 shales in Cangdong Sag (Bohai Bay Basin) and a alyzed 50 samples to reveal adsorbed and free oil contents, which primarily resided i >20 nm pores, controlled by organic matter abundance and thermal maturity not por volume. High-S1 organic-rich shales were optimal exploration targets, guiding Ek2 sha development prioritization.

Contribution 10 proposed the Salmon Salar Optimization algorithm, mimickin salmon social behavior for high-precision searches. It performed better than benchmark and resolved deep-sea probe design constraints, proving effective for complex engineerin systems, thus advancing resource development efficiency.

Contribution 11 improved fracture permeability assessment for South China Se granite reservoirs by integrating Darcy–Poiseuille laws with fracture parameters (angl aperture, tortuosity). Confirmed by core experiment data, the model effectively evaluate single fractures but required refinement for intersecting cases, advancing permeabilit prediction in complex unconventional reservoirs.

Contribution 12 evaluated Eocene–Oligocene source rock potential in China's norther Yinggehai Basin, where tectonic activity drove east-to-south sedimentary trough migratio favoring lacustrine, shallow marine, and deltaic facies. The Early Oligocene's warm–humi climate, reducing conditions, and high productivity optimized organic matter preservatio The established sedimentary models showed significant hydrocarbon exploration potentia in this region.

Contribution 13 established a rapid spontaneous imbibition method using thin shal samples to assess pore connectivity and ultimate imbibed porosity. The results from th Qingshankou Formation (Fm), Shahejie Fm, and Liushagang Fm showed distinct average of connectivity and porosity. By standardizing sample dimensions and reducing testin time, it enhanced accuracy in assessing pore connectivity and ultimate imbibed porosit enabling efficient sweet-spot evaluation within a day.

Acknowledgments: As Guest Editors of the Special Issue "High-Efficient Exploration and Development of Oil & Gas from Ocean—2nd Edition", we wish to extend our sincere gratitude to all the authors whose valuable contributions made the publication of this Special Issue possible.

Conflicts of Interest: The authors declare no conflicts of interest.

List of Contributions

1. Cai, J.; Guo, R. Early Jurassic Gypsum within Eastern African Continental Marginal Basins and Its Significance for Gas Play. *J. Mar. Sci. Eng.* **2024**, *12*, 93. https://doi.org/10.3390/jmse12010093.
2. Huang, W.; Gao, K.; Feng, Y. Predicting Stick-Slips in Sheared Granular Fault Using Machine Learning Optimized Dense Fault Dynamics Data. *J. Mar. Sci. Eng.* **2024**, *12*, 246. https://doi.org/10.3390/jmse12020246.
3. Liu, P.; Zhang, W.; Mao, S.; Su, P.; Chen, H.; Hu, L. Study on the Mechanism of Natural Gas Hydrate Decomposition and Seabed Seepage Triggered by Mass Transport Deposits. *J. Mar. Sci. Eng.* **2024**, *12*, 646. https://doi.org/10.3390/jmse12040646.
4. Li, C.; Guo, S.; Zhou, Q.; Xu, C.; Chen, G. Lower Limits of Petrophysical Properties Allowing Natural Gas Accumulation in Marine Sandstones: An Example from the Qiongdongnan Basin, Northern South China Sea. *J. Mar. Sci. Eng.* **2024**, *12*, 735. https://doi.org/10.3390/jmse12050735.
5. Liu, R.; Wang, G.; Wang, Y.; Hao, X.; Qin, F.; Fang, X.; Meng, W.; Liu, G. Characteristics and Reservoir Development Model of the Unconformity Caused by Huaiyuan Movement in Bohai Bay Basin, China: A Case Study of Chengdao-Zhuanghai Buried Hill in Jiyang Depression. *J. Mar. Sci. Eng.* **2024**, *12*, 804. https://doi.org/10.3390/jmse12050804.
6. Wang, Y.; Feng, Y. Experimental and Numerical Simulation Investigation of Cement Sheath Integrity during Multi-Stage Fracturing in Offshore Tight Oil Reservoir. *J. Mar. Sci. Eng.* **2024**, *12*, 814. https://doi.org/10.3390/jmse12050814.
7. Tang, M.; Xiong, S.; Zhang, Q.; Hong, R.; Peng, C.; Xie, R. A Novel Method for Analyzing Sandbar Distribution in Shelf-Type Tidal Deltas Using Sediment Dynamic Simulation. *J. Mar. Sci. Eng.* **2024**, *12*, 1102. https://doi.org/10.3390/jmse12071102.
8. Nie, Y.; Li, C.; Zhou, Y.; Yu, Q.; Zuo, Y.; Meng, Y.; Xian, C. Intelligent Prediction of Sampling Time for Offshore Formation Testing Based on Hybrid-Driven Methods. *J. Mar. Sci. Eng.* **2024**, *12*, 1348. https://doi.org/10.3390/jmse12081348.
9. Ma, B.; Hu, Q.; Pu, X.; Yang, S.; Wang, X.; Han, W.; Wen, J. Occurrence Mechanism and Controlling Factors of Shale Oil from the Paleogene Kongdian Formation in Cangdong Sag, Bohai Bay Basin, East China. *J. Mar. Sci. Eng.* **2024**, *12*, 1557. https://doi.org/10.3390/jmse12091557.
10. Guo, J.; Yan, Z.; Sato, Y.; Zuo, Q. Salmon Salar Optimization: A Novel Natural Inspired Metaheuristic Method for Deep-Sea Probe Design for Unconventional Subsea Oil Wells. *J. Mar. Sci. Eng.* **2024**, *12*, 1802. https://doi.org/10.3390/jmse12101802.
11. Guo, J.; Gu, B.; Lv, H.; Zhu, Z.; Zhang, Z. Improved Fracture Permeability Evaluation Model for Granite Reservoirs in Marine Environments: A Case Study from the South China Sea. *J. Mar. Sci. Eng.* **2024**, *12*, 1868. https://doi.org/10.3390/jmse12101868.
12. Pei, J.; Hu, G.; Huo, Z.; Chen, Z.; Chen, Y.; Fu, X.; Wang, W.; Liu, H.; Wang, Y.; Luo, J.; et al. Geological Conditions and Sedimentary Models of Oligocene and Eocene Effective Source Rocks in the Northern Yinggehai Basin. *J. Mar. Sci. Eng.* **2025**, *13*, 100. https://doi.org/10.3390/jmse13010100.
13. Hong, Z.; Meng, M.; Deng, K.; Bao, J.; Wang, Q.; Liu, X. A Quick Method for Appraising Pore Connectivity and Ultimate Imbibed Porosity in Shale Reservoirs. *J. Mar. Sci. Eng.* **2025**, *13*, 174. https://doi.org/10.3390/jmse13010174.

References

1. Luo, Q.; Goodarzi, F.; Zhong, N.; Qiu, N.; Wang, X.; Suchý, V.; Khan, I.; Zheng, X.; Liu, B.; Ardakani, O.H.; et al. Dispersed organic matter from pre-Devonian marine shales: A review on its composition, origin, evolution, and potential for hydrocarbon prospecting. *Earth Sci. Rev.* **2025**, *261*, 105027. [CrossRef]
2. Meng, M.; Hu, Q.; Wang, Q.; Hong, Z.; Zhang, L. Effect of initial water saturation and water film on imbibition behavior in tight reservoirs using nuclear magnetic resonance technique. *Phys. Fluids* **2024**, *36*, 056603. [CrossRef]

3. Wang, Q.; Li, Y.; Utley, J.E.P.; Gardner, J.; Liu, B.; Hu, J.; Shao, L.; Wang, X.; Gao, F.; Liu, D.; et al. Terrestrial dominance of organi carbon in an Early Cretaceous syn-rift lake and its correlation with depositional sequences and paleoclimate. *Sediment. Geol.* **202** *455*, 106472. [CrossRef]
4. Sun, Q.; Wang, Q.; Shi, F.; Alves, T.; Gao, S.; Xie, X.; Wu, S.; Li, J. Runup of landslide-generated tsunamis controlled b paleogeography and sea-level change. *Commun. Earth Environ.* **2022**, *3*, 244. [CrossRef]
5. Wang, X.; Kneller, B.; Sun, Q. Sediment waves control origins of submarine canyons. *Geology* **2023**, *51*, 310–314. [CrossRef]
6. Luo, Q.; Fariborz, G.; Zhong, N.; Wang, Y.; Qiu, N.; Skovsted, C.B.; Suchý, V.; Hemmingsen Schovsbo, N.; Morga, R.; Xu, Y.; et a Graptolites as fossil geo-thermometers and source material of hydrocarbons: An overview of four decades of progress. *Earth Sc Rev.* **2020**, *200*, 103000. [CrossRef]
7. Wang, D.; Dong, Y.; Sun, D.; Yu, B. A three-dimensional numerical study of hydraulic fracturing with degradable divertin materials via CZM-based FEM. *Eng. Fract. Mech.* **2020**, *237*, 107251. [CrossRef]

Journal of
Marine Science and Engineering

MDPI

rticle

:arly Jurassic Gypsum within Eastern African Continental ʌarginal Basins and Its Significance for Gas Play

ın Cai [1,2,3,*] and Rong Guo [1,2]

1 Cooperative Innovation Center of Unconventional Oil and Gas, Yangtze University (Ministry of Education & Hubei Province), Wuhan 430100, China; grcjdx@163.com

2 Key Laboratory of Exploration Technologies for Oil and Gas Resources, Ministry of Education, Yangtze University, Wuhan 430100, China

3 Key Laboratory of Polar Geology and Marine Mineral Resources (China University of Geosciences, Beijing), Ministry of Education, Beijing 100083, China

* Correspondence: caijun.jz@163.com

Abstract: Although the eastern African continental marginal basins have discovered giant gas fields, it is not clear whether the Early Jurassic gypsum associated with the main source rocks controls the gas play. In this paper, we use well logging, seismic reflection, and organic geochemistry data to synthesize the distribution and origin of the gypsum deposits and their control over the gas play. The results show that from 201 Ma to 183 Ma, a thick suite of mudstone with thin-layered gypsum began to form in the southern gulf-like sea due to water stratification. With the sea level falling since 183 Ma, a thick suite of gypsum was deposited in some grabens, where the sedimentary environment changed to lagoons. The gypsum increases the maturity threshold depth of Lower Jurassic source rocks and delays the peak time of gas generation, resulting in the ongoing filling of East African natural gas reservoirs and the formation of giant reserves. In addition, the gypsum not only controls the distribution of conventional gas but also makes the Lower Jurassic shale, where the faults are not developed, an important place for shale gas exploration in the future due to the good sealing properties of the gypsum.

Keywords: gypsum; depositional model; shale gas; conventional gas; eastern African continental margin

itation: Cai, J.; Guo, R. Early Jurassic ypsum within Eastern African ontinental Marginal Basins and Its ıgnificance for Gas Play. *J. Mar. Sci. ng.* **2024**, *12*, 93. https://doi.org/).3390/jmse12010093

cademic Editor: George ontakiotis

eceived: 10 November 2023
evised: 22 December 2023
ccepted: 22 December 2023
ublished: 2 January 2024

1. Introduction

Gypsum deposits have always been one of the hot research topics due to their usage in many industries (e.g., chloralkali industry, fertilizer industry) [1]. For a few decades, gypsum has gained much attention in the petroleum industry because of its controls on oil and gas generation and accumulations worldwide [2]. The Bohai Bay Basin [3,4], the Ordos Basin [5,6], the Jianghan Basin [7], the Songliao Basin [8,9], the Lower Congo Basin [10], the Persian Gulf [11,12], the Pricaspian [13], the African passive continental margin [14,15], the Brazilian deep-water basins [16,17], and the Tarim Basin [18], all have significant petroleum plays.

Despite the multiple achievements mentioned above, there have been relatively few publications on the gypsum deposited in the eastern African continental marginal basins, which has been considered to be the world's largest increase in natural gas reserves in the last decade. It is currently known that the gypsum is associated with the main source rocks of the eastern African continental marginal basins, which are considered to be the Early Jurassic mudstone [19]. Unfortunately, the distribution range of the gypsum still needs to be solved, which further limits the understanding of controlling its effects on gas play. In this paper, we attempt to clarify the distribution range, establish an appropriate sedimentary model for the Early Jurassic gypsum deposits, and finally discuss its controlling effects on gas play.

2. Geological Background

The eastern African continental margin is composed of several basins stretching fro north to south, including the Somalia Basin, Lamu Basin, Tanzania Basin, Ruvuma Basi Mozambique Basin, Mulongdawa Basin, and Mazenga Basin, with a total sedimentary are of about 3.7×10^6 km^2 (Figure 1). Our research area focuses on the Rovuma, Tanzania, an Lamu Basins due to abundant seismic data availability.

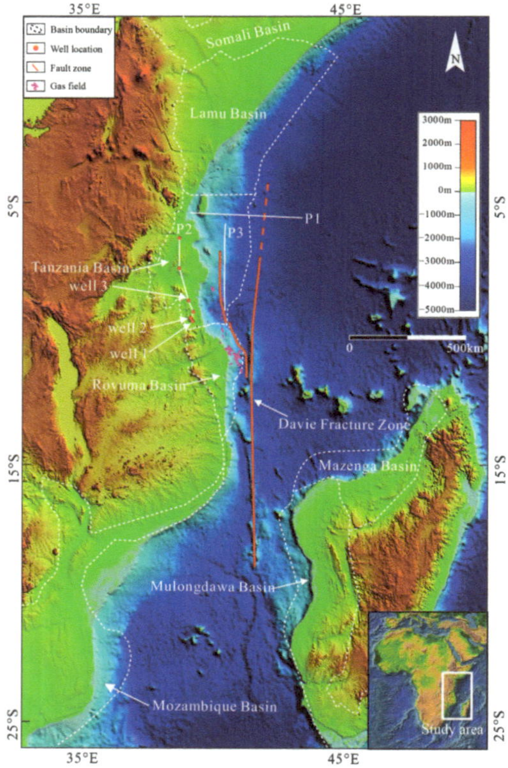

Figure 1. The map shows the locations of the Somalia Basin, Lamu Basin, Tanzania Basin, Ruvum Basin, Mozambique Basin, Mulongdawa Basin, and Mazenga Basin. The basin locations were referre to by [20].

The eastern African continental marginal basins are developed in the basement c the Carboniferous crystalline rock, with a basinal depositional thickness of more tha 10 km. Its tectonic evolution can be divided into three stages: the intracontinental rift stag the intercontinental rift stage, and the passive continental margin stage. The first stag initiates from the late Carboniferous to the Triassic. Along with the gradual formatio of the Gondwana continent, a strong "mantle plume" rose beneath the southeast of th continent, causing regional crustal uplifting, faulting, and volcanic activity, forming widely distributed intracontinental basin where a set of fluvial, lacustrine, and marsh facies were formed [21]. Since the strata deposited during the intercontinental riftin are best preserved in the Kaloo basin within South Africa, this period is also known a the Kaloo Rift Valley and is represented by the Kaloo Group deposits as a tectonic ever affecting the entire Gondwana continent [22]. The profile reveals that the structural style c this period is dominated by a wide range of horsts, grabens, and half-grabens (Figure 2a

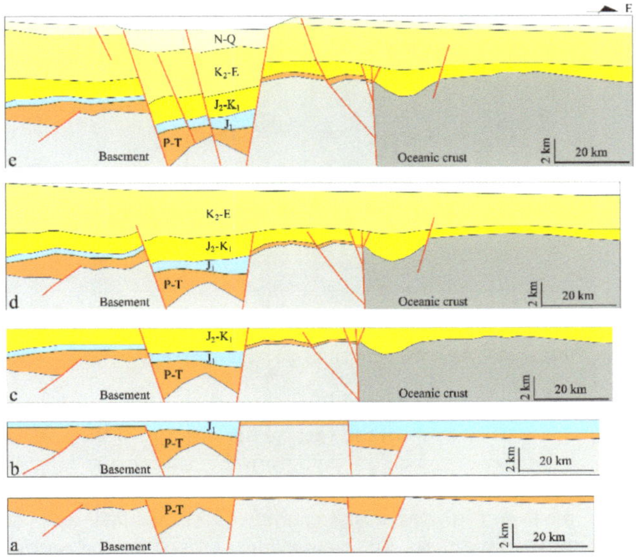

Figure 2. (**a**)The diagram presents the tectonic structure and sedimentary fill of the eastern African continental margin at the intracontinental rift stage; (**b**) The diagram presents the tectonic structure and sedimentary fill at the intercontinental rift stage; (**c**) The diagram presents the tectonic structure and sedimentary fill from the Middle Jurassic to the Early Jurassic; (**d**) The diagram presents the tectonic structure and sedimentary fill from the Upper Jurassic to the Paleogene; (**e**) The diagram presents the tectonic structure and sedimentary fill from the Neogene to the present. Abbreviations: P-T, Permian to Tiassic; J_1: Lower Jurassic; J_2-K_1: Middle Jurassic to Lower Cretaceous; K_2-E: Upper Cretaceous to Paleogene; N-Q: Neogene to Quaternary.

The second stage begins with the Early Jurassic. The Gondwana continent began to break up into several different blocks from northwest to southeast, where seafloor spreading and drifting are limited to the northeast, resulting in large-scale rifting and subsidence occurring in present-day coastal areas of Somalia, Kenya, Tanzania, and Madagascar, causing seawater flooding from the northeast and the formation of narrow bays similar to the Red Sea [23]. The seismic profiles reveal that the structural style is dominated by strata filling, but a small number of preexisting faults have been reactivated under the regional extensional stress field, resulting in graben formation (Figure 2a,b), showing that this period is dominated by thermal subsidence with limited normal faulting over most of the basins.

In the middle Jurassic, the Gondwana continent was split into two continents, the East and the West. The entire East African Sea began to prograde into the passive continental margin stage [24], which could be further divided into three secondary stages according to the sub-tectonic events. From the middle Jurassic to the early Cretaceous, the continents of Madagascar and East Gondwana broke away from the African continent and drifted southward along the Davie fracture zone characterized by dextral strike-slipping, resulting in a dextral shear continental margin on the southern margin of East Africa [25]. From the late Cretaceous to the Eocene, the tectonic activity of the East African continent margin was weak (Figure 2d), with a set of shallow sea and delta deposits [26]. From the Oligocene, due to the activity of the Afar mantle plume in northern East Africa, with accelerated subsidence in the central part of the eastern African continental margin and relative uplift of the eastern Davie fracture zone (Figures 2e and 3), the East African Rift began to form [27].

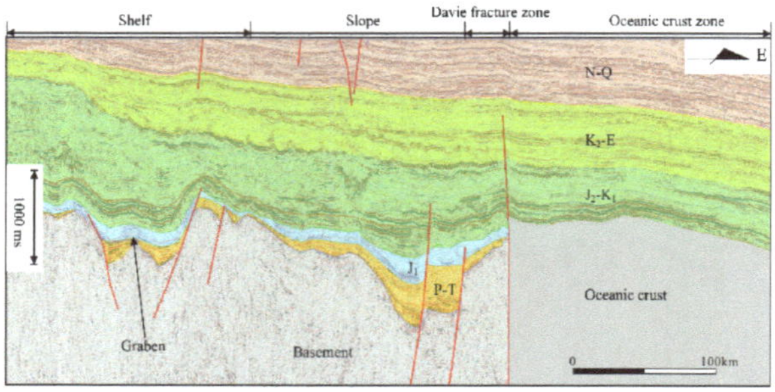

Figure 3. The interpreted seismic profile reveals the tectonic framework of the Lower Jurassic within the eastern African continental margin. The red line presents the fault. See P1 for the location.

3. Materials and Methods

The analysis of the vertical distribution of the Early Jurassic gypsum deposits were primarily based on the data from the drilling and logging of six wells. Considering the heterogeneous distribution of the rock and the few drillings in the deep-water encounter of the Lower Jurassic, a large amount of seismic data were used to classify seismic sequences and seismic facies. The main method is to obtain the petrophysical parameters of the Lower Jurassic, including the densities, interval velocities, and wave impedance characteristics of mudstone, sandstone, and gypsum deposits, based on logging data, which could better utilize seismic reflections in seismic profiles across key wells to establish connections with lithology. Then, in areas without wells, the planar distribution of the gypsum can be inferred based on similar seismic facies. Notably, during our interpretation of the seismic data, we reviewed strata and structural models from unpublished research reports and published literature to obtain biostratigraphic and chemostratigraphic constraints on the age of seismic stratigraphy from key selected wells. In addition, in discussing the sedimentary environment, more than 100 mudstone samples were collected to measure the values of the total organic carbon (TOC) and hydrogen index (HI), and the detailed analytical procedure and accuracy follow [28].

4. Results

4.1. Vertical Distribution of the Gypsum Deposits

Drilling records show that the Lower Jurassic succession could be divided into the lower and upper parts based on the difference in lithology. The lower part is dominated by a thick suite of mudstone with thin-layered gypsum. The upper part is dominated by a thick suite of gypsum deposits with thin layers of mudstone. For example, the lower part of the Lower Jurassic succession of well 1, located within the southern Tanzanian Basin toward land, are mainly mudstone with scarce gypsum layers based on the data from drilling and logging (Figure 4). The gamma-ray curve is finger-shaped, while the acoustic curve is at middle amplitude, which could be considered to be the shallow marine deposits. The gamma-ray curve of the upper part of the Lower Jurassic succession is in the shape of a gearbox and low-amplitude funnel, which is significantly different from the lower part of the Jurassic succession. Combined with data from drilling and logging, the lithology within this part are dominated by a thick suite of gypsum with thin layers of mudstone. The gypsum percentage increases, while the mudstone component decreases. Well 2 has similar characteristics to well 1 (Figure 4), the Lower Jurassic of which can also be divided in two parts vertically, the lower and the upper part of the Lower Jurassic sequence. The lower part of the Lower Jurassic succession is shallow marine-thick siliciclastic mudstone with thin layers of gypsum, while the upper part of the Lower Jurassic succession is presented by thick gypsum deposits. It should be noted that red gypsum deposits appear in the upper part of the Lower Jurassic succession.

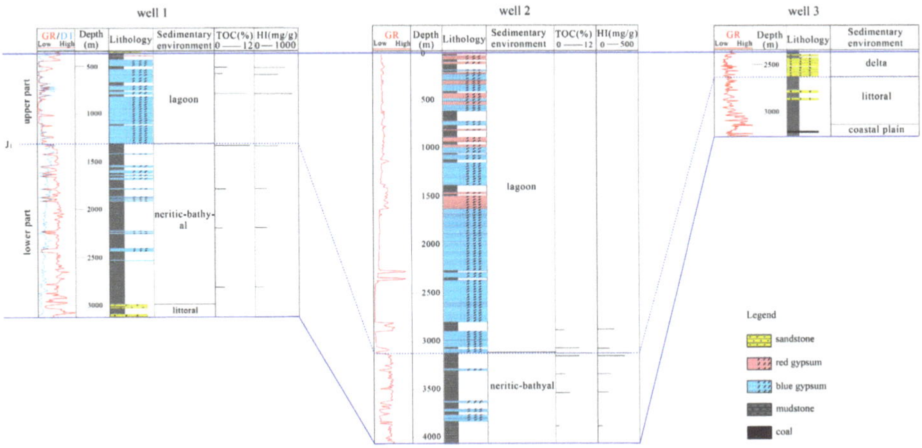

Figure 4. Regional Lower Jurassic stratigraphy across well 1, well 2, and well 3 within the eastern African continental margin. GR: gamma curve; DT: acoustic curve; TOC: total organic carbon; HI: hydrogen index. See P2 for the locations of well 1, well 2, and well 3.

4.2. Planar Distribution of the Gypsum Deposits

More drilling records reveal the heterogeneity of gypsum distribution. For example, according to well 3 (Figure 4), in the Lower Jurassic, with a 19 m-thick black coal seam at the bottom, a thick suite of mudstone upward, and sandstone at the top, gypsum was not observed within well 3 (Figure 4). The results show that the values of acoustic logging of the Lower Jurassic sandstone are mainly concentrated around 225 μs/m, with mudstone around 315 μs/m and gypsum around 200 μs/m (Figure 5), indicating that mudstone exhibits low-speed characteristics, while sandstone and gypsum exhibit high-speed characteristics. The results from logging data show that the average velocity in sandstone are 2620 kg/m^3, with gypsum being 2070 kg/m^3, and mudstone being 2420 kg/m^3 (Table 1). The wave impedance of rock is equal to the product of longitudinal wave velocity and density. The calculation results show that the wave impedance of gypsum is about 900 × 10^4 kg/(s·m), with sandstone about 1164 × 10^4 kg/(s·m), and mudstone about 768.35 × 10^4 kg/(s·m) (Table 1), indicating that the wave impedance of mudstone is low, the wave impedance of sandstone is high, and the wave impedance of gypsum is between mudstone and sandstone. If it is a thick suite of layered gypsum with thin-layered mudstone, the seismic facies would be characterized by high amplitude and continuous reflectance. If it is a thick suite of mudstone with thin-layered gypsum, it would be characterized by continuous parallel weak reflection or bland reflection. Due to the significant wave impedance difference between mudstone and sandstone, interbedded sandstone-mudstone would be characterized by strong reflection.

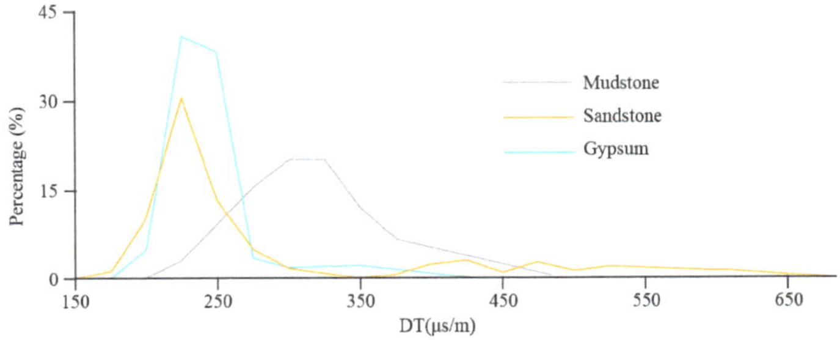

Figure 5. The diagram shows the distribution frequency of interval transit time in different Lower Jurassic lithologies.

Table 1. The average rock mechanics parameters of the Lower Jurassic.

Lithology	DT (μs/m)	Velocity (m/s)	Density (kg/m^3)	Wave Impedance (kg/(s·m))
sandstone	225	4444	2620	1164.33×10^4
gypsum	230	4348	2070	900×10^4
mudstone	315	3175	2420	768.35×10^4

The seismic profiles through the key wells confirm the above speculation. For exampl the seismic profile across well 1 indicates that the whole Lower Jurassic seismic facies coul be divided into three parts: the lowest part is characterized by weakly continuous stron reflection corresponding to the sandstone-mudstone intercalation at the bottom of th Lower Jurassic, the middle part is characterized by weak reflection corresponding to th thick suite of mudstone with thin layers of gypsum, and the upper part is characterize by continuous strong reflection corresponding to the thick suite of gypsum deposits wit thin layers of mudstone (Figure 6). The more seismic data show that most Lower Jurassi succession are characterized by weak reflection (Figure 7), while the Lower Jurassic i the nearshore and the northern edge of the Tanzania Basin, as well as the Lamu Basin, i characterized by weakly continuous strong reflections. It is notable that in some nearshor grabens, the Lower Jurassic succession is characterized by weak reflections in the lowe part and continuous, strong reflections in the upper part.

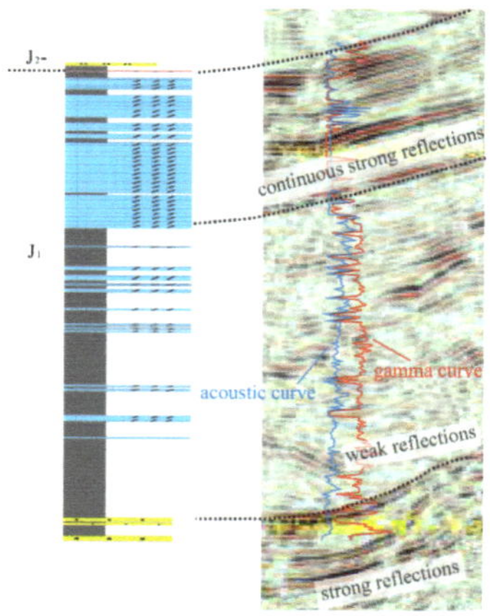

Figure 6. The through-well 1 seismic profile indicates that the whole Lower Jurassic seismic facie could be divided into three parts.

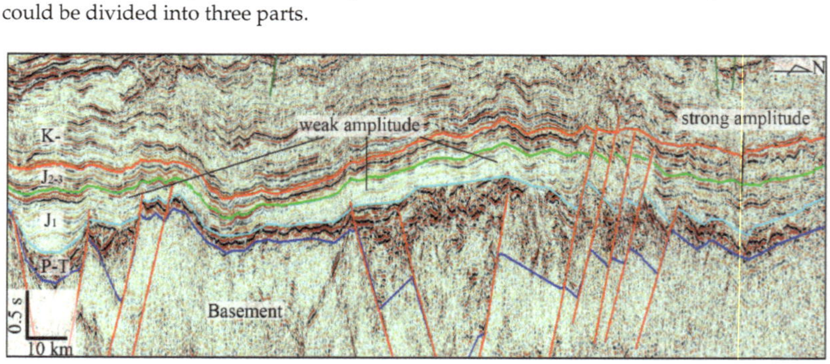

Figure 7. The seismic profile shows that the Lower Jurassic is characterized by weak reflection while the northern part is characterized by weakly continuous, strong reflections. Different colo lines present stratigraphic boundaries and faults. See P3 for the location.

4.3. Variations of the TOC and HI

Although the TOC values change greatly, as could be seen from Figure 8, the major TOC values within the northern wells range from 1.3% to 2.2%, while the TOC values within the southern wells range from 5.1% to 6.1%, which shows that the TOC values within the southern wells are much higher than those within the northern wells. The HI values have a phenomenon similar to the TOC values. The average HI values within the northern wells range from 96 mg/g to 133 mg/g, while the average HI values within the southern part range from 276 mg/g to 836 mg/g (Figure 8).

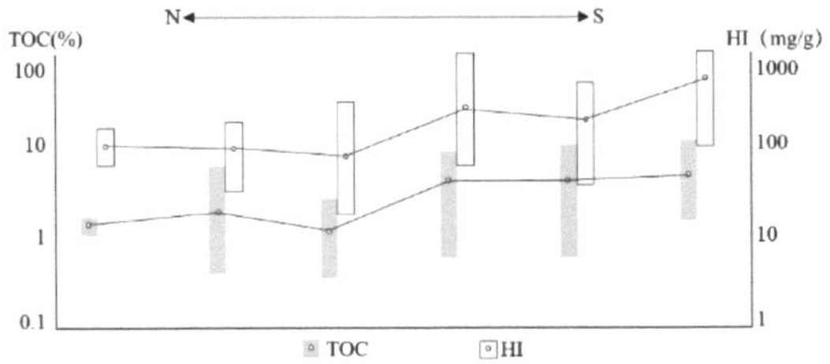

Figure 8. Box- and whisker plots show the range and average values of the TOC values and the HI values. See Figure 1 for the location of the wells.

5. Discussion

5.1. Sea Level Changes

Two main reasons could account for the differential distribution of gypsum deposits: the plastic rheology of gypsum and the differential sedimentary environment, respectively. Considering that there is no evidence of large-scale plastic rheology of gypsum on the seismic profiles within the eastern African continental marginal basins, the difference in sedimentary environment is the main controlling factor of the differential gypsum distribution. Because sea level change plays an important role in controlling the sedimentary environment, it is necessary to discuss sea level change within the eastern African continental marginal basins to clarify the sedimentary environment. The gamma-ray curve of the Lower Jurassic within well 3 is mainly bell-shaped and funnel-shaped, and a 19 m-thick black coal seam at the bottom of the Lower Jurassic reflects that the initial environment may be a coastal plain, followed by the littoral sea and delta inferred by the upward-changing lithology. Environmental changes show a sea level rise during the early stage of the Early Jurassic, followed by a sea level fall during the late stage of the Early Jurassic. The presumption could also be evidenced by seismic stratigraphy. The stacking geometry of the Lower Jurassic in seismic profiles also enables a two-fold division. The lower is composed of a wedge-shaped body onlaping and converging to the east in a direction to the Davie fracture ridge (Figures 1 and 3). This retrogradational stacking pattern and reversal of the convergence direction are attributed to transgressive deposits that accumulate mainly in areas adjacent to the shoreline, resulting in a cut-off of sediment supply to the marine environment. Therefore, it indicates a rapid sea level rise that creates coastal accommodation faster than sediment can fill it [29,30]. The overlying section is defined by a series of offlapping clinoforms that step down into the basin (Figure 9). Such a basinward shift and the lack of topset aggradation may be caused by progradation during a relative sea-level fall.

The above analysis suggests that the eastern African continental margin experienced a rapid sea level rise during the early stage of the Early Jurassic, followed by a sea level fall during the late stage of the Early Jurassic. Both changes coincide with those of the global sea level change [31], which can be seen as the global sea level rise from 201 Ma to 183 Ma, followed by the sea level fall. Thus, we interpret that the Early Jurassic strata within this continental margin experienced a sea level rise of about 18 Ma and a sea level fall of about 12 Ma.

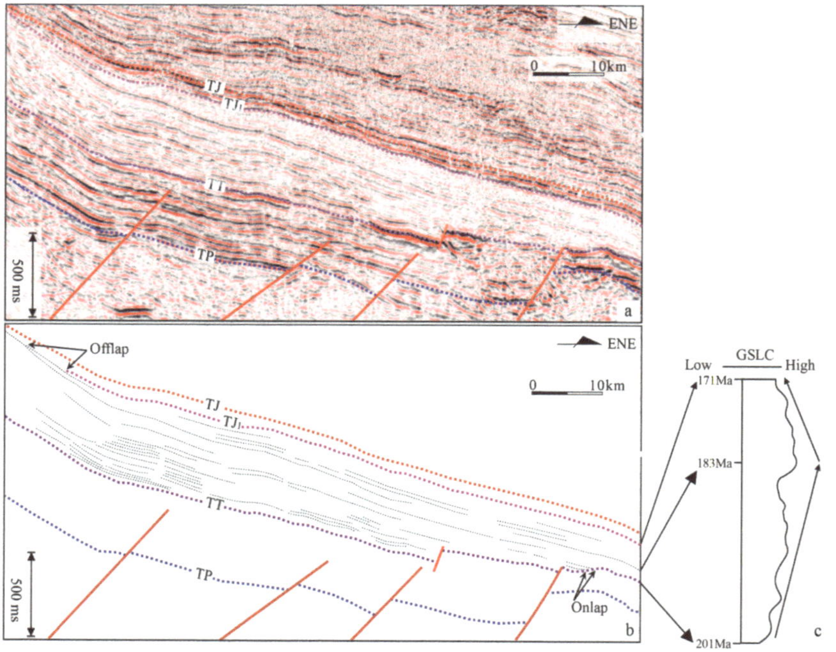

Figure 9. (**a**) Seismic profile and (**b**) geological profile showing the retrogradation and marine onla at the early stage of the Early Jurassic and the offlap pattern at the late stage. TP: top of Permia TT: top of Triassic; TJ_1: top of Lower Jurassic; TJ: top of Jurassic. (**c**) Shows the global sea level chang curve during the Early Jurassic [31].

5.2. Changes in the Sedimentary Environment through the Early Jurassic

During the Early Jurassic, with the break-up of the eastern Gondwana continen and the western Gondwana continent in a "V" shape from the northeast (Figure 10a the Paleo-Tethys Ocean began to form from northeast to southwest, forming a gulf-lik sea [22,23,32,33]. Northern Tanzania and Lamu Basins were close to the open sea, wher they had a strong capacity for water exchange, while southern Tanzania and Rovum Basins were relatively far away from the open sea environment with weak water circulatior resulting in an increase in reducibility from northeast to southwest.

The organic geochemical data related to source rocks supports the above opinion. Th TOC could reflect the degree of the organic matter enrichment. Statistics indicate that th overwhelming majority of global organic matter is degraded during sedimentation, an the proportion preserved is less than 0.5% [34]. Thus, the preservation condition of organi matter is an important main controlling factor of organic matter enrichment [24,35,36 Demaison & Moore [36] calculated the primary productivities of the global ocean an the TOC distributions in seabed sediments and found that the source rock distributio was critically affected by the redox degree of the water. Additionally, the organic-ric source rocks developed in the deep sea. Such environments existed in the lower Cretaceou of the Atlantic Ocean and the Pacific Ocean. In such environments, there is evidence o reductive conditions and the preservation of organic matter [37]. The Kivu Lake in easterr Africa [38], the Black Sea [39], the Bohai Bay Basin in eastern China [40], and the Bayingeb Basin in eastern China [41]. Wei & Jiang [41] also suggest that the preservation conditio is the main controlling factor of organic matter enrichment. Moreover, the preservatio condition of modern sedimentary settings is also a major factor controlling organic matte enrichment [37,42]. The content of organic carbon in a reducing environment is generall 1.0% to 20.0%, while the TOC values in an oxic environment are lower, ranging from 0.2% to 4.0% [37]. For example, although Albert Lake exhibits the highest productivity amon the four lakes within the western part of the Eastern African Rift Valley, the TOC value in the surface sediments is the lowest [42]. The reason is that Albert Lake is in an oxic environment with frequent water turbulence, where it is difficult for organic matter to be buried [42]. If the reducibility increases from the northeast of the gulf-like sea to the southwest, the preservation of organic matter within the southwestern part of the gulf-like sea is much better than that within the northeastern part. Thus, it can be speculated that the average TOC values within the southwestern part are higher than those within the northeastern part, which is in line with the variations of the measured TOC values (Figure 8). In addition, the HI is the ratio of the amount of pyrolytic hydrocarbon to the total organic carbon, which is generally used to reflect the organic matter type and can also indicate the capacity of preservation of organic matter after deposition [43]. The high value of HI represents that the bottom of the water with a strong capacity for organic matter preservation is in a reducing environment, while the low value of HI represents a relative oxic environment [43]. Considering the consistent organic matter type of the Early Jurassic source rock within the Tanzania Basin [19], the variations in the HI values are likely controlled by differences in preservation conditions. The measured results of the HI values also suggest that the reducibility within the southern part is better than that within the northern part (Figure 8).

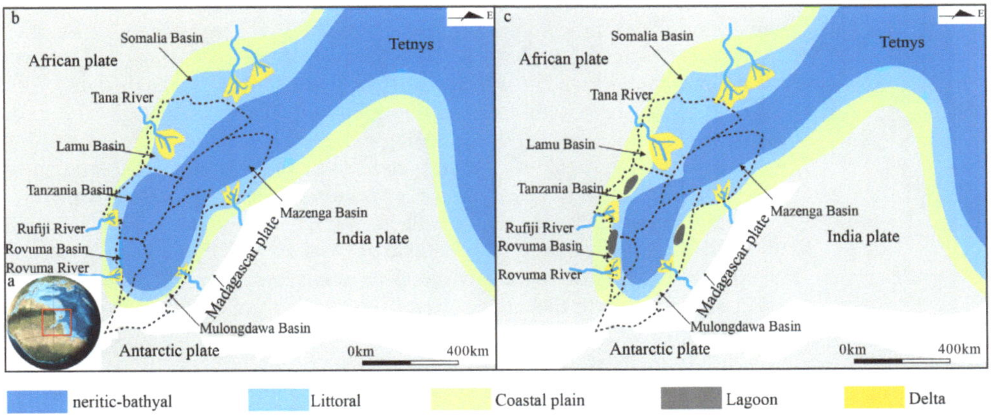

Figure 10. (**a**) A global paleogeographical map showing the paleographic reconstruction during the Early Jurassic. (**b**) The map shows the sedimentary environment at the early stage of the Early Jurassic; (**c**) The map shows the sedimentary environment at the late stage of the Early Jurassic. The reconstructed paleo-plate pattern and paleo-sedimentary environment refer to [22,23,32,33].

The Early Jurassic of the eastern African continental marginal basins consists of several lithological changes. The succession begins with a thick suite of mudstone with thin-layered gypsum, followed by a thick suite of gypsum deposits with thin layers of mudstone. TOC values increase gradually from bottom to top, and the HI values have similar characteristics (Figure 4), reflecting the possibility of gradually increasing reducibility. For the thick suite of gypsum deposits with thin layers of mudstone, the values of both TOC and HI remain at a relatively high level initially, with a decrease at the top, probably indicating strong reducibility in the early stage and later oxidation enhancement, which corresponds to the increase of red gypsum deposits within the upper part of the Lower Jurassic succession (e.g., well 1 and well 2).

Based on the analysis above, combined with data from drillings and seismic, unpublished research reports, and published literature [20,23], four types of sedimentary environments from 201 Ma to 183 Ma are identified, including delta, coastal plain, littoral, and neritic-bathyal environments (Figure 10b). During the period of sea level rise, the lagoon appears (Figure 10c).

5.3. Depositional Model

There are presently two main genetic models accounting for the formation of gypsu deposits. The first is the shallow water genetic model, which suggests that in a relativel closed sedimentary environment under arid climate conditions, the water is concentrate and gypsum is precipitated due to prevailing water evaporation over water influx [1,44,4 The other is the deep water model [46–50], which considers water stratification resulting i the formation of gyspsum deposits [51]. A dominant control on whether water bodies a well stratified or mixed is depth. Brainerd and Gregg [51] propose that waves generated b winds at the air-water surface will mix downward to the bottom of the waves, forming mixed layer. The maximum depth of a mixed layer was marked by the storm-wave bas which is about 50 m in dense brines, as proposed by some scholars [47,52,53]. However, th volume within deep water basins is generally huge below the wave base [53], which tenc to have large vertical gradients in density in hypersaline settings, resulting in the upp mixed layer and the bottom water being commonly stably stratified [51]. For example, fo the deep-water gypsum deposition in the Dead Sea during meromictic periods, the wat column remains stably stratified throughout the year.

Although paleo-water depth, as a significant parameter for paleomorpholog restoration and tectonic subsidence history analysis in the basin, has always been value by geologists, its recovery is difficult in international frontier research [54]. Howeve considering that the geological structure of the Early Jurassic gulf-like sea is simila to that of the present Red Sea [23], both of which are in the stage of intercontinent rifting, the water depth of the former may be in the same range as the latter, which has a average depth of 558 m and a maximum depth of 3039 m. Although such a compariso does not take into account other geological factors, such as sea-level rise and fall, th water depth of the Early Jurassic gulf-shaped sea may have far exceeded the storm-wav base level of the dense brine. Sufficient water depth makes the stratification of brin in the bottom water stable, which reflects that the deep-water genetic model is suitabl for explaining the formation mechanism of gypsum rock in the Early Jurassic. It shoul be pointed out that this interpretation still needs more evidence, such as geochemistr and paleoecology, in the future. In addition, the shallow water genetic model need the water evaporation to be much larger than the influx, meaning that the water dept must continue to decline, which is inconsistent with the continuous sea level rise fror 201 Ma to 183 Ma. Thus, the deep water model for the gypsum within the East Africa continental margin basins from 201 Ma to 183 Ma could be established. In this mode (Figure 11a), with the break-up of the Gondwana continent, the global sea level bega to rise. The Paleo-Tethys Ocean first invaded the rifting basin from the northeast t form a gulf-like sea. At this time, the Ruvuma River, the Rufiji River, and the othe river systems provided clastic material to the basins, forming a sedimentary syster including delta, coastal plain, littoral, and neritic-bathyal environments. As the sea leve continued to rise, it exceeded the storm-wave base for dense brines, which tend to hav large vertical gradients in density. The bottom water and upper water within the centra of the basins, which corresponds to the central of the neritic-bathyal facies, were stabl stratified, allowing thin gypsum layers to precipitate. Notably, thin-layered gypsun deposits could be precipitated at the edge of the neritic-bathyal facies due to the satiabl depth created by the activities of extensional faults. However, the strong water exchang in the northeastern part impeded stratification, resulting in little precipitation of th gypsum, which is why gypsum deposits are lacking in the Lamu Basin. Vertically, a the sea level continues to rise, the values of TOC and HI increase. The more stable th bottom water layer becomes, the more reducibility becomes stronger.

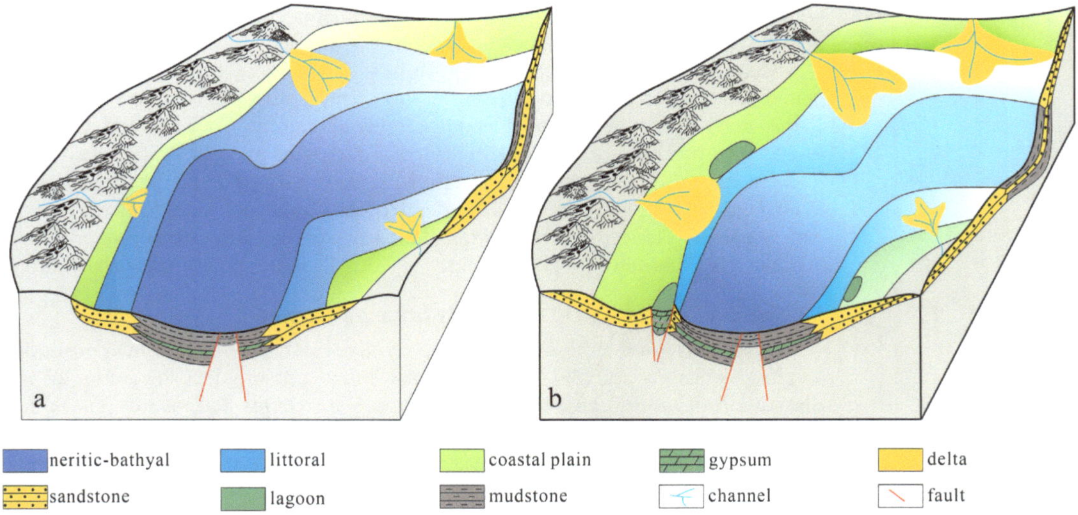

Figure 11. (**a**) A schematic diagram shows the sedimentary model during sea level rise. (**b**) Schematic diagram showing the sedimentary model during the seal level drawdown.

The sea level within the East Africa continental margin basins began to fall around 183 Ma ago, and the area of the neritic-bathyal environment gradually became smaller. Considering that brine stratification existed in deeper water during this period, the area of gypsum precipitation became limited to the center of the neritic-bathyal environment. However, in terms of the distribution range of a thick suite of gypsum far from the center of the basins, the deep water genetic model is herein difficult to account for its formation. It can be found that the thick gypsum deposits are mainly located in grabens formed by extensional faulting. The grabens provided accommodation space for subsequent thick gypsum deposition, and the rise of the areas adjacent to the grabens limited the influx of seawater, which further resulted in the formation of the lagoons (Figure 11b). As the sea level continued to decline, the brine within lagoons became more concentrated, and then the thick gypsum deposits precipitated.

Although gypsum-bearing basins are widely developed in the world, the model of the Early Jurassic gypsum within the East African continental marginal basins cannot be mechanically applied to other basins. It is necessary to comprehensively consider whether the target basin has a similar tectonic and sedimentary background as the Early Jurassic East continental marginal basins. For example, whether the gypsum deposits are mainly formed in the intercontinental rift period or whether the gypsum deposit formation period has similar sea level changes. Only by meeting these characteristics can we consider learning from the depositional model of the Early Jurassic gypsum deposits within the East African continental marginal basins.

5.4. Geological Significance for Gas Play

5.4.1. Expanding the Scope of the Hydrocarbon Generation Window

Since the discovery of the first large gas field in the deep-water area of East Africa in 2010, the deep-water areas of Tanzania and Rovuma basins have become one of the hot spots for global deep-water oil and gas exploration due to their huge proven gas reservoir reserves [21]. Geochemical evidence suggests that the Lower Jurassic mudstone is the main source rock for gas reservoirs in the Rovuma and Tanzania Basins [19]. It should be pointed out that conventional thermal history simulations of source rocks reached a peak of gas generation during the Paleocene, followed by a gradual decline in gas production [21], which makes it difficult to explain the ongoing filling of East African natural gas reservoir and the formation of giant reserves.

Considering that the gypsum deposits have high thermal conductivity and a low hea generation rate [55], they can transmit ground temperature easily from the deep stratun to the shallow stratum. This may lead to a lower temperature gradient and an abnormall low temperature of the source rock in the lower part of the gypsum stratum than in the nor gypsum stratum. The abnormal temperatures in the lower part of the gypsum stratum resul in an increase in the maturity threshold depth of the source rocks and delay the peak time c gas generation. This can effectively expand the scope of the hydrocarbon generation windov and may be the reason why giant amounts of natural gas are still continuously accumulatin

5.4.2. Sealing Property and Potential Hydrocarbon Traps

Irrespective of their thickness and geodynamic setting, the gypsum deposits acted a an effective sealing unit [56]. The widely distributed gypsum deposits hinder the upwar dissipation of natural gas. Although the main targets for deep-water gas exploratio currently in the Rovuma and Tanzania Basins are the Late Cretaceous and Paleogen turbidite sandstones, most of the gas fields are located near the large-scale fault zone connecting the source and the reservoirs, indicating that gas generated from Lower Jurassi source rocks vertically migrated along the large-scale fault zones and accumulated i the turbidite sandstones (Figure 12). Considering that both the Rovuma and Tanzani Basins were dominated by thermal subsidence during the Early Jurassic period, a smal number of preexisting faults have been reactivated under the regional extensional stres field, resulting in the formation of some horst-graben structures. The Triassic fluvial an lacustrine sandstones [21] in the horst structures can be used as the pointing area for th near-source dominant migration of natural gas, forming "new source, old reservoir" typ buried hills, which should be given more attention in further gas exploration (Figure 12).

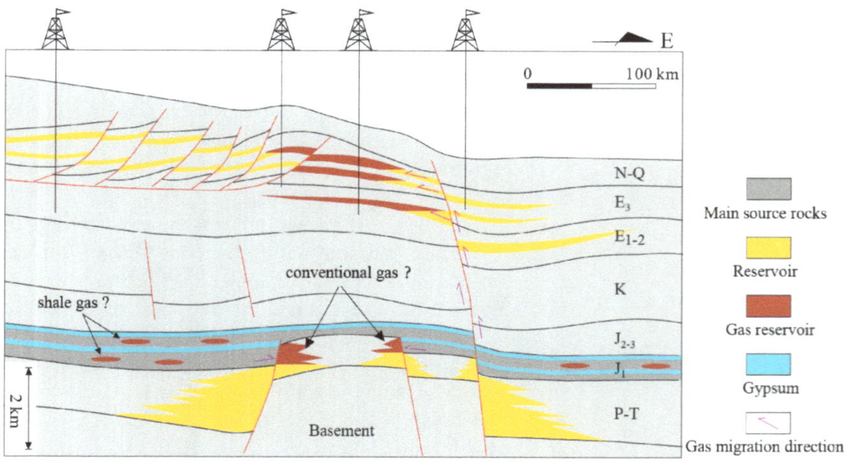

Figure 12. The model shows gas migration and accumulation within the Eastern African continenta marginal basins.

Because of the good sealing properties of gypsum deposits, most of the discovered ga fields are located near the large-scale fault zones connecting the source and the reservoir In addition, the Lower Jurassic shale where the faults are not developed may be rich in a large amount of undischarged shale gas, and the place with a high brittle mineral conten will be an important place for shale gas exploration in the future.

6. Conclusions

(1) The thick suite of mudstone with thin-layered gypsum deposits is widely distributed in the Lower Jurassic within the Rovuma and southern Tanzania Basins. The thick suite of gypsum deposits with thin layers of mudstone is mainly distributed in the upper part of the Lower Jurassic, located in grabens formed by extensional faulting.
(2) With the break-up of the Gondwana continent, the eastern African continental margin basins experienced a sea level rise from 201 Ma to 183 Ma, with a gradual increasing reducibility from northeast of the gulf-like sea to southwest, and then began to fall until the end of the Early Jurassic with a gradual decreasing reducibility.
(3) The deep-water gypsum genetic model could account for the distribution of the thick suite of mudstone with the thin layers of gypsum deposited during the entire Early Jurassic. The model of the lagoon could account for the limited distribution of the thick suite of gypsum deposts with thin layers of mudstone deposited during the late stage of the Early Jurassic.
(4) The high thermal conductivity and low heat generation rate of gypsum deposits result in increasing the maturity threshold depth of Lower Jurassic source rocks and delaying the peak time of gas generation. The good sealing of gypsum makes the discovered Late Cretaceous and Paleocene gas fields mainly located near the large-scale fault zones connecting the source and the reservoirs, and the traps of Triassic fluvial and lacustrine sandstones in the horst structures could be potential conventional gas exploration targets, while the Lower Jurassic shale where the faults are not developed may be an important place for shale gas exploration in the future.

Author Contributions: J.C.: Conceptualization, methodology, writing, and funding acquisition. R.G.: data curation and formal analysis. All authors have read and agreed to the published version of the manuscript.

Funding: This research was financially supported by the National Natural Science Foundation of China (No. 42002150), the Open Foundation of Cooperative Innovation Center of Unconventional Oil and Gas, Yangtze University (Ministry of Education & Hubei Province) (No. UOG2024-12), the Open Foundation Project of the Key Laboratory of Polar Geology and Marine Mineral Resources (China University of Geosciences, Beijing, China), Ministry of Education (No. PGMR-2023-201), and the National Key Research Program of China (No. 2017ZX05032-002).

Data Availability Statement: Data is contained within the article.

Acknowledgments: We sincerely thank three anonymous reviewers for their critical comments and constructive suggestions, which have greatly improved the quality of this paper.

Conflicts of Interest: The authors declare no conflict of interest.

References

Warren, J.K. Evaporites through time: Tectonic, climatic and eustatic controls in marine and nonmarine deposits. *Earth-Sci. Rev.* **2010**, *98*, 217–268. [CrossRef]

Yu, Y.; Tang, L.; Yang, W.; Qiu, N.; Li, W. Salt structures and hydrocarbon accumulations in the Tarim Basin, northwest China. *AAPG Bull.* **2014**, *98*, 135–159. [CrossRef]

Wang, X.; Fei, Q.; Zhang, J. Cenozoic Diapiric Traps in Eastern China. *AAPG Bull.* **1985**, *69*, 2098–2109. [CrossRef]

Zhu, C.; Jiang, F.; Zhang, P.; Zhao, Z.; Chen, X.; Wu, Y.; Chen, Y.; Wang, W.; Song, Z.; Hu, T.; et al. Effect of petroleum chemical fraction and residual oil content in saline lacustrine organic-rich shale: A case study from the Paleogene Dongpu Depression of North China. *Petrol. Sci.* **2023**, *20*, 649–669. [CrossRef]

Fu, L.; Li, J.; Xu, W.; Guo, W.; Li, N.; Zhang, Y.; Song, W.; Sun, Y. Characteristics and main controlling factors of Ordovician deep subsalt reservoir in central and eastern Ordos Basin, China. *J. Nat. Gas Geosci.* **2021**, *6*, 13–25. [CrossRef]

Meng, M.; Ge, H.; Shen, Y.; Ji, W.; Wang, Q. Rock fabric of tight sandstone and its influence on irreducible water saturation in Eastern Ordos Basin. *Energy Fuel* **2023**, *37*, 3685–3696. [CrossRef]

Zhou, J.; Wu, G.; Geng, Y.; Guo, Y.; Chang, X.; Peng, C.; Ai, C. Laboratory study of the factors affecting hydraulic fracturing effect for inter-salt oil shale layers, Qianjiang Depression, China. *Petrol. Sci.* **2023**, *20*, 1690–1706. [CrossRef]

Meng, M.; Ge, H.; Shen, Y.; Wang, L. Influence of rock fabric on salt ion diffusion behavior in upper cretaceous lacustrine shale from Songliao Basin. *J. Petrol. Sci. Eng.* **2022**, *208*, 109355. [CrossRef]

9. Meng, M.; Peng, J.; Ge, H.; Ji, W.; Li, X.; Wang, Q. Rock Fabric of Lacustrine Shale and Its Influence on Residual Oil Distribution the Upper Cretaceous Qingshankou Formation, Songliao Basin. *Energy Fuel* **2023**, *37*, 7151–7160. [CrossRef]
10. Liang, C.; Ding, W.; Liu, Y.; Li, M. The Late Cretaceous—Miocene supra-salt structures in Block M, Lower Congo Basin (Cong West Africa) and the controls on channel development. *Geol. J.* **2022**, *57*, 4167–4182. [CrossRef]
11. Edgell, H.S. Salt tectonism in the Persian Gulf basin. *Geol. Soc. Lond. Spec. Publ.* **1996**, *100*, 129–151. [CrossRef]
12. Hassanpour, J.; Yassaghi, A.; Muñoz, J.A.; Jahani, S. Salt tectonics in a double salt-source layer setting (Eastern Persian Gulf, Irar Insights from interpretation of seismic profiles and sequential cross-section restoration. *Basin Res.* **2021**, *33*, 159–185. [CrossRe
13. Volozh, Y.; Talbot, C.; Ismail-Zadeh, A. Salt structures and hydrocarbons in the Pricaspian basin. *AAPG Bull.* **2003**, *87*, 313–33 [CrossRef]
14. Jackson, M.P.; Hudec, M.R.; Jennette, D.C.; Kilby, R.E. Evolution of the Cretaceous Astrid thrust belt in the ultradeep-water Low Congo Basin, Gabon. *AAPG Bull.* **2008**, *92*, 487–511. [CrossRef]
15. Quirk, D.G.; Schødt, N.; Lassen, B.; Ings, S.J.; Hsu, D.; Hirsch, K.K.; Von Nicolai, C. Salt tectonics on passive margins: Exampl from Santos, Campos and Kwanza basins. *Geol. Soc. Lond. Spec. Publ.* **2012**, *363*, 207–244. [CrossRef]
16. Adam, J.; Ge, Z.; Sanchez, M. Salt-structural styles and kinematic evolution of the Jequitinhonha deepwater fold belt, centr Brazil passive margin. *Mar. Petrol. Geol.* **2012**, *37*, 101–120. [CrossRef]
17. Alves, T.M.; Fetter, M.; Lima, C.; Cartwright, J.A.; Cosgrove, J.; Gangá, A.; Queiroz, C.L.; Strugale, M. An incomplete correlatic between pre-salt topography, top reservoir erosion, and salt deformation in deep-water Santos Basin (SE Brazil). *Mar. Petrol. Ge* **2017**, *79*, 300–320. [CrossRef]
18. Qin, P.; Zhong, D.; Su, C.; Mo, T.; Tang, Y. Effect of the internal plastic deformation of salt structures on the lithologic successic of evaporites: A case study on the Palaeogene Kumugeliemuqun formation, Kelasu Thrust Belt, Kuqa depression, Tarim Basi *J. Petrol. Sci. Eng.* **2022**, *208*, 109559. [CrossRef]
19. Sun, Y.; Sun, T.; Xu, Z. Source rock characteristics and oil-gas origins in Tanzania basin, East Africa coast. *China Offshore Oil G* **2016**, *28*, 13–19. [CrossRef]
20. Liang, J.; Kong, L.; Qiu, C.; Li, H.; Jia, S.; Long, X. Gas accumulation mechanism in East Africa coastal key basins. *Earth Sci.* **202** *46*, 2919–2933. [CrossRef]
21. Bird, D.E.; Burke, K.; Hall, S.A.; Casey, J.F. Gulf of Mexico tectonic history: Hotspot tracks, crustal boundaries, and early sa distribution. *AAPG Bull.* **2005**, *89*, 311–328. [CrossRef]
22. Catuneanu, O.; Wopfner, H.; Eriksson, P.G.; Cairncross, B.; Rubidge, B.S.; Smith RM, H.; Hancox, P.J. The Karoo basins south-central Africa. *J. Afr. Earth Sci.* **2005**, *43*, 211–253. [CrossRef]
23. Deng, Y. Discusion on the control effect of gulf on marine petroleum. *Acta Pet. Sin.* **2018**, *39*, 1–11. [CrossRef]
24. Reeves, C. Re-examining the evidence from plate-tectonics for the initiation of Africa's passive margins. In Proceedings of th Geological Society of Houston/Petroleum Exploration Society of Great Britain, London, UK, 9–10 September 2009.
25. Geiger, M.; Clark, D.N.; Mette, W. Reappraisal of the timing of the breakup of Gondwana based on sedimentological and seism evidence from the Morondava Basin, Madagascar. *J. Afr. Earth Sci.* **2004**, *38*, 363–381. [CrossRef]
26. Key, R.M.; Smith, R.A.; Smelror, M.; Sæther, O.M.; Thorsnes, T.; Powell, J.H.; Njange, F.; Zandamela, E.B. Revised lithostratigraph of the Mesozoic-Cenozoic succession of the onshore Rovuma Basin, northern coastal Mozambique. *S. Afr. J. Geol.* **200** *111*, 89–108. [CrossRef]
27. Mougenot, D.; Recq, M.; Virlogeux, P.; Lepvrier, C. Seaward extension of the East African Rift. *Nature* **1986**, *321*, 599–60 [CrossRef]
28. Cai, J.; Luo, S.; Guan, Y.; Dong, P. Main controlling factors of organic matter enrichment of the Mesoproterozoic source rocks fror the North China Craton. *Geol. J.* **2021**, *56*, 422–433. [CrossRef]
29. Neal, J.; Abreu, V. Sequence stratigraphy hierarchy and the accommodation succession method. *Geology* **2009**, *37*, 779–78 [CrossRef]
30. Zhang, M.; Zhang, J.; Xu, F.; Li, J.; Liu, J.; Hou, G.; Zhang, P. Paleocene sequence stratigraphy and depositional systems in th Lishui Sag, East China Sea Shelf Basin. *Mar. Petrol. Geol.* **2015**, *59*, 390–405. [CrossRef]
31. Haq, B.U.; Hardenbol, J.; Vail, P.R. Chronology of fluctuating sea levels since the Triassic. *Science* **1987**, *235*, 1156–1167. [CrossRe
32. Phethean, J.J.; Kalnins, L.M.; van Hunen, J.; Biffi, P.G.; Davies, R.J.; Mccaffrey, K.J. Madagascar's escape from A frica: A higl resolution plate reconstruction for the Western Somali Basin and implications for supercontinent dispersal. *Geochem. Geophy Geosyst.* **2016**, *17*, 5036–5055. [CrossRef]
33. Scotese, C.R. Jurassic and Cretaceous plate tectonic reconstructions. *Palaeogeogr. Palaeoclimatol. Palaeoecol.* **1991**, *87*, 493–50 [CrossRef]
34. Schoepfer, S.D.; Shen, J.; Wei, H.; Tyson, R.V.; Ingall, E.; Algeo, T.J. Total organic carbon, organic phosphorus, and biogenic bariur fluxes as proxies for paleomarine productivity. *Earth-Sci. Rev.* **2015**, *149*, 23–52. [CrossRef]
35. Arthur, M.A.; Dean, W.E.; Pratt, L.M. Geochemical and climatic effects of increased marine organic carbon burial at th Cenomanian/Turonian boundary. *Nature* **1988**, *335*, 714–717. [CrossRef]
36. Hedges, J.I.; Keil, R.G. Sedimentary organic matter preservation: An assessment and speculative synthesis. *Mar. Chem.* **199** *49*, 81–115. [CrossRef]
37. Demaison, G.J.; Moore, G.T. Anoxic environments and oil source bed genesis. *Org. Geochem.* **1980**, *2*, 9–31. [CrossRef]

38. Degens, E.T.; von Herzen, R.P.; Wong, H.; Deuser, W.G.; Jannasch, H.W. Lake Kivu: Structure, chemistry and biology of an East African rift lake. *Geol. Rundsch.* **1973**, *62*, 245–277. [CrossRef]
39. Mazzini, A.; Ivanov, M.K.; Nermoen, A.; Bahr, A.; Bohrmann, G.; Svensen, H.; Planke, S. Complex plumbing systems in the near subsurface: Geometries of authigenic carbonates from Dolgovskoy Mound (Black Sea) constrained by analogue experiments. *Mar. Petrol. Geol.* **2008**, *25*, 457–472. [CrossRef]
40. Jin, Q.; Zhu, G.; Wang, J. Deposition and distribution of high-potential source rocks in saline lacustrine environments. *J. China Univ. Pet.* **2008**, *32*, 19–23. [CrossRef]
41. Wei, H.; Jiang, X. Early Cretaceous ferruginous and its control on the lacustrine organic matter accumulation: Constrained by multiple proxies from the Bayingebi Formation in the Bayingebi Basin, inner Mongolia, NW China. *J. Petrol. Sci. Eng.* **2019**, *178*, 162–179. [CrossRef]
42. Katz, B.J. Controls on distribution of lacustrine source rocks through time. *AAPG Mem.* **1990**, *50*, 132–139.
43. Talbot, M.R.; Livingstone, D.A. Hydrogen index and carbon isotopes of lacustrine organic matter as lake level indicators. *Palaeogeogr. Palaeoclimatol. Palaeoecol.* **1989**, *70*, 121–137. [CrossRef]
44. Hsü, K.J. Origin of saline giants: A critical review after the discovery of the Mediterranean evaporite. *Earth-Sci. Rev.* **1972**, *8*, 371–396. [CrossRef]
45. Hsü, K.J.; Ryan, W.B.; Cita, M.B. Late Miocene desiccation of the Mediterranean. *Nature* **1973**, *242*, 240–244. [CrossRef]
46. Brongersma-Sanders, M. Origin of major cyclicity of evaporites and bituminous rocks: An actualistic model. *Mar. Geol.* **1971**, *11*, 123–144. [CrossRef]
47. Christeleit, E.C.; Brandon, M.T.; Zhuang, G. Evidence for deep-water deposition of abyssal Mediterranean evaporites during the Messinian salinity crisis. *Earth Planet. Sci. Lett.* **2015**, *427*, 226–235. [CrossRef]
48. Debenedetti, A. The problem of the origin of the salt deposits in the Mediterranean and of their relations to the other salt occurrences in the Neogene formations of the contiguous regions. *Mar. Geol.* **1982**, *49*, 91–114. [CrossRef]
49. Schmalz, R.F. Deep-water evaporite deposition: A genetic model. *AAPG Bull.* **1969**, *53*, 798–823. [CrossRef]
50. Schmalz, R.F. The Mediterranean salinity crisis: Alternative hypotheses. *Carbonate Evaporite* **1991**, *6*, 121–126. [CrossRef]
51. Brainerd, K.E.; Gregg, M.C. Surface mixed and mixing layer depths. *Deep Sea Res. Part I Oceanogr. Res. Pap.* **1995**, *42*, 1521–1543. [CrossRef]
52. Dietz, R.S. Wave-base, marine profile of equilibrium, and wave-built terraces: A critical appraisal. *Geol. Soc. Am. Bull.* **1963**, *74*, 971–990. [CrossRef]
53. Sonnenfeld, P. *Models of Upper Miocene Evaporite Genesis in the Mediterranean Region*; Springer: New York, NY, USA, 1985; pp. 323–346.
54. Plint, A.G.; Tyagi, A.; Hay, M.J.; Varban, B.L.; Zhang, H.; Roca, X. Clinoforms, paleobathymetry, and mud dispersal across the Western Canada Cretaceous foreland basin: Evidence from the Cenomanian Dunvegan Formation and contiguous strata. *J. Sediment. Res.* **2009**, *79*, 144–161. [CrossRef]
55. O'Brien, J.J.; Lerche, I. The influence of salt domes on paleotemperature distributions. *Geophysics* **1984**, *49*, 2032–2043. [CrossRef]
56. Liu, W.; Zhao, H.; Liu, Q.; Zhou, B.; Zhang, D.; Wang, J.; Lu, L.; Luo, H.; Meng, Q.; Wu, X. Significance of gypsum-salt rock series for marine hydrocarbon accumulation. *Pet. Res.* **2016**, *37*, 1451–1462. [CrossRef]

Journal of
Marine Science and Engineering

MDPI

Article

Predicting Stick-Slips in Sheared Granular Fault Using Machine Learning Optimized Dense Fault Dynamics Data

Weihan Huang [1], Ke Gao [1,2,*] and Yu Feng [3]

1 Department of Earth and Space Sciences, Southern University of Science and Technology, Shenzhen 518055, China; 11930742@mail.sustech.edu.cn
2 Guangdong Provincial Key Laboratory of Geophysical High-Resolution Imaging Technology, Southern University of Science and Technology, Shenzhen 518055, China
3 School of Civil Engineering, Sun Yat-sen University, Zhuhai 519082, China; fengy253@mail.sysu.edu.cn
* Correspondence: gaok@sustech.edu.cn

Abstract: Predicting earthquakes through reasonable methods can significantly reduce the damage caused by secondary disasters such as tsunamis. Recently, machine learning (ML) approaches have been employed to predict laboratory earthquakes using stick-slip dynamics data obtained from sheared granular fault experiments. Here, we adopt the combined finite-discrete element method (FDEM) to simulate a two-dimensional sheared granular fault system, from which abundant fault dynamics data (i.e., displacement and velocity) during stick-slip cycles are collected at 2203 "sensor" points densely placed along and inside the gouge. We use the simulated data to train LightGBM (Light Gradient Boosting Machine) models and predict the gouge-plate friction coefficient (an indicator of stick-slips and the friction state of the fault). To optimize the data, we build the importance ranking of input features and select those with top feature importance for prediction. We then use the optimized data and their statistics for training and finally reach a LightGBM model with an acceptable prediction accuracy (R^2 = 0.94). The SHAP (SHapley Additive exPlanations) values of input features are also calculated to quantify their contributions to the prediction. We show that when sufficient fault dynamics data are available, LightGBM, together with the SHAP value approach, is capable of accurately predicting the friction state of laboratory faults and can also help pinpoint the most critical input features for laboratory earthquake prediction. This work may shed light on natural earthquake prediction and open new possibilities to explore useful earthquake precursors using artificial intelligence.

Keywords: disasters prediction; stick-slip; earthquake; machine learning; combined finite-discrete element method (FDEM)

Citation: Huang, W.; Gao, K.; Feng, Y. Predicting Stick-Slips in Sheared Granular Fault Using Machine Learning Optimized Dense Fault Dynamics Data. *J. Mar. Sci. Eng.* **2024**, *12*, 246. https://doi.org/10.3390/jmse12020246

Academic Editor: Markes E. Johnson

Received: 15 January 2024
Revised: 26 January 2024
Accepted: 27 January 2024
Published: 30 January 2024

1. Introduction

Reliable prediction of earthquakes, a long-standing goal of earthquake research, is a prerequisite for reducing losses from earthquake disasters. Although many efforts have been made in the past, earthquake prediction is still in its infancy due to the inherent complexity of the underlying earthquake physics. In recent years, artificial intelligence, especially machine learning (ML), has become the mainstream tool for data analysis in various fields. The ML approaches can explore inner patterns from large data volumes and fit nonlinear mapping relationships between different variables in high dimensions. While various earthquake sources and propagation models have been developed, the intricate physics from earthquake nucleation to dynamic rupture poses tremendous challenges for accurate and reliable modelling and forecasting. In particular, integrating various heterogeneous fault data and extracting complex hidden patterns beyond simplified physical laws demands advanced data-driven approaches. Data collected from natural earthquakes typically includes seismic catalogues, earthquake waveforms, surface deformation, electromagnetic fields, temperature fields, gravity fields, and observations of changes in underground fluids and geochemical compositions [1]. Therefore, the ML algorithms, with a solid ability to capture underlying unknown nonlinear patterns, may thus be appropriate for earthquake prediction exploration.

There is evidence that machine learning has been used for the prediction of disasters such as earthquakes and tsunamis. Asencio-Cortés et al. [2] studied the sensitivity of 19 earthquake catalogue features to machine learning prediction models and explored the impact of different types of earthquake catalogues on the prediction target. Asim et al. [3] expanded the earthquake catalogue to include 60 entries by calculating statistical features within a fixed time window. The study used 60 earthquake characteristics and the mRMR criterion to extract relevant features. A prediction system was developed using SVR and HNN, with EPSO optimizing weights. This system, applied to Hindu Kush, Chile, and Southern California, showed improved prediction performance compared to previous studies. Allen and Melgar [4] constructed various decoders for seismic data using deep neural networks to predict earthquake magnitudes in the next seven days. Brykov et al. [5] employed different machine learning classifiers to predict events in specific areas with magnitudes equal to or greater than a certain threshold, showcasing strong predictive capabilities. Furthermore, Corbi et al. [6] effectively trained a machine learning prediction model capable of detecting sliding events by analysing recorded fault deformation data. In addition to the features based on earthquake catalogues mentioned above, machine learning models for natural earthquake prediction are also utilized to explain the earthquake process, such as earthquake magnitude changes analysis [7,8] and investigation of the underlying processes driving earthquake occurrence [9,10]. In this type of work, input data from catalogues of past events are crucial since they may directly affect how well ML models will perform. However, due to the limited number of periodic cycles of historical events recorded from many natural faults, reliable prediction of natural earthquakes using ML is still challenging.

As an alternative, it has been recognized that laboratory rock shearing experiments exhibiting stick-slips can generate patterns similar to the intermittent dynamics of natural earthquakes [11]. However, in contrast to natural earthquakes, laboratory earthquakes have the advantages of controllable seismogenic conditions, easy monitoring of frictional dynamics, highly repeatable earthquake processes, and abundant periodic cycles, and thus have been extensively employed to explore the physics of earthquakes [12–16]. The frictional dynamics data obtained in the stick-slip experiments can enable ML applications in laboratory earthquake prediction. For example, Rouet-Leduc et al. [17] revealed a correlation between the acoustic emission signal released in a laboratory shear test and the macroscopic shear stress based on a random forest model; Bolton et al. [18] classified the acoustic emission signals in laboratory shear experiments by an unsupervised K-cluster method and predict the occurrence time, duration, and magnitude of laboratory earthquakes using the data trained with classified features. Wang et al. [19] used a deep-learning transformer model to predict fault characteristics like displacement and friction using acoustic emission signals. The ML-based laboratory earthquake prediction may pave the way for developing appropriate approaches and procedures for natural earthquake prediction. However, due to the limitations of monitoring equipment, some laboratory experiments can only collect limited types of signals with restricted quantities. Particularly, the local friction dynamic information (displacement, velocity) inside the fault is difficult to acquire directly during the experiment because of the restriction of sensor location in the physical model. As a result, the datasets utilized in the existing ML work are usually small, i.e., using no more than a few sensor points, and thus the generalization ability of the ML models in these applications may be limited.

Complementary to physical laboratory experiments, numerical simulations of laboratory-scale sheared granular fault experiments have also been conducted to explore earthquake physics in recent years [20–27]. Setting up dense monitoring points both on the shearing plate and inside the gouge in the numerical model allows for collecting sufficie macroscopic and microscopic friction dynamics data during stick-slip cycles with spati resolutions that are practically impossible in physical experiments. Numerical simulatio can enable a more detailed investigation of laboratory earthquake predictions with variou seismic features and also the interaction between these features leading to recurring sli events. For example, Ren et al. [28] used the rigid particle-based discrete element metho (DEM) to simulate a sheared stick-slip experiment containing a granular gouge and, thereb collected friction dynamics data of different particles to train a series of XGBoost [29] mo els for predicting the system's macroscopic friction. They found that the fault dynami responses at different locations have entirely different contributions to the overall stick-sli behaviour and incorporating more statistical features of the friction dynamics signals i training can improve prediction accuracy. However, only the velocity signals from one c two particles in the DEM simulation are utilized in each training. Ma et al. [30] also traine an XGBoost model using data and their statistics from a DEM-simulated sheared granul fault gouge to investigate the relations between the microslips and macroscopic frictio and suggested that the trained ML model can well distinguish the recharge and drop stag of the entire system.

Very recently, the state-of-the-art combined finite-discrete element method (FDEM has been applied as an improvement of the traditional rigid particle-based DEM for sheare granular fault simulations [31–33]. The FDEM inherits the merits of both the finite elemer method (FEM) and the DEM and has proven to be a powerful tool for simulating granul materials (see the introduction of the FDEM in Text S1 of Supplementary Material) [34]. I an FDEM realization of the sheared granular fault system, the DEM module is responsib for processing the particle–particle and particle–plate interactions, and the FEM modu allows for simulating the deformation of both particles and shearing plates. A combinatio of the FEM and the DEM is superior to traditional DEM simulations where only rigid par cles and plates are allowed (see Figure S1 in Supplementary Material for the compariso between the FDEM and the DEM). Gao et al. [31] mentioned that compared to the DEM the FDEM could easily realize the explicit representation of granular fault systems and thu better capture more detailed and realistic fault microscopic dynamics data in both particle and shearing plates during stick-slip cycles. In our previous work, we have thoroughl validated the appropriateness of the FDEM for simulating laboratory earthquakes [31 Based on our FDEM simulated sheared granular fault data, Wang et al. [35] trained convolutional encoder-decoder using the kinematic energy of the numerical model an successfully transferred the trained ML model to predict the frictions of physical labor tory experiments. The transfer learning practice demonstrates great potential for trainin laboratory earthquake ML prediction models using FDEM simulations.

In this work, continuing with the FDEM simulated 2D granular fault system unde stick-slip shearing, we further probe into the fault dynamics data collected at 2203 "senso points (i.e., displacements and velocities in both x and y directions at each sensor poin densely distributed both on the shearing plates and inside the gouge. We train and te LightGBM [36] models using the time-series data obtained at these sensor points as inpu features and the gouge-plate normalized shear stress (shear stress divided by the applie constant normal stress) as the label (i.e., prediction target). The main purpose of this stud is to investigate whether the regular fault dynamics data (displacements and velocitie contain effective information on the instantaneous shear stress of the fault and to explor how the input features could be optimized to better predict the fault friction state when large number of fault dynamics data are available.

This paper is organized as follows. Section 2 briefly introduces the FDEM mode the LightGBM, and SHAP value approaches. In Section 3, we first determine the optima number of input features that are representative of the fault system through a series of M training and testing using different numbers of sensor data. We then filter the 8812 inpu features (2203 × 4) to alleviate the possibility of feature correlation based on the obtaine optimal number and the importance ranking of input features. We train the LightGBM model using the optimized (filtered) dataset and test its prediction performance. Following this, we also expand the optimized feature data by calculating their statistics to form a new input feature dataset and check the performance of the newly trained LightGBM model. In Section 4, the SHAP value analysis is carried out as a preliminary trial to discuss the contribution of input features to the prediction results. The conclusions are given at the end.

2. Methods

2.1. FDEM Simulation of Stick-Slips

In this study, a laboratory-scale sheared granular fault system is simulated using the FDEM to produce fault dynamics data and macroscopic shear stress fluctuations associated with stick-slip processes. The FDEM model, which was first referenced by Gao et al. [31], consists of two shearing plates and one granular fault gouge and is realized according to the laboratory photoelastic experiment conducted by Geller et al. [37] (Figure 1). The gouge, confined between two identical deformable plates, consists of 2817 bidisperse circular particles with diameters of either 1.2 or 1.6 mm. The material and numerical simulation parameters are shown in Table S1 of the Supplementary Material. All particles are randomly generated and placed between the two plates. Each plate is 570 mm × 250 mm in width and height, and to induce friction, numerous half-circular "teeth" are created on the plates along the gouge-plate boundary. We apply the shear loading and normal pressure through two stiff bars attached to the top end of the upper plate and the bottom end of the lower plate, respectively. The bottom stiff bar is fixed in the shear direction (x direction), so it can only move in the vertical direction (y direction) under a constant normal pressure P = 28 kPa. The granular gouge is sheared by pushing the top stiff bar horizontally towards the right-hand side (x direction) at a constant velocity $V = 5.0 \times 10^{-4}$ m/s with its vertical movement fixed. A detailed illustration of the model geometry and parameter selection is presented in Text S2 of the Supplementary Material. As in our previous work [31], we calibrated the FDEM simulation by comparing its equivalent seismic moments with those obtained from the laboratory experiments by Geller et al. [37]. Figure S2 (Supplementary Material) shows that the FDEM simulation generates slip events with magnitudes adequately following the Gutenberg-Richter distribution [38]. These prove that the FDEM model can simulate the stick-slip behaviour in sheared granular gouges with sufficient accuracy.

To track the frictional dynamics signals during shearing and to avoid the edge effect, the 2203 "sensor" points are placed at the centres of the particles and the half-circular teeth that cover the middle segment of the granular gouge layer with a span of ~340 mm in the x direction (Figure 1a). We thus collect a total of 8812 (2203 × 4, i.e., V_x, V_y, D_x, and D_y at each sensor point) items of time-series data. The sensors on the two plates (i.e., on the half-circular teeth) are numbered first starting from 0 to 285, followed by the ones on the particles. The sensors on the bottom plate are numbered with even numbers (i.e., 0, 2, ..., 284), and the sensors on the top plate are numbered with odd numbers (i.e., 1, 3, ..., 285). The sensor numbers on the particles are then followed in a sequence from left to right and bottom to top. We set a time step of 10^{-7} s and run the model for almost 3.0×10^{8} time steps with a total running time of ~30 s. The model reaches a steady state after the first 3 s, and the data collected after 5 s are used here for ML analysis. The normalized gouge-plate shear stress and the x and y velocity (V_x, V_y) and displacement (D_x, D_y) at the 2203 sensor points are recorded every 1 ms. Figure 1b presents the time series of the normalized shear stress (NSS) and the exemplar displacement and velocity data recorded at a selected sensor at No. 382, manifesting the repetitive generation of slip events and recharging processes.

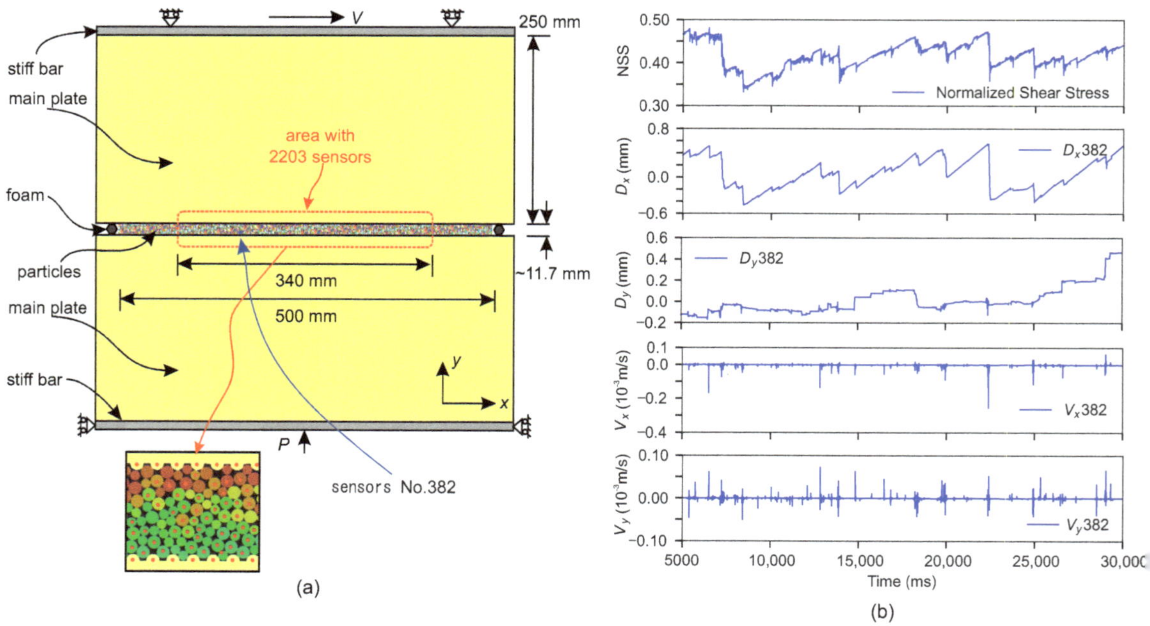

Figure 1. (**a**) Schematic of the FDEM model and the sensor locations. (**b**) Exemplar fault dynamics dat such as (from top to bottom) the normalized gouge-plate shear stress, and the x and y displacement and velocities collected at sensor No. 382. "NSS" is short for normalized shear stress.

2.2. LightGBM (Light Gradient Boosting Machine) Approach

The dense microscopic fault dynamics data collected at different sensor locations in the granular fault system may reflect the macroscopic stick-slip behaviour differently. Some of these data likely significantly contribute to the ML model's prediction performance while others are merely redundant information due to potential feature correlations [39,40]. Therefore, filtering important input features and alleviating feature correlation are crucial in training ML prediction models, especially when abundant data are available.

This study employs the LightGBM (Light Gradient Boosting Machine) approach [36] to select important input features and train prediction models. LightGBM is built based on the Gradient Boosting Decision Tree (GBDT) and is one of the most versatile ML algorithms. Unlike other GBDT algorithms adopting a pre-sorting method for node splitting, LightGBM uses the histogram of input features to substantially reduce its memory consumption and computational cost [36]. Before training the model, we first convert the input feature data at each time instant into a histogram in the form of $\{(x_i, y_i),\ i = 1, 2, \ldots .n\}$, where x_i is the bin vector of input features and y_i is the corresponding bin values. Then, the K additive tree functions in the LightGBM algorithm are represented as

$$\begin{aligned} \hat{y}_i &= \sum_{k=1}^{K} f_k(x_i),\ \ f_k \in F \\ f(x) &= w_{q(x)} \end{aligned} , \quad (1$$

where f_k represents the classification and regression tree (CART) model; $w_{q(x)}$ is the score of the leaf node; $q_{(x)}$ is the specific leaf node; $\hat{y}_i$ is the prediction value; F is the set space that contains all trees. We utilize the root mean square error (RMSE) loss function during training. RMSE is a widely used evaluation metric that quantifies the disparity between predicted values and actual observations. By minimizing RMSE, our objective is to enhance the model performance in predicting the label. The mathematical expression of RMSE can be illustrated as

$$\text{RMSE} = \sqrt{\frac{1}{n}\sum_{i=1}^{n}(y_i - \hat{y}_i)^2}, \tag{2}$$

where i, from 1 to n, denotes the index of input instances, n is the number of instances, and y_i and $\hat{y}_i$ are the ground truth and prediction, respectively. Consequently, the objective function of all K additive trees can be simplified as

$$\sum_{t=1}^{K}\sqrt{\frac{1}{n}\sum_{i=1}^{n}(y_i - \hat{y}_i)^2} + \sum_{t=1}^{K}\Omega(f_t),\ f_t \in F, \tag{3}$$

where t, from 1 to K, denotes the index of the tth tree, and $\Omega(f_t)$ is the regularization value of the tth tree. To train the model, residuals are calculated in each iteration relative to the previous tree. The prediction for the tth iteration is the summation of the previous tree's prediction, i.e.,

$$\hat{y}_i^{(t)} = \sum\nolimits_t f_t(x_i) = \hat{y}_i^{(t-1)} + f_t(x_i), \tag{4}$$

where $\hat{y}^{(t-1)}$ is the summation of all the predicted values before the tth tree; $f_t(x_i)$ is the predicted value of the tth tree. One base decision tree is sequentially chosen and added to the ensemble model to reduce each iteration's residual by minimizing the objection function $F_{\text{obj}^{(t)}}$ during each iteration. The objection function of the tth iteration is

$$F_{\text{obj}^{(t)}} = \sqrt{\frac{1}{n}\sum_{i=1}^{n}\left(y_i - \left(\hat{y}^{(t-1)} + f_t(x_i)\right)\right)^2} + \Omega(f_t), \tag{5}$$

where $\Omega(f_t)$ is the regularization value of the tth tree. The LightGBM model can be continuously optimized through the growth process of the tree model for constructing feature splits. A detailed explanation of LightGBM is presented in Text S3 (Supplementary Material), and the typical structure of LightGBM is shown in Figure 2.

To fully use the input data and consider the commonly used training data proportion in ML, we clip the first 80% of each item of the collected time-series data (5–25 s) as the training dataset and the remaining 20% (25–30 s) for testing. In order to find the optimal hyperparameters, we divide the training dataset into five non-overlapping folds, i.e., using the first four folds for training and the last one for validation. The hyperparameters of the LightGBM model are tuned using grid search [41] and Bayesian optimization [42,43]. We first employ the Bayesian optimization to narrow the range of hyperparameters from a large solution space to a certain degree, and then use the grid search method to search for optimal hyperparameters in a small solution space. The optimal hyperparameters are selected from the group with the lowest training loss in cross-validation, as shown in Table S2 in the Supplementary Material. We use R^2, a normalized parameter ranging from 0 to 1 (from bad to good in terms of prediction accuracy), to calculate the gap between the prediction results and ground truth, i.e.,

$$R^2 = 1 - \frac{\sum_{i=1}^{n}(y_i - \hat{y}_i)^2}{\sum_{i=1}^{n}(y_i - \overline{y})^2}, \tag{6}$$

where $\overline{y}$ denotes the mean of the ground truth. The R^2 is a relative indicator of a machine learning model's prediction performance. Since we are comparing different machine learning models' performance on the same testing dataset in the present paper, the R^2 can satisfy our needs.

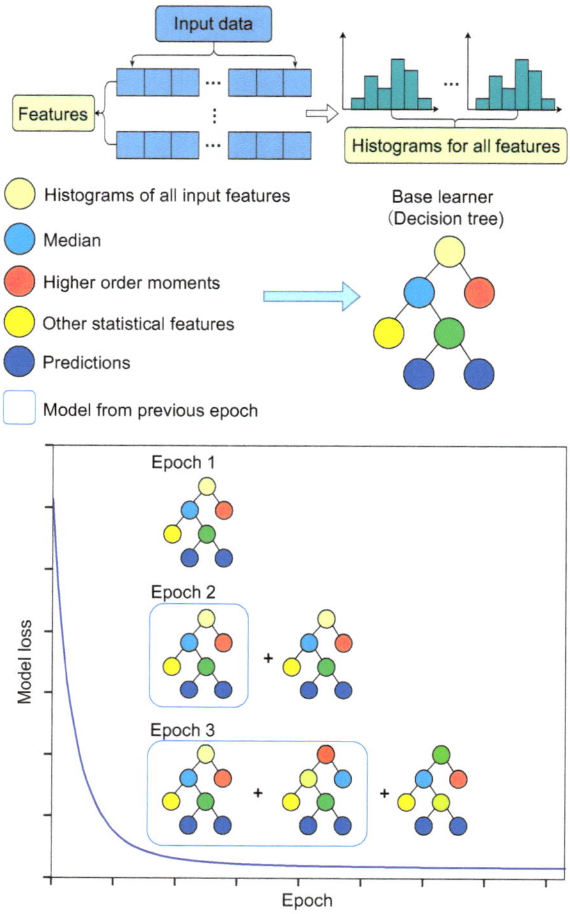

Figure 2. Schematic of LightGBM based on the gradient boosting decision tree (GBDT).

It is worth noting that for time-series data, the LightGBM works differently from deep-learning approaches, where a consecutive segment of time-series data from each input feature is often used. Here, we make a "now" prediction, and instead, the LightGBM models are trained step by step in temporal space, i.e., all input feature data points from the same time instant are collected and trained in each time step using a histogram-based algorithm to predict the normalized shear stress at the same time instance. Thus, temporal relationships between adjacent data points in each time-series feature are not considered in our LightGBM model. To achieve decent results, appropriate feature engineering should be conducted for time-series data to improve the prediction accuracy of LightGBM model. This can be completed by filtering the input feature data based on their importance ranking according to the information gained in training, which measures the predictive ability of a feature on the target by counting the number of times an input feature has been used in node splitting in a decision tree [44]. Additionally, the LightGBM approach can effectively preserve each feature's physical meaning, allowing for straightforward interpretations of prediction models compared to deep learning. Upon acquiring a decent LightGBM model, the SHAP value approach [44], as will be illustrated next, can be employed to quantitatively trace back and interpret the contribution of input features to the prediction results.

2.3. SHAP (Shapley Additive exPlanation) Value

Interpretation of input features is a crucial part of the applications of ML models to real-world problems since it facilitates our understanding of the underlying physics in the trained model and also helps evaluate whether the model could meet the initial expectation. After obtaining the best ML model, the contribution of each input feature can be quantified using the SHAP value approach, a method developed from cooperative game theory [44]. The SHAP value is an additive feature attribution method. Specifically, at a time instant, the summation of the SHAP values of all input features is equal to the predicted value minus the mean predicted value of all input features, i.e.,

$$\begin{aligned}\sum_{j=1}^{p} \phi_j(\hat{f}) &= \sum_{j=1}^{p} (\beta_j x_j - E(\beta_j X_j)) \\ &= (\beta_0 + \sum_{j=1}^{p} \beta_j x_j) - (\beta_0 + \sum_{j=1}^{p} E(\beta_j X_j)) \\ &= \hat{f}(x) - E(\hat{f}(X))\end{aligned} \tag{7}$$

Here, j, from 1 to p, denotes the index of input feature; $\hat{f}(x)$ represents the model prediction of input instant x; $\phi_j(\hat{f})$ represents the contribution of model prediction $\hat{f}(x)$ from input feature j; β_j is the weight of input feature j; x_j denotes the contribution of input feature j; X_j represents the set of model prediction of feature j; $E(\beta_j X_j)$ is the average contribution of prediction for feature j. By calculating the SHAP values, the evolution of the contributions of each input feature on the prediction results with respect to time can be obtained.

3. Predictions

This section demonstrates how the input features can be optimized to achieve better prediction performance. The purpose is to identify the appropriate procedure upon which the information in input features can be fully excavated. With the appropriate procedure, the laboratory slip events can be accurately and efficiently predicted when abundant frictional dynamics data are available.

3.1. Input Feature Data Optimization

As mentioned earlier, the fault dynamics data collected at 2203 densely distributed sensor points may contain redundant and correlated information. Here, we filter the data by first exploring the optimal number of input features that are representative of the fault system from an ML prediction viewpoint. This is achieved by training and testing a series of LightGBM models using data from an increasing number of sensor points, say, 1, 5, 10, 20, 50, 100, 200, 500, 1000, 1500, and 2203. For each of these numbers, we randomly draw data from all the 2203 sensor points using the specified number of sensors and then use the selected data to train a LightGBM model and test its performance. The procedure is repeated ten times for each number (except 2203 since all data are used) to alleviate uncertainty. Note that for each sensor, all four items of time-series data, i.e., V_x, V_y, D_x, and D_y, are used in the training and testing.

The testing performance of each trained LightGBM model in terms of R^2 and the required number of epochs for convergence in training are presented in Figure 3a,b. It is manifest that the prediction accuracy of the trained LightGBM models is poor when the number of sensor data used is less than 10. The accuracy significantly improves when the sensor number exceeds 20 (Figure 3a). As the number continues to increase, the R^2 mainly hovers around 0.8, while the models' training expense significantly increases (Figure 3b). Particularly, we reach an R^2 of 0.82 for the LightGBM model trained using all the 2203 sensor data without screening (Figure 3a); however, the time needed in training is very demanding, i.e., the training takes an epoch of ~3300 upon convergence (Figure 3b). These results demonstrate that the dense fault dynamics data at various sensor points on the shearing plates and inside the gouge do contain effective information regarding the shear stress of the fault, and the LightGBM model trained using a decent amount of sensor data can adequately capture the complex relations between microscopic frictional dynamic in the sheared granular fault system and the macroscopic slip events. In the meantime, th also confirms that the data from the 2203 sensors indeed have redundant information.

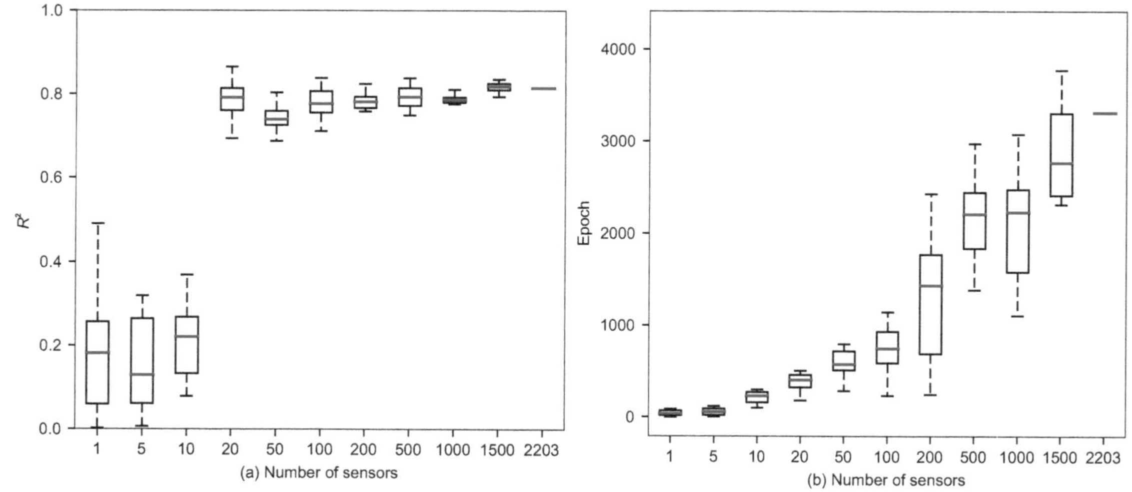

Figure 3. Box plots of the LightGBM model performance trained using the data from an increasin number of sensors. The training and testing are repeated ten times for each sensor number to obtai the statistics. (**a**) Statistics of the testing R^2 of the ten LightGBM models trained under each sens number. (**b**) Statistics of the required training epochs for convergence for the ten LightGBM model trained under each sensor number. Note that the model trained using all the 2203 sensor data is no repeated, and the results are indicated by a single blue bar.

Compromising the training efficiency and prediction accuracy, we choose an optima number of sensor points of ~20 for feature filtering, corresponding to ~80 (~20 × 4) inpu features. To optimize the 8812 input features, we calculate and rank the importance value of each input feature based on the LightGBM model trained using all the 2203 senso data. Finally, we choose a list of 88 features with top rankings as the optimized featur dataset. In contrast, the remaining features mainly have importance values in more tha one order smaller than the largest one. Note that the exact number of sensors related t these 88 selected features is more than 22 since maybe only 1–2 items of time-series dat from one sensor have been selected to form the optimized feature dataset. Due to spac limitations, only the top 50 optimized input features are presented in Figure 4.

3.2. Prediction Using Optimized Data

We proceed by retraining and testing a LightGBM model using the 88 optimize input features. As shown in Figure 5a, the ML model trained on the optimized feature generates an R^2 of 0.90 and RMSE of 0.0045 for testing, outperforming the previous mode trained using all 8812 input features (Figure 3a). This demonstrates that although onl 88 input features are used here, the trained model can achieve a better prediction accurac than the one trained using all data. In other words, the fault dynamics information in th 88 optimized input features is sufficient to represent the dynamics of the entire sheare granular fault system; in contrast, the remaining unused features are mainly redundan to the prediction of normalized shear stress. It is also worth noting here that the trainin efficiency is significantly improved, as evidenced by the fact that the RMSE converge after only ~100 epochs (see inset of Figure 5a), which contrasts the ~3300 epochs needed t converge in the previous training where all input features are used (Figure 3b). Additionall the RMSE evolution on the testing set shows no significant degradation, indicating n sign of overfitting. In general, training LightGBM models on the optimized feature dat can reduce the chance of model overfitting, improve prediction accuracy, and enhance training efficiency.

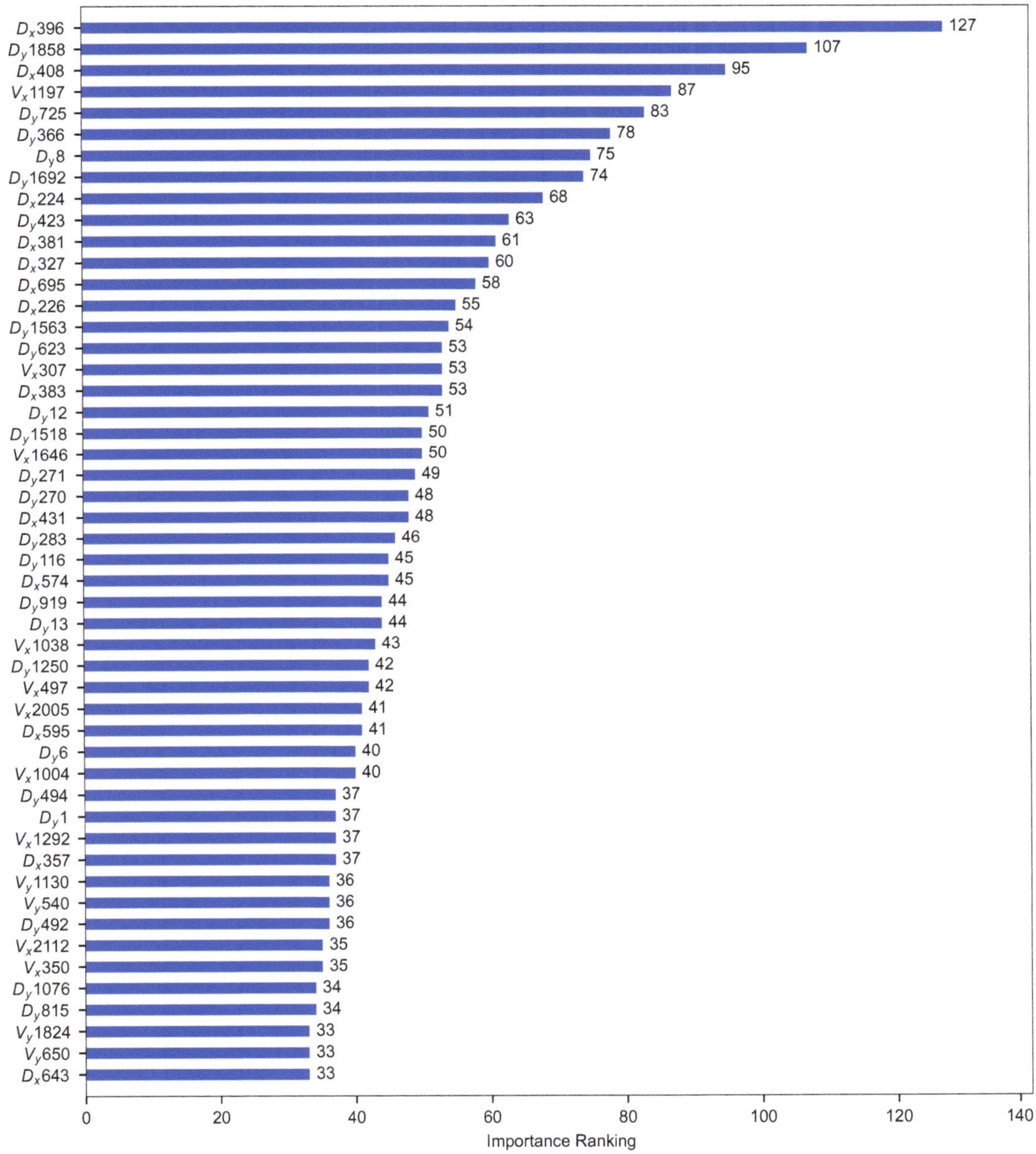

Figure 4. The top 50 optimized input features with higher importance values. For the names of these input features located on the left side of the horizontal bar, the symbols V_x, V_y, D_x, and D_y represent the velocity and displacement in the x and y directions, and the number following the symbol indicates the sensor ID mentioned in Section 2.1. The importance value of each input feature is given on the right side of the horizontal bar.

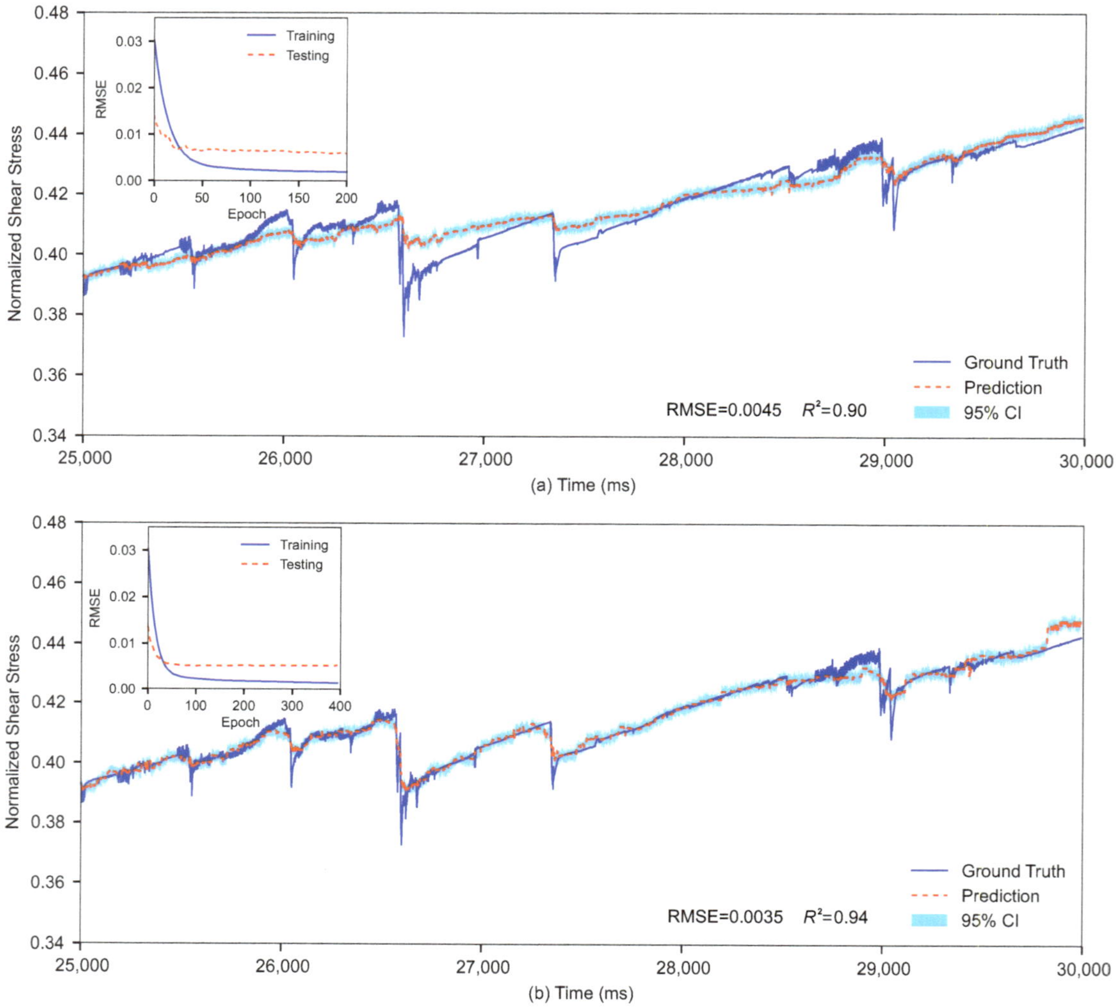

Figure 5. The prediction results of the two trained LightGBM models, together with the 95% conf dence intervals (light blue shaded region). (**a**) Performance of the model trained with the optimize sensor data on the testing set. (**b**) Performance of the model trained with the optimized sensor dat and their statistics on the testing set. The inset in each subfigure presents the evolution of RMSE fo the corresponding model during training and testing.

3.3. *Prediction Using Optimized Data and Their Statistics*

In laboratory-earthquake-related ML practices, to fully exploit the underlying informa tion from the obtained fault dynamics data and to achieve better prediction performanc the statistics of raw data are also commonly used as input features to train the predictio models [17,28,45]. Here, on top of the 88 optimized time-series data, we use the slidin time window approach to calculate 8 statistical parameters, such as the mean, varianc skewness, kurtosis, quartile, 1st percentile, 91st percentile, and median of the data locate in each window, for each item of the time-series data and comprises a total of 792 inpu features (88 $\times$ 9) for ML training and testing. The size of the sliding window used t calculate the statistics is 300 ms, which is chosen based on the fact that the periods of mos stick-slip cycles are within the range of 300 ms in our FDEM simulated results. Adjacen sliding windows overlap each other by 299 ms. Again, we train and test a LightGBM mode using the 792 new input features and obtain an improved R^2 = 0.94 and RMSE = 0.0035 for testing. As shown in Figure 5b, the predicted normalized shear stress shows a very good match with the actual values. Although the training takes more epochs (~200) before reaching stable results (inset of Figure 5b) than the previous training using only the 88 optimized feature data (inset of Figure 5a), compared with the first case using all the 8812 input features (Figure 3), the result of this model is notably superior in terms of both prediction accuracy and training efficiency.

Through the above progressive adjustment of the input features from raw data to the optimized data and to their statistics, the prediction performance of the trained LightGBM models has been gradually improved. Particularly, the statistics-based input features provide an in-depth excavation of the raw data, which helps to better capture the distribution characteristics and variation trends of the data, thereby improving the model's performance. We also provide a precise visualization of the predictive performance of the two LightGBM models (models related to Figure 5a,b) on the testing dataset in Figure S3 of the Supplementary Material.

One may easily argue that when statistics-based features are used, there are cases where slip events are predicted prior to actual events (e.g., the slip event at 29.0 s in Figure 5b), i.e., an artificial smoothing effect. However, as shown in Figure 5a, without using statistics-based features, there are also predicted slips before the actual slip events, e.g., the same slip event at 29.0 s. Therefore, we would attribute the smoothing largely to prediction inaccuracy. Given the complexity of our FDEM-simulated irregular stick-slip cycles, a precise prediction is challenging. Nevertheless, the model trained using the statistics-based features indeed improves the prediction accuracy, as is clearly evidenced by the increased R^2 from Figure 5a to Figure 5b.

4. Discussion

The above ML analyses based on the dense fault dynamics data extracted from the FDEM simulated sheared granular fault system demonstrate that upon appropriate optimization of input features, we can achieve acceptable performance and accuracy for laboratory earthquake prediction using LightGBM. However, ML models are often called "black boxes", which means that although we may reach accurate predictions, the mechanism behind the models remains elusive. To probe into the working mechanism of the LightGBM prediction model, we evaluate the role each input feature plays in the prediction using the SHAP value approach, which quantifies the direct contribution of an input feature to the predicted value at each instant of time. Specifically, the positive SHAP value of an input feature results in an increase in the predicted normalized shear stress at that time and vice versa. Based on the last LightGBM model with the highest prediction accuracy—the one trained using the optimized data and their statistics (i.e., 792 input features), we calculate the SHAP values of each input feature at each instant of time.

Due to space limitations, we first plot in Figure 6 the top 50 input features with large cumulative magnitudes of the yielded SHAP values (in descending order). The cumulative magnitude of SHAP values is the summation of the magnitudes (absolute values) of all yielded SHAP values for an input feature, indicating a feature's overall contribution to prediction at all times. The actual SHAP value versus the corresponding feature value (relative value from low to high, denoted by the colour bar) at each instant of time is presented in Figure 6. An enlarged version listing the top 200 input features is presented in Figure S4 in the Supplementary Material. These plots show that the different microscopic fault dynamics data types have completely distinct contributions to the predicted normalized shear stress.

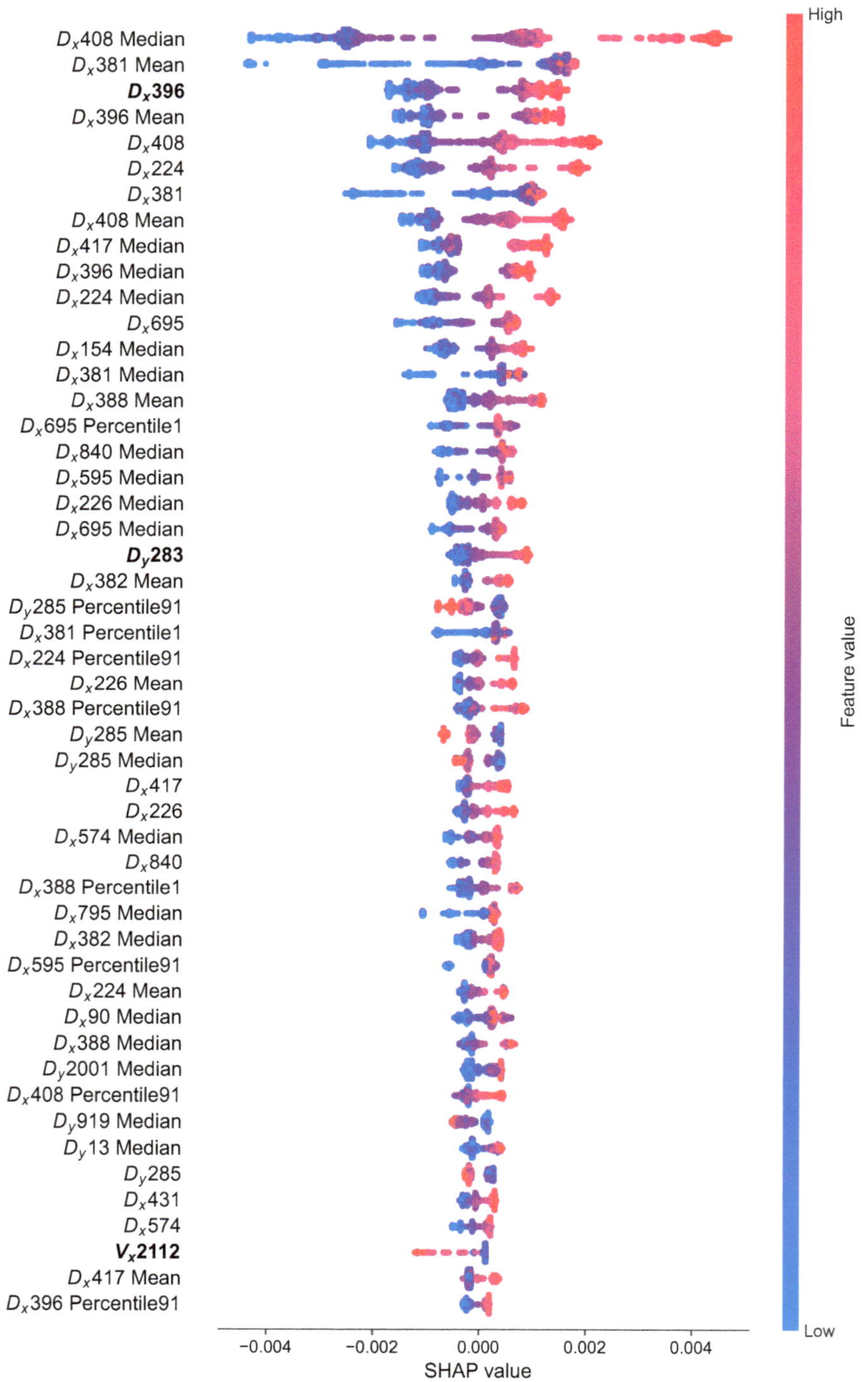

Figure 6. The SHAP value versus feature value at each instant of time for the top 50 input features with relatively larger cumulative magnitudes of the yielded SHAP values. The results are based on the last LightGBM model trained using the optimized input features and their statistics. Typical input features analysed in Figure 7 are marked in bold.

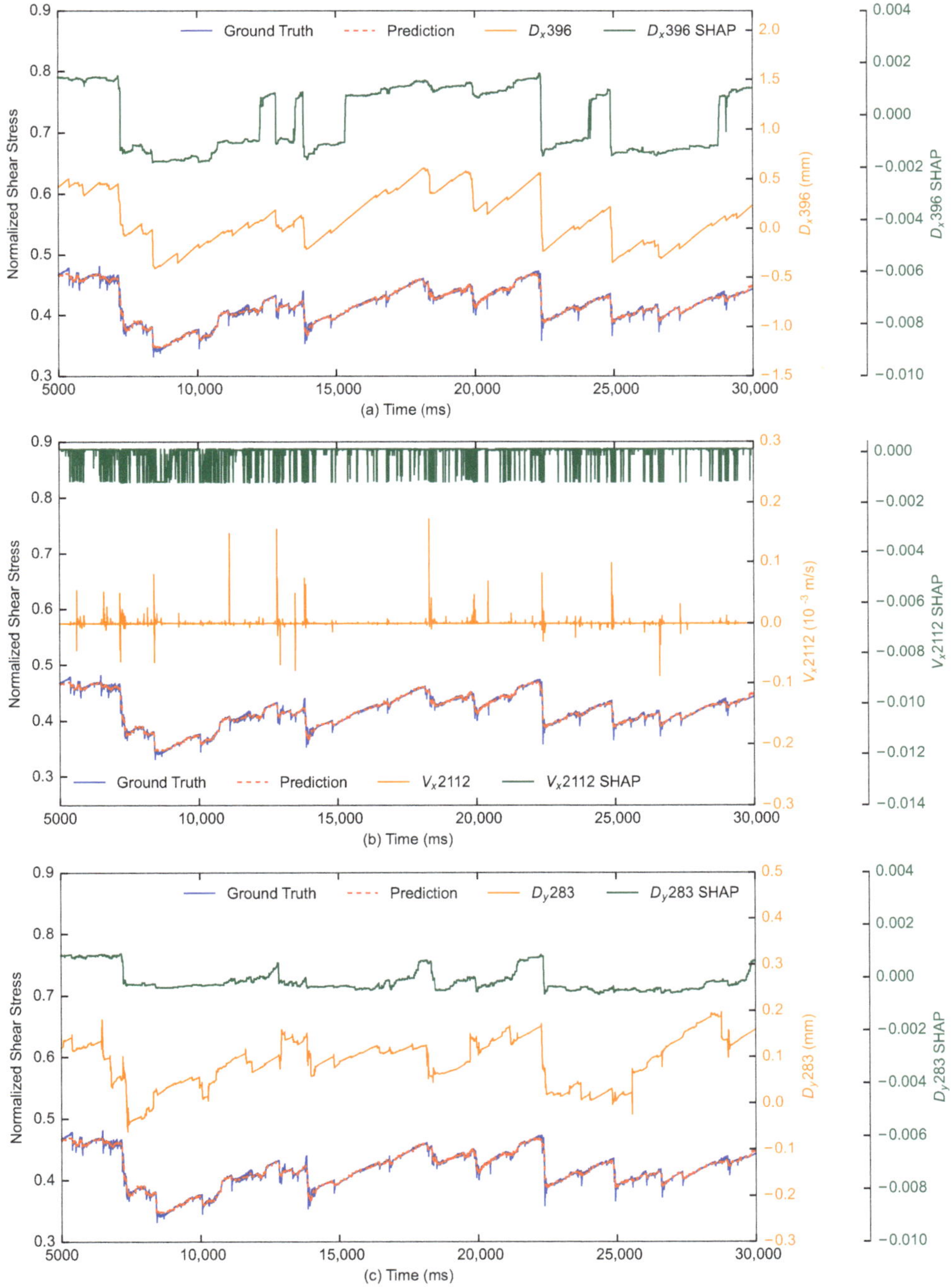

Figure 7. The evolution of SHAP values of typical features with respect to time for the LightGBM model trained using the optimized data and their statistics (i.e., 792 input features). (**a**) The evolution of SHAP values of D_x at sensor No. 396, where the increase/decrease in SHAP value has a similar trend to D_x. (**b**) The evolution of SHAP values of V_x at sensor No. 2112, where spikes occur at slips and cause a sharp drop in its SHAP value. (**c**) The evolution of SHAP values of D_y at sensor No. 283, where D_y shows a slight increase during the stick phases and sudden drop at slips; however, it mainly only causes a sharp increase in its SHAP value when a slip occurs.

The top input features shown in Figure 6 (e.g., the top 10–15) yield both positive an negative SHAP values. The behaviour of this type of input feature is further detaile in Figure 7a by the temporal evolution of the SHAP value of an exemplar D_x at sensc No. 396, alongside the change of D_x and the normalized shear stress for comparison. A shown in Figure 7a, the changing trend of the SHAP value is roughly consistent with th feature value, i.e., during the stick stages, both the feature value and SHAP value increas and sudden drops occur at slips for both. Moreover, the larger the feature value is, th higher the generated SHAP value will be (although imprecisely). The rough consistenc between the input feature and the corresponding SHAP value in both the stick and sli stages is commonplace for the majority of the top input features, especially the one related to D_x and its statistics, e.g., the data collected at sensors with No. 224, 408, an 417, among many others (Figure S4). This agrees with our intuitive understanding tha features with better prediction capability should have similar evolution patterns and pac to the prediction target. However, not all D_x-related features exhibit similar evolutio patterns to the predicted target. As shown in Figure S5 of the Supplementary Material, th correlation between the listed D_x features and the normalized shear stress is considerabl weak. Nevertheless, the SHAP value provides a powerful approach that helps filter th D_x-related features with high correlations and contributions to the prediction target.

There are also input features that mainly yield negative SHAP values, e.g., the V_x c sensor No. 2112 (Figure 6). As detailed in Figure 7b regarding its temporal evolutio versus the SHAP value and normalized shear stress, when a slip occurs, a negative SHA value will be generated, together with a spike in V_x (can be positive or negative). Howeve during the stick phases, V_x is very small, and the yielded SHAP value is mainly aroun zero, thus having no distinct contribution to the prediction. These demonstrate that thi type of input feature (V_x) mainly contributes to the prediction results at slips by bringin down the predicted normalized shear stress. Similar phenomena can also be found in th V_x of a few other sensors, e.g., the ones with No. 406, 1004, and 1888 (Figure S4).

On the contrary, a small number of input features behave the opposite, e.g., the D_y c sensor No. 283 (Figure 6). Specifically, when a slip occurs, this type of feature mainl generates positive SHAP values and pushes up the predicted normalized shear stres (Figure 7c). During stick phases, although we can also witness an increase in these featur values (similar to D_x), they mainly have minor contributions to the prediction result i.e., the yielded SHAP values are mainly around zero or of a small negative value. Thes input features include the D_y of sensors with No. 8, 12, and 285 (Figure S4). It is wort noting that the V_y of sensors barely has notable contributions to the prediction results, a there are only 4 input features related to V_y on the top 200 feature list shown in Figure S4

The SHAP values of input features can help uncover local friction dynamics' rol in laboratory earthquake prediction. From the above analyses, we find that the inpu features related to D_x mainly occupy the top of the list shown in Figure 6 (or Figure S4 simply because these top features evolve at a similar pace to the normalized shear stres in both the stick and slip phases. This is followed by the input features related to D However, compared with D_x, the magnitudes of the yielded SHAP values are relativel smaller. We suspect this may be partially caused by the fact that the input features relate to D_y only contribute to the prediction results at slips. We can observe a small numbe of input features related to V_x on the list in Figure 6 (or Figure S4), and they also onl contribute to the prediction results at slips. We barely see input features related to V_y i Figure S4. Details of the sensor numbers and their locations are presented in Figure Locations of the sensors related to the top 200 input features listed in Figure S4 are als presented in Figure 8 (coloured). All these demonstrate that, in general, the input feature related to displacements contribute more to the prediction results than those associate with velocity. In addition, the features related to the shearing direction (i.e., the x directior can better reflect the friction dynamics of the system and thus have higher contributions t the prediction results than those in the y direction.

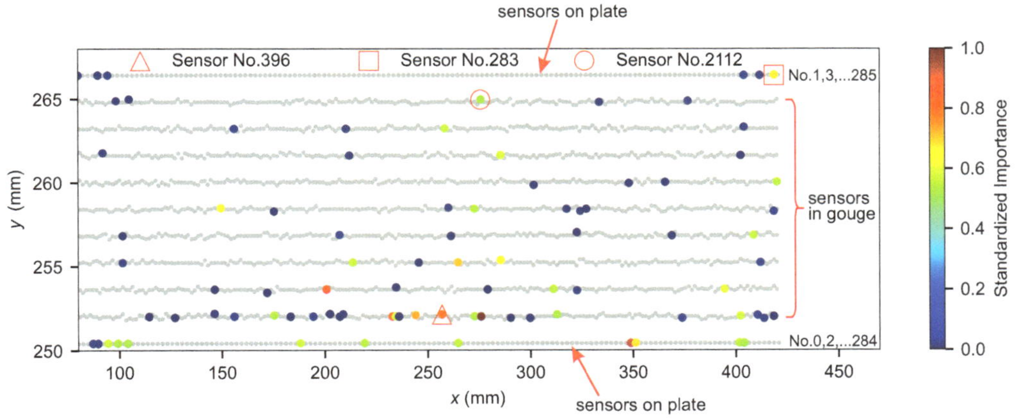

Figure 8. Sensor locations and their numbers in the granular gouge fault system. The sensors related to the top 200 input features presented in Figure S4 are coloured according to their normalized importance values.

Through the SHAP value analysis, accurate evaluations of the contribution of each feature to the predicted normalized shear stress can be obtained. Therefore, the LightGBM prediction model combined with the SHAP value analysis can quantify the impact of microscopic fault dynamics data on the macroscopic stick-slip response in a sheared granular fault system. The above analysis also implies that for natural earthquakes, and even tsunami prediction, the displacement data (or equivalent data) monitored near the fault may be more valuable than velocity data, and the data resolved along the fault shearing direction are more valuable than those perpendicular to the fault. With the SHAP value analysis, we can further optimize the earthquake prediction. The arrangement of natural earthquake sensor positions is a key factor. By introducing our model, we can optimize this arrangement. Specifically, we can select the positions corresponding to the most important features based on the analysis results of the model and arrange sensors at these positions. In this way, we can make the most of limited resources and improve the accuracy and efficiency of earthquake prediction.

It is worth noting that due to the large dimension of the dense fault dynamics data used in the present research (say, 8812), dimensionality reduction on the input features is necessary. The LightGBM approach employed here could achieve our goal of effectively filtering the redundant information from a large number of fault dynamics data and preserving the original physical meanings of the final optimized data, which facilitates the interpretability of the prediction model. Therefore, the LightGBM model together with the SHAP value analysis can efficiently and accurately predict laboratory earthquakes based on dense fault dynamics data and may also help explore valuable precursors for upcoming slip events.

5. Conclusions

In this paper, we have simulated a sheared granular fault system using the FDEM to collect fault dynamics data (V_x, V_y, D_x, and D_y) at 2203 "sensor" points during stick-slip cycles and trained a series of LightGBM models by optimizing and excavating the input features. The prediction performance of these models is tested upon the generation of stick-slips characterized by the normalized shear stress between the shearing plates and gouge. The purpose is to explore the appropriate ML model training procedure and achieve better laboratory earthquake predictions when abundant fault dynamics data are available. The SHAP values of input features are also calculated to quantify the contribution of each input feature to the prediction results.

The ML study uses the sensor data as input features and the gouge-plate normalize shear stress as the label, and the first 80% of the time-series proportion is reserved fc training and the remaining 20% for testing. We first optimize the 8812 input feature by randomly selecting an increasing number of sensor data and training and testing th LightGBM models' performance. When the number of sensors used in training exceeds 2 the R^2 mainly hovers around 0.8, while the models' training expense significantly increase We choose an optimal number of sensor points of ~20 by considering both the model' prediction accuracy and training efficiency. Then, 88 input features with top importanc are selected as the optimized dataset based on the LightGBM model trained using all th 8812 input features. Following this, the LightGBM model trained using the optimize dataset yields an R^2 of 0.90; however, the training efficiency is remarkably enhanced. The to fully use the underlying fault dynamics information, by calculating the 8 statistica parameters of the 88 optimized features, we compose an enlarged dataset with 792 inpu features and retrain a LightGBM model with a significantly improved prediction accurac of $R^2 = 0.94$. Based on the final trained LightGBM model, we extract the SHAP values c each input feature and compare them with the normalized shear stress. We find that th input features related to displacements contribute more to the prediction results than thos associated with velocity, and the ones in the x direction (fault shearing direction) can bette reflect the friction dynamics of the system than those in the y direction.

The analyses demonstrate that the dense fault dynamics data obtained from the goug and plates contain the necessary information to train appropriate ML models so that th fault friction state can be adequately predicted. However, the dense fault dynamics dat may contain redundant information and increase training expenses. Using ML approache to screen out correlated information and optimize the data can facilitate the trainin process and achieve better prediction results. Additionally, appropriate utilization of th statistics of the fault dynamics data may help extract important hidden information fror the data and enhance the prediction performance. Finally, the SHAP values obtained wit the trained LightGBM model can quantify the contribution of each input feature to th prediction results.

The data collection process from earthquake sensors is vital. Aim to gather informa tion from sensors that are rich in earthquake catalogues and closely tied to earthquak occurrences. Trying to gather data from sensors deployed in locations that the machin learning model considers to be of high importance will significantly aid in earthquak prediction. This study offers useful guidance for this process. Constructing a predictio model requires diligent data screening, feature extraction, and model optimization. Thes steps help to create a reliable representation of complex physical processes. This researc provides an effective example of how to create an efficient prediction framework. Th SHAP value analysis emphasizes the importance of features such as displacement and di placement statistics. This insight offers a fresh viewpoint for exploring potent earthquak precursor signals and setting up early warning systems. Our work has broad application While it can predict laboratory earthquakes, its framework, methods, and conclusion provide a solid base for forecasting modelling processes to natural faults. This broadens th potential for employing machine learning in natural earthquake prediction. This work ca shed light on natural earthquake prediction in terms of selecting valuable monitoring dat and training appropriate ML models and also opens new possibilities to explore valuabl precursors for earthquake prediction.

Supplementary Materials: The following supporting information can be downloaded at: https:/ www.mdpi.com/article/10.3390/jmse12020246/s1. Text S1: Brief introduction of FDEM; Text S2 Supplementary explanation of FDEM model setup; Text S3: Light Gradient Boosting Machin (LightGBM); Figure S1: Numerical representation of the granular fault system using DEM and FDEM (a) In DEM, the plates are simplified by a set of bonded particles, and the gouge layer is composed o a series of rigid particles. Therefore, both the gouge particles and the shearing plates cannot deforn (b) In FDEM, the plates are explicitly represented, and both plates and particles are further discretize into finite elements to capture their detailed deformation and movement; Figure S2: Probabilit density distribution of the seismic moment of all slip events, where the detailed calculation of the moment is explained by Gao et al. [31]. The results agree with the physical experiment data collected in Geller et al. [37]. The probability density distribution is consistent with the Gutenberg-Richter distribution and is predicted to scale as $M^{-3/2}$ (the power $-3/2$ is within the range of -1.4 to -1.8 observed in natural earthquakes; Figure S3: Performance of the two LightGBM models trained with different datasets. (a) Training with optimized sensor data. (b) Training with optimized sensor data and their statistics; Figure S4: The SHAP value versus the feature value at each instant of time for the top 200 input features with relatively larger cumulative magnitudes of the yielded SHAP values. The results are based on the last LightGBM model trained using the optimized dataset and their statistics; Figure S5: Evolutions of Dx for several sensors and their comparison with the normalized shear stress. (a) Normalized shear stress. (b) Dx of sensor No. 95. (c) Dx of sensor No. 295. (d) Dx of sensor No. 1832; Table S1: Material and numerical simulation parameters; Table S2: The optimal hyperparameters of the LightGBM model trained using 8812 features; References [31–42,45–60] are citied in the Supplementary Materials.

Author Contributions: W.H.: conceptualization, methodology, software, investigation, formal analysis, writing original draft. K.G.: conceptualization, resources, funding acquisition, supervision, writing review and editing. Y.F.: validation, writing review and editing. All authors have read and agreed to the published version of the manuscript.

Funding: This work is supported by the National Natural Science Foundation of China (42374070), the National Key R&D Program of China (2022YFF0800601), and the Guangdong Provincial Key Laboratory of Geophysical High-resolution Imaging Technology (2022B1212010002).

Institutional Review Board Statement: Not applicable.

Informed Consent Statement: Not applicable.

Data Availability Statement: The data used in the paper will be available upon request to the corresponding author. The Supplementary Material includes an expanded explanation of the FDEM method and model setup and a brief introduction to the LightGBM approach.

Acknowledgments: We sincerely thank two anonymous reviewers for their critical comments and constructive suggestions, which have greatly improved the quality of this paper.

Conflicts of Interest: The authors declare no conflict of interest.

References

1. Meng, M.; Ge, H.; Shen, Y.; Ji, W.; Wang, Q. Rock Fabric of Tight Sandstone and Its Influence on Irreducible Water Saturation in Eastern Ordos Basin. *Energy Fuels* **2023**, *37*, 3685–3696. [CrossRef]
2. Asencio-Cortés, G.; Martínez-Álvarez, F.; Morales-Esteban, A.; Reyes, J. A Sensitivity Study of Seismicity Indicators in Supervised Learning to Improve Earthquake Prediction. *Knowl.-Based Syst.* **2016**, *101*, 15–30. [CrossRef]
3. Asim, K.M.; Idris, A.; Iqbal, T.; Martínez-Álvarez, F. Earthquake Prediction Model Using Support Vector Regressor and Hybrid Neural Networks. *PLoS ONE* **2018**, *13*, e0199004. [CrossRef] [PubMed]
4. Allen, R.M.; Melgar, D. Earthquake Early Warning: Advances, Scientific Challenges, and Societal Needs. *Annu. Rev. Earth Planet. Sci.* **2019**, *47*, 361–388. [CrossRef]
5. Brykov, M.N.; Petryshynets, I.; Pruncu, C.I.; Efremenko, V.G.; Pimenov, D.Y.; Giasin, K.; Sylenko, S.A.; Wojciechowski, S. Machine Learning Modelling and Feature Engineering in Seismology Experiment. *Sensors* **2020**, *20*, 4228. [CrossRef] [PubMed]
6. Corbi, F.; Bedford, J.; Sandri, L.; Funiciello, F.; Gualandi, A.; Rosenau, M. Predicting Imminence of Analog Megathrust Earthquakes with Machine Learning: Implications for Monitoring Subduction Zones. *Geophys. Res. Lett.* **2020**, *47*, e2019GL086615. [CrossRef]
7. Asim, K.M.; Martínez-Álvarez, F.; Basit, A.; Iqbal, T. Earthquake Magnitude Prediction in Hindukush Region Using Machine Learning Techniques. *Nat. Hazards* **2016**, *85*, 471–486. [CrossRef]
8. Mousavi, S.M.; Beroza, G.C. A Machine-Learning Approach for Earthquake Magnitude Estimation. *Geophys. Res. Lett.* **2020**, *47*, e2019GL085976. [CrossRef]
9. Beroza, G.C.; Segou, M.; Mousavi, S.M. Machine Learning and Earthquake Forecasting-Next Steps. *Nat. Commun.* **2021**, *12*, 4761. [CrossRef]
10. Johnson, C.W.; Johnson, P.A. Learning the Low Frequency Earthquake Activity on the Central San Andreas Fault. *Geophys. Res. Lett.* **2021**, *48*, e2021GL092951. [CrossRef]
11. Brace, W.F.; Byerlee, J.D. Stick-Slip as a Mechanism for Earthquakes. *Science* **1966**, *153*, 990–992. [CrossRef] [PubMed]
12. Tinti, E.; Scuderi, M.M.; Scognamiglio, L.; Di Stefano, G.; Marone, C.; Collettini, C. On the Evolution of Elastic Properties During Laboratory Stick-Slip Experiments Spanning the Transition from Slow Slip to Dynamic Rupture. *J. Geophys. Res. Solid Earth* **2016**, *121*, 8569–8594. [CrossRef]

13. Leeman, J.R.; Saffer, D.M.; Scuderi, M.M.; Marone, C. Laboratory Observations of Slow Earthquakes and the Spectrum of Tecton Fault Slip Modes. *Nat. Commun.* **2016**, *7*, 11104. [CrossRef] [PubMed]
14. Rouet-Leduc, B.; Hulbert, C.; Bolton, D.C.; Ren, C.X.; Riviere, J.; Marone, C.; Guyer, R.A.; Johnson, P.A. Estimating Fault Frictic from Seismic Signals in the Laboratory. *Geophys. Res. Lett.* **2018**, *45*, 1321–1329. [CrossRef]
15. Bolton, D.C.; Shreedharan, S.; Rivière, J.; Marone, C. Acoustic Energy Release During the Laboratory Seismic Cycle: Insights c Laboratory Earthquake Precursors and Prediction. *J. Geophys. Res. Solid Earth* **2020**, *125*, e2019JB018975. [CrossRef]
16. Bolton, D.C.; Shreedharan, S.; McLaskey, G.C.; Rivière, J.; Shokouhi, P.; Trugman, D.T.; Marone, C. The High-Frequency Signatu of Slow and Fast Laboratory Earthquakes. *J. Geophys. Res. Solid Earth* **2022**, *127*, e2022JB024170. [CrossRef]
17. Rouet-Leduc, B.; Hulbert, C.; Lubbers, N.; Barros, K.; Humphreys, C.J.; Johnson, P.A. Machine Learning Predicts Laborato Earthquakes. *Geophys. Res. Lett.* **2017**, *44*, 9276–9282. [CrossRef]
18. Bolton, D.C.; Shokouhi, P.; Rouet-Leduc, B.; Hulbert, C.; Rivière, J.; Marone, C.; Johnson, P.A. Characterizing Acoustic Signals ar Searching for Precursors During the Laboratory Seismic Cycle Using Unsupervised Machine Learning. *Seismol. Res. Lett.* **2019**, 9 1088–1098. [CrossRef]
19. Wang, K.; Johnson, C.W.; Bennett, K.C.; Johnson, P.A. Predicting Future Laboratory Fault Friction through Deep Learnir Transformer Models. *Geophys. Res. Lett.* **2022**, *49*, e2022GL098233. [CrossRef]
20. Hazzard, J.F.; Mair, K. The Importance of the Third Dimension in Granular Shear. *Geophys. Res. Lett.* **2003**, *30*, 1708. [CrossRef
21. Abe, S.; Mair, K. Grain Fracture in 3D Numerical Simulations of Granular Shear. *Geophys. Res. Lett.* **2005**, *32*, L05305. [CrossRe
22. Mair, K.; Hazzard, J.F. Nature of Stress Accommodation in Sheared Granular Material: Insights from 3D Numerical Modelin *Earth Planet. Sci. Lett.* **2007**, *259*, 469–485. [CrossRef]
23. Mair, K.; Abe, S. 3D Numerical Simulations of Fault Gouge Evolution During Shear: Grain Size Reduction and Strain Localizatio *Earth Planet. Sci. Lett.* **2008**, *274*, 72–81. [CrossRef]
24. Griffa, M.; Ferdowsi, B.; Guyer, R.A.; Daub, E.G.; Johnson, P.A.; Marone, C.; Carmeliet, J. Influence of Vibration Amplitude o Dynamic Triggering of Slip in Sheared Granular Layers. *Phys. Rev. E* **2013**, *87*, 012205. [CrossRef]
25. Ferdowsi, B. Discrete Element Modeling of Triggered Slip in Faults with Granular Gouge. Application to Dynamic Earthquak Triggering. Ph.D. Thesis, ETH-Zürich, Zürich, Switzerland, 2014.
26. Dorostkar, O.; Guyer, R.A.; Johnson, P.A.; Marone, C.; Carmeliet, J. On the Micromechanics of Slip Events in Sheared, Flui Saturated Fault Gouge. *Geophys. Res. Lett.* **2017**, *44*, 6101–6108. [CrossRef]
27. Wang, C.; Elsworth, D.; Fang, Y. Influence of Weakening Minerals on Ensemble Strength and Slip Stability of Faults. *J. Geophy Res. Solid Earth* **2017**, *122*, 7090–7110. [CrossRef]
28. Ren, C.X.; Dorostkar, O.; Rouet-Leduc, B.; Hulbert, C.; Strebel, D.; Guyer, R.A.; Johnson, P.A.; Carmeliet, J. Machine Learnin Reveals the State of Intermittent Frictional Dynamics in a Sheared Granular Fault. *Geophys. Res. Lett.* **2019**, *46*, 7395–740 [CrossRef]
29. Chen, T.; He, T.; Benesty, M.; Khotilovich, V.; Tang, Y.; Cho, H.; Chen, K. Xgboost: Extreme Gradient Boosting. *R Package Versio 0.4-2* **2015**, *1*, 1–4.
30. Ma, G.; Mei, J.; Gao, K.; Zhao, J.; Zhou, W.; Wang, D. Machine Learning Bridges Microslips and Slip Avalanches of Sheare Granular Gouges. *Earth Planet. Sci. Lett.* **2022**, *579*, 117366. [CrossRef]
31. Gao, K.; Euser, B.J.; Rougier, E.; Guyer, R.A.; Lei, Z.; Knight, E.E.; Carmeliet, J.; Johnson, P.A. Modeling of Stick-Slip Behavic in Sheared Granular Fault Gouge Using the Combined Finite-Discrete Element Method. *J. Geophys. Res. Solid Earth* **2018**, 12 5774–5792. [CrossRef]
32. Gao, K.; Guyer, R.; Rougier, E.; Ren, C.X.; Johnson, P.A. From Stress Chains to Acoustic Emission. *Phys. Rev. Lett.* **2019**, *123*, 04800 [CrossRef]
33. Gao, K.; Guyer, R.A.; Rougier, E.; Johnson, P.A. Plate Motion in Sheared Granular Fault System. *Earth Planet. Sci. Lett.* **202** *548*, 116481. [CrossRef]
34. Munjiza, A.; Lei, Z.; Divic, V.; Peros, B. Fracture and Fragmentation of Thin Shells Using the Combined Finite-Discrete Elemer Method. *Int. J. Numer. Methods Eng.* **2013**, *95*, 478–498. [CrossRef]
35. Wang, K.; Johnson, C.W.; Bennett, K.C.; Johnson, P.A. Predicting Fault Slip Via Transfer Learning. *Nat. Commun.* **2021**, *12*, 731 [CrossRef] [PubMed]
36. Ke, G.; Meng, Q.; Finley, T.; Wang, T.; Chen, W.; Ma, W.; Ye, Q.; Liu, T.-Y. Lightgbm: A Highly Efficient Gradient Boosting Decisio Tree. In Proceedings of the 31st International Conference on Neural Information Processing Systems, Long Beach, CA, USA, 4– December 2017; pp. 3149–3157.
37. Gutenberg, B.; Richter, C.F. Magnitude and Energy of Earthquakes. *Nature* **1955**, *176*, 795. [CrossRef]
38. Khosravikia, F.; Clayton, P. Machine Learning in Ground Motion Prediction. *Comput. Geosci.* **2021**, *148*, 104700. [CrossRef]
39. Ren, C.X.; Peltier, A.; Ferrazzini, V.; Rouet-Leduc, B.; Johnson, P.A.; Brenguier, F. Machine Learning Reveals the Seismic Signatur of Eruptive Behavior at Piton De La Fournaise Volcano. *Geophys. Res. Lett.* **2020**, *47*, e2019GL085523. [CrossRef] [PubMed]
40. Bergstra, J.; Bengio, Y. Random Search for Hyper-Parameter Optimization. *J. Mach. Learn. Res.* **2012**, *13*, 281–305.
41. Martinez-Cantin, R. Bayesopt: A Bayesian Optimization Library for Nonlinear Optimization, Experimental Design and Bandit *J. Mach. Learn. Res.* **2014**, *15*, 3735–3739.
42. Victoria, A.H.; Maragatham, G. Automatic Tuning of Hyperparameters Using Bayesian Optimization. *Evol. Syst.* **2021**, 1 217–223. [CrossRef]

3. Lundberg, S.M.; Erion, G.G.; Lee, S.-I. Consistent Individualized Feature Attribution for Tree Ensembles. *arXiv* **2018**, arXiv:1802.03888.
4. Johnson, P.A.; Rouet-Leduc, B.; Pyrak-Nolte, L.J.; Beroza, G.C.; Marone, C.J.; Hulbert, C.; Howard, A.; Singer, P.; Gordeev, D.; Karaflos, D.; et al. Laboratory Earthquake Forecasting: A Machine Learning Competition. *Proc. Natl. Acad. Sci. USA* **2021**, *118*, e2011362118. [CrossRef]
5. Dorostkar, O.; Guyer, R.A.; Johnson, P.A.; Marone, C.; Carmeliet, J. On the role of fluids in stick-slip dynamics of saturated granular fault gouge using a coupled computational fluid dynamics-discrete element approach. *J. Geophys. Res. Solid Earth* **2017**, *122*, 3689–3700. [CrossRef]
6. Geller, D.A.; Ecke, R.E.; Dahmen, K.A.; Backhaus, S. Stick-Slip Behavior in a Continuum-Granular Experiment. *Phys. Rev. E Stat. Nonlin. Soft Matter Phys.* **2015**, *92*, 060201. [CrossRef]
7. Euser, B.; Rougier, E.; Lei, Z.; Knight, E.E.; Frash, L.P.; Carey, J.W.; Viswanathan, H.; Munjiza, A. Simulation of Fracture Coalescence in Granite via the Combined Finite–Discrete Element Method. *Rock Mech. Rock Eng.* **2019**, *52*, 3213–3227. [CrossRef]
8. Lei, Q.; Gao, K. Correlation between Fracture Network Properties and Stress Variability in Geological Media. *Geophysical Research Letters* **2018**, *45*, 3994–4006. [CrossRef]
9. Lei, Z.; Rougier, E.; Knight, E.E.; Munjiza, A.A.; Viswanathan, H. A generalized anisotropic deformation formulation for geomaterials. *Comput. Part. Mech.* **2016**, *3*, 215–228. [CrossRef]
0. Lei, Z.; Rougier, E.; Munjiza, A.; Viswanathan, H.; Knight, E.E. Simulation of discrete cracks driven by nearly incompressible fluid via 2D combined finite-discrete element method. *Int. J. Numer. Anal. Methods Geomech.* **2019**, *43*, 1724–1743. [CrossRef]
1. MiDi, G. On dense granular flows. *Eur. Phys. J. E* **2004**, *14*, 341–365. [CrossRef]
2. Munjiza, A.; Andrews, K. NBS contact detection algorithm for bodies of similar size. *Int. J. Numer. Methods Eng.* **1998**, *43*, 131–149. [CrossRef]
3. Munjiza, A.; Rougier, E.; John, N.W.M. MR linear contact detection algorithm. *Int. J. Numer. Methods Eng.* **2006**, *66*, 46–71. [CrossRef]
4. Munjiza, A.A. Discrete Elements in Transient Dynamics of Fractured Media. Ph.D. Thesis, Swansea University, Swansea, UK, 1992.
5. Munjiza, A.A. *The Combined Finite-Discrete Element Method*; John Wiley & Sons: Hoboken, NJ, USA, 2004.
6. Munjiza, A.A.; Knight, E.E.; Rougier, E. *Computational Mechanics of Discontinua*; John Wiley & Sons: Hoboken, NJ, USA, 2011.
7. Munjiza, A.A.; Rougier, E.; Knight, E.E. *Large Strain Finite Element Method: A Practical Course*; John Wiley & Sons: Hoboken, NJ, USA, 2014.
8. Okubo, K.; Bhat, H.S.; Rougier, E.; Marty, S.; Schubnel, A.; Lei, Z.; Knight, E.E.; Klinger, Y. Dynamics, radiation and overall energy budget of earthquake rupture with coseismic off-fault damage. *J. Geophys. Res. Solid Earth* **2019**, *124*, 11771–11801. [CrossRef]
9. Rougier, E.; Munjiza, A.; Lei, Z.; Chau, V.T.; Knight, E.E.; Hunter, A.; Srinivasan, G. The combined plastic and discrete fracture deformation framework for FDEM. *Int. J. Numer. Methods Eng.* **2019**, *121*, 1020–1035. [CrossRef]
0. Tatone, B.S.A.; Grasselli, G. A calibration procedure for two-dimensional laboratory-scale hybrid finite–discrete element simulations. *Int. J. Rock Mech. Min. Sci.* **2015**, *75*, 56–72. [CrossRef]

Journal of
Marine Science and Engineering

MDPI

Article

Study on the Mechanism of Natural Gas Hydrate Decomposition and Seabed Seepage Triggered by Mass Transport Deposits

Pengqi Liu [1,2], Wei Zhang [1,2,*], Shuang Mao [3,4,*], Pibo Su [1,2], Huaizhen Chen [5] and Liguo Hu [6]

1 Sanya Institute of South China Sea Geology, Guangzhou Marine Geological Survey, Sanya 572025, China; liupengqi@mail.cgs.gov.cn (P.L.); spb_525@sina.com (P.S.)
2 Academy of South China Sea Geological Science, China Geological Survey, Sanya 572025, China
3 Economics & Technology Research Institute of Liaohe Oilfield Company, PetroChina, Panjin 124000, China
4 School of Geosciences, Yangtze University, Wuhan 430100, China
5 State Key Laboratory of Marine Geology, School of Ocean and Earth Science, Tongji University, Shanghai 200092, China; huaizhenchen@tongji.edu.cn
6 PetroChina Liaohe Oilfield Company, Panjin 124010, China; hulg@petrochina.com.cn
* Correspondence: zwei_a@mail.cgs.gov.cn (W.Z.); ms1991@petrochina.com.cn (S.M.)

Abstract: Previous studies indicate that mass transport deposits are related to the dynamic accumulation of natural gas hydrates and gas leakage. This research aims to elucidate the causal mechanism of seabed seepage in the western region of the southeastern Qiongdongnan Basin through the application of seismic interpretation and attribute fusion techniques. The mass transport deposits, bottom simulating reflector, submarine mounds, and other phenomena were identified through seismic interpretation techniques. Faults and fractures were identified by utilizing variance attribute analysis. Gas chimneys were identified using instantaneous frequency attribute analysis. Free gas and paleo-seepage points were identified using sweetness attributes, enabling the analysis of fluid seepage pathways and the establishment of a seepage evolution model. Research has shown that in areas where the mass transport deposits develop thicker layers, there is a greater uplift of the bottom boundary of the gas hydrate stability zone, which can significantly alter the seafloor topography. Conversely, the opposite is true. The research indicates that the upward migration of the gas hydrate stability zone, induced by the mass transport deposits in the study area, can result in the rapid decomposition of gas hydrates. The gas generated from the decomposition of gas hydrates is identified as the principal factor responsible for inducing seabed seepage. Moderate- and low-speed natural gas seepage can create spiny seamounts and domed seamounts, respectively.

Keywords: Qiongdongnan Basin; gas hydrate; seafloor seepage mechanism; attribute analysis technology; the mass transported deposits; spiny seamounts; domed seamounts

Citation: Liu, P.; Zhang, W.; Mao, S.; Su, P.; Chen, H.; Hu, L. Study on the Mechanism of Natural Gas Hydrate Decomposition and Seabed Seepage Triggered by Mass Transport Deposits. *J. Mar. Sci. Eng.* **2024**, *12*, 646. https://doi.org/10.3390/jmse12040646

Academic Editor: Atilla Incecik

Received: 10 February 2024
Revised: 28 March 2024
Accepted: 2 April 2024
Published: 12 April 2024

1. Introduction

Submarine gas seepage represents a ubiquitous and dynamic geological phenomenon within the marine environment. It involves the migration of various fluids, such as shallow gas, submarine groundwater, cold seep fluid, hydrothermal fluid, and liquefied fine-grained sediments. These fluids flow through predominant pathways, like faults, fractures, mud volcanoes, or gas chimneys beneath the seabed interface, eventually reaching their discharge or eruption on the ocean floor [1]. Such seepage occurrences have been documented across various marine environments, spanning passive and active continental margins, including the South China Sea, North Sea, Baltic Sea, Black Sea, North Atlantic, Gulf of Mexico, Sea of Japan, Japan Sea Trench, Mediterranean Sea, Indian Ocean, and Southwest Pacific [2–8].

The manifestation of submarine gas seepage is intricately linked to geological processes and is influenced by factors, such as rapid sedimentation, tectonic uplift, erosion, faulting, and hydrate decomposition [9–12]. During periods of rapid sedimentation, substantial sediment accumulation can result in the formation of thick subsurface layers. This can lead to elevated pore water pressure and the subsequent migration of pre-existing subsurface fluids due to compaction-related compression. Tectonic uplift and erosion have complex effects on the migration of subsurface fluids, altering flow paths and velocities, and impacting storage and migration processes [13,14]. Faulting can disrupt upper fluid seals, which can facilitate vertical fluid migration [15]. Submarine gas seepage serves as an indicator of the presence of natural gas hydrates and hydrocarbon accumulations within sedimentary layers. In regions with intense gas seepage, high gas hydrate saturation is often observed on or below the seabed [16]. Additionally, gas hydrates may contribute to geological formation fracturing, further promoting submarine seepage occurrences [17,18]. Such seepage events significantly modify seafloor topography and geomorphology, giving rise to various submarine micro-landform types, including pockmarks, authigenic carbonate crusts, spiny seamounts, domed seamounts, and mud volcanoes [19].

The research focus area is located within the deepwater region of the northern South China Sea continental slope. Previous studies have investigated submarine micro-landforms and fluid seepage in this area, uncovering various types of micro-landforms, such as pockmarks, spiny seamounts, and domed seamounts. Marine geological investigations over several years have established a close correlation between submarine micro-landform formation, gas hydrate formation, and decomposition, in addition to the widespread development of the mass transport deposit (MTD) systems [20]. However, current research inadequately examines the causal mechanisms of submarine seepage involving hydrates and MTDs. This study aims to explore the interrelationships among natural gas hydrates, submarine micro-landforms, and MTD sedimentary systems, elucidating their impacts on submarine gas seepage mechanisms, particularly focusing on the influence of MTDs on hydrate decomposition and seafloor seepage intensity.

Methodologically, this study employs variance attribute volume, sweetness attribute volume, and instantaneous frequency attribute volume fusion analysis techniques to identify geological phenomena, such as MTDs, bottom simulating reflectors (BSRs), spiny seamounts, fault systems, gas chimneys, and paleo-seepage points. Through relational analysis, a unique model for the evolution of seafloor fluid seepage specific to the study area is proposed. This contributes to a deeper understanding of seafloor geomorphology formation, evolution, and fluid seepage mechanisms.

This study utilizes seismic attribute fusion technology to assess the influence of MTDs on the gas hydrate stability zone, uncovering their effects on submarine natural gas seepage and micro-landforms. Additionally, a unique fluid seepage genesis model specific to the study area is proposed, expanding upon previous research that focused solely on the impact of MTDs on subsurface fluids. References remain unchanged.

2. Geologic Setting, Data, and Method

The Qiongdongnan Basin (QDNB) is located in the northwestern part of the South China Sea, to the southeast of Hainan Island, and extends in the NEE direction. It is considered a Cenozoic extensional basin developed over strongly thinned continental crust and is a significant oil and gas, as well as natural gas, hydrate basin [21]. The study area is situated in a tectonically gentle region between the continental slope and the uplifted platform, and the topography is relatively flat (Figure 1).

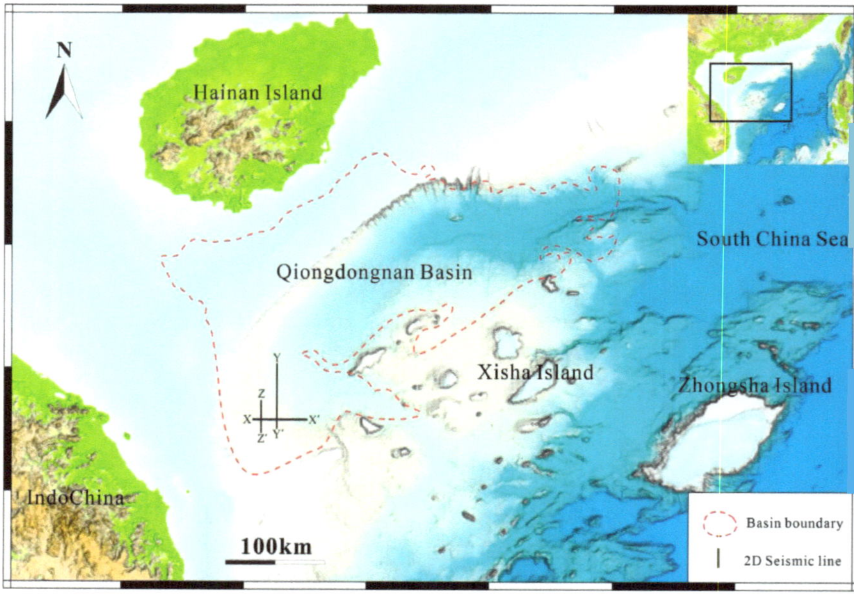

Figure 1. Regional location map of the study area. The red dashed line represents the boundary c the QDNB. The black line segments represent the positions of seismic profiles.

The sedimentary fill during the tectonic rifting period in the study area comprise three stratigraphic units: the Eocene, Yacheng Formation, and Lingshui Formation. Th sedimentary transition from continental to shallow marine facies is characterized by th development of fan-deltas and deltas. During the faulting period, the sedimentary fil includes the Miocene Lingshui Formation Member 1–2, the Lower-Middle Miocene Sany Formation, and the Meishan Formation. The sedimentation gradually transitions fron shallow marine to semi-deep marine, featuring the development of deltas, fan-deltas, an platform-margin reefs, among other features. During the subsidence stage, the sedimentar fill includes the Upper-Middle Miocene Huangliu Formation, Upper Miocene Yinggeha Formation, and Quaternary Ledong Formation. These sediments were deposited in semi-deep marine environment and are characterized by gravity flow deposits, such a debris flows, turbidite fans, slope apron fans, and canyon channels (See Figure 2) [22,23].

The target interval of this study is the Quaternary Leidong Formation (Figure 2 Within this formation, there is significant variation in sediment thickness. The sedimer tary thickness is greater in the northwest direction, with a maximum sedimentation rat reaching up to 550 m per million years. In contrast, the sedimentary thickness is thinne in the southeast direction, with a sedimentation rate of only 50 m per million years. Th sedimentary strata in the northwest direction are influenced by the influx of materials fron the continental slope. During the lowstand and transgression periods, debris flow and tu bidite deposits are developed, mainly comprising slope fans, basin floor fans, channel-leve systems, and MTDs that rapidly accumulate [24]. During the highstand system tract, th depositional environment transitions to a low-energy, fine-grained settling sedimentatio in a semi-deep sea setting. In the southeast direction, the sedimentary strata are mainl influenced by slow sedimentation during the subsidence stage, resulting in semi-dee marine deposition. The hydrodynamic conditions are relatively weak, creating a stabl semi-deep marine depositional environment. The study area contains four sets of SR pod the Yacheng Formation, the LingShui Formation Member 3, the LingShui Formation Mem ber 1–2, the Meishan Formation, and the Huangliu Formation. Among them, the LingShu Formation Member 3 and the LingShui Formation Member 1–2 in the Meishan Formatio are the main gas source layers for gas hydrates [25].

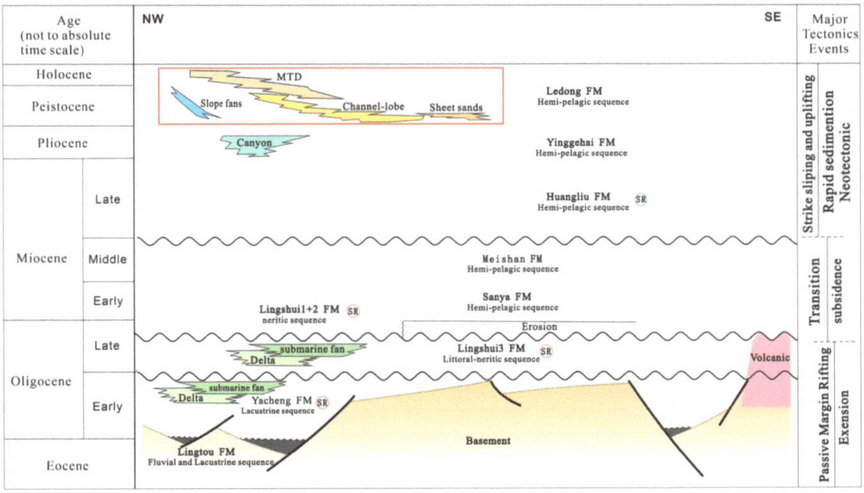

Figure 2. The simplified diagram illustrating the sedimentary facies, lithology, and regional tectonic movements from the Paleogene to the Quaternary in the study area. The red rectangular box represents the target interval of the study. Within this interval, the red circles labeled as "SR" denote thermogenic gas source rocks, while the green circles labeled as "SR" represent biogenic gas source rocks.

3. Data and Method

The seismic volume covers a surface area of 2900 km^2 and has a total trace length of 7 s TWT. The bin spacing is 12.5 m in both directions with a sample rate of 2 ms TWT. The seismic volume is a full-offset, post-stack time migrated (PSTM) volume. Processing steps included signal noise reduction, corrections, and velocity analysis using Kirchhoff PSTM. Noise removal is carried out within the frequency range of 10 Hz to 20 Hz. The resolution of seismic data depends on the dominant frequency of the seismic signal and the interval velocity in the zone of interest. The interval velocity of the sediment package containing mounds and domes is approximately 0.7 km/ms, as determined from velocity data obtained through the well-to-seismic correlation within the 2D seismic volume.

When the subsurface contains fluids, it can affect the amplitude, frequency, and energy of seismic waves, thereby causing post-stack attribute anomalies in seismic data. To identify and analyze geological factors, such as free gas, faults, and fractures, a combination of attributes, like instantaneous frequency, sweetness, and variance, can be selected. The instantaneous frequency refers to the derivative of the instantaneous phase with respect to time [26,27]. The main factors causing variations in instantaneous frequency values include changes in lithology, properties of the fillings within the strata, and lateral heterogeneity in distribution [28]. The presence of free gas in pores within the strata can lead to high-frequency attenuation and low-frequency enhancement effects. In this study, the instantaneous frequency within the frequency range of 13 Hz to 28 Hz was chosen as a sensitive attribute to reflect the spatial distribution characteristics of gas-bearing formations. The reason for choosing this range is that it effectively reflects the developmental range of subsurface fluids. The sweetness attribute is the ratio of seismic amplitude to the square root of the instantaneous frequency. Methane fluid seepage is often accompanied by phenomena, such as authigenic carbonate crusts and chemosynthetic biogenic shells. These occurrences result in high-amplitude anomalies, leading to high values of the sweetness attribute. This attribute plays a crucial role in identifying both present-day seepage sites and ancient seepage points on the seabed [29]. The variance attribute is used to describe the lateral heterogeneity of seismic properties, such as lithology and stratigraphy, by utilizing the similarity between adjacent seismic traces [30]. It can reveal the discontinuous information within the data volume and characterize fault systems. This study primarily utilizes th multi-attribute fusion technique to analyze the source-sink system of natural gas hydrat mineralization. This approach reduces the uncertainties caused by interpreting a single dat volume and highlights abundant geological information, thereby enhancing the reliabilit of geological understanding. The accuracy of individual seismic attributes in the geologica interpretation and comprehensive evaluation may be subject to uncertainties. Conductin geological interpretation through the simultaneous use of multiple attributes significantl reduces ambiguity, thereby enhancing the reliability of the geological understanding.

4. Results

In Figure 3A, on the left side, there are submarine spiny seamounts developed on th seafloor, while on the right side, there is a submarine domed seamount. On the whole, th Huangliu Formation, Yinggehai Formation, and Ledong Formation, situated below the tw geological bodies, exhibit significant low-frequency anomalies, along with the presence o gas chimney structures. It is inferred that there is a substantial amount of gas in the strat with the spiky mounds corresponding to a higher gas content. Coherent body anomalie can indicate the development of faults. From the figure, it can be observed that the are corresponding to the spiny mounds shows the development of faults throughout the seabe to the Huangliu Formation, with the fractures being most pronounced in the Yinggaoha Formation. The domed seamount also exhibits faults extending from the seabed to th Yinggaohai Formation, but the fracture density is lower compared to the spiny seamount These fractures serve as conduits for gas migration. The spiny seamounts exhibit anomalie in the sweetness attribute on the seabed, while sporadic anomalies are observed in th sweetness attribute of the underlying strata of the MTDs. The domed seamount show no anomalies of sweetness attributes on the seabed, but distinct scattered anomalies ar present in the underlying strata of the MTDs. The Yinggaohai Formation exhibits localize anomalies in sweetness attributes. These anomalies could be attributed to the presence o gas-rich formations or the occurrence of seafloor seepage points (both current and ancient which might have led to the formation of authigenic carbonate concretions and chemo-bi shells in the vicinity. From the local zoom-in of Figure 3A (Figure 3B), it can be observe that sporadic BSR occurrences develop near the base of the gas hydrate stability zon (BGHSZ) corresponding to the domed seamount, while no apparent BSR phenomenon i observed near BGHSZ corresponding to the spiny seamounts. Profile YY′ intersects wit profile XX′ at point N. In the vicinity of the intersection point on profile YY′ (within th yellow box in Figure 3B), the BSR phenomenon is observed near the BGHSZ. However, i the vicinity of the intersection point on profile XX′ (within the yellow box in Figure 3D the BSR is even more pronounced. Profile YY′ intersects with profile ZZ′ at point M. Bot profiles show clear BSR phenomena in the vicinity of the intersection point within th BHSZ (Figure 3C,D). At the intersection point on profile ZZ′, there is a distinct presence o a domed seamount on the seafloor. The Ledong Formation contains two stages of MTD (MTD1 and MTD2). As shown in Figure 3B, both stages of massive sulfide deposits (MTDs exhibit thicker accumulations on the spiny seamounts compared to the domed seamount Based on the paleostatic conditions of the study area, MTD1 and MTD2 correspond to th paleo-base of the gas hydrate stability zone (Pelo-BGHSZ1 and Pelo-BGHSZ2, respectivel as indicated in the figure). Furthermore, the paleo-BHSZ intersects with anomalies in th sweetness attribute. The base map in Figure 3E displays the combined thickness of MTD and MTD2. The white dashed line delineates the development zone of spiny seamounts while the black dashed line marks the development zone of domed seamounts. It can b observed that the spiny seamount development zone exhibits thicker MTDs, while th domed seamount development zone shows thinner MTDs.

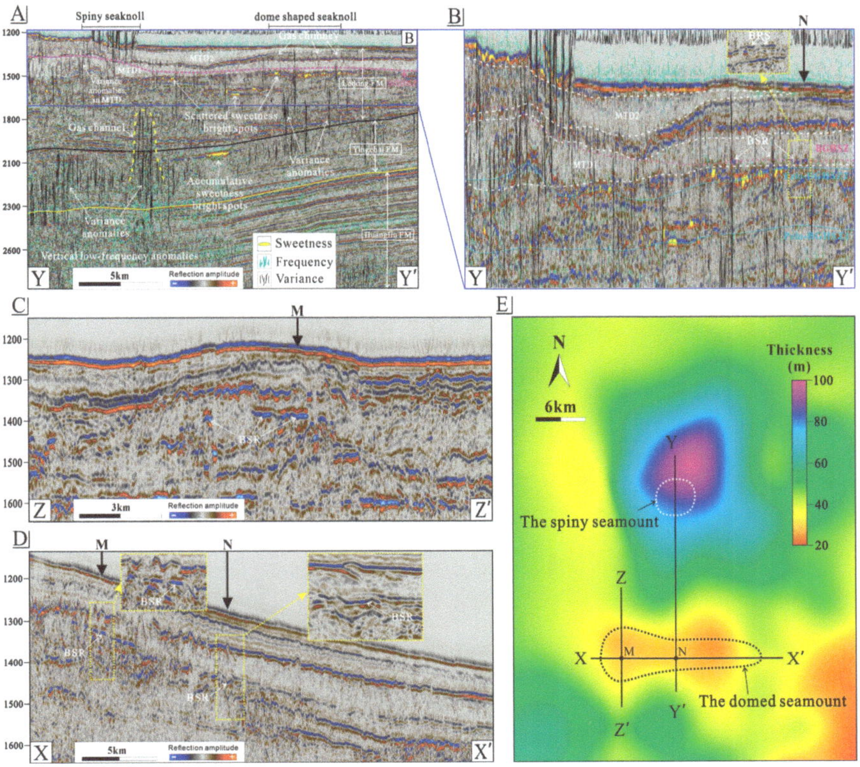

Figure 3. (**A**) The superimposed seismic profiles of the seismic data, instantaneous dominant frequency attribute, variance attribute, and sweetness attribute in the research area. The white dashed circles indicate MTDs, the yellow dashed lines represent gas chimney outlines, the pink dashed line represents the present-day BGHSZ, the bright green color represents the instantaneous dominant frequency attribute values in the range of 13 to 28 Hz, the black vertical lines indicate high-value anomalies in the variance attribute, and the yellow color represents data with sweetness attribute values above 5000. (**B**) The enlarged view of the blue-boxed region in (**A**). The light blue dashed line represents the paleo base of the gas hydrate stability zone. (**C**) The seismic profile at position Z-Z′ in (**E**). (**D**) The seismic profile at position X-X′ in (**E**). (**E**) The map indicating the positions of the seismic profiles. The base map color represents the sum of MTD1 and MTD2 thicknesses, the white dashed line represents the development range of spiny seamounts, the black dashed line represents the development range of domed seamount, and M and N points are the intersections of seismic lines.

5. Discussions

5.1. Mechanism of Hydrate Decomposition Induced by MTDS

The formation of gas hydrates requires appropriate temperature and pressure conditions. Lower temperatures and higher pressures facilitate the formation of hydrogen bonds between water molecules and natural gas molecules, leading to the formation of stable gas hydrate structures that prevent the escape of natural gas molecules. Additionally, the formation of gas hydrates involves nucleation and growth processes. Under suitable temperature and pressure conditions, water and gas molecules can form initial gas hydrate structures through a process known as nucleation. Over time, more water and gas molecules are adsorbed and incorporated into the existing nuclei, forming larger gas hydrate crystals, a process known as growth. This process requires suitable temperature and pressure conditions to ensure adequate energy and stability.

MTDs are widely distributed in various settings, such as outer continental shelve upper continental slopes, canyons, uplifts (volcanoes, ridges, salt domes), and sidewal of channels. They are commonly found in deepwater basins globally and typically for during sea-level falls. Previous studies suggest that the rapid deposition of MTDs can cau rapid changes in the temperature and pressure conditions within their coverage areas. Th leads to a significant uplift in the gas hydrate stability zone boundary, resulting in the rapi decomposition of pre-existing gas hydrates.

Submarine mound features in the study area are primarily distributed in regions wit high Quaternary sedimentation rates and thick accumulations of MTDs. On seismic profile these features exhibit distinct "spike" shapes. These development areas of submarir mound features are accompanied by extensive seafloor sediment folding phenomen Submarine domed seamounts are mainly distributed in areas with smaller Quaternar sediment thickness and smaller thicknesses of MTDs. On seismic profiles, they exhib gentle positive uplift slopes. As the thickness of MTDs increases, their impact on the ga hydrate stability zone boundary becomes more significant. Therefore, compared to dome seamounts, the influence of MTDs corresponding to spiny seamounts is greater on th stability zone boundary, leading to a higher intensity of gas hydrate decomposition.

The influence of MTDs on the temperature and pressure conditions of gas hydrate can result in the upward movement of the gas hydrate stability zone boundary, leadin to the decomposition of pre-existing gas hydrates. Subsequently, the decomposed ga can reform gas hydrate deposits within new suitable stability zones, thereby affecting th distribution patterns and formation mechanisms of gas hydrate deposits.

5.2. *The Geological Effects Caused by Hydrate Decomposition*

The decomposition of natural gas hydrates yields a significant volume of gas, whic rises and creates fractures. These fractures act as preferential conduits for the upwar migration of free gas. In regions abundant in gas supply, these fractures continue t propagate upwards, facilitating gas seepage toward the seafloor.

The formation of hydrates induces sediment expansion and carbonate precipitatio around seafloor seepage sites, leading to the formation of submarine spiny seamount These seamount features arise from the accumulation of free gas within sedimentary laye The gas displaces water in the upper sediment pores, causing localized expansion an forming positive topographic uplifts [31]. An analysis of the instantaneous frequenc attribute in Figure 3A reveals that the development of submarine spiny seamount area signifies a higher gas supply, which can stimulate persistent and rapid gas seepage on th seafloor, consequently leading to ongoing destabilization of the shallow sediment. Thi phenomenon likely contributes significantly to the chaotic seismic imaging and prominer manifestation of seafloor gas seepage in the underlying sedimentary layers of the submarin mound features.

The thickness of MTDs is relatively thin in the wing areas, resulting in a comparativel smaller uplift of the BGHSZ in the MTDs' wing coverage regions. The gas formed by th decomposition of natural gas hydrates can be rapidly trapped within the newly forme hydrate stability zone. This process leads to the creation of new natural gas hydrat deposits, which, in turn, hinder the upward development of fractures. In these region strong seabed seepage is usually not present. The sediment expansion caused by th formation of hydrates can create submarine domed seamount features on the seabed. Th formation of submarine domed seamounts is similar to that of spiny seamount structure Both result from the displacement of water in the pore space of overlying sediment b leaking gas, causing localized expansion of the sediment volume. In most cases, submarin domed seamounts are situated directly above gas-rich layers and typically only rise few meters above the normal seabed, but their spans can extend up to several hundre meters [32].

In their study on the seabed gas seepage characteristics in the northern Gulf of Mexic continental slope, Roberts et al. found that moderate-velocity natural gas seepage can lea to the formation of gas hydrate mounds, isolated authigenic carbonate mounds, and small cold seeps on the seafloor [33]. Low-velocity natural gas seepage, on the other hand, is more likely to result in the formation of authigenic carbonate mounds, hardgrounds, and nodules on the seafloor. Therefore, we deduce that the morphology of these mound-like structures may represent varying degrees of seepage intensity. The process of moderate natural gas seepage can lead to the formation of seafloor spiny seamounts. Gas accumulates within the shallow subsurface hydrate stability zone, and under suitable temperature and pressure conditions, it is prone to forming blocky or fissure-filling natural gas hydrates [34]. Low-speed natural gas seepage processes can lead to the formation of domed seamounts. Domed seamounts are often associated with carbonate rock substrates. Due to their well-developed fractures, they are prone to forming diffusive-type natural gas hydrates. The spiny seamounts associated with hydrates typically display an oval or concentric positive topography when viewed from above. The formation of these features is a result of the accumulation of free gas within sedimentary layers. This process involves gas generated by hydrate dissociation replacing the water in the pores of upper sediment layers, causing localized expansion and resulting in a positive topographic feature [35].

5.3. Potential Seafloor Seepage Patterns in the Study Area

In summary, this study proposes a seafloor seepage model for the southern QDNB (Figure 4).

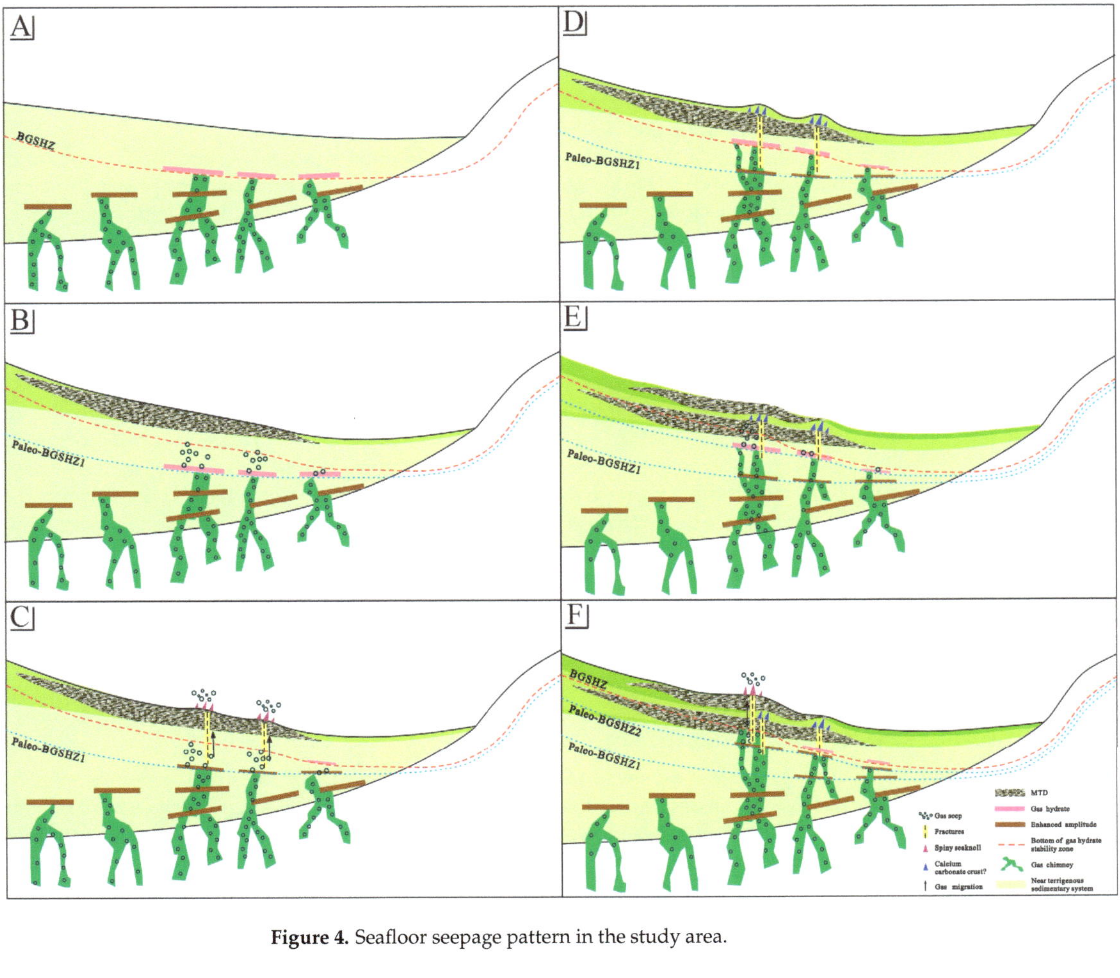

Figure 4. Seafloor seepage pattern in the study area.

Stage A. Influenced by tectonic activities since the Middle-Neogene, deep-seate fluids are released upward along tectonic weak zones characterized by developed fracture forming vertical migration pathways primarily composed of gas chimneys. The Huangli Formation offers favorable conditions for biogenic gas generation, as sedimentary roc layers contain adequate organic matter abundance, appropriate maturity, and temperatur conditions. Biogenic gas migrates toward the hydrate stability zone through gas chimney The continuous biogenic gas supply, combined with the accumulation of in-situ microbia methane gas within the stability zone, leads to the formation of methane hydrates whe the methane concentration reaches saturation. Simultaneously, as biogenic gas continues it upward migration and encounters highly permeable sedimentary layers, it can accumulat to form shallow gas reservoirs.

Stage B. With the rapid deposition of the first-phase MTDs, significant changes i temperature and pressure occur within the stability zone formed during Stage A in th MTD coverage area, leading to the decomposition of the original gas hydrate bodies and th release of a large amount of methane gas. This results in the decomposition of pre-existin hydrate deposits formed in Stage A and the release of a significant amount of methan gas. The central part of MTDs exhibits greater thickness, leading to a notable uplift of th BGHSZ and the release of a larger amount of methane gas. Conversely, the thinner flanks o MTDs experience less uplift of the BGHSZ, leading to the reduced release of methane ga

Stage C. When the methane gas rapidly released from hydrate decomposition and th continuously supplied deep-seated fluids reach a certain level, they undergo overpressur release (Figure 4C). At this point, microfractures carrying fluids and sediment penetrat through the seafloor, resulting in the upward arching of the strata, spiny seamounts, an seafloor sediment folding. Due to the ongoing gas leakage, self-generated carbonate crust and chemosynthetic biogenic shells form near the leakage points, leading to an expansio in the distribution and development range of spiny seamounts.

Stage D. As the high-level depositional regime experiences stable, uniform, and grad ual subsidence, the temperature and pressure within the stability zone gradually stabiliz resulting in a slowdown in the rate of hydrate decomposition. With increasing sedimen thickness, the BGHSZ gradually moves upward until it reaches a relatively stable positio At this point, the continuously supplied fluids will slowly form new hydrate bodies nea the BGHSZ.

Stage E. With the rapid deposition of the latest MTDs, there is another rapid change i temperature and pressure within the stability zone. The BGHSZ formed during Stage D experiences rapid uplift, leading to the rapid decomposition of natural gas hydrate bodie

Stage F. When the thickness of the most recent MTDs is substantial, the correspondin alteration of the BGHSZ is significant, leading to a higher rate of decomposition of hydrat bodies and stronger overpressure release. During this period, microfractures carryin fluids and sediment pierce the seafloor, alongside the ongoing gas leakage and expansio of gas hydrate mineralization. These factors contribute to the formation of more distinc underwater mounds and upward arching in the geological layers. On the other hand, whe the thickness of the latest MTDs is thinner, there is relatively minimal alteration in th BGHSZ. After the decomposition of hydrates, the gas formed quickly re-mineralizes withi a shorter vertical fissure in the new stable zone, without venting or leaking to the seafloo Consequently, this process leads to the development of domed seamounts.

6. Conclusions

6.1. Impacts of MTDs on Seafloor Gas Seepage

The rapid sedimentation of MTDs leads to the swift alteration of the BGHSZ, triggerin the substantial decomposition of hydrates and the release of a significant amount of natura gas. This gas migrates towards the seafloor and punctures through it, constituting th primary cause of varying degrees of seafloor seepage in the study area. In regions wher MTDs are thicker, the uplift of the BGHSZ is more pronounced. Consequently, the speed o natural gas hydrate decomposition accelerates, resulting in a higher gas seepage rate and greater damage to the seafloor micro-features. Conversely, the opposite holds true.

6.2. Association of Seafloor Seamounts with Gas Seepage

It has been discovered that seafloor spiny seamounts and domed seamounts are closely associated with seafloor gas seepage. Spiny seamounts represent moderate-speed natural gas seepage in the shallow overburden, while domed seamounts signify slow-speed natural gas seepage in the same layer. The zones with dense development of fractures in the underlying formations of both types of mounds are ideal sites for leaky gas hydrates. They could potentially serve as crucial indicators for identifying leaky gas hydrates during exploration activities on the northern slope of the South China Sea.

Author Contributions: Writing—original draft preparation, P.L.; methodology, W.Z. and L.H.; validation, S.M. and H.C.; data curation, P.S. All authors have read and agreed to the published version of the manuscript.

Funding: We gratefully acknowledge the financial support provided by the National Science Foundation of China (42176215), the Guangzhou Science and Technology Planning Project (202201011434), the First Batch of the "Nanhai New Star" project (NHXXRCXM202357), the Guangdong Basic and Applied Basic Research Foundation (2019B030302004), Project of Sanya Yazhou Bay Science and Technology City (SCKJ-JYRC-2023-02), and Sanya Science and Technology Innovation Project (2022KJCX14).

Data Availability Statement: The data presented in this study are available on request from the corresponding author. The data are not publicly available due to legal restrictions.

Conflicts of Interest: Author Pengqi Liu was employed by Sanya Institute of South China Sea Geology, Guangzhou Marine Geological Survey. The remaining authors declare that the research was conducted in the absence of any commercial or financial relationships that could be construed as a potential conflict of interest.

References

1. Wang, B. Research on Typical Seabed Fluid Flows in China Marginal Seas. Ph.D. Dissertation, Chinese Academy of Sciences, Qingdao China, 2018.
2. Judd, A.; Hovland, M. *Seabed Fluid Flow Impact of Geology, Biology and the Marine Environment*; Great Britain at the Alden Press: Oxford, UK, 2007.
3. Ho, S.; Cartwright, J.A.; Imbert, P. Vertical evolution of fluid venting structures in relation to gas flux, in the Neogene-Quaternary of the Lower Congo Basin, Offshore Angola. *Mar. Geol.* **2012**, *332*, 40–55. [CrossRef]
4. Wang, J.; Wu, S.; Kong, X.; Ma, B.; Li, W.; Wang, D.; Gao, J.; Chen, W. Subsurface fluid flow at an active cold seep area in the Qiongdongnan Basin, northern South China Sea. *J. Asian Earth Sci.* **2018**, *168*, 17–26. [CrossRef]
5. Moore, G.F.; Saffer, D.; Studer, M.; Costa Pisani, P. Structural restoration of thrusts at the toe of the Nankai Trough accretionary prism off Shikoku Island, Japan: Implications for dewatering processes. *Geochem. Geophys. Geosyst.* **2011**, *12*. [CrossRef]
6. Goshovskyi, S. Gas seeps as the black sea hydrocarbon resource. *Int. Multidiscip. Sci. GeoConf. SGEM* **2019**, *19*, 877–882.
7. Bondarenko, V.; Ganushevych, K.; Sai, K.; Tyshchenko, A. Development of gas hydrates in the Black Sea. *Tech. Geoinformational Syst. Min.* **2011**, 55–59. Available online: https://www.researchgate.net/publication/330310759_Development_of_gas_hydrates_in_the_Black_sea (accessed on 4 March 2024).
8. Klymenko, V.; Ovetskyi, S.; Martynenko, V.; Vytiaz, O.; Uhrynovskyi, A. An alternative method of methane production from deposits of subaquatic gas hydrates. *Min. Miner. Depos.* **2022**, *16*, 11–17. [CrossRef]
9. Yang, C.; Li, X.; Yao, Y.; Chang, X.; Wu, J.; Huang, L.; Ju, D. The subsurface fluid-flow systems and their genetic mechanism in the southwestern barents sea. *Mar. Geol. Quat. Geol.* **2015**, *35*, 135–143.
10. Mark, C.; Srikumar, R.; Conor, O.; Annika, C.; Ronan, O.; Ruth, P. Geological settings and controls of fluid migration and associated seafloor seepage features in the north Irish Sea. *Mar. Pet. Geol.* **2021**, *123*, 104762.
11. Dugan, B.; Flemings, P.B. Overpressure and Fluid Flow in the New Jersey Continental Slope: Implications for Slope Failure and Cold Seeps. *Science* **2000**, *289*, 288–291. [CrossRef] [PubMed]
12. Crutchley, G.J.; Kroeger, K.F.; Pecher, I.A.; Gorman, A.R. How tectonic folding influences gas hydrate formation: New Zealand's Hikurangi subduction margin. *Geology* **2019**, *47*, 39–42. [CrossRef]
13. Doré, A.G.; Jensen, L.N. The impact of late Cenozoic uplift and erosion on hydrocarbon exploration: Offshore Norway and some other uplifted basins. *Glob. Planet. Chang.* **1996**, *12*, 415–436. [CrossRef]

14. Pape, T.; Blumenberg, M.; Reitz, A.; Scheeder, G.; Schmidt, M.; Haeckel, M.; Wallmann, K.; Bohrmann, G. Oil and gas seepage offsho Georgia (Black Sea)–Geochemical evidences for a Paleogene-Neogene hydrocarbon source rock. *Mar. Pet. Geol.* **2021**, *128*, 1049 [CrossRef]
15. Chen, J.; Song, H.; Guan, Y.; Pinheiro, L.M.; Geng, M. Geological and oceanographic controls on seabed fluid escape structures the northern Zhongjiannan Basin, South China Sea. *J. Asian Earth Sci.* **2018**, *168*, 38–47. [CrossRef]
16. Bouchaala, F.; Guennou, C. Estimation of viscoelastic attenuation of real seismic data by use of ray tracing software: Applicati to the detection of gas hydrates and free gas. *C. R. Geosci.* **2012**, *344*, 57–66. [CrossRef]
17. Liang, J.Q.; Fu, S.Y.; Chen, F.; Su, P.; Shang, J.; Lu, H.; Fang, Y. Characteristics of methane seepage and gas hydrate reservoir in th northeastern slope of South China Sea. *Nat. Gas Geosci.* **2017**, *28*, 761–770.
18. Waite, W.F.; Santamarina, J.C.; Cortes, D.D.; Dugan, B.; Espinoza, D.N.; Germaine, J.; Jang, J.; Jung, J.W.; Kneafsey, T. Shin, H.; et al. Physical properties of hydrate-bearing sediments. *Rev. Geophys.* **2009**, *47*. [CrossRef]
19. Chen, L.; Song, H. Geophysical features and identification of natural gas seepage in marine environment. *Prog. Geophys.* **2005**, 2 1067–1073.
20. Wan, Z.; Chen, C.; Liang, J.; Zhang, W.; Huang, W.; Su, P. Hydrochemical characteristics and evolution mode of cold seeps in th Qiongdongnan Basin, South China Sea. *Geofluids* **2020**, *2020*, 4578967. [CrossRef]
21. Zhao, Z.; Sun, Z.; Sun, L.; Wang, Z.; Sun, Z. Cenozoic tectonic subsidence in the Qiongdongnan basin, northern South China Se *Basin Res.* **2018**, *30*, 269–288. [CrossRef]
22. Hongfang, G.; Xin, N.; Weidong, L. Source to sink analysis of a sea basin: The Quaternary deepwater turbidite fan system in Pea River Valley-Northwest subbasin, Northern South China Sea. *Mar. Geol. Quat. Geol.* **2021**, *41*, 1–12.
23. Lei, Z.; Su, M.; Zhang, L.; Shuai, Q.; Sun, M.; Liu, J.; Yang, R. Sediment sources and transport patterns of Qiongdongnan Basin i northern slope of South China Sea since late Miocene. *J. Mar. Sci.* **2016**, *34*, 35–42.
24. Pang, X. Sequence stratigraphy configuration of deepwater gravity-flow sediments and its controls: A line of thinking in sequen stratigraphy of gravity-flow sediments in Baiyun deepwater area, the northern South China Sea. *China Offshore Oil Gas* **201** *24*, 1–8.
25. Mao, S.; Hu, G.; Hu, L.; Liu, P.; Chen, X.; Qin, M.; Cheng, F. Evaluation of Natural Gas Hydrate Fault System: A Case from a Sa in Deep-Water Slope Area of the Northern South China Sea. *Lithosphere* **2022**, *2022*, 9049312. [CrossRef]
26. Taner, M.T.; Koehler, F.; Sheriff, R.E. Complex seismic trace analysis. *Geophysics* **1979**, *44*, 1041–1063. [CrossRef]
27. Bouchaala, F.; Ali, M.Y.; Matsushima, J.; Bouzidi, Y.; Takougang, E.T.; Mohamed, A.A.; Sultan, A.A. Scattering and intrins attenuation as a potential tool for studying of a fractured reservoir. *J. Pet. Sci. Eng.* **2019**, *174*, 533–543. [CrossRef]
28. Huang, Y.P.; Geng, J.H.; Zhong, G.F.; Guo, T.L.; Pu, Y.; Ding, K.Y.; Ma, J.Q. Seismic attribute extraction based on HHT and i application in a marine carbonate area. *Appl. Geophys.* **2011**, *8*, 125–133. [CrossRef]
29. Hart, B.S. Channel detection in 3-D seismic data using sweetness. *AAPG Bull.* **2008**, *92*, 733–742. [CrossRef]
30. Khan, N.; Khan, S.; Rehman, F.; Sajid, M. Structural and seismic attribute analysis of the Paleocene carbonate reservoir from th Balkassar Field, Potwar Plateau, Pakistan. *J. Pet. Res. Stud.* **2024**, *14*, 18–35.
31. Musgrave, R.J.; Bangs, N.L.; Larrasoaña, J.C.; Gràcia, E.; Hollamby, J.A.; Vega, M.E. Rise of the base of the gas hydrate zone sinc the last glacial recorded by rock magnetism. *Geology* **2006**, *34*, 117–120. [CrossRef]
32. Zhao, T.; Zhang, X.; Feng, J. Acoustic detection techniques for seabed hydrocarbon seepage. *Mar. Geol. Quat. Geol.* **2010**, 3 149–156.
33. Roberts, H.H.; Hardage, B.A.; Shedd, W.W.; Hunt, J., Jr. Seafloor reflectivity—An important seismic property for interpretin fluid/gas expulsion geology and the presence of gas hydrate. *Lead. Edge* **2006**, *25*, 620–628. [CrossRef]
34. Shang, J.; Wu, L.; Liang, J.; Sha, Z.B. The microtopographic features and gas seep model on the slope in the northeastern Sout China Sea. *Mar. Geol. Quat. Geol.* **2014**, *34*, 129–136.
35. Paull, C.K.; Normark, W.R.; Ussler III, W.; Caress, D.W.; Keaten, R. Association among active seafloor deformation, moun formation, and gas hydrate growth and accumulation within the seafloor of the Santa Monica Basin, offshore California. *Ma Geol.* **2008**, *250*, 258–275. [CrossRef]

Journal of
Marine Science and Engineering

MDPI

Article

Lower Limits of Petrophysical Properties Allowing Natural Gas Accumulation in Marine Sandstones: An Example from the Qiongdongnan Basin, Northern South China Sea

Chao Li [1,2], Shuai Guo [3], Qianshan Zhou [1,2,*], Chaochao Xu [1,4,*] and Guojun Chen [1,2]

[1] Northwest Institute of Eco-Environment and Resources, Chinese Academy of Sciences, Lanzhou 730000, China; lichaomails@lzb.ac.cn (C.L.); gjchen@lzb.ac.cn (G.C.)
[2] Key Laboratory of Petroleum Resources Exploration and Evaluation, Lanzhou 730000, China
[3] Research Institute of China National Offshore Oil Corporation, Beijing 100028, China; guoshuai@cnooc.com.cn
[4] University of Chinese Academy of Sciences, Beijing 100049, China
* Correspondence: zhouqianshan@nieer.ac.cn (Q.Z.); xuchaochao1998@163.com (C.X.)

Abstract: The lower limits of petrophysical properties for an effective reservoir are among the key parameters for assessing hydrocarbon reserves and are therefore directly related to hydrocarbon exploration and development strategies. However, the lower limits for marine sandstone gas reservoirs are still not clear and the impact factors also remain to be discussed. This study analysed the lower petrophysical property limits of an effective sandstone reservoir in the Qiongdongnan Basin using porosity, permeability and gas testing. The results showed that the lower porosity and permeability limits of effective reservoirs developed in the deltas are 8.9% and 1.2×10^{-3} μm^2, respectively, and 11.3% and 4.0×10^{-3} μm^2 in the submarine canyons and fans, respectively. Sedimentary facies, sediment transport distance, grain size and burial depth of sandstone significantly influence the lower physical property limits. The lower porosity and permeability limits increase with the increase in sediment transport distance as well as the decrease in sandstone grain size and burial depth. Sediment sources and sedimentary facies determine whether sandstone can become an effective reservoir in the Qiongdongnan Basin. Specifically, the sediment source dramatically influences the petrophysical properties of sandstone. The sandstone sourced from the Red River has higher porosity and permeability, followed by the sandstone sourced from the Hainan Uplift, and the sandstone sourced from the palaeo-uplift within the basin has the lowest porosity and permeability. The feldspar dissolution by CO_2 and organic acid is the primary formation mechanism of the effective reservoir in the Lingshui Formation, whereas the dissolution of glauconite is more common in the sandstone reservoirs of the Sanya and Meishan formations.

Keywords: effective reservoir; lower petrophysical properties limits; formation mechanism; Qiongdongnan Basin; South China Sea

Citation: Li, C.; Guo, S.; Zhou, Q.; Xu, C.; Chen, G. Lower Limits of Petrophysical Properties Allowing Natural Gas Accumulation in Marine Sandstones: An Example from the Qiongdongnan Basin, Northern South China Sea. *J. Mar. Sci. Eng.* **2024**, *12*, 735. https://doi.org/10.3390/jmse12050735

Academic Editor: Dimitris Sakellariou

Received: 28 March 2024
Revised: 25 April 2024
Accepted: 25 April 2024
Published: 28 April 2024

1. Introduction

Recently, tight oil or gas exploration has received increasing attention [1–6]. Due to the high technical requirements for tight oil or gas exploration, its use costs much more than conventional oil or gas [7–12]. Efficient tight oil or natural gas exploration is one way to reduce costs, making oil and gas explorers consider the lower physical property limits [13–17]. The definition of the lower limits of physical properties for effective reservoirs is the minimum porosity and minimum permeability required for fluid accumulation in rocks [13–15]. The lower porosity and permeability limits of the reservoirs have become fundamental factors for assessing oil or natural gas reserves; they are directly related to oil or natural gas exploration strategies [18–22]. Similarly, deepwater oil or gas exploration also faces this dilemma [23,24]. In addition to considering the buried depth of the reservoirs, offshore drilling pays attention to the current seawater depth because the requirements of deepwater oil or gas exploration for drilling technology are far higher than those of shallow-wate areas [23,24]. Deepwater oil or gas exploration costs much more than shallow-water oil c natural gas exploration [23–25]. Therefore, when exploring deepwater oil or natural ga researchers must also consider the lower porosity and permeability limits of the reservoir However, the lower porosity and permeability limits of the reservoirs distributed in marin deepwater areas have rarely been studied.

Previous studies have determined the lower porosity and permeability limits of e fective reservoirs using empirical statistics, oil testing, oil-bearing analysis and othe methods [13–15,17,20,22,26–29]. The results show that the lower porosity and permeabi ity limits of the tight sandstone of the Ordos, Junggar and Songliao basins, China, ar 3–12% and 0.02–1 × 10^{-3} μm^2 [13–15,17,30]. Source rock, buried depth and hydrocarbo generation kinetics are the primary factors influencing the lower porosity and permeabi ity limits of tight reservoirs [17,18,20,26,28,29,31–33]. The development and evolution c sandstone reservoirs are jointly controlled by the basin tectonic background, sedimentar facies, diagenesis, palaeoclimate, palaeotemperature and other factors [20,27,31–33]. Th sedimentary facies forms the basis for reservoir development, and the later reconstructio of a reservoir formed by diagenesis is key to reservoir performance [20,31,33]. Pressure an cementation determine the degree of compaction of a reservoir, while dissolution is the ke to the development of secondary pores in a reservoir [31,33]. Therefore, the sedimentar facies and diagenesis are vital for determining reservoir properties [20,31,33]. Howeve the effects of sedimentary facies, sediment transport distance and grain size on the lowe porosity and permeability limits of effective reservoirs are yet to be discussed.

The objectives of this study were to (1) investigate the lower porosity and permeabilit limits of effective reservoir and its influence factors and (2) evaluate the roles of sedimen source, sediment transport distance, sedimentary facies and dissolution diagenesis on th formation of an effective reservoir. The results are of practical significance for guiding th selection of natural gas exploration zones in the Qiongdongnan Basin (QDNB).

2. Geological Background

The QDNB is in the northern area of the South China Sea (see Figure 1A). NE-, NNE NW- and EW-trending faults controlled the development of the Songxi Sag (SXS), Songdon Sag (SDS), Yabei Sag (YBS), Yanan Sag (YNS) and Songtao Uplift (STU) in the northern depression; the Changchang Sag (CCS), Songnan–Baodao Sag (SN–BDS), Lingshui Sa (LSS), Ledong Sag (LDS) and Songnan Uplift (SNU) in the central depression and th Beijiao Sag (BJS) in the southern depression (see Figure 1B). The western QDNB contain YBS, YNS, LDS and LSS, and the eastern QDNB contains SXS, SDS, SN–BDS and CCS (se Figure 1B). The tectonic evolution of the QDNB includes the rifting (35–23 Ma) and post-rif subsidence stages (23–0 Ma) [34–36].

The QDNB is developed on Mesozoic basement and contains a thickness of 6000–12,000 n Tertiary to Quaternary sediments (see Figure 2) [37,38]. Rifting commenced in the Eocen and finished around the late Oligocene, leaving a series of sags that are filled with lacustrin sediments (see Figure 2) [37,38]. During the early Oligocene, the Yacheng Formatio (YCF) comprised neritic and coastal-plain coal-bearing sediments (see Figure 2) [37,38 Immediately above the Yacheng Formation is the littoral to neritic Lingshui Formatio (LSF) (see Figure 2) [37,38]. Following the rifting stage, the QDNB experienced post-rif thermal subsidence to date and was filled with a thick sequence of marine sediment (see Figure 2) [37,38]. In the post-rift subsidence stage, marine sediments dominated b mudstones with occasional turbidite channels and deepwater fan sandstone were deposite in the Miocene Sanya Formation (SYF), Meishan Formation (MSF), Huangliu Formatio (HLF) and the Pliocene Yinggehai Formation (YGHF) (see Figure 2) [37,38]. The primar strata in this study are the LSF, SYF and MSF.

The sediments source of the QDNB includes the palaeo-uplifts within the basin, th Hainan Uplift and the Red River [39–42]. These sediment sources were dominant i different periods [43]. The basin was in the rifting stage during the deposition of the YC and LSF, and each sag was an independent sedimentary unit [34–36]. The sediments of each sag during this period primarily originated from the palaeo-uplifts adjacent to the sag, and the transport distance of these sediments was short [41]. During the deposition of the Miocene SYF and MSF, the basin was in the post-rift subsidence stage, and the sediments from the Hainan Uplift entered the basin widely and occupied a dominant position [34,36,44]. The transport distance of these sediments is longer than that of the Oligocene sediments. During the late Miocene HLF and Pliocene YGHF depositions, the sediment transported by the Red River entered the basin and deposited as a large submarine fan and canyon with the longest transportation distance [42,45,46].

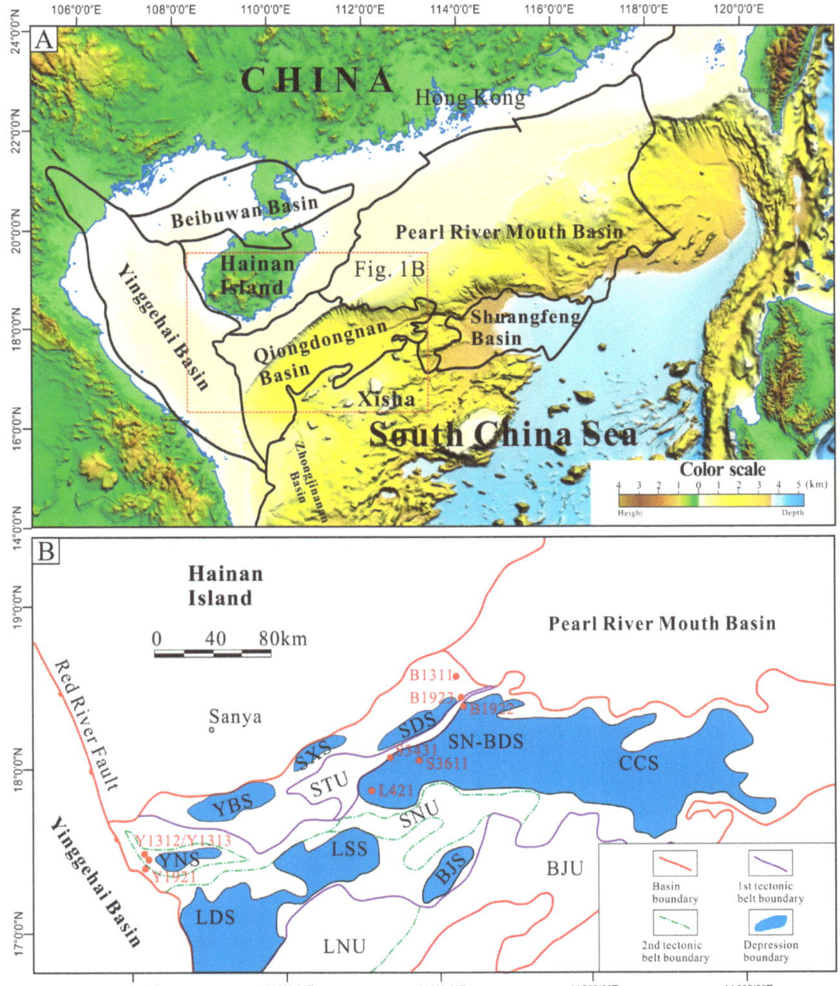

Figure 1. Map showing the location of the Qiongdongnan Basin (**A**) and the sags of the Qiongdongnan Basin (**B**). A 3D topographic map in (**A**), modified after Yang [47]. Legends: YBS: Yabei Sag; SXS: Songxi Sag; SDS: Songdong Sag; YNS: Yanan Sag; LDS: Ledong Sag; LSS: Lingshui Sag; SN–BDS: Songnan–Baodao Sag; CCS: Changchang Sag; BJS: Beijiao Sag; STU: Songtao Uplift; SNU: Songnan Uplift; LNU: Lingnan Uplift.

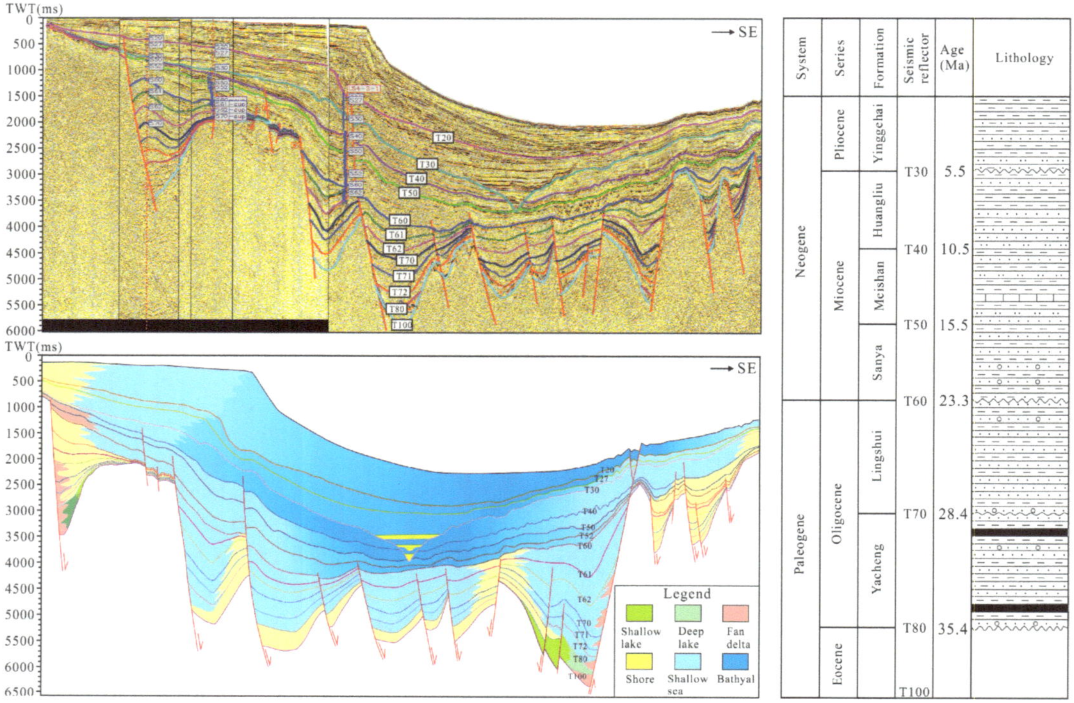

Figure 2. The **left** seismic profile and interpretation profile shows the internal structure, tectoni stage, sedimentary facies and evolution of the Qiongdongnan Basin. The **right** sketch map shows th sequence classification, seismic reflector, geologic age and lithologic characteristics of the Qiongdon nan Basin. The ages of the sequence boundaries and formations in the Qiongdongnan Basin wer provided by the Research Institute of China National Offshore Oil Corporation.

3. Materials and Methods

3.1. Samples and Their Petrography, Porosity and Permeability

Systematic observations were conducted on cores from the QDNB, covering appro imately 200 m of the cores from 10 wells. The China National Offshore Oil Corporatio (CNOOC) provided 60 core samples from the MSF, SYF and LSF sandstone. Sixty sample were prepared as cast thin sections to investigate the mineralogical compositions, diager esis and pores. A Nicon LV100 polarizing petrographic microscope (Nikon Corporatio Tokyo, Japan) was used to observe thin sections under different magnifications. Quant tative petrographic analysis was performed by counting at least 300 points/thin section to differentiate between detrital composition, cements, foraminifera and types of pore Scanning electron microscope (SEM) analysis was performed on nineteen samples usin an FEI Merlin Compact SEM (Carl Zeiss AG, Oberkochen, Germany). The purpose of th SEM analyses was to determine the type and morphology of the foraminifera. Approx mately 406 mineralogical compositions of the MSF, SYF and LSF sandstone were used, c which 110 mineralogical compositions refer to different samples that were point-counte through thin sections, 162 mineralogical compositions were obtained from the CNOOC, an the other 134 mineralogical compositions from Gao [48] and Zhong [49]. Approximatel 2438 porosity and permeability values of the LSF, SYF, MSF, HLF and YGHF sandston were used, of which 1147 porosities and permeabilities were obtained from the CNOO and the other 1291 from Zhao [50], You [51] and Su [52,53].

3.2. The Lower Porosity and Permeability Limits of Effective Reservoir Analysis

The lower porosity and permeability limits of the effective reservoir were investigated using 423 measured porosities and permeabilities and interpreted gas saturations from 44 wells (including the prementioned 10 wells of which samples were prepared as cast thin sections to investigate the pores). Depending on whether the reservoir contains natural gas and water, the reservoirs were interpreted as a gas layer, gas and water layer, gas-bearing water layer, water layer and dry layer [13–15]. The CNOOC provided the porosity, permeability and gas- or water-layer interpretation results. Here, we define the gas-bearing water layer, gas and water layer, gas layer and water layer as the geological fluid-bearing layer. On the cross-plot with porosity and permeability as abscissa and ordinate, respectively, the samples of those interpreted as the dry layer are frequently distributed in different regions, with the samples interpreted as a geological fluid-bearing layer, and the porosity and permeability boundaries between them are considered as the lower porosity and permeability limits of the effective reservoir [13–15].

4. Results

4.1. Sandstone Petrography

4.1.1. Late Oligocene LSF

The LSF sandstone includes four lithofacies: conglomeratic coarse sandstone, coarse sandstone, fine sandstone and siltstone (see Figure 3). The LSF sandstone comprises subfeldsarenite, sublitharenite, litharenite, feldspathic litharenite, lithic arkose and feldsarenite (see Figure 4A). The average quartz content is 59.5%, of which the single-crystal quartz content is 45.9%, the polycrystalline quartz content is 13.6% and the secondary growth of quartz is generally developed (see Figure 5A). The average feldspar content is 16.3% and its type is primarily potassium feldspar. The average rock fragment content is 7.7% and the type is predominantly igneous and metamorphic rock fragments (see Figure 5A). The average matrix content is 10.4%, and the cement is dominated by calcite, with a content of 0.2–8.6%, followed by iron dolomite (3.5%) and iron calcite (2.1%). The grain size is typically fine to medium with a few coarse grains present, well-to-moderately sorted and subangular to round (see Figure 5A).

4.1.2. Lower Miocene SYF

The SYF sandstone includes two lithofacies, primarily comprising siltstone and fine sandstone (see Figure 3). The SYF sandstone contains subfeldsarenite and sublitharenite and some quartz arenite and litharenite (see Figure 4B). The average quartz, feldspar and rock fragment contents are 60.3%, 6.9% and 6.4%, respectively (see Figure 5B). The feldspar and rock fragments are mainly potassium feldspar and metamorphic and extrusive rock fragments, with average contents of 6.7%, 2.3% and 2.6%, respectively. The argillaceous matrix content is 2.7–34.6%. The cement is mainly iron calcite (avg. 4.5%), followed by dolomite (2.8%) and iron dolomite (2.8%). The grain size is silt to fine grain, poorly to moderately sorted and angular to subangular (see Figure 5B).

4.1.3. Middle Miocene MSF

The MSF sandstone includes two lithofacies, primarily comprising siltstone and muddy siltstone (see Figure 3). The MSF sandstone contains subfeldsarenite, sublitharenite, litharenite, feldspathic litharenite and some quartz arenite (see Figure 4C). The quartz content ranges from 33.0% to 74.0% (avg. 51.8%) (see Figure 5C). Quartz primarily comprises single-crystal quartz; the secondary growth of quartz is undeveloped. The feldspar content is 2.5–6.5% (avg. 5.5%) and its type is mainly potassium feldspar. The rock fragment content is 1.0–2.5% (avg. 1.6%) and the type is primarily metamorphic rock fragments (avg. 1.5%). The argillaceous matrix content in sandstone ranges from 2% to 26% (avg. 7.3%). The cement is mainly iron calcite (5.3%), followed by calcite (1.3%), dolomite (1.1%) and iron dolomite (0.6%). The sandstone texture of the MSF is similar to that of the SYF, with the silt grain size being poorly sorted and angular to subangular (see Figure 5C).

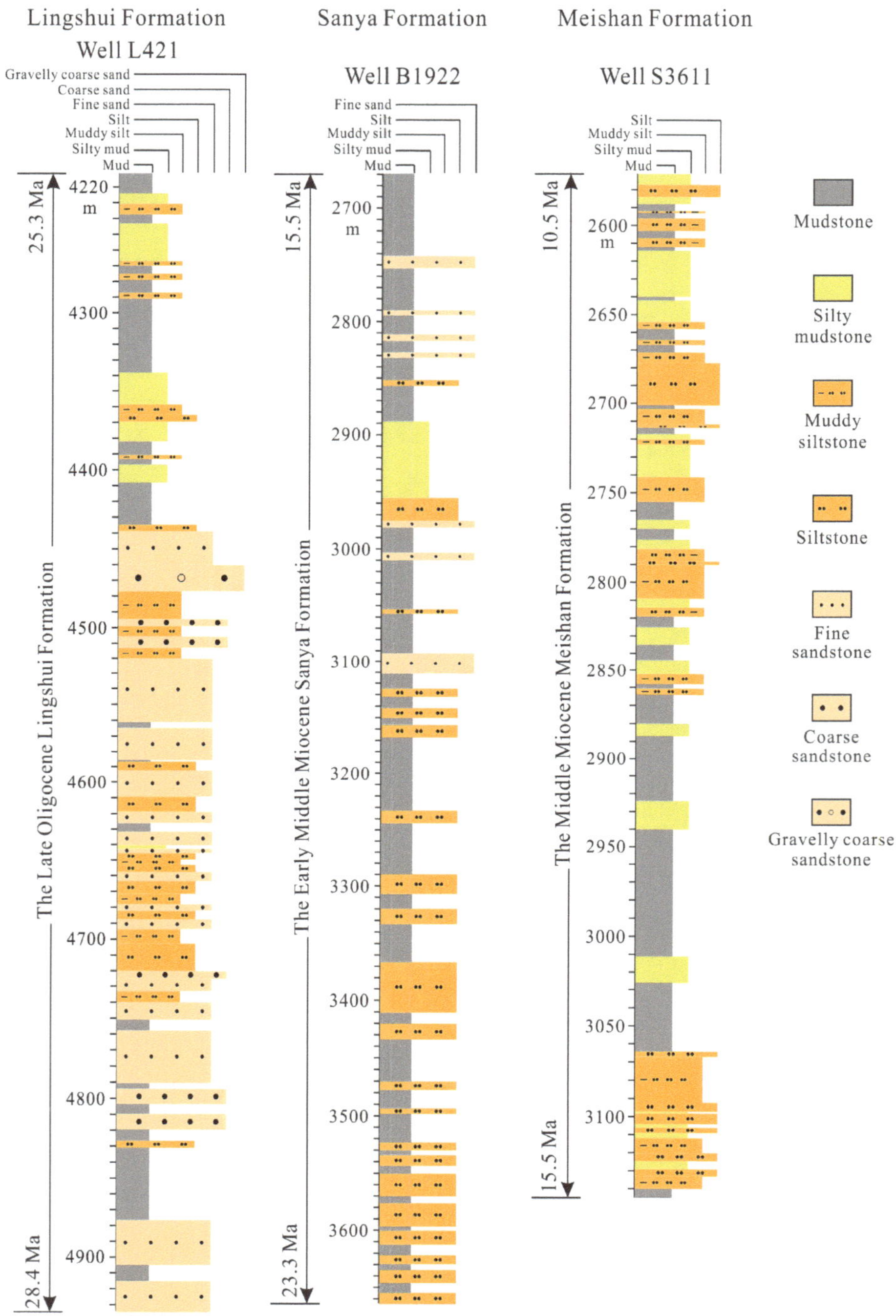

Figure 3. Well L421 shows that the Lingshui Formation sandstone comprises siltstone, fine sandstone coarse sandstone and conglomeratic coarse sandstone; well B1922 shows that the Sanya Formation sandstone comprises siltstone and fine sandstone; and well S3611 shows that the Meishan Formation sandstone comprises siltstone and muddy siltstone.

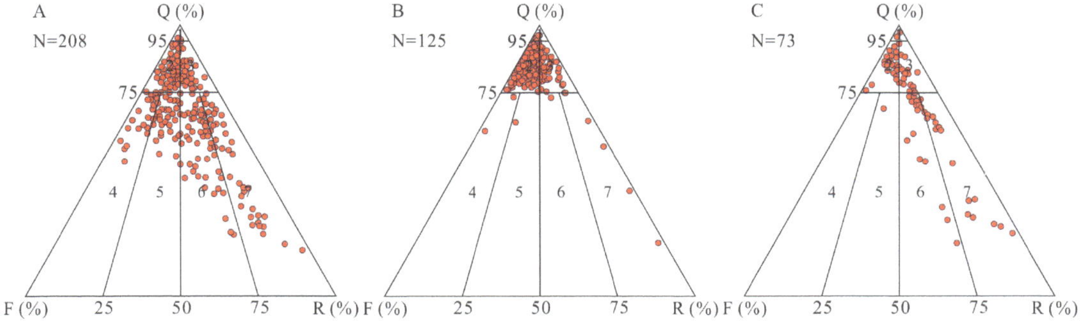

Figure 4. The mineralogical compositions of (**A**) the Lingshui Formation, (**B**) the Sanya Formation and (**C**) the Meishan Formation sandstones are plotted in the triangular chart (after Folk [54]). 1: quartz arenite; 2: subfeldsarenite; 3: sublitharenite; 4: feldsarenite; 5: lithic arkose; 6: feldspathic litharenite; 7: litharenite.

Figure 5. Lithology and pore characteristics of the Lingshui, Sanya and Meishan Formation sandstones. (**A**) The thin-section micrograph shows that the Lingshui Formation (LSF) sandstone is fine–medium grain, moderately well-sorted and subangular to subrounded; well Y1312, 3883.15 m.

(**B**) The Sanya Formation (SYF) sandstone is silt to fine grain, poorly to moderately sorted and angul to subangular; well S3431, 2575.20 m. (**C**) The Meishan Formation sandstone is silt grain, moderate sorted and angular to subangular; well B1923, 2426.50 m. (**D**) A few primary intergranular an many intergranular dissolution pores are developed in the LSF sandstone; well Y1921, 3827.90 r (**E**) Feldspar grains dissolved to form intragranular pores in the LSF sandstone; well Y1313, 3816.10 r (**F**) Secondary dissolution pores are widely developed in the SYF sandstone; well S3431, 2576.00 r (**G**) Feldspar grains dissolved to form intragranular pores in the SYF sandstone; well S3431, 2572.90 (**H**) Glauconite grains dissolved to form intergranular and intragranular pores in the SYF sandston Well S3431, 2566.00 m. (**I**) Secondary dissolution pores are widely developed in the Meishan Form tion sandstone; well B1311, 1582.10 m. Legend: quartz (Q), quartz overgrowth (QO), feldspar (F dissolved felspar (DF), glauconite (G), dissolved glauconite (DG), primary porosity (PP), secondar pores (SP).

4.2. Pore Systems

4.2.1. Primary Intergranular Pore

Only a few primary intergranular pores are preserved in the LSF, SYF and MSF san stones. The minerals are typically inline or in uneven contact because the LSF sandston has undergone strong compaction (see Figure 5A). The lesser development of cemer allows a few primary intergranular pores to be retained (see Figure 5D,E). These pores ar usually triangular or elongated in thin sections. The SYF and MSF sandstones experienc inadequate compaction and the minerals are typically in point contact (see Figure 5F– Because of strong carbonate cementation and the generous argillaceous filling, the primar intergranular pores in the SYF and MSF sandstones were not preserved. Only when th MSF sandstone is well-sorted and the argillaceous matrix is low can the residual primar intergranular pores be observed between the quartz grains.

4.2.2. Secondary Dissolution Pore

Secondary dissolution pores widely exist in the LSF, SYF and MSF sandstones (se Figure 5E–I). The dissolution pores comprise intergranular dissolved, intragranular di solved, mould and dissolution-enlarged pores (see Figure 5E–I).

The intergranular dissolution pores in the LSF, SYF and MSF sandstones are primaril formed by the dissolution of the feldspar grain edges (see Figure 5E,F,I). The morpholog of the feldspar intergranular dissolution pores is highly irregular, frequently showing a ba like shape. Furthermore, the intergranular dissolution pores related to the dissolution c the glauconite grain edges are developed in the SYF and MSF sandstones (see Figure 5H, Since glauconite is a unique mineral in the SYF and MSF sandstones, the intergranula dissolution pores related to glauconite are unique to the SYF and MSF sandstones (se Figure 5H,I). The morphology of glauconite intergranular dissolution pores is much mor regular than that of feldspar intergranular dissolution pores, frequently similar to the shap of the glauconite particles (see Figure 5H).

The feldspar and glauconite intragranular dissolution primarily form the intragranula dissolution pores in the LSF, SYF and MSF sandstones (see Figure 5E,G,H). The intragran lar dissolution of feldspar is widespread in the LSF sandstone (see Figure 5E). However, th intragranular dissolution of glauconite is more common in the SYF and MSF sandstone (see Figure 5H). The intragranular dissolution pores related to glauconite in the SYF an MSF sandstones are more developed than the intergranular dissolution pores related t feldspar. Feldspar intragranular dissolution is frequently carried along the cleavage c feldspar to the periphery to form irregular intragranular dissolution pores (see Figure 5E The dissolution pores in glauconite grains are primarily regular, continuous flakes (se Figure 5H).

Mould pores are widely developed in the SYF and MSF sandstones and are forme when the glauconite particle is almost entirely dissolved, and only the outline of th grain remains (see Figure 5H). The size and shape of the glauconite grains determine th size and morphology of these pores. The glauconite in the SYF and MSF sandstones ar mainly stripes, triangles and ellipses; therefore, the related mould pores are also stripes, triangles and ellipses. The complete dissolution of feldspar to form a mould pore is rarely observed. However, after several continuously arranged feldspars undergo intergranular or intragranular dissolution, dissolution-enlarged pores are formed in the LSF, SYF and MSF sandstones.

4.2.3. Foraminiferal Cavities

Foraminifera are widely developed in the SYF sandstone and are occasionally observed in the MSF sandstone (see Figure 6). The most common foraminifera in the SYF sandstone are in the *Operculina* sp. It contains a few fossil fragments of the *Cycloclypeus* sp. and *Miogypsina* sp., with a shell diameter of 2–6 mm (see Figure 6). Other biological debris included minimal Globigerina and bivalves. Foraminiferal shells in thin sections typically appear in a directional arrangement. The size of foraminiferal bodies buried in the same position is similar, the shells are intact and many foraminiferal cavities are preserved. Foraminiferal cavities are isolated pores separated by partition walls, and only some are partially connected due to the destruction of the shell structure (see Figure 6).

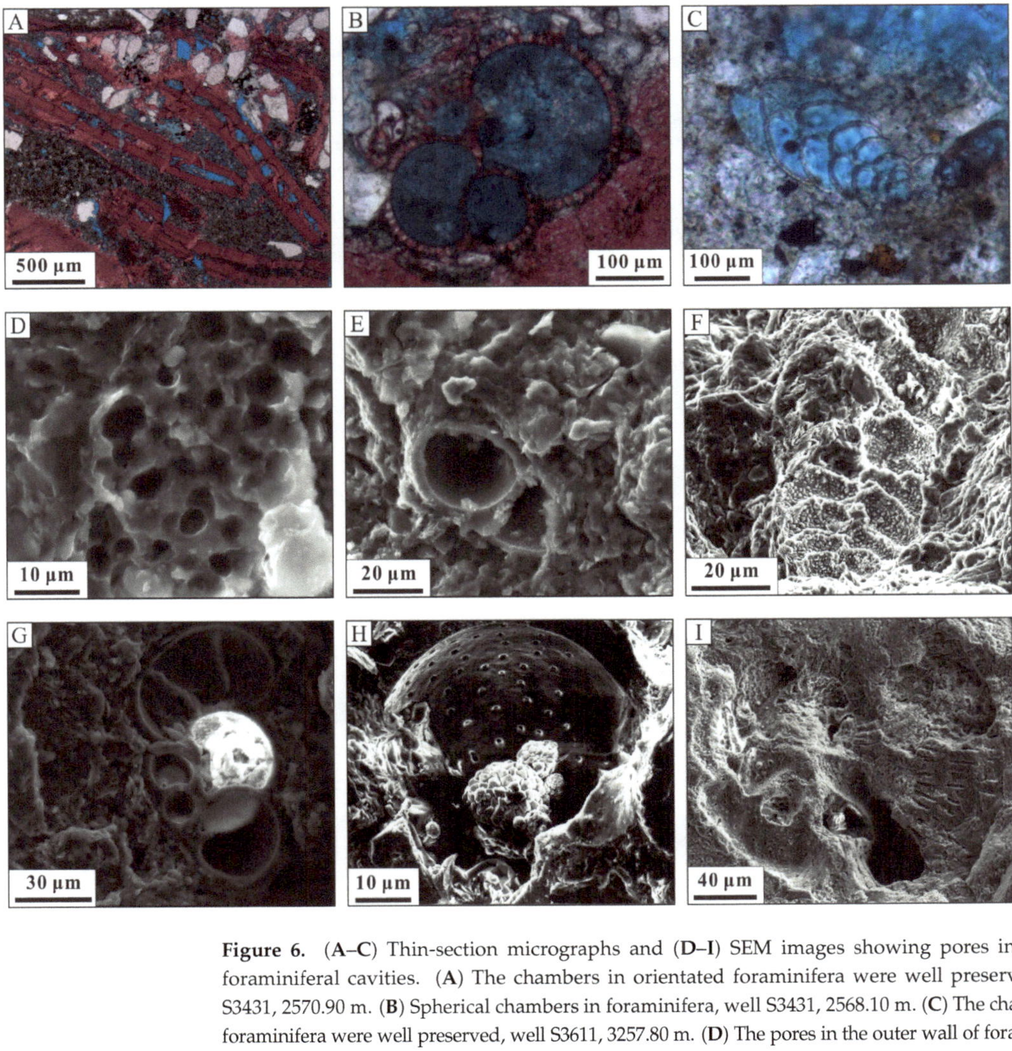

Figure 6. (**A–C**) Thin-section micrographs and (**D–I**) SEM images showing pores in various foraminiferal cavities. (**A**) The chambers in orientated foraminifera were well preserved, well S3431, 2570.90 m. (**B**) Spherical chambers in foraminifera, well S3431, 2568.10 m. (**C**) The chambers in foraminifera were well preserved, well S3611, 3257.80 m. (**D**) The pores in the outer wall of foraminifera,

well S3611, 3257.80 m. (**E**) Spherical chambers in foraminifera, well S3611, 3257.80 m. (**F**) Th chambers in foraminifera, well S3611, 2813.20 m. (**G**) Spherical chambers in foraminifera, well S361 3257.80 m. (**H**) The pores in the outer wall of foraminifera, well B1331, 1968.90 m. (**I**) The chambers i foraminifera, well Y9, 2221.57 m.

4.3. Porosity and Permeability

In the western QDNB, the LSF sandstone has a porosity of 2.1–29.35% and a perm ability of 0.07–2549 × 10^{-3} μm^2 (see Figure 7A,B). Three abnormally high petrophysica property zones developed at depths of 2700–2900 m, 3650–4150 m and 4450–4800 m wit maximum porosities of approximately 30%, 25% and 20%, respectively (see Figure 7A). I the eastern QDNB, the amount of LSF sandstone with a burial depth of less than 3500 r porosity is generally high, ranging from 10% to 30%, and that of LSF with a burial dept greater than 3500 m is rapidly reduced to 2–15% (see Figure 7A). Two abnormally hig porosity zones developed at depths of 3750–4300 m and 4800–5250 m, with maximur porosities of approximately 17% and 13% (see Figure 7A).

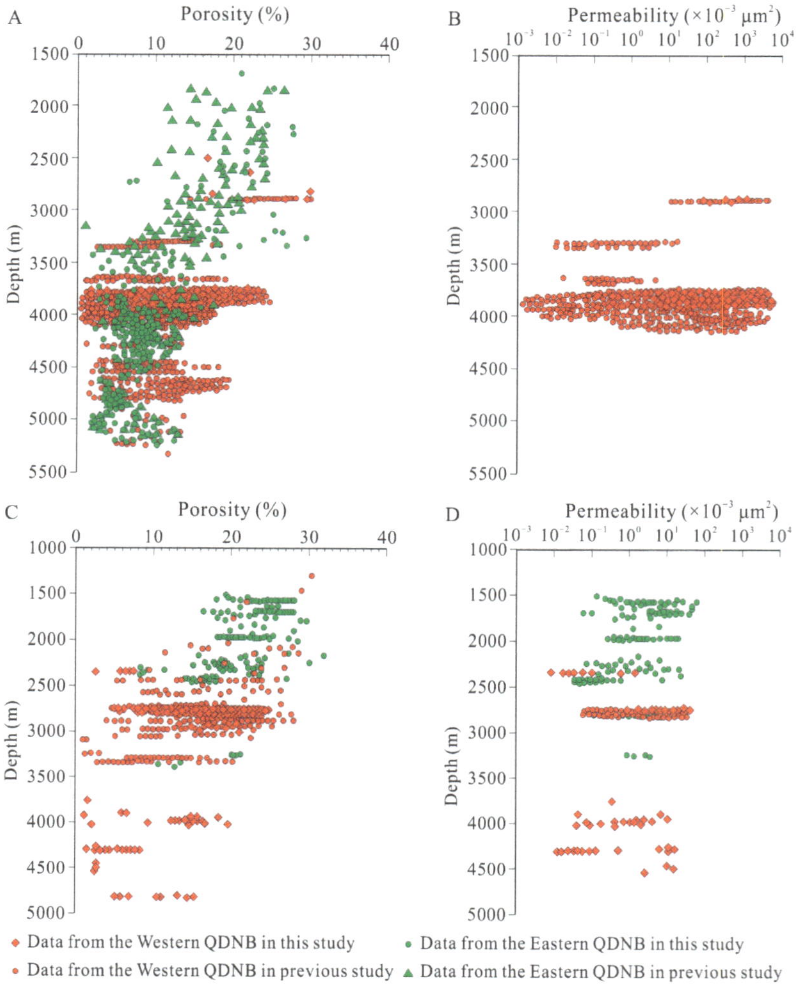

Figure 7. (**A**) Porosity and (**B**) permeability of the Lingshui Formation sandstones. (**C**) Porosity an (**D**) permeability of the Sanya and Meishan Formation sandstones.

The burial depths of the SYF and MSF sandstones in the western QDNB are typically greater (>2300 m) than those in the eastern QDNB. The burial depth of the SYF and MSF sandstones in the western QDNB is generally greater than 2300 m, whereas that of the eastern QDNB is usually less than 2500 m. The difference in the SYF and MSF sandstones' burial depths in the eastern QDNB ensures a higher porosity of 8–32% and permeability of 0.05–93.8 $\times 10^{-3}$ μm^2, whereas, in the western QDNB, they have a low porosity of 1–28% and permeability of 0.01–20 $\times 10^{-3}$ μm^2 (see Figure 7C,D). The SYF and MSF sandstones in the western QDNB developed one abnormally high porosity zone at 4700–4800 m deep (see Figure 7C).

5. Discussion

5.1. The Lower Porosity and Permeability Limits of the Effective Reservoir

5.1.1. Sandstone in Different Sedimentary Facies

Various sedimentary facies are developed in the QDNB, including delta, submarine canyon and submarine fan [35,40,42,46]. The lower porosity and permeability limits of effective reservoirs in different sedimentary facies is an exciting topic. Considering that the submarine fan data are few in this study, the submarine fan and submarine canyon data are taken as the sample for statistical analysis. Statistical analysis shows that the porosity and permeability of sandstone developed in the delta are greater than 8.9% and 1.2 $\times 10^{-3}$ μm^2, respectively, and the sandstone might contain water or natural gas, thus becoming an effective reservoir (see Figure 8A). For sandstone developed in submarine canyons and fans, the lower porosity and permeability limits of sandstone that can become an effective reservoir are 11.3% and 4.0 $\times 10^{-3}$ μm^2, respectively (see Figure 8B). The result indicates that with the increase in the sediment transport distance, the lower porosity and permeability limits of the reservoir will increase. The possible reason for this result is that, with the increase in the sediment transport distance, the sandstone will be better sorted and rounded. Intergranular pore development is more common, resulting in the sandstone having higher porosity and permeability. When geological fluids enter these sandstone layers, they preferentially accumulate into a reservoir with higher porosity and permeability. This process improves the lower porosity and permeability limits of whether the sandstone can contain geological fluids.

5.1.2. Sandstone with Different Grain Sizes

Argillaceous siltstone, siltstone and fine and coarse sandstone are enriched in the QDNB [40–42]. The pore structures and petrophysical parameters of sandstone with different grain sizes vary considerably [49–53]. Therefore, it is necessary to analyse the lower porosity and permeability limits of effective reservoirs with varying grain sizes. Statistical analysis shows that only when the porosity and permeability of coarse sandstone, fine sandstone, siltstone and argillaceous siltstone are greater than 9.9% and 7.7 $\times 10^{-3}$ μm^2, 10.2% and 1.9 $\times 10^{-3}$ μm^2, 12.3% and 9.9 $\times 10^{-3}$ μm^2 and 12.2% and 3.9 $\times 10^{-3}$ μm^2, respectively (see Figure 8C–F), the sandstone might contain water or natural gas, thus becoming effective reservoirs. The result indicates that with the decreasing grain size, the lower porosity limit for sandstone to be an effective reservoir will increase and no significant correlation exists between the grain size and lower permeability limit because, with the decrease in grain size, the porosity and permeability of the reservoir decrease. It is challenging for geological fluids to enter reservoir with poor porosity and permeability, requiring higher porosity and permeability limits for siltstones to be effective reservoirs.

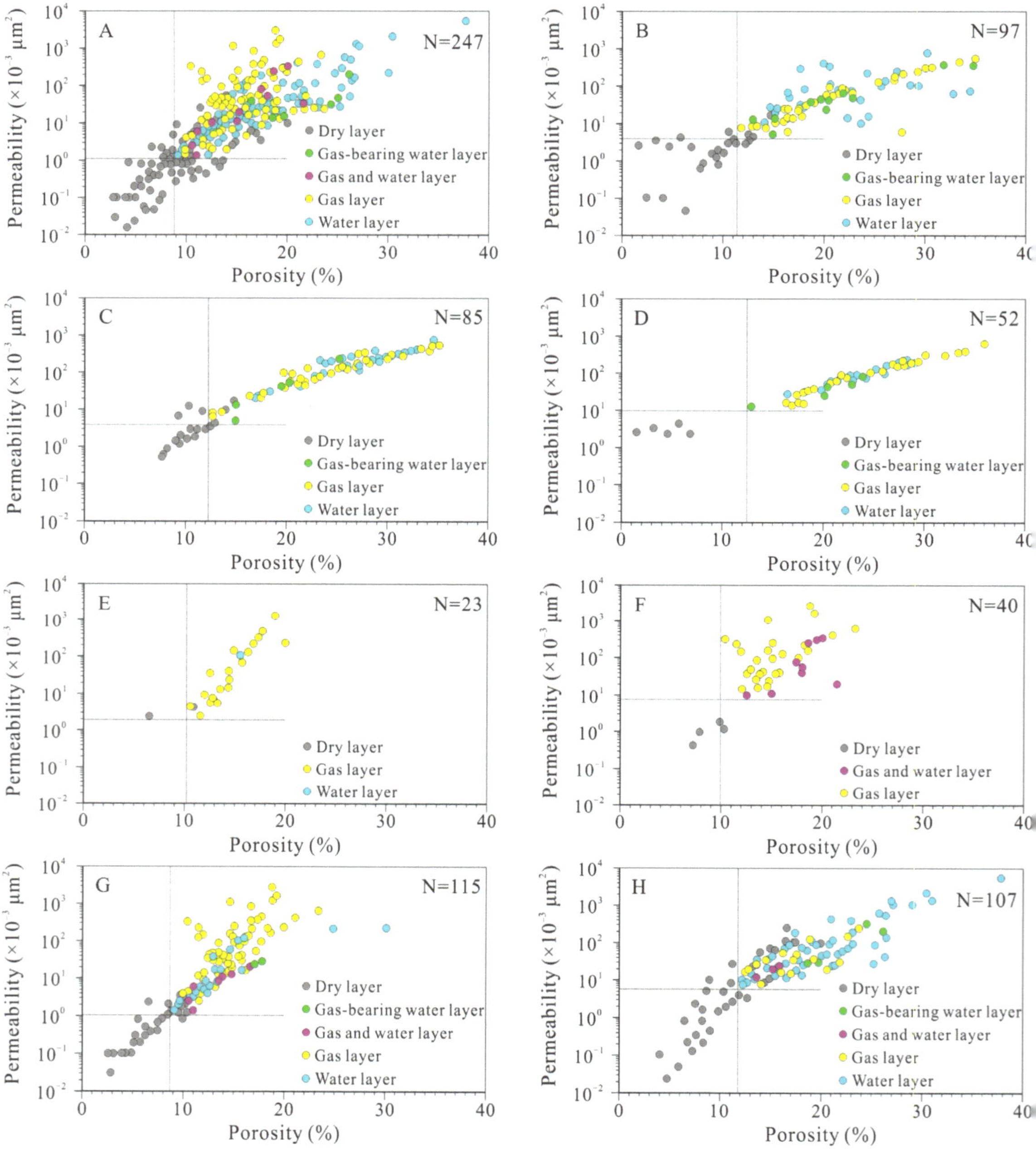

Figure 8. The porosity, permeability and gas testing relation graphs of the sandstones developed i the (**A**) deltas, (**B**) submarine canyons and fans, (**C**) coarse sandstone, (**D**) fine sandstone, (**E**) siltston (**F**) argillaceous siltstone, (**G**) Lingshui Formation and (**H**) Sanya and Meishan formations.

5.1.3. Sandstone in Different Strata

The burial depth of the sandstone influences its diagenetic process, which will chang the petrophysical parameters of the sandstone, causing the reservoir in the deep stratum t frequently have a lower porosity and permeability [49–51,55]. We discuss the variation i the lower porosity and permeability limits of effective reservoirs in different formations. Considering the small thickness and few data of the MSF (primarily 100–250 m thick), the MSF and the SYF data are taken as samples for statistics. Statistical analysis shows that the porosity and permeability of sandstone developed in the late Oligocene LSF are greater than 8.8% and 1.0×10^{-3} μm^2, respectively, and the sandstone might contain water or natural gas, thus becoming an effective reservoir (see Figure 8G). For sandstone developed in the Miocene MSF and SYF, the lower porosity and permeability limits of sandstone that can become effective reservoirs are 11.8% and 5.7×10^{-3} μm^2 (see Figure 8H). The result indicates that with the decrease in the stratum period, the lower porosity and permeability limits of the effective reservoir will increase because, in shallower buried formations, sandstone tends to experience weaker compaction and cementation, resulting in higher porosity and permeability. When geological fluids enter the shallower buried sandstone, they preferentially accumulate into sandstone with high porosity and permeability. This process improves the lower porosity and permeability limits of whether the shallower buried sandstone can contain geological fluids.

5.2. Formation Mechanism of the Effective Reservoir

5.2.1. Effect of Sediment Source and Transport Distance

Sediment source and transport distance are fundamental factors determining the petrophysical parameters of the QDNB sandstone [41,49,51]. In particular, the sediment transport distance significantly influences the petrophysical parameters of sandstones [41,49,50]. A petrophysical parameter comparative analysis shows that the sandstone sourced from the Red River drainage has experienced the longest transport distance [42,45,46], and the porosity and permeability are the highest (see Figure 9A). The sandstone reservoir sourced from the Hainan Uplift experienced a medium transport distance [34,36,44], and the porosity and permeability are in the middle of the three provenances (see Figure 9A). The sandstone reservoir sourced from the palaeo-uplift within the basin experienced the shortest transport distance [41], and the porosity and permeability were minor in the three provenances (see Figure 9A). Furthermore, the petrophysical parameters of sandstones from the palaeo-uplift within the QDNB and the Hainan Uplift vary widely, indicating that only a portion of the sandstone from these two provenances can be effective reservoirs (see Figure 9A).

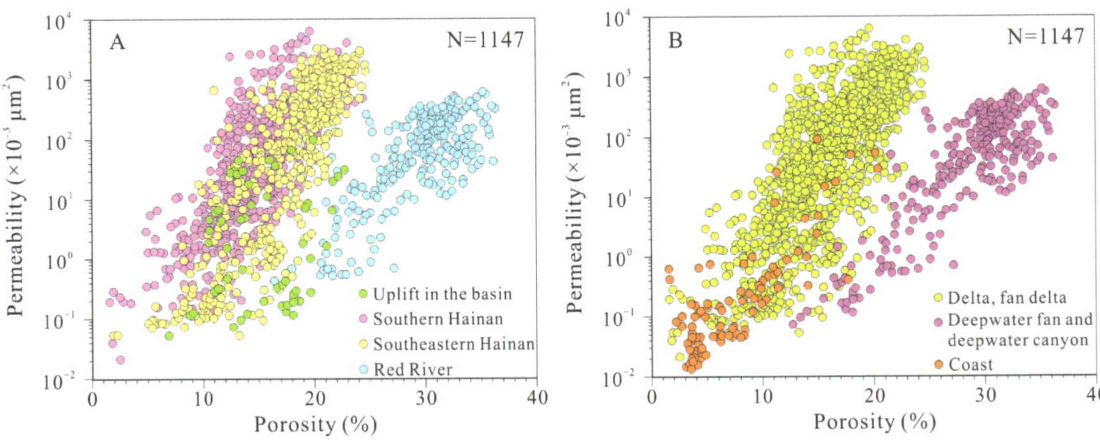

Figure 9. The relationship between porosity, permeability and (**A**) sediment source and (**B**) sedimentary facies of the sandstones in the Qiongdongnan Basin.

5.2.2. Effect of Sedimentary Facies

Reservoir quality is influenced by depositional process [55] and sedimentary facies are therefore another major factor determining the petrophysical parameters of the QDNB sandstone [49–51]. A comparative analysis of the petrophysical parameters of sandstone developed in different sedimentary facies shows that the sandstones developed in dee water fans and submarine canyons have higher porosity and permeability, followed b deltas and fan deltas. The sandstone developed in the coastal area has the lowest porosit and permeability (see Figure 9B). The porosity and permeability of sandstone developed i deepwater fans, submarine canyons, fan deltas and deltas of the QDNB vary widely. Th deepwater sedimentary sandstone sourced from the Red River system has higher porosit and permeability, whereas that from the southeast Hainan Uplift has a lower porosity an permeability (see Figure 9). The sandstones developed in fan deltas and deltas have a wid range of porosity and permeability because the sediment sources are primarily from th palaeo-uplift within the basin and the Hainan Uplift (see Figure 9).

5.2.3. Dissolution of Various Minerals during Diagenesis

(1) Dissolved minerals

Feldspar dissolution is an important diagenetic reaction affecting reservoir qualit evolution in sandstones with detrital feldspars [56]. Feldspar particles are the primar dissolved minerals in the LSF sandstone reservoir of the QDNB (see Figure 5D,E). Th average feldspar content in the LSF sandstone is 16.3%, of which 85% have suffered varyin degrees of dissolution. Some feldspar particles are only dissolved at the outer edge, an some are inside the particles (see Figure 5D,E). Most feldspar particles in the LSF sandston were dissolved, indicating that the feldspar content determines the development degree o dissolution pores. The feldspar content in the LSF sandstone is proportional to the grai size of the sandstone [48,49]. In sandstone reservoirs with coarser grain sizes, the feldspa content is higher and the dissolution pores of feldspar are more developed than those wit finer grain sizes [48,49].

In addition to feldspar, glauconite dissolution is more common in the SYF and MS sandstone reservoirs (see Figure 5H). The average feldspar contents in the SYF and MS sandstone reservoirs are 6.9% and 5.5%, respectively, which is much lower than the feldspa content in the LSF sandstone reservoir. However, the glauconite content in the SYF an MSF sandstone reservoirs reached 3–15%, of which 95% have suffered different degree of dissolution (see Figure 5H). Moreover, most dissolution occurs inside the glauconit particles, which significantly contributed to the dissolution pore development in the SY and MSF sandstones (see Figure 5H).

(2) Dissolved acidic geological fluids

Carbonic acid is one of the primary acidic fluids in the QDNB [57]. The CO_2 conten in the natural gas reservoir of the QDNB is 0.5–97.2%, and, in particular the CO_2 con tent in the natural gas reservoir distributed in the northern SN–BDS is typically highe than 80% [53,58]. The carbon isotope value of CO_2 in the QDNB is between −7.50‰ an 3.90‰, the value of $^3He/^4He$ is 3.46–8.75 × 10^{-6} and the R/Ra value is between 2.47 an 7.87 [53,58]. Such recorded values indicates that the CO_2 in the QDNB is primarily o mantle-derived inorganic origin. Mantle-derived CO_2 in the QDNB enters sandstone rese voirs mainly through magmatic diapirs and eruptive volcanoes [58]. The fluid inclusion show that the CO_2 charging periods are the Pliocene and Quaternary [53,58]. Simulatio experiments show that CO_2-rich deionised water can have a strong dissolution effect o potash feldspar and albite [57].

The organic acid released from the source rock is another essential acidic fluid i the QDNB [53]. The source rocks in the QDNB are developed in the YCF [58]. Th YCF in shallow-water areas comprises 483–910 m mudstones and 5–13.5 m carbonaceou mudstones and coals [58]. The total organic carbon (TOC) value is 0.4–98.5% [58]. Th YCF in deepwater areas comprises shales, carbonaceous shales and coals, with TOC value of 0.4–21% [58]. The YCF source rocks primarily contain types IIb and III kerogens [58 The YCF source rocks in the QDNB reached a mature stage at 22 Ma and entered the ga generation window at 10 Ma [58]. Thermal simulation experiments show that the YC source rocks can generate many organic acids, with an average yield ratio of organic acid t kerogen of 3.8 mg/g; moreover, the primary component is oxalic acid, which has a strong dissolution ability [53]. When the organic acids generated by the YCF source rocks enter the adjacent sandstone reservoir of the LSF, SYF and MSF along the fault system, they dissolve the soluble minerals in the sandstone reservoir and form dissolution pores [56].

5.3. Implication for Deepwater Natural Gas Exploration

The LSF effective reservoir in the deepwater area of the QDNB is mainly distributed in the northern margin of the LDS and LSS, the periphery of the SNU and the northern margin of the SNS. The sedimentary facies of the sand bodies in the northern area of the LDS and LSS are deepwater fans [41]. The sediments of these deepwater fans are mainly sourced from the deltas developed in the Yacheng area, where the sediments are primarily medium- and coarse-grained sandstone [41]. Therefore, deepwater fan sandstone developed in the northern margin of the LDS and LSS is also considered coarse grained [41]. Furthermore, dissolution pores are widely developed in the LSF sandstone in this area. The facies of the sand body developed on the periphery of the SNU is a beach bar, and no drilling has encountered this set of sand bodies. From the sedimentary environment, it is inferred that the sandstone should be pure. It is adjacent to the source rock and prone to organic acid dissolution. The sedimentary facies of the sand bodies developed in the northern margin of the SNS are fan deltas and coastal seas and are prone to organic acid dissolution.

The SYF effective reservoir in the deepwater area of the QDNB is mainly distributed in the northern margin of the LDS and LSS. The sedimentary facies of the sand bodies developed in the northern margin of the LDS and LSS are deepwater fans [41]. However, the grain size and argillaceous content of the sandstone vary [41]. The delta sediments developed on the northern shelf of the LDS and LSS have a coarse grain size, making the deepwater fans developed in this area have a coarser grain size and lower argillaceous content than in other areas [41]. Furthermore, the development of dissolution pores results in the deepwater fan sand bodies in this area having a higher porosity and permeability.

The MSF effective reservoir in the deepwater area of the QDNB is primarily distributed in the central and northern regions of the LDS and LSS and the northern margin of the SNS and CCS. The depositional facies of the sandstone are deepwater fans [40]. The sand bodies developed in the LDS and LSS are sheet sand but differ from the sheet sand in the northern margin of the SNS [40]. The sheet sand developed in the northern area of the LDS and LSS has a coarser grain size and lower argillaceous content [40]. Simultaneously, due to the development of dissolution pores, the sheet sand developed in the LDS and LSS has high porosity and permeability. The sheet sand developed in the northern margin of the SNS has multiple stages of evolution [40]. The deepwater fans developed in the early stage are small, are dominated by siltstone and rich in argillaceous [40]. The deepwater fans developed when the sea level decreased to the lowest during the middle Miocene has a large scale and a fine grain size [40]. Moreover, no magmatic diapirs are developed in the development area of the late-stage deepwater fan, which can avoid CO_2 filling [57]. The sedimentary facies of the sand bodies developed in the northern margin of the CCS are deepwater channels [40]. The lithology of the deepwater channel sandstone is well-sorted fine sandstone [40] with a high porosity and permeability.

6. Conclusions

This study analysed the lower petrophysical property limits and their influence factors of the effective marine sandstone gas reservoirs in the Qiongdongnan Basin using porosity, permeability and gas testing. Then, the roles of sediment source, sediment transport distance, sedimentary facies and dissolution diagenesis on the formation of an effective sandstone reservoir were evaluated, and finally the favourable reservoir development zone was proposed.

- A few primary intergranular and many secondary dissolution pores are preserved in the LSF, SYF and MSF sandstones of the QDNB. Foraminifera were widely preserved in the SYF sandstone. In the western QDNB, three abnormally high porosity and permeability zones developed in the LSF at depths of 2700–2900 m, 3650–4150 and 4450–4800 m, whereas there are two abnormally high porosity zones in the eas ern QDNB at depths of 3750–4300 m and 4800–5250 m. The SYF and MSF san stones in the eastern QDNB have a high porosity of 8–32% and a permeability 0.05–93.8 × 10^{-3} μm^2, whereas the SYF and MSF sandstones in the western QDN have a low porosity of 1–28% and a permeability of 0.01–20 × 10^{-3} μm^2. The SY and MSF sandstones in the western QDNB developed abnormally high porosity 4700–4800 m deep.

- The lower porosity and permeability limits of effective reservoirs developed in th deltas are 8.9% and 1.2 × 10^{-3} μm^2, whereas the lower porosity and permeability limi of those developed in submarine canyons and fans are 11.3% and 4.0 × 10^{-3} μm The lower porosity and permeability limits of the effective reservoirs with coars fine, silty and argillaceous silty grain sizes are 9.9% and 7.7 × 10^{-3} μm^2, 10.2% an 1.9 × 10^{-3} μm^2, 12.3% and 9.9 × 10^{-3} μm^2 and 12.2% and 3.9 × 10^{-3} μm^2, respective The lower porosity and permeability limits of effective reservoirs developed in th late Oligocene LSF are 8.8% and 1.0 × 10^{-3} μm^2, whereas the lower porosity an permeability limits of effective reservoirs developed in the Miocene MSF and SYF a 11.8% and 5.7 × 10^{-3} μm^2.
- The sandstones from the Red River have higher porosity and permeability, followe by those from the Hainan Uplift. The palaeo-uplift within the basin presents th lowest porosity and permeability. The sandstone reservoirs developed in deepwat fans and submarine canyons have a higher porosity and permeability, followed b deltas and fan deltas, and the sandstone reservoir developed in coastal areas has th lowest porosity and permeability. Dissolution of the feldspars by CO_2 and organi acid, resulting in dissolution pores, is considered the primary mechanism for increase porosity of the effective LSF reservoir. Glauconite particle dissolution is common i the SYF and MSF sandstone reservoirs.

Author Contributions: Conceptualization, C.L. and G.C.; methodology, C.L. and S.G.; formal analysi C.L. and C.X.; resources, S.G.; writing—original draft preparation, C.L.; writing—review and editin Q.Z.; visualization, C.X.; funding acquisition, G.C. All authors have read and agreed to the publishe version of the manuscript.

Funding: This research was funded by "Youth Innovation Promotion Association CAS" (Grant N 2022431), and the National Science and Technology Major Project of the Ministry of Science an Technology of China (Grant No. 2016ZX05026-007-05).

Institutional Review Board Statement: Not applicable.

Informed Consent Statement: Not applicable.

Data Availability Statement: The data that support the findings of this study are available withi the article.

Acknowledgments: We are grateful to the Research Institute of the China National Offshore O Corporation for their technical support, sampling assistance and granting permission to publish th results of this study.

Conflicts of Interest: Author S.G. was employed by the Research Institute of China National Offshor Oil Corporation. The remaining authors declare that the research was conducted in the absence c any commercial or financial relationships that could be construed as a potential conflict of interest.

References

1. Behmanesh, H.; Hamdi, H.; Clarkson, C.R.; Thompson, J.M.; Anderson, D.M. Analytical modeling of linear flow in single-phas tight oil and tight gas reservoirs. *J. Pet. Sci. Eng.* **2018**, *171*, 1084–1098. [CrossRef]
2. Hu, S.Y.; Zhu, R.K.; Wu, S.T.; Bai, B.; Yang, Z.; Cui, J.W. Exploration and development of continental tight oil in China. *Pet. Explo Dev.* **2018**, *45*, 790–802. [CrossRef]

3. Milad, M.; Junin, R.; Sidek, A.; Imqam, A.; Alusta, G.A.; Augustine, A.; Abdulazeez, M.A. Experimental investigation of bypassed-oil recovery in tight reservoir rock using a two-step CO_2 soaking strategy: Effects of fracture geometry. *Upstream Oil Gas Technol.* **2023**, *11*, 100093. [CrossRef]
4. Sun, L.D.; Zou, C.N.; Jia, A.L.; Wei, Y.S.; Zhu, R.K.; Wu, S.T.; Guo, Z. Development characteristics and orientation of tight oil and gas in China. *Pet. Explor. Dev.* **2019**, *46*, 1073–1087. [CrossRef]
5. Syed, F.I.; Dhaghi, A.K.; Muther, T. Laboratory to field scale assessment for EOR applicability in tight oil reservoirs. *Pet. Sci.* **2022**, *19*, 2131–2149. [CrossRef]
6. Meng, M.M.; Zhang, Y.X.; Yuan, B.; Li, Z.J.; Zhang, Y. Imbibition behavior of oil-saturated rock: Implications for enhanced oil recovery in unconventional reservoirs. *Energy Fuels* **2023**, *37*, 13759–13768. [CrossRef]
7. Altawati, F.; Emadi, H.; Pathak, S. Improving oil recovery of Eagle Ford shale samples using cryogenic and cyclic gas injection methods—An experimental study. *Fuel* **2021**, *302*, 121170. [CrossRef]
8. Du, J.H.; Liu, H.; Ma, D.S.; Fu, J.H.; Wang, Y.H.; Zhou, T.Y. Discussion on effective development techniques for continental tight oil in China. *Pet. Explor. Dev.* **2014**, *41*, 217–224. [CrossRef]
9. Lange, S.S.; Shrestha, L.; Nnoli, N.; Aniagu, S.; Rawat, S.; McCant, D. Do shale oil and gas production activities impact ambient air quality? Acomprehensive study of 12 years of chemical concentrations and well production data from the Barnett Shale region of Texas. *Environ. Int.* **2023**, *175*, 107930. [CrossRef]
10. Solarin, S.A.; Gil-Alana, L.A.; Lafuente, C. An investigation of long range reliance on shale oil and shale gas production in the U.S. market. *Energy* **2020**, *195*, 116933. [CrossRef]
11. Wang, H.; Liao, X.; Lu, N.; Cai, Z.; Liao, C.; Dou, X. A study on development effect of horizontal well with SRV in unconventional tight oil reservoir. *J. Energy Inst.* **2014**, *87*, 114–120. [CrossRef]
12. Wei, J.G.; Zhou, X.F.; Zhou, J.M.; Li, J.T.; Wang, A.L. Recovery efficiency of tight oil reservoirs with different injection fluids: An experimental investigation of oil-water distribution feature. *J. Pet. Sci. Eng.* **2020**, *195*, 107678. [CrossRef]
13. Bai, Y.B.; Zhao, J.Z.; Wu, W.T. Methods to determine the upper limits of petrophysical properties in tight oil reservoirs: Examples from the Ordos and Songliao Basins. *J. Pet. Sci. Eng.* **2021**, *196*, 107983. [CrossRef]
14. Cui, H.Y.; Zhong, N.N.; Li, J.; Wang, D.L.; Li, Z.S.; Hao, A.S.; Liang, F. Study on the lower limits of petrophysical parameters of the Upper Paleozoic tight sandstone gas reservoirs in the Ordos Basin, China. *J. Nat. Gas Geosci.* **2017**, *2*, 21–28. [CrossRef]
15. Pang, H.; Ding, X.G.; Pang, X.Q.; Geng, H. Lower limits of petrophysical parameters allowing tight oil accumulation in the Lucaogou Formation, Jimusaer Depression, Junggar Basin, Western China. *Mar. Pet. Geol.* **2019**, *101*, 428–439. [CrossRef]
16. Shakya, S.; Li, B.X.; Etienne, X.L. Shale revolution, oil and gas prices, and drilling activities in the United States. *Energy Econ.* **2022**, *108*, 105877. [CrossRef]
17. Zheng, D.Y.; Pang, X.Q.; Zhou, L.M.; You, X.C.; Liu, X.H.; Guo, F.X.; Li, W. Critical conditions of tight oil charging and determination of the lower limits of petrophysical properties for effective tight reservoirs: A case study from the Fengcheng Formation in the Fengcheng area, Junggar Basin. *J. Pet. Sci. Eng.* **2020**, *190*, 107135. [CrossRef]
18. Qiao, J.C.; Zeng, J.H.; Cai, J.C.; Jiang, S.; An, T.; Xiao, E.Z.; Zhang, Y.C.; Feng, X.; Yang, G.Q. Pore-scale heterogeneity of tight gas sandstone: Origins and impacts. *J. Nat. Gas Sci. Eng.* **2021**, *96*, 104248. [CrossRef]
19. ŠLiaupa, S.; Lozovskis, S.; Lazauskienė, J.; Šliaupienė, R. Petrophysical and mechanical properties of the lower Silurian perspective oil/gas shales of Lithuania. *J. Nat. Gas Sci. Eng.* **2020**, *79*, 103336. [CrossRef]
20. Wang, J.; Cao, Y.C.; Xiao, J.; Liu, K.Y.; Song, M.S. Factors controlling reservoir properties and hydrocarbon accumulation of the Eocene lacustrine beach-bar sandstones in the Dongying Depression, Bohai Bay Basin, China. *Mar. Pet. Geol.* **2019**, *99*, 1–16. [CrossRef]
21. Zhang, F.; Jiang, Z.X.; Sun, W.; Li, Y.H.; Zhang, X.; Zhu, L.; Wen, M. A multiscale comprehensive study on pore structure of tight sandstone reservoir realized by nuclear magnetic resonance, high pressure mercury injection and constant-rate mercury injection penetration test. *Mar. Pet. Geol.* **2019**, *109*, 208–222. [CrossRef]
22. Zhang, L.C.; Song, X.J.; Du, Y.J.; Lu, S.F.; Xiao, D.S.; Jiang, S.; Chen, X.L.; Zhang, R.; Yu, R.Y. The upper and lower limits and grading evaluation of the Shahezi tight gas reservoirs in the Xujiaweizi Rift, northern Songliao Basin: Implications from microscopic pore structures. *J. Pet. Sci. Eng.* **2022**, *212*, 110224. [CrossRef]
23. Randolph, M.F.; Gaudin, C.; Gourvenec, S.M.; White, D.J.; Boylan, N.; Cassidy, M.J. Recent advances in offshore geotechnics for deep water oil and gas developments. *Ocean Eng.* **2011**, *38*, 818–834. [CrossRef]
24. Skogdalen, J.E.; Vinnem, J.E. Quantitative risk analysis of oil and gas drilling, using Deepwater Horizon as case study. *Reliab. Eng. Syst. Saf.* **2012**, *100*, 58–66. [CrossRef]
25. Huo, J.H.; Zhang, R.Z.; Yu, B.S.; Che, Y.J.; Wu, Z.S.; Zhang, X.; Peng, Z.G. Preparation, characterization, investigation of phase change micro-encapsulated thermal control material used for energy storage and temperature regulation in deep-water oil and gas development. *Energy* **2022**, *239*, 122342. [CrossRef]
26. Epuh, E.E.; Joshua, E.O. Modeling of porosity and permeability for hydrocarbon Exploration: A case study of Gongola arm of the Upper Benue Trough. *J. Afr. Earth Sci.* **2020**, *162*, 103646. [CrossRef]
27. Farahani, M.; Aghaei, H.; Masoumi, H. Effect of pore type on porosity, permeability and pore volume compressibility of geological formations due to in-situ stress change. *J. Pet. Sci. Eng.* **2022**, *218*, 110986. [CrossRef]

28. Wang, J.J.; Wu, S.H.; Li, Q.; Zhang, J.J.; Guo, Q.H. Characterization of the pore-throat size of tight oil reservoirs and its control o reservoir physical properties: A case study of the Triassic tight sandstone of the sediment gravity flow in the Ordos Basin, Chin *J. Pet. Sci. Eng.* **2020**, *186*, 106701. [CrossRef]
29. Xia, Y.X.; Tian, Z.H.; Xu, S.; Wei, W.; Cai, J.C. Effects of microstructural and petrophysical properties on spontaneous imbibition i tight sandstone reservoirs. *J. Nat. Gas Sci. Eng.* **2021**, *96*, 104225. [CrossRef]
30. Meng, M.; Ge, H.; Shen, Y.; Ji, W.; Wang, Q. Rock fabric of tight sandstone and its influence on irreducible water saturation i Eastern Ordos Basin. *Energy Fuels* **2023**, *37*, 3685–3696. [CrossRef]
31. Er, C.; Zhao, J.Z.; Li, Y.Y.; Si, S.H.; Bai, Y.B.; Wu, W.T.; Han, Q.Y. Relationship between tight reservoir diagenesis and hydrocarbo accumulation: An example from the early Cretaceous Fuyu reservoir in the Daqing oil field, Songliao Basin, China. *J. Pet. Sc Eng.* **2022**, *208*, 109422. [CrossRef]
32. Liu, K.; Wang, R.; Shi, W.Z.; Zhang, W.; Qi, R.; Qin, S.; Xu, L.T. Tectonic controls on Permian tight gas accumulation: Constrain from fluid inclusion and paleo-structure reconstruction in the Hangjinqi area, northern Ordos Basin, China. *J. Nat. Gas Sci. En* **2020**, *83*, 103616. [CrossRef]
33. Ren, D.Z.; Zhou, D.S.; Liu, D.K.; Dong, F.J.; Ma, S.W.; Huang, H. Formation mechanism of the Upper Triassic Yanchang Formatio tight sandstone reservoir in Ordos Basin—Take Chang 6 reservoir in Jiyuan oil field as an example. *J. Pet. Sci. Eng.* **2019**, *17* 497–505. [CrossRef]
34. Hu, B.; Wang, L.S.; Yan, W.B.; Liu, S.W.; Cai, D.S.; Zhang, G.C.; Zhong, K.; Pei, J.X.; Sun, B. The tectonic evolution of th Qiongdongnan Basin in the northern margin of the South China Sea. *J. Asian Earth Sci.* **2013**, *77*, 163–182. [CrossRef]
35. Franke, D.; Savva, D.; Pubellier, M.; Steuer, S.; Mouly, B.; Auxietre, J.-L.; Meresse, F.; Chamot-Rooke, N. The final rifting evolutio in the South China Sea. *Mar. Pet. Geol.* **2014**, *58*, 704–720. [CrossRef]
36. Morley, C.K. Major unconformities/termination of extension events and associated surfaces in the South China Seas: Review an implications for tectonic development. *J. Asian Earth Sci.* **2016**, *120*, 62–86. [CrossRef]
37. Gong, C.L.; Wang, Y.M.; Hodgson, D.M.; Zhu, W.L.; Li, W.G.; Xu, Q.; Li, D. Origin and anatomy of two different types o mass–transport complexes: A 3D seismic case study from the northern South China Sea margin. *Mar. Pet. Geol.* **2014**, *54*, 198–21 [CrossRef]
38. Wang, Z.F.; Sun, Z.P.; Zhang, D.J.; Zhu, J.T.; Li, X.S.; Huang, B.J.; Guo, M.G.; Jiang, R.F. Geology and hydrocarbon accumulation in the deepwater of the northwestern South China Sea—With focus on natural gas. *Acta Oceanol. Sin.* **2015**, *34*, 57–70. [CrossRef
39. Clift, P.D.; Carter, A.; Wysocka, A.; Van Hoang, L.; Zheng, H.; Neubeck, N. A Late Eocene-Oligocene through-flowing Rive between the Upper Yangtze and South China Sea. *Geochem. Geophys. Geosystems* **2020**, *21*, e2020GC009046. [CrossRef]
40. Li, C.; Chen, G.J.; Zhou, Q.S.; Yang, H.Z.; Lyu, C.F.; Guo, S.; Sun, R.; Ma, M. Seismic geomorphology of three types of deepwate fans and their relationship with slope morphology: Qiongdongnan Basin, northern South China Sea. *Mar. Pet. Geol.* **202** *124*, 104814. [CrossRef]
41. Lyu, C.F.; Li, C.; Chen, G.J.; Zhang, G.C.; Ma, M.; Zhang, Y.; Sun, Z.T.; Zhou, Q.S. Zircon U–Pb age constraints on the provenanc of Upper Oligocene to Upper Miocene sandstones in the western Qiongdongnan Basin, South China sea. *Mar. Pet. Geol.* **202** *126*, 104891. [CrossRef]
42. Wang, Y.M.; Xu, Q.; Li, D.; Han, J.H.; Lü, M.; Wang, Y.F.; Li, W.G.; Wang, H.R. Late Miocene red river submarine fan, northweste South China Sea. *Chin. Sci. Bull.* **2011**, *56*, 1488–1494. [CrossRef]
43. Zhao, M.; Shao, L.; Liang, J.S.; Li, Q.Y. No Red River capture since the late Oligocene: Geochemical evidence from the Northweste South China Sea. *Deep. Sea Res. Part II Top. Stud. Oceanogr.* **2015**, *122*, 185–194. [CrossRef]
44. Savva, D.; Pubellier, M.; Franke, D.; Chamot-Rooke, N.; Meresse, F.; Steuer, S.; Auxietre, J.L. Different expressions of rifting on th South China Sea margins. *Mar. Pet. Geol.* **2014**, *58*, 579–598. [CrossRef]
45. Liang, C.; Liu, C.Y.; Xie, X.N.; Yu, X.H.; He, Y.L.; Chen, H.; Zhou, Z.; Tian, D.M.; Lu, B.Y.; Mi, H.G.; et al. The role of large-scal mass wasting processes in changing the sediment dispersal pattern in the deep-water Central Canyon of the northwestern Sout China Sea. *Mar. Pet. Geol.* **2020**, *122*, 104693. [CrossRef]
46. Su, M.; Wu, C.H.; Chen, H.; Li, D.F.; Jiang, T.; Xie, X.N.; Jiao, H.J.; Wang, Z.F.; Sun, X.M. Late Miocene provenance evolution at th head of Central Canyon in the Qiongdongnan Basin, Northern South China Sea. *Mar. Pet. Geol.* **2019**, *110*, 787–796. [CrossRef]
47. Yang, S.; Qiu, Y.; Zhu, B. *Atlas of Geology and Geophysics of the South China Sea*; China Navigation Publications: Tianjin, China, 201
48. Gao, Y.; Qu, X.Y.; Yang, X.B.; You, L.; Zhong, J.; Dong, X.F.; Cao, Y.Q.; Wang, Y.P. Characteristics of fluid inclusions an accumulation period of Miocene reservoir in Ledong-Lingshui Sag of Qiongdongnan Basin. *Mar. Orig. Pet. Geol.* **2018**, *23*, 83–9 (In Chinese with English Abstract)
49. Zhong, J.; Yang, X.B.; Zhu, P.Y.; Xu, S.L.; Deng, X.L.; Tuo, L.; Li, X.; Song, P. Porosity evolution differences of the Lingshu Formation reservoir between Baodao and Changchang Sag, Qiongdongnan Basin. *Earth Sci.* **2019**, *44*, 2665–2676. (In Chinese wit English Abstract)
50. Zhao, X.Q. *Diagenesis and the Favourable Reservoir Prediction of the Target Layer in the Qiongdongnan Basin*; Northeast Petroleu University: Daqing, China, 2010. (In Chinese with English Abstract)
51. You, L.; Li, W.; Li, C.; Zhao, Z.J. Main factors affecting physical properties of deep burial reservoir in Baodao area of Southeas Hainan Basin. *Spec. Oil Gas Reserv.* **2014**, *21*, 37–40. (In Chinese with English Abstract)

2. Su, A.; Chen, H.H.; He, C.; Lei, C.; Lei, M.Z.; Liu, Y.H. Digenesis controlling development of abnormal high porosity zones: A case from Yacheng area in the western Qiongdongnan basin, South China Sea. *J. China Universtiy Min. Technol.* **2017**, *46*, 345–355. (In Chinese with English Abstract)
3. Su, A.; Du, J.M.; Chen, H.H.; Yu, Y.; Lei, M.Z. Diagenetic fluid type and activity history of controlling the development of abnormal pore zone: Taking the north margin of Baodao Sag, Qiongdongnan Basin as an example. *Nat. Gas Geosci.* **2016**, *27*, 1837–1847. (In Chinese with English Abstract)
4. Folk, R.L. *Petrology of Sedimentary Rocks*; Hemphills Publishing Company: Sutton, UK, 1974.
5. Sun, N.L.; Zhong, J.H.; Ge, Y.Z.; van Loon, A.T. Reservoir quality of the Middle-Late Triassic Yanchang Formation (Ordos Basin) as controlled by sedimentology and diagenesis. In *The Ordos Basin*; Elsevier: Amsterdam, The Netherlands, 2022; pp. 421–460.
6. Yuan, G.H.; Cao, Y.C.; Gluyas, J.; Xiaoyan, L.; Xi, K.L.; Wang, Y.Z.; Jia, Z.Z.; Sun, P.; Oxtoby, N. Feldspar dissolution, authigenic clays, and quartz cements in open and closed sandstone geochemical systems during diagenesis: Typical examples from two sags in Bohai Bay Basin, East China. *AAPG Bull.* **2015**, *99*, 2121–2154. [CrossRef]
7. Li, C.Z.; Chen, G.J.; Li, C.; Tian, B.; Sun, R.; Su, L.; Lu, Y.X.; Wang, L.J. Experimental Study on Water-rock Reactions with CO_2 Fluid in a Deep Sandstone Formation under High Temperature and Pressure. *Acta Geol. Sin.-Engl. Ed.* **2021**, *95*, 268–279. [CrossRef]
8. Huang, B.J.; Tian, H.; Li, X.S.; Wang, Z.F.; Xiao, X.M. Geochemistry, origin and accumulation of natural gases in the deepwater area of the Qiongdongnan Basin, South China Sea. *Mar. Pet. Geol.* **2016**, *72*, 254–267. [CrossRef]

Journal of
Marine Science and Engineering

MDPI

Article

Characteristics and Reservoir Development Model of the Unconformity Caused by Huaiyuan Movement in Bohai Bay Basin, China: A Case Study of Chengdao-Zhuanghai Buried Hil in Jiyang Depression

Ruijuan Liu [1,2], Guozhi Wang [3,4,*], Yongshi Wang [5], Xuefeng Hao [2], Feng Qin [2], Xianxu Fang [6], Wei Meng [2] and Gang Liu [1]

[1] Colleges of Earth Sciences, Chengdu University of Technology, Chengdu 610059, China; liuruijuan2004@126.com (R.L.)
[2] Research Institute of Petroleum Exploration and Development, Shengli Oilfield, SINOPEC, Dongying 25700, China
[3] National Key Laboratory of Oil and Gas Reservoir Geology and Exploitation, Chengdu University of Technology, Chengdu 610059, China
[4] Institute of Sedimentary Geology, Chengdu University of Technology, Chengdu 610059, China
[5] Sinopec Shengli Oilfield Company, Dongying 25700, China; wangysh623@sina.com
[6] School of Geosciences, China University of Petroleum (East China), Qingdao 266580, China
* Correspondence: wangguozhi66@163.com; Tel.: +86-133-8817-8798

Abstract: It is beneficial in terms of the theoretical significance and application prospects to defin the structure and reservoir development model of the lower Paleozoic unconformity in the Jiyan Depression of Bohai Bay Basin, China, for oil and gas exploration of unconformity in carbonat strata. Geological and geochemical evidence shows that a regional unconformity formed during th Huaiyuan Movement in the lower Paleozoic strata of the Jiyang Depression. Along the top of th regional unconformity between the Yeli Liangjiashan Formation and Fengshan Formation, variou types of karst breccia have developed, showing prominent characteristics of development and vertica karst zonation. The paleokarst zone can be divided into the vadose zone and the underflow zon and there are apparent differences between the two zones in terms of the mode of karst activit and type of reservoir space. Primitive sedimentary microfacies, dolomitization, and superger karstification controlled the reservoirs of the Fengshan Formation and Yeli-Liangjiashan Formatio There are significant differences in the original physical properties due to the differences in th original sedimentary microfacies. The pore development of granular dolomite of high-energy beac facies has the best reservoir performance. In the later period, the superposition of dolomitization an supergene karstification resulted in apparent differences in karst development mode, developmer intensity, reservoir type, and reservoir physical properties. Among them, the granular dolomit reservoir has the best physical properties and has developed a cavity-type reservoir that has a plana distribution along an unconformity surface.

Keywords: Chengdao-Zhuanghai buried hill; the lower Paleozoic; Huaiyuan movement; unconformit structure; supergene karstification; reservoir development model

Citation: Liu, R.; Wang, G.; Wang, Y.; Hao, X.; Qin, F.; Fang, X.; Meng, W.; Liu, G. Characteristics and Reservoir Development Model of the Unconformity Caused by Huaiyuan Movement in Bohai Bay Basin, China: A Case Study of Chengdao-Zhuanghai Buried Hill in Jiyang Depression. *J. Mar. Sci. Eng.* **2024**, *12*, 804. https://doi.org/10.3390/jmse12050804

Academic Editor: Dmitry A. Ruban

Received: 5 April 2024
Revised: 9 May 2024
Accepted: 9 May 2024
Published: 11 May 2024

1. Introduction

Statistics show that nearly one-third of the world's large carbonate oil and gas field are related to unconformities [1–3]. Previous studies have shown that due to the influenc of tectonic movement, carbonate strata are exposed to the surface, and under the influenc of atmospheric freshwater leaching, supergene karstification occurs to form unconformit and a large number of dissolution pores and caves that are developed inside the unconfo mity can be used as good reservoirs and essential channels for oil and gas migration [4–13

Therefore, this critical study predicts the distribution of oil and gas reservoirs in carbonate strata to clarify the characteristics and development model of unconformity. Different scholars have effectively identified unconformity interfaces from the perspective of stratigraphic cycles, C and O isotopes, trace element data, and foraminiferal biostratigraphy [14–20]. On this basis, the karst profile structure and horizontal distribution controlled by paleokarst under the unconformity have been deeply analyzed by combining present and past methods. The zonation and intensity of karst vertically and horizontally in carbonate reservoirs with unconformities are vital factors controlling the formation and distribution of secondary reservoir spaces [17,21–27].

The lower Paleozoic strata in the Jiyang Depression of Bohai Bay Basin, China, experienced multiple stages of tectonic uplift, weathering and leaching, and hydrothermal transformation; these activities occurred in the Huaiyuan Movement (early Caledonian Movement), late Caledonian Movement, early Hercynian Movement, Indosinian Movement, Yanshan Movement, and Himalayan Movement [26]. The early Alpine stage, mainly the Mesozoic, is called the Yanshan Stage in China. The narrow Alpine Age, called the "Himalayan Age" in Asia, happened in the Cenozoic. Theoretically, there may be multistage unconformities and multiple zones of karst reservoirs. Previous studies have focused mainly on Himalayan unconformities and paleokarstification [28,29]. Although some petroleum geologists have proposed that the Huaiyuan Movement affected the Jiyang Depression [30], they believe that the reservoirs under the tectonic surface were not affected or controlled by supergenetic activities during the Huaiyuan Movement but rather by multistage uplift and dissolution that affected the surficial lake environment in the Paleogene faulted basin [26,31]. In addition to the lack of direct geological and geochemical evidence, the record of the Huaiyuan Movement has been a controversial and unresolved issue. With the continuous increase in oil and gas exploration in buried hills in the Jiyang Depression, the discovery of large-scale oil reservoirs in the Yeli Formation and Liangjiashan Formation, and the accumulation of a large amount of drilling data, petroleum geologists have begun to understand the Huaiyuan Movement.

The Huaiyuan Movement, first named by Li Siguang (1939) [32], is an important regional and multistage "curtain" tectonic movement that occurred in the early Paleozoic on the North China platform; the movement lasted from the end of the Zhangxia Stage in the middle Cambrian to the Liangjiashan Stage in the Early Ordovician. Frequently exposed strata between the Majiagou Formation and the Liangjiashan Formation and within the Yeli and Liangjiashan Formations were denuded from the Cambrian to the Ordovician. The unconformity between the late Liangjiashan and Majiagou stages is the prominent unconformity formed by this movement, and the duration of this depositional discontinuity is 3~18 Ma [32,33]. The Jiyang Depression is one of the essential oil-bearing basins on the North China Platform, and the whole tectonic evolution of the Huaiyuan Movement had an essential influence on the formation of the lower Paleozoic reservoir and oil and gas migration and accumulation in the Jiyang Depression. At present, because the duration of exposure and denudation in this period was shorter than that during the Caledonian Movement, the evidence of the record of this movement is relatively hidden, and it has not attracted extensive or in-depth attention from geologists [30,31]. What are the characteristics of the Huaiyuan Movement in the Jiyang Depression? Did it control the reservoir? Is this reservoir widely developed throughout the depression? What are the development patterns of the reservoir? All of these questions need to be answered. Therefore, the lower Paleozoic Chengdao-Zhuanghai buried hill in the Jiyang Depression is selected as the research area. Through field outcrop surveys, core observations, hand sample observations, geochemical analyses, and seismic and drilling data, evidence of unconformities, the characteristics of the development of unconformity structures, and reservoir development models during the Huaiyuan Movement are revealed. This study provides a new direction for research on lower Paleozoic reservoir prediction in the Jiyang Depression. It significantly contributes to the study of unconformity in carbonate strata worldwide.

2. Geological Background

The Chengdao-Zhuanghai buried hill is located to the northeast of the Jiyang Depression in the Bohai Bay Basin, eastern China (Figure 1). It is located in the central and eastern parts of the North China Platform, adjacent to three hydrocarbon-generating depressions (Huanghe Kou Sag, Bozhong Sag, and Chengbei Sag) on three sides, and connected to the Chengbei Sag on the southwest, the Bozhong Sag on the northeast, and the Huanghe Kou Sag on the east. The oil source conditions are excellent (Figure 2a).

The multi-stage tectonic movement influenced the formation of the Chengdao-Zhuanghai buried hill. During the Caledonian tectonic movement, the Lower Paleozoic carbonate platform as a whole was compressed and uplifted, the strata suffered denudation, and the Upper Ordovician series, Silurian Series, Devonian series, and Lower Carboniferous Series were missing, forming sedimentary discontinuities as long as 120 Ma. In the Hercynian tectonic movement stage, the Upper Paleozoic strata developed the cratonic sedimentary stage of land–sea intersection. In the late period, the whole was uplifted to land, suffered denudation, and the tectonic activity was quiet, and the fault was not developed. During the Indosinian tectonic movement, the early and Middle Triassic were stable deposits, while the NE–SW compression uplift and denudation occurred in the late Triassic. In the stage of Yanshan tectonic movement, the early to Middle Jurassic was in the process of filling and replenishing with weak fault activity, and the Late Jurassic experienced overall uplift, the Early Cretaceous NE–SW extensional fault depression, and the Late Cretaceous compressive uplift and strata denudation. In the stage of the Xishan tectonic movement, the S–N intense extensional fault depression occurred in the Paleogene, and the depression settlement occurred from the Neogene to the Quaternary, and the late sediments overlay the earlier sedimentary strata, forming the present buried hill [34].

Chengdao-Zhuanghai buried hill developed from the bottom up: Archean, Lower Paleozoic Cambrian, Ordovician, Upper Paleozoic Carboniferous, Permian, Mesozoic Jurassic and Cretaceous. The lower Paleozoic strata of the Chengdao-Zhuanghai buried hill are Cambrian and Ordovician in age (Figure 2b), formed in a marine environment, and are a set of stable carbonate strata. Influenced by the uplift and denudation of the Caledonian Movement, Hercynian Movement, Indosinian Movement, Yanshan Movement, Himalayan Movement, and other tectonic movements in the study area, the transformation and destruction of supergenetic karstification in multiple stages occurred, forming multistage unconformities and multiple paleokarst interfaces (Wang et al. [35]). In different parts of the Chengdao-Zhuanghai buried hill, parallel unconformities in the Cenozoic, Mesozoic, or some of the upper Paleozoic strata cover the lower Paleozoic Badou Formation, Majiagou Formation, or Yeli-Liangjiashan Formation. The late Cambrian and Ordovician strata were mainly affected by the Huaiyuan Movement (early Caledonian Movement), the late Caledonian Movement–early Hercynian Movement, and the Indosinian Movement, while the Yanshan Movement and the Himalayan Movement had relatively weak influences on the lower Paleozoic strata. Current exploration practices have shown that there are good reservoir spaces and migration channels in the lower Paleozoic strata in the Chengdao-Zanghai buried hill, resulting in oil and gas enrichment in several Formations (the Badou Formation, Majiagou Formation, Yeili Liangjiashan Formation, and Fengshan Formation) [35,36].

The lithologies of the Yeli-Liangjiashan Formation in the study area are dominantly dolomitic karst breccia, crystalline dolomite, and granular dolomite. Crystalline limestone, calcareous dolomite, dolomitic limestone, argillaceous dolomite, and mudstone are also present in local areas, with total formation thicknesses of 90~150 m. The Fengshan Formation is mainly composed of dolomitic or calcareous karst breccia, crystalline limestone, and crystalline dolomite, and some areas contain muddy limestone. The total formation thicknesses are 100~110 m. The Majiagou Formation, overlying the Yeli-Liangjiashan Formation, is dominated by limestone and dolomitic limestone interbedded with several sets of micritic dolomite and salt-rich rocks, with thicknesses ranging from 150 to 600 m (Figure 2b) [35]. In terms of sedimentary facies evolution, the Cambrian–Early Ordovician was mainly divided into three stages: the continuous transgression stage from the early Cambrian

Mantou stage to the middle Cambrian Xuzhuang stage, which was dominated by a tidal flat environment; the middle Cambrian Zhangxia to late Cambrian Gushan Changshan stage, during which the maximum extent of the ocean occurred and was dominated by subtidal high-energy shoals (Zhangxia stage), as well as open sea; and the late Fengshanian and Early Ordovician stage, during which the water was shallower, and the Yeli-Liangjiashan and Fengshan Formations formed in tidal flat and restricted sea environments. Granular beach facies, evaporative platform dolomite facies, restricted platform limestone facies, and limestone-dolomite facies are of equal occurrence in the Chengdao-Zhuanghai area in the northeast Jiyang Depression (Figure 3) [26,37].

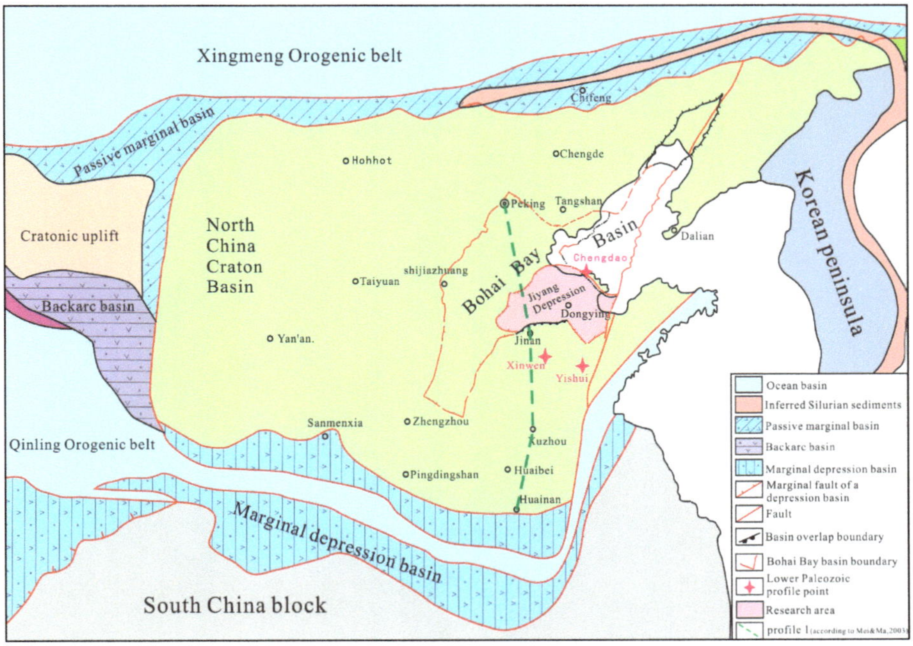

Figure 1. Early Paleozoic tectonic-paleogeography diagram of North China Plate (modified according to Ma Shuai, 2023 [37]).

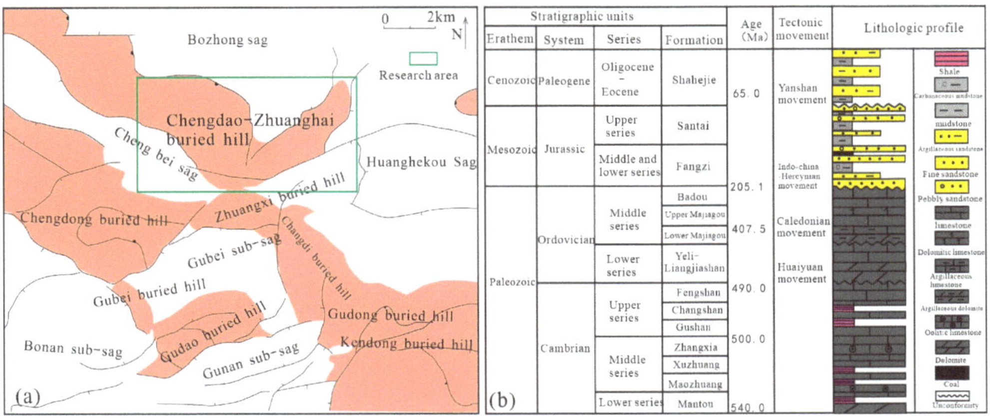

Figure 2. (**a**) Structural location map of Chengdao-Zhuanghai buried hill area. (**b**) Stratigraphic composite column chart of Chengdao-Zhuanghai buried hill (CB302).

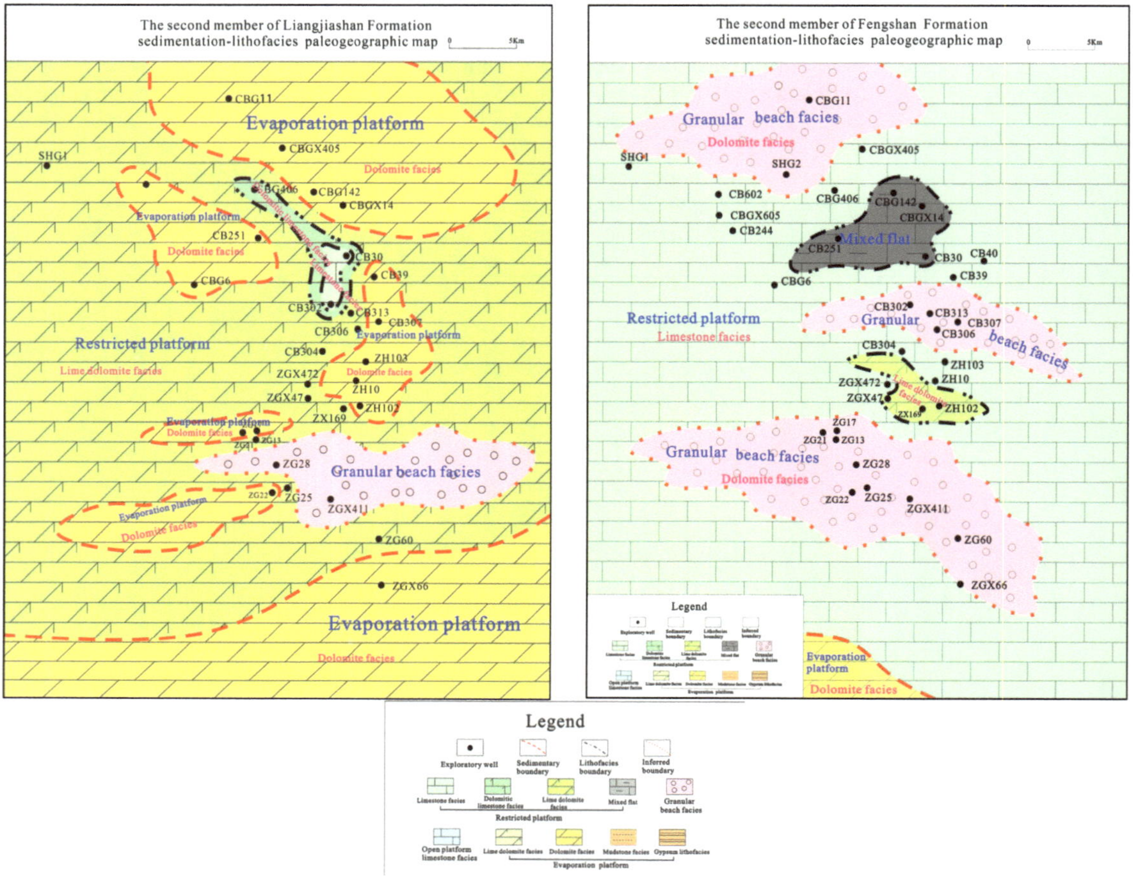

Figure 3. Yeli-Liangjiashan Formation and Fengshan Formation. sedimentation-lithofacies paleoge-graphic map in Chengdao and Zhuanghai buried hill (according to Sinopec Shengli oilfield, 2022).

3. Materials and Methods

In this study, the evidence, structural characteristics and reservoir development mod of unconformities in the Yeli-Liangjiashan Formation and Fengshan Formation are studie by means of outcrop exploration and core observations. Two observation areas, the Xinwe area of Xintai city and the Yishui area of Linyi city, were selected for the field surve The specific survey sites are shown in Figure 4. The characteristics of lithology, solutio holes (pores), and cracks in the outcrop directly show the longitudinal characteristics o an unconformity structure. Cores of 12 wells with a total length of 150 m were selecte from the Chengdao-Zhuanghai buried hill for detailed observation. The characteristics o rock types, cracks, and solution holes (pores) in the scale range of hand specimens wer obtained. Sixty representative samples were selected from the cores to prepare thin section The rock types, reservoir space types, and intercalation characteristics were studied i detail with a ZEISS Imager.A2 m microscope (Carl Zeiss, Oberkochen, Germany). Th characteristics of cracks, solution holes (pores), and cementation of karst reservoirs on microscopic scale were obtained. On this basis, 47 representative samples were selecte and C and O isotopes ($\delta^{13}C_{V\text{-}PDB}/‰$ and $\delta^{18}O_{V\text{-}PDB}/‰$) were analyzed by a MAT253 ga stable isotope mass spectrometer ($\delta^{13}C$, with 0.0037‰ and $\delta^{18}O$, with 0.013‰). Twenty-tw pieces of carbon and oxygen isotope analysis data were obtained, and five pieces of carbo and oxygen isotope analysis data were obtained from literature research to complete the study of fluid types in the paper. Sr isotope analysis ($\delta^{87}Sr_{V\text{-}PDB}$/‰, $\delta^{86}Sr_{V\text{-}PDB}$/‰) of the cement was performed with a MAT261 solid stable isotope mass spectrometer (accuracy $^{87}Sr/^{86}Sr < 10$ ppm). Eighteen sets of analytical data were obtained that could be used to distinguish the period of occurrence of supergene karstification. Core observations and sampling were completed in the core bank at the Shengli Oilfield, and thin-section observations and geochemical analysis were completed at the Academy of Geosciences, Chengdu University of Technology Laboratory.

4. Results

4.1. Peripheral Outcrop Characteristics

The points where the outcrop was observed are located in the Matouya area, Xinwen city, Shandong Province (Area 1), and in the Yangzhuang area, Yishui County, Linyi city, Shandong Province (Area 2), both of which are dominated by the exposed Yeli-Liangjiashan Formation, with a small amount of exposed Fengshan Formation strata. The sedimentary characteristics of the Yeli-Liangjiashan Formation and Fengshan Formation are similar to those of the whole North China Platform in the late Cambrian and early Ordovician. This paper focuses on the Yeli-Liangjiashan Formation in Area 1. Regionally, the thickness of the Majiagou Formation overlying the Yeli-Liangjiashan Formation is approximately 600 m.

In the Yeli-Liangjiashan Formation in Area 1, the strata have a strike of NW20°, dip direction of 30°, and dip angle of 35°. The exposed strata are distributed in thick layers (the thickness of a single layer reaches 1 m), and the thickness of a single layer in the section is uniform overall. The lithology is mainly dolomite, and the lithology is uniform both vertically and laterally (Figure 4a,b,i). Vertically, the reservoir space exhibits apparent zonation, which can be roughly divided into an upper fractured vuggy zone (Zone 1) and a lower cavity zone (Zone 2). High-angle cracks are developed in the upper fractured vuggy zone, with a maximum length of up to 2 m and a density of up to 5 bars/m. The cracks are curved, and dissolution expansion holes and other holes are often present along the edges of the cracks. Most of the holes are connected to the cracks (Figure 4c). Stratified dissolution pores developed in the cavity zone and were interconnected. The diameters of the pores range from 5 to 15 cm. The dissolution pores are filled with different degrees of multistage secondary calcite, and the boundaries between secondary calcite at different periods are clearly distinguishable (Figure 4d,e). Karst breccias are present in Zone 1 and Zone 2, and the karst breccias are angular to subangular in shape, with the largest diameter reaching 20 cm and the most minor diameter reaching 1 cm. Calcite veins filled and developed along rock fractures and interlayer fractures, and a large number of clay impurities filled the karst breccias (Figure 4f). Elliptic chert nodules are present in the dolomites of the Yeli-Liangjiashan Formation. The nodules are approximately 18 cm in size and are distributed along bedding planes (Figure 4g). The outcrop of the Fengshan Formation in Area 1 (Figure 4h) shows the development of bamboo leaf-shaped dolomites, some of which show disorderly accumulation, some of which show bedding distribution characteristics, and fractures and dissolution cavities are less common than those in the Yeli-Liangjiashan Formation.

Figure 4. Geological characteristics of the Yeli-Liangjiashan Formation and Fengshan Formation i Matouya area, Xinwen City and Yangzhuang area, Yishui County, Linyi City, Shandong Provinc (**a**) The dolomites of Yeli-Liangjiashan Formation are distributed in thick layers; (**b**) Developmen of dissolved pores and cavities, thick layer distribution of dolomite in Yeli-Liangjiashan Formatior (**c**) Vertical cracks and holes are developed in Liangjiashan Formation; (**d**) The profile of Liangjiasha Formation shows the development of layered dissolved pores; (**e**) Stratified dissolved pores ar developed and connected in the Yeli-Liangjiashan Formation; (**f**) Development of karst breccia i Yeli-Liangjiashan Formation; (**g**) Chert nodules are developed in dolomite of Liangjiashan Forma tion; (**h**) Bamboo leaf-shaped dolomite is developed in Fengshan Formation. (**i**) The dolomites o Liangjiashan Formation are distributed in thick layers. Except for photo (**i**) of Yangzhuang area, othe photos are of Matouya area.

4.2. Core Characteristics

(1) Rock types and developmental characteristics

The lithology of the Yeli-Liangjiashan Formation and Fengshan Formation is mainl dolomite, and the most common rock type is crystalline dolomite, followed by granula dolomite and dolomitic breccia with different thicknesses. Among them, the crystallin dolomite is usually gray–gray-white, sugar-like, primarily fine and mesocrystalline, an massive in structure, with dissolution cavity development and sometimes visible crack filled with calcite or dolomite veins (Figure 5a). Under the microscope, the dolomite i mostly microcrystalline to mesocrystalline (mainly fine to mesocrystalline dolomite), an it is mainly composed of hemiidiomorphic to idiomorphic crystals (Figure 6a–c,e,g,h,l The crystal type is uniform, and the boundaries between particles are clear, showing poin and line contacts. Granular dolomite with a circular appearance can be observed in the core, and it is mainly gray in color, intergranular and with calcite cement (Figure 5e). The original granular outline of granular dolomite is visible under the microscope; the pores between particles are relatively developed, and no cemented filling is observed (Figure 6k).

Dolomitic breccia mainly developed at the tops of the Yeli-Liangjiashan Formation and Fengshan Formation. According to their genetic types and development locations, two subtypes can be further delineated: karst breccia related to percolation (type 1 karst breccia) and karst breccia related to underflow action (type 2 karst breccia). The characteristics of the two types of karst breccia are significantly different.

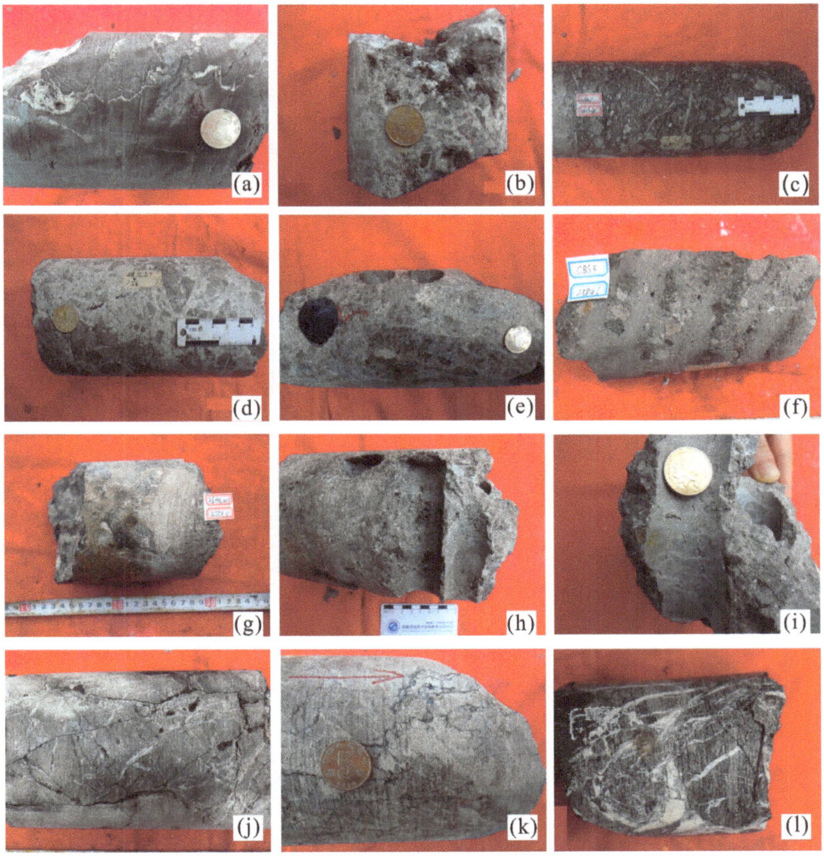

Figure 5. Core characteristics of Yeli-Liangjiashan Formation in Chengdao-Zhuanghai buried hill. (**a**) Well CBG100, 2562 m, granular (fine crystalline) dolomite, the high angle crack is connected with the dissolved pore; (**b**) Well CBG5, 2586 m, dolomitic breccia, dissolved pores developed, partially filled with calcite; (**c**) Well CBG5, 2602 m, the dolomitic breccia with good roundness is arranged in a directional manner; (**d**) Well CBG5, 2585.4 m, dolomitic breccia, high degree of calcite vein filling; (**e**) Well CB302, 4002.16 m, intergranular dolomite filled with calcite; (**f**) Well CBG5, 2584.6 m, granular dolomite, subrounded breccia, layered dissolved pores are developed; (**g**) Well CB403, 2759.5 m, the dolomitic karst breccia is interspersed with percolating sand; (**h**) Well CB302, 4001.26 m, limy karst breccia with dissolved pores developed and a small part filled with soil; (**i**) Well CB302, 4001.86 m, limy karst breccia with unfilled pore development; (**j**) Well CBG5, 2601.5 m, limy dolomite cracks are developed, and the cracks are filled with mud; (**k**) Well CBG100, 2560.47 m, dolomite limestone, high angle crack development; (**l**) Well CB244, 2921.27 m, granular (fine crystalline) dolomite, network cracks filled with calcite veins.

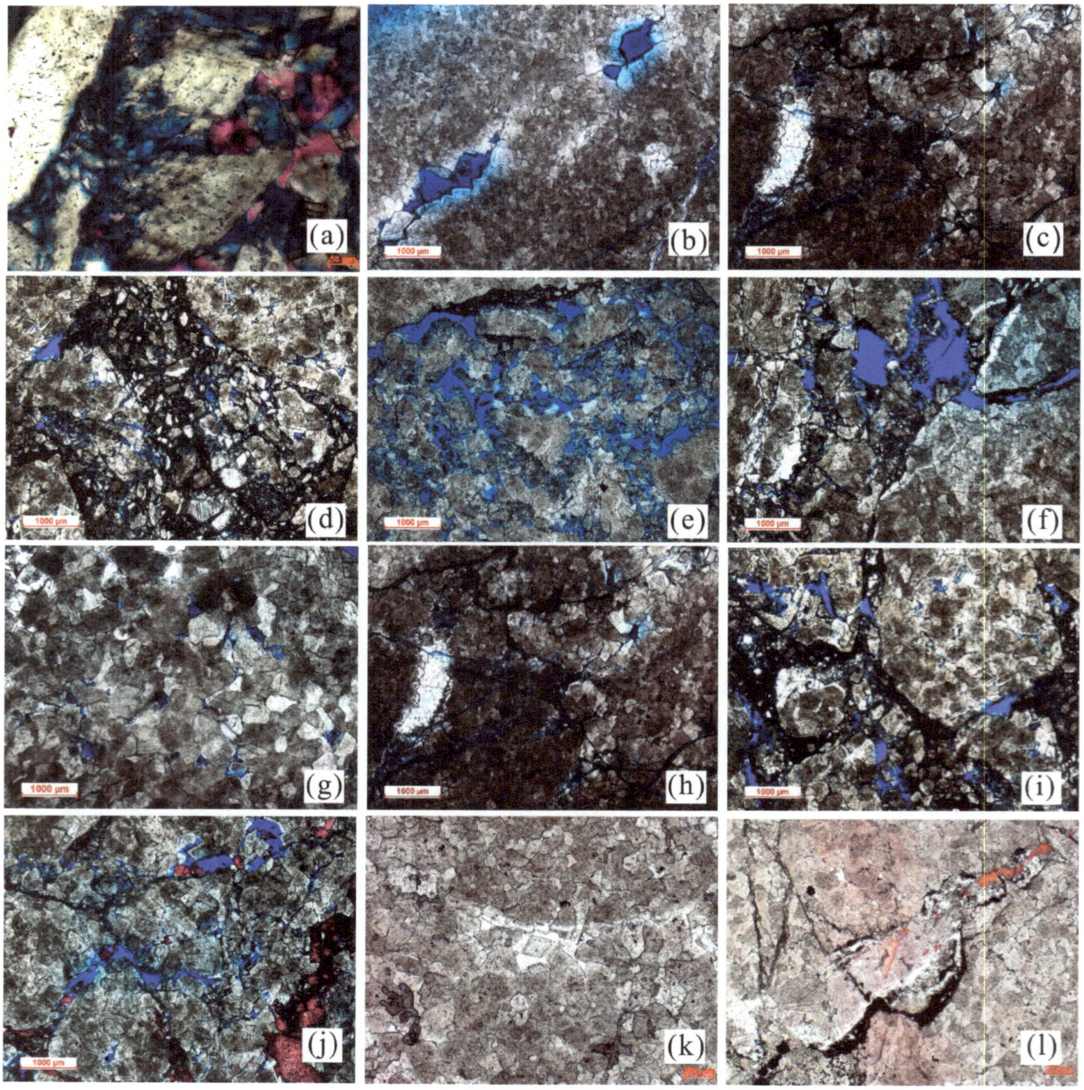

Figure 6. Observation features of Yeli-Liangjiashan Formation in Chengdao-Zhuanghai buried hi under microscope. (**a**) Well CB302, 4001.26 m, dolomitic breccia, calcite is filled in the primary pores c dolomite grains (-); (**b**) Well CB244, 2904.67 m, mesocrystalline dolomite and intercrystalline pore an crack (-); (**c**) Well CB244, 2914.37 m, mesocrystalline dolomite and cracks are developed (-); (**d**) We CB302, 4008.26 m, the dolomitic karst breccia is filled with protolith debris and the dissolved por develops (-); (**e**) Well CB302, 4005.35 m, dolomitic karst breccia, intergranular pores are developed, an the breccia is filled with percolating sand (-); (**f**) Well CB302, 4008.26 m, dissolved pores in granula dolomite (-); (**g**) Well CB244, 2925 m, the primary pore in mesocrystalline dolomite is not filled (- (**h**) Well CB244, 2910 m, the dissolved pore and crack development of mesocrystalline dolomit (-); (**i**) Well CB302, 4008.26 m, dolomitic karst breccia, residual granular dolomite, dissolved por development (-); (**j**) Well CB302, 4006.5 m, dolomitic karst breccia, intergranular and intragranula pores are filled with calcite (-); (**k**) Well CB244, 2920.6 m, granular dolomite, the particles are relativel complete, the pores developed in the early stage and were partially filled by cement in the later stag (-); (**l**) Well ZG28, 4122.92 m, granular dolomite, the crack is filled with asphalt (-).

Type 1 karst breccia mainly developed in atmospheric freshwater vadose zones. The composition of the breccia is dolomitic or lime-rich, and the breccia has a relatively loose structure, angular shape, and different clast sizes, often showing the characteristics of chaotic and mixed accumulation. The breccia is often filled with mud, sand from percolation, etc., and the dissolution cavities are developed both in the karst breccia and in the interstitials between the clasts in breccia. This breccia type is composed of crystalline dolomite, granular dolomite, granular limestone, and lime dolomite. This type of breccia is found in well CB302 (4001.26–4009.06 m) and well CBG403 (2759.5 m) (Figure 6g–i). In well CB302, this breccia type developed at a depth of nearly 480 m from the paleokarst unconformity interface (3520 m) between the overlying lower Paleozoic and Mesozoic strata. Most clasts are approximately 0.5 cm in diameter, and the clasts are disorderly and unoriented. The breccia has obvious silicification, which is limited to the breccia, or it is filled with mud, dolomite debris, quartz, etc., and no silicification is found. Sometimes, late calcite veins cut through the breccia, dissolution holes are present between the clasts, cracks are present, and oil immersion tests were conducted. The oil immersion test results show that the interval contains oil. Under the microscope, some of the dissolved pores in this type of breccia are filled with calcite, the calcite crystal form is good, some of them are dissolved in the later stage, the boundaries between the breccia particles are clear, and the breccia is filled with argillaceous intercalations, sand from percolation and calcite (Figure 6d,f,i,j).

Type 2 karst breccia is mainly found in the underflow zone. This breccia type is dolomitic or lime-rich, and the clasts are subangular to subrounded. The breccia usually shows an excellent directional arrangement or vertical grain order and generally shows characteristics indicating transport and river redeposition. Well CBG5 contains this type of karst breccia at 2602 m, and CB244 contains this type at 2923.7 m. In these wells, subrounded dolomitic breccia can be observed in a stratified arrangement (Figure 5c,f). Intermediate-like solution pores are present in the breccia, and their sizes are uneven. The breccia is filled with dolomitic debris, etc. Under the microscope, the development of solution pores and microcracks between clasts can be observed. Most of them are effective storage spaces, and some of the particles are filled with calcite (Figure 6b,c,h,k). In the whole core section, there is a noticeable change in the grain size in the longitudinal direction, this change in the grain size has multiple rhythmic characteristics (Figure 5f), and oil immersion tests were conducted on this core section.

(2) Crack types and developmental characteristics

In the core, there are two kinds of cracks with different occurrences in the Yeli-Liangjiashan and Fengshan Formations. The first crack type (crack 1) is usually a curved and high-angle fracture. Soil filling is evident from 2560 to 2562 m in the CBG100 well, and calcite filling occurs. Dissolved pores are often present in the upper and lower parts of such cracks, and the dissolved pores are often connected to the cracks in a bead-like shape. The pores are often filled with clay or saturated sand (Figure 5a,k,j). Under the microscope, these cracks appear to be dissolved and expanded; some of the cracks are filled with mud, dolomite debris, and saturated sand (Figure 6d,f,i,j); and some of the cracks present adequate storage space. The second type of crack (crack 2) is present at 2921.27 m in well CB244 and at 2601.5 m in well CBG5. These cracks usually present medium- and low-angle characteristics, and the edges are relatively straight, showing the feature of cutting each other in a network. The interior is often filled with sparite, dolomite, and other veins. Some calcite grains with good crystal shapes are visible (Figure 5e,l); dissolved pores often develop on both sides of the cracks, and some of them are filled with calcite. These cracks are readily apparent under the microscope in a semifilled-filled state, and some microcracks are filled with bitumen or calcite (Figure 6c,h,l).

(3) Types and developmental characteristics of dissolved pores (holes)

In the core, the solution pores in the Yeli-Liangjiashan Formation and Fengshan Formation can be divided into three categories according to their occurrence. The sizes of the first type of dissolved pore (pore 1) are relatively large. The CBG100 well (2562 m),

CBG5 well (2601.5 m), and ZH101 well (145.75 m) are well developed in many places and have 2–15 mm diameters. The pores are filled with clay or saturated sand, often associated with the first crack type or located below it. The connectivity with the first type of crack (crack 1) is more frequent (Figure 5a,k,j). The second type of dissolved pore (pore 2), with diameters of 1–3 mm, can be observed at 2584.6 m, 2602 m, and other locations in the CBG5 well. The prominent feature of pore 2 is that the dissolved pores are distributed in layers and connected, and their diameters can reach 8–15 mm. The dissolved pores are filled and semifilled, and they are often filled with saturated sand. This type is often found in aligned subangular–subrounded breccias (Figure 5e). The third type of dissolved pore (pore 3), which is present at 4145.75 m in the ZH101 well and at 2586 m in the CBG5 well, varies in size, with the largest being 5 mm and the smallest being 1 mm. This type is filled with a large amount of calcite; some are unfilled, and many contain oil (Figure 5i).

As observed under the microscope, there are many dissolved pores and cracks in the dolomites of the Yeli-Liangjiashan and Fengshan Formations. The edges of the dolomite crystals are irregular and harbor shapes; the cracks are dissolved and expanded, and large dissolution pores are formed along the dissolution–expansion cracks (Figure 6f). The phenomenon of cracks or dissolved pores filled with asphalt can be observed in some sections (Figure 6l). The reservoir space consists of various dissolved pores, dissolution expansion cracks, and intergranular pores of residual granular dolomite, among which the dissolved pores include secondary pores, such as integrated–dissolved pores, mold pores, intergranular dissolved pores, and intragranular dissolved pores. The intragranular dissolved pores are mainly present in the karst breccia dolomites (Figure 6b), and the pore sizes are relatively small, generally between 0.05 and 0.1 mm. Intergranular dissolution pores are mainly present between the granular dolomite and breccia within the residual granular dolomite and dolomitic karst breccias (Figure 6c,d,i), showing varying sizes and uneven distributions under the microscope. The pore sizes are generally 0.2–0.8 mm with a maximum of 2 mm, and some of the pores are filled or half-filled by calcite idiomorphic crystals (Figure 6d). The mold pores are mainly present in the dolomitic karst breccias, and relatively small clasts are entirely dissolved, leaving only their appearance in the remaining breccia (Figure 6f). Intergranular and intragranular dissolved pores are mainly present in fine-grained and medium-grained dolomites, with relatively small pore sizes. Generally, they are approximately 0.1 mm long, and some of them are filled with gypsum (Figure 6g,h).

At both the core scale and the hand sample scale, there are many types of reservoir spaces in the dolomites of the Yeli-Liangjiashan Formation and Fengshan Formation. They are interconnected and interleaved, forming a composite space-like network in the reservoir, with crack-cavity type (Figure 5a,j), cavity type (Figure 5b,f), and crack type spaces (Figure 5l,k). Among them, the crack-cavity and crack types are present in reservoirs where matrix pores are not present or near faults. The cavity type is relatively prevalent in the dolomite reservoirs of the Yeli-Liangjiashan and Fengshan Formations or inside the faults.

Due to the variety of carbonate reservoir types in the Yeli-Liangjiashan Formation and Fengshan Formation, their reservoirs are highly heterogeneous, and their reservoir parameters, such as porosity and permeability, vary widely. Based on the existing routine core analysis data, 24 samples from three core wells in the Yeli-Liangjiashan Formation and Fengshan Formation in the study area were statistically analyzed. The maximum porosity was 12.2%, the minimum was 1.1%, the maximum permeability was 266.659 $\times 10^{-3}$ μm^2 and the minimum permeability was 0.025 $\times 10^{-3}$ μm^2 (Table 1). The wide variation range in porosity and permeability fully indicates the heterogeneity and complexity of reservoir types in space.

(4) Longitudinal zoning features

Through observation and analysis of the core characteristics of several wells, according to the characteristics of karst breccia development, the reservoir in the study area has the characteristics of karst zonation in the vertical direction, that is, the upper vadose zone and the lower underflow zone. Statistics show that the vadose zone and underflow zone can be present independently, either simultaneously or separately in the Yeli-Liangjiashan Formation or Fengshan Formation, (Figures 7 and 8).

Table 1. Statistical table of porosity and permeability of some wells in the Yeli-Liangjiashan Formation.

Well	Formation	Porosity (%)				Permeability/$\times10^{-3}$ μm^2			
		Minimum Value	Maximum Value	Sample Quantity	Mean Value	Minimum Value	Maximum Value	Sample Quantity	Mean Value
CB244	Yeli-Liangjiashan Formation	1.1	12.2	18	4.32	0.075	266.659	16	21.535
CB39	Yeli-Liangjiashan Formation	2.3	4.6	4	3.15	0.025	8.01	4	2.043
CBG6	Yeli-Liangjiashan Formation	9.4	2.5	4	4.48	1.545	1.545	1	1.545

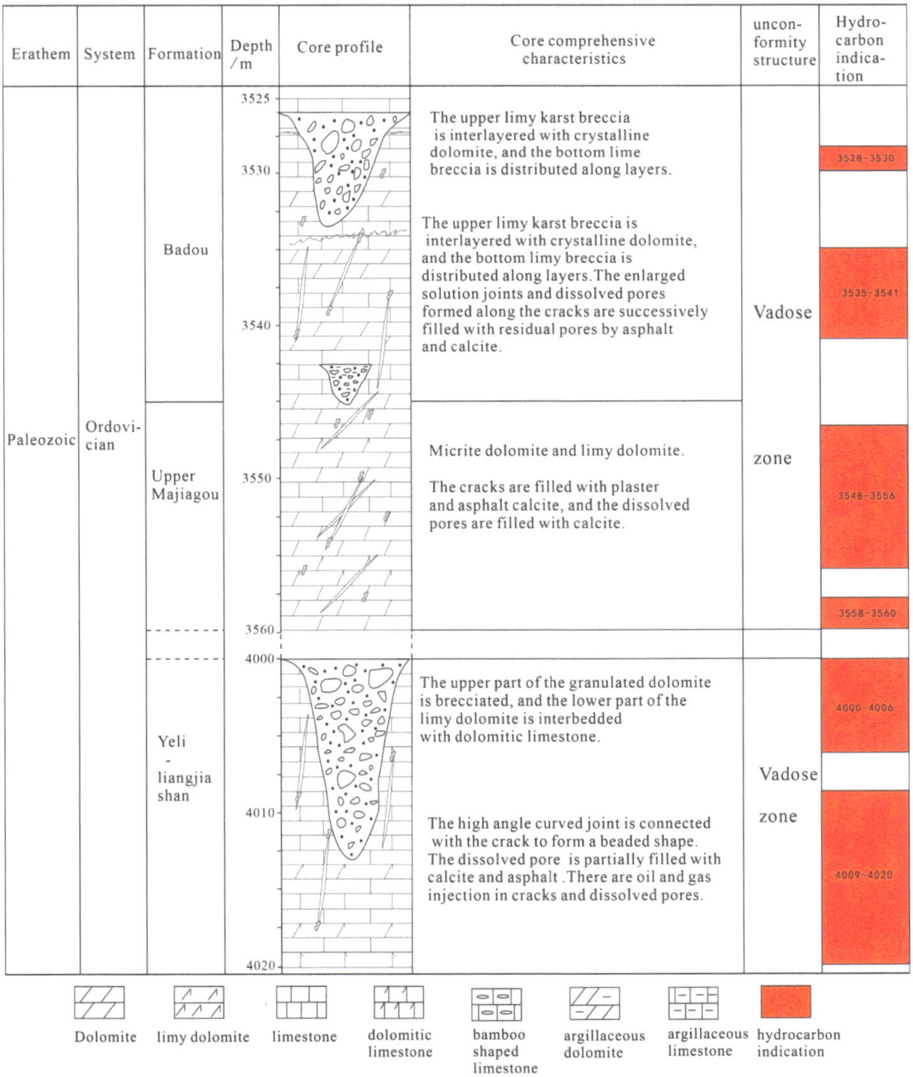

Figure 7. Well CB302 core–karst composite column diagram.

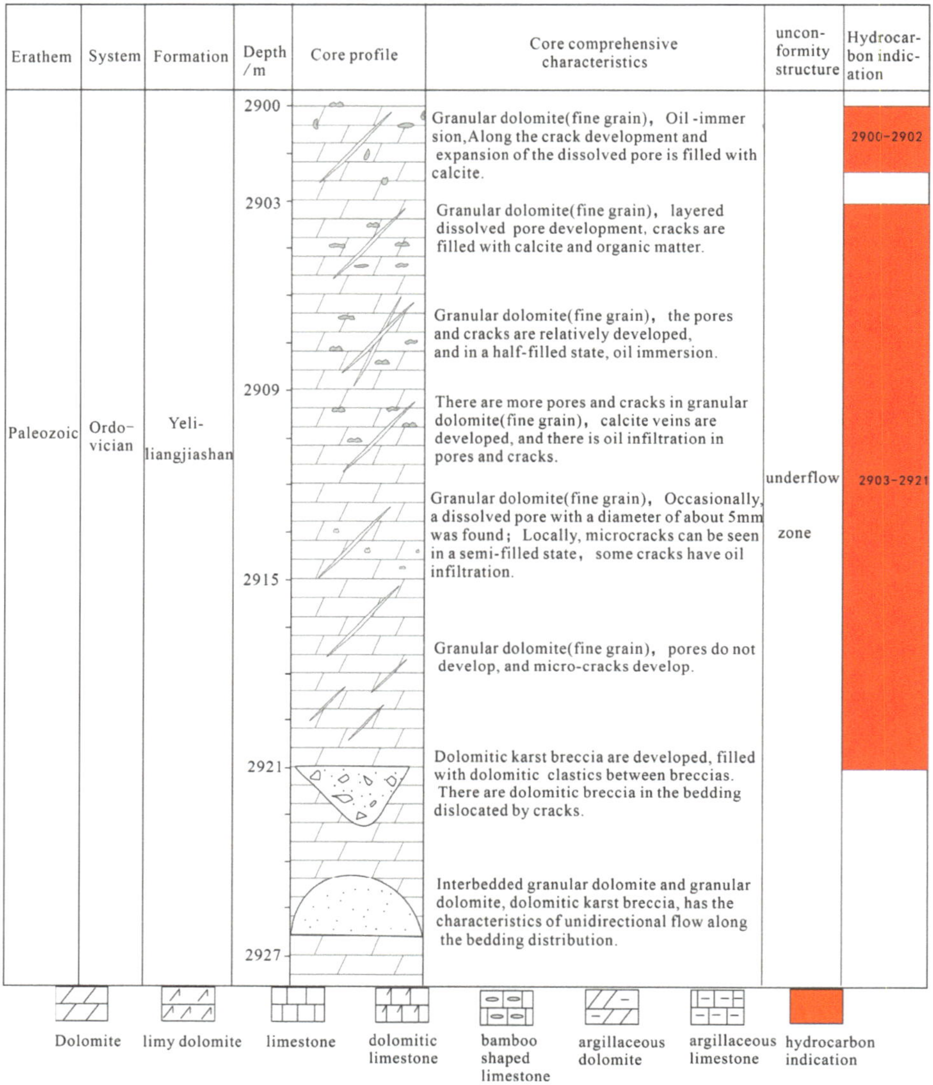

Figure 8. Well CB244 core–karst composite column diagram.

Type 1 karst breccia is present in the vadose zone, and the longitudinal reservoi development characteristics of the Yeli-Liangjiashan Formation in the CB302 well are th most significant (Figure 7). Type 1 dolomitic karst breccia, limy dolomite and dolomiti limestone are present in the well, and type 1 breccia is present in the 4001.26–4009.06 m segment. In this zone, high-angle curved cracks (crack 1) are mainly present, and dissolve pores (pore 1) are present in the upper or lower parts of the cracks in the longitudina direction of their development. They are connected with cracks (crack 1). The mos significant pores reach 15 mm in diameter, and the pores are filled with mud or calcit but there are still residual pores (pore 3) and oil that is present. In addition to high-angl curved cracks (crack 1), low-angle network cracks (crack 2) are also present but are no unique. According to the drilling data, leakage of 113 m^3 occurred when this zone of th karst breccia section was drilled.

Karst breccia type 2 and layered dissolved pores are present in the underflow zone. This zone is revealed by the longitudinal reservoir development characteristics of well CB244 (Figure 8). The lithologies are mainly granular (fine-grained) dolomite and granular (sand-dust) dolomite, with network cracks (crack 2) and layered dissolved pores (pore 2). Cracks (crack 2) and dissolved pores (pore 2 and pore 3) are half-filled with asphalt and calcite filling materials. In the 2921.37–2923.67 m segment, dolomitic karst breccia is present in a subangular shape, with pore 3 in the breccia, and the pores are filled with calcite. In the 2923.67–2925.27 m section, the granular (sand debris) dolomite and micritic dolomite are interstratified, and the dolomitic karst breccia is forward-distributed. The clasts are subangular, with excellent roundness, and they show the characteristics of unidirectional water flow. The breccia is filled with sparry calcite cement.

4.3. Isotopic Geochemical Characteristics

According to the geochemical characteristics of C and O isotopes in rock samples from the Yeli-Liangjiashan Formation and Fengshan Formation (Table 2), the isotopic variations in C and O in the Yeli-Liangjiashan Formation are greater, with $\delta^{13}C$ values varying from −9.92‰ to −0.14‰ and $\delta^{18}O$ values varying from −15.46‰ to −7.97‰. The $\delta^{13}C$ and $\delta^{18}O$ values range from −0.56‰ to 1.34‰ and from −14.23‰ to −6.24‰, respectively, in the Fengshan Formation. According to the Sr isotope geochemical characteristics of the Liangjiashan Formation and Fengshan Formation samples (Table 3), the $^{87}Sr/^{86}Sr$ values of the Yeli-Liangjiashan Formation have extensive variation, with values ranging from 0.7103 to 0.7145. The $^{87}Sr/^{86}Sr$ values of the Fengshan Formation vary between 0.7103 and 0.7144.

Table 2. Geochemical characteristics of C and O isotopes of Yeli-Liangjiashan Formation and Fengshan Formation.

Serial Number	Sample Number	Lithology	Stratum	Sampling Depth (m)	$\delta^{13}C_{V\text{-}PDB}$ (‰)	$\delta^{18}O_{V\text{-}PDB}$ (‰)	Remark
1	HB19-1	dolomite	Yeli-Liangjiashan Formation	-	−9.92	−8.84	According to Zhang, Y.F., and Wang, Q.C., 2007 [38]
2	HB19	dolomite	Yeli-Liangjiashan Formation	-	−7.68	−7.97	
3	HB12	dolomite	Yeli-Liangjiashan Formation	-	−5.24	−9.97	
4	HB09-1	dolomite	Yeli-Liangjiashan Formation	-	−7.89	−8.92	
5	HB09	dolomite	Yeli-Liangjiashan Formation	-	−7.45	−9.51	
6	CB302-10	dolomite	Yeli-Liangjiashan Formation	4003.3	−1.29	−13.18	This study
7	CB302-4	dolomite	Yeli-Liangjiashan Formation	4006.5	−1.28	−12.39	
8	CB302-8	dolomite	Yeli-Liangjiashan Formation	4005.4	−0.14	−9.78	
9	CB302-1	dolomite	Yeli-Liangjiashan Formation	4008.3	−0.46	−9.56	
10	CB302-21	dolomitic breccia	Yeli-Liangjiashan Formation	4002	−0.96	−10.42	
11	CB302-15	dolomite	Yeli-Liangjiashan Formation	4003.34	−2.43	−13.24	
12	ZH101-2	dolomite	Yeli-Liangjiashan Formation	4076.21	−2.3	−15.46	
13	ZH101-3	dolomitic breccia	Yeli-Liangjiashan Formation	4076.91	−2.59	−14.87	
14	CBG100-1	dolomite	Fengshan Formation	2561.68	−0.02	−7.31	

Table 2. *Cont.*

Serial Number	Sample Number	Lithology	Stratum	Sampling Depth (m)	$\delta^{13}C_{V\text{-}PDB}$ (‰)	$\delta^{18}O_{V\text{-}PDB}$ (‰)	Remark
15	CBG100-2	dolomite	Fengshan Formation	2561.68	−0.01	−6.59	This study
16	CBG100-3	dolomite	Fengshan Formation	2561.68	−0.38	−6.73	
17	CBG100-4	dolomite	Fengshan Formation	2561.68	−0.05	−6.24	
18	CBG100-5	dolomite	Fengshan Formation	2566.27	−0.32	−6.88	
19	CBG100-6	dolomite	Fengshan Formation	2566.17	−0.22	−8.1	
20	CBG100-7	dolomite	Fengshan Formation	2566.17	−0.53	−6.59	
21	CBG100-8	dolomite	Fengshan Formation	2566.17	−0.56	−7.41	
22	ZG39-1	limestone	Fengshan Formation	4338.2	0.27	−10.07	
23	ZG39-2	limestone	Fengshan Formation	4341.1	0.21	−10.6	
24	ZG39-3	limestone	Fengshan Formation	4343.3	0.44	−10.19	
25	ZG39-4	dolomitic limestone	Fengshan Formation	4345.9	0.7	−14.23	
26	ZG39-5	limestone	Fengshan Formation	4349.3	1.17	−12.57	
27	ZG39-6	limestone	Fengshan Formation	4351.8	1.34	−10.9	

Table 3. Sr isotope geochemical characteristics of Yeli-Liangjiashan Formation and Fengshan Formation.

Serial Number	Sample Number	Lithology	Stratum	Sampling Depth (m)	$^{87}Sr/^{86}Sr$
1	CB302-1A	dolomite	Yeli-Liangjiashan Formation	4001.2	0.7103
2	CB302-2A	dolomite	Yeli-Liangjiashan Formation	4001.46	0.7133
3	CB302-3A	dolomite	Yeli-Liangjiashan Formation	4004.45	0.7145
4	CB302-4A	dolomite	Yeli-Liangjiashan Formation	4008.5	0.7138
5	CB302-5A	dolomite	Yeli-Liangjiashan Formation	4001.5	0.7105
6	SHG 2-1A	dolomite	Yeli-Liangjiashan Formation	2404	0.7097
7	SHG2-2A	dolomite	Yeli-Liangjiashan Formation	2407.9	0.7093
8	SHG2-3A	dolomite	Yeli-Liangjiashan Formation	2408	0.7116
9	ZG39-1	dolomite	Fengshan Formation	4338.2	0.7108
10	ZG39-2	dolomite	Fengshan Formation	4341.1	0.7107
11	ZG39-3	dolomite	Fengshan Formation	4343.3	0.7106
12	ZG39-4	dolomitic limestone	Fengshan Formation	4345.9	0.7144
13	ZG39-5	dolomite	Fengshan Formation	4349.3	0.7103
14	ZG39-6	dolomite	Fengshan Formation	4351.8	0.7116
15	ZH10	dolomite	Fengshan Formation	4658.14	0.7113
16	ZG39-7	dolomite	Fengshan Formation	4359.2	0.7108
17	ZG39-8	dolomite	Fengshan Formation	4361.8	0.7117
18	ZG39-9	dolomite	Fengshan Formation	4358.1	0.7126

5. Discussion

5.1. Evidence of the Record of Huaiyuan Movement

(1) Geological evidence of the Huaiyuan Movement

Regionally, different parts of the Yeli-Liangjiashan Formation and Lower Majiagou Formation overlie different strata on the North China Platform, resulting in regional parallel unconformities between the Yeli-Liangjiashan Formation and Fengshan Formation, and the same is true between the Lower Majiagou Formation and Yeli-Liangjiashan Formation. For example, Osleger (1991) [39] noted that at the end of the Cambrian period, a global sea level decline occurred at the bottom interface of the Ordovician strata. Liu et al. (1997) [40]

proposed a paleokarst-related unconformity at the top of the Ordovician Liangjiashan Formation in the North China Platform. The unconformity is parallel, and there is an apparent change in lithology at the boundary. Underly the boundary are medium-fine crystalline dolomites with chert bands, which have different degrees of karstification. According to Mei and Ma (2003) [41], on the basis of the contact relationship between the Cambrian and overlying strata in different regions of the North China Platform, from the Xishan section in Beijing to the Zhang Xia section in Tai'an Shandong, the Jiawang section in Xuzhou, and the Huainandongshan section (see profile 1 in Figure 1 for the specific location), there was a short period of exposure to different degrees between the Cambrian and the Ordovician, especially in Shandong Province. There was more prolonged exposure between the deposition of the Fengshan Formation and the Yeli-Liangjiashan Formation than between that of the Xuzhou and Zhunnan sections, thus forming an unconformity between them.

Through field observations of the Yeli-Liangjiashan Formation in the Matouya area, Xinwen city, Shandong Province, and Yangzhuang, Yishui County, Linyi city, Shandong Province, many karst breccias are found in the Yeli-Liangjiashan Formation. The vadose zone and underflow zone can be delineated (Figure 4), indicating that after the deposition of the Yeli-Liangjiashan Formation, many karst breccias formed in the Yeli-Liangjiashan Formation. It was briefly uplifted to the surface and weathered and denuded. According to the seismic data in the study area, different parts of the Lower Majiagou Formation overlie the Fengshan Formation and Yeli-Liangjiashan Formation, resulting in regional unconformities between the Lower Majiagou Formation and the Fengshan Formation and Yeli-Liangjiashan Formation. For example, in the CB271-CBG403-CBG4 North–South seismic profile (Figure 9, see profile in Figure 10 for the specific location), the bottom seismic reflection axes of the lower Paleozoic Badou Formation, Upper Majiagou Formation, Lower Majiagou Formation, Yeli-Liangjiashan Formation, and Fengshan Formation are visible. In the northeastern area of the CBG4 bulge, the Cambrian and Ordovician Yeli-Liangjiashan Formation forms wedges from southwest to northeast. The Formation gradually thins in the northeast direction and truncates and is overlapped by the Majiagou Formation. This situation is more common in the lower Paleozoic strata of the Chengdao-Zhuanghai buried hill, indicating that tectonic uplift and exposed denudation occurred during the Huaiyuan Movement [42]. In the Chengdao-Zhuanghai buried hill, there are three overlapping regional unconformity surfaces at the top of the Ordovician System, the bottom of the Majiagou Formation, and the top of the Cambrian System. In the past, people thought that the unconformity in this area formed during the late Caledonian Movement [25]. However, it may be the unconformity of the Huaiyuan Movement (early Caledonian Movement) and the late Caledonian Movement superimposed together.

According to the observations of solid drill cores, karst breccias of different sizes have been found in the Fengshan Formation and Yeli-Liangjiashan Formation. Saturated sand structures between breccias are present, and karst breccias are the most direct manifestation of the strata's exposure to atmospheric freshwater leaching [17]. The existence of many karst breccias in the Yeli-Liangjiashan Formation and Fengshan Formation indicates that the studied strata must have been subjected to atmospheric freshwater leaching for some time after deposition, resulting in the formation being subjected to dissolution. During the longitudinal leaching and dissolution of atmospheric freshwater, the mud and sand on the weathering crust formed by weathering permeated downward into the cracks and filled in the karst breccia.

The vertical zonation of karst indicates the record of the Huaiyuan Movement. Many studies on modern and ancient karst regions have been conducted by different geologists. According to the extent of karst system development and modes of groundwater movement and karst action, geologists have suggested that there are zones of hydrodynamic action in vertical karst activity, and these can be delineated from top to bottom: surface karst zone–vadose karst zone–runoff karst zone–underflow karst zone [43]; surface karst zone–vertical percolation zone–runoff solution zone–underflow solution zone [44];

or surface karst zone–vertical percolation zone–horizontal underflow zone–deep slo flow zone [21]. The drill core shows that similar zones of karst hydrodynamic action ar visible in wells CB244, CB302, CB307, and ZH102 in the study area. The vertical vados zone and horizontal underflow zone can be divided from top to bottom (Figures 7 and 8 The horizontal underflow zone is affected by the mode of horizontal flow. Horizont layered dissolved pores and underground rivers are present, and subrounded karst brecci is present and displays vertical characteristics of normally graded bedding (Figure 5 These data show the characteristics of long-distance transport and river deposition. Th zonation results from formation uplift and denudation after deposition and the long-ter transformation of atmospheric freshwater.

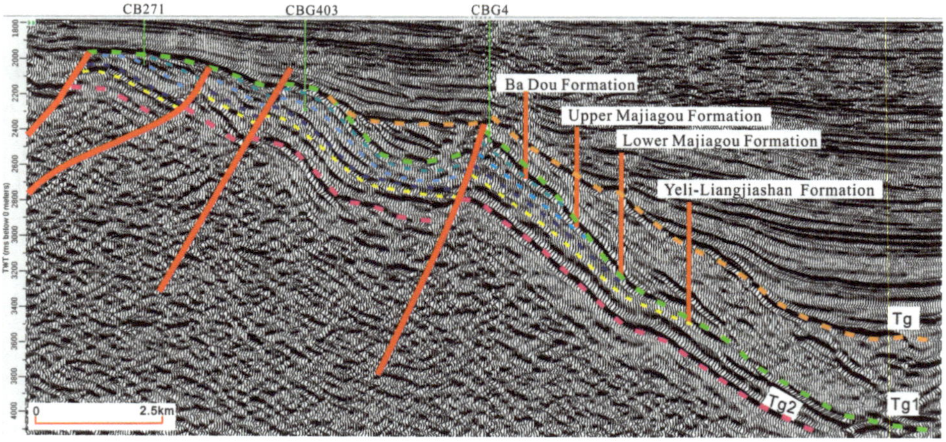

Figure 9. Seismic reflection characteristics of each section in the lower Paleozoic in Chengda Zhuanghai buried hill (Tg—Top of the Upper Paleozoic,Tg1—Top of the Lower Paleozoic, Tg2—To of Mantou Formation).

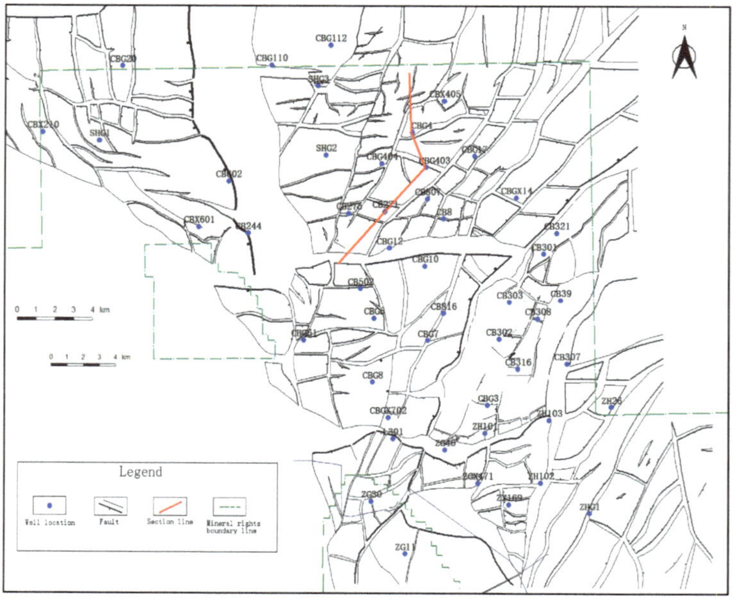

Figure 10. Structure map of top of the Lower Paleozoic of Chengdao-Zhuanghai buried hill.

Observing the outcrop and drill core, the unconformity formed during the Huaiyuan movement exists in the study area, consistent with previous studies. Further compared with previous studies, it can be seen from the seismic and tectonic characteristics of the study area and the zonation of core karstification that there were at least two significant periods of supergene karstification during the Huaiyuan movement, which resulted in unconformity between the Fengshan Formation and Yeli-Liangjiashan Formation, and between the Yeli-Liangjiashan Formation and overlying strata.

(2) Geochemical evidence of the Huaiyuan Movement

Isotopic analysis of C, O, and Sr in the Yeli-Liangjiashan Formation reveals that $\delta^{13}C$ values range from −9.92‰ to −0.14‰, $\delta^{18}O$ values range from −15.46‰ to −7.97‰, and $^{87}Sr/^{86}Sr$ values range from 0.7103 to 0.7145. Studies have shown that in Ordovician seawater, $\delta^{13}C$ values vary from approximately −3‰ to 3‰, $\delta^{18}O$ values vary from approximately −11‰ to −3.6‰ [45], and the average $^{87}Sr/^{86}Sr$ value is 0.7085 [46]. The $\delta^{13}C$ values of some carbonate rocks in the Yeli-Liangjiashan Formation are lower than those of seawater during the same period, and the abnormally low $\delta^{13}C$ values (−5.24‰–9.92‰) are mainly related to atmospheric freshwater [47]. The $\delta^{18}O$ values of the Yeli-Liangjiashan Formation are greater than those of seawater from the same period, and the $^{87}Sr/^{86}Sr$ values are more enriched than the $^{87}Sr/^{86}Sr$ values in seawater from the same period (Figure 10), indicating that the Yeli-Liangjiashan Formation has been affected by alteration by atmospheric freshwater, which is more enriched in $^{87}Sr/^{86}Sr$ than seawater [48–50]. The diagram showing the Sr–O isotope relationship (Figure 11) reveals that some samples have Sr–O isotope characteristics that are similar to those of unmodified marine carbonate rocks; others have abnormally high and abnormally low values, which may be equivalent to the Sr–O isotope characteristics of atmospheric freshwater, and some samples display values between the two. These results show that this part of the sample is an intermediate product of the water–rock reaction.

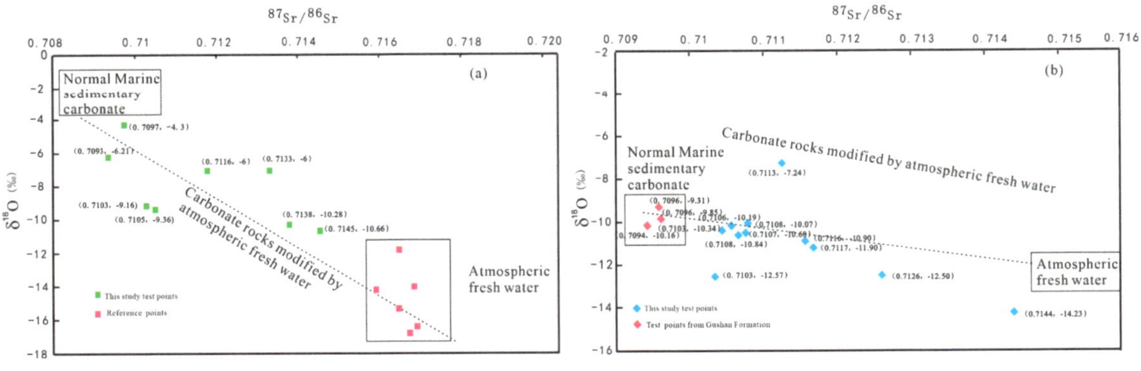

Figure 11. Sr–O isotope correlation between Yeli-Liangjiashan Formation (**a**) and Fengshan Formation (**b**).

The results show that the $\delta^{13}C$ values range from −0.56‰ to 1.34‰, the $\delta^{18}O$ values range from −14.23‰ to −6.24‰, and the $^{87}Sr/^{86}Sr$ values range from 0.7103–0.7144. The literature shows that the $^{87}Sr/^{86}Sr$ values of Cambrian seawater are between 0.7081 and 0.7093 [45,46], the $\delta^{13}C$ values vary from −2‰ to 0‰ (Veizer et al., 1999) [45], and the mean $\delta^{18}O$ value is −9.8‰. The $\delta^{18}O$ value of the Fengshan Formation carbonate rocks is more negative than that of seawater from the same period, and the $^{87}Sr/^{86}Sr$ values are more enriched than the $^{87}Sr/^{86}Sr$ values of seawater from the same period, indicating that the Fengshan Formation was also undergoing alteration by atmospheric freshwater [49–51]. The Sr–O isotopic correlation diagram (Figure 11) shows that the Sr–O isotopic relationship characteristics are similar to those of the Yeli-Liangjiashan Formation, and some samples show that the values are the product of water–rock transformation.

According to the geochemical characteristics of C, O, and Sr isotopes mentione above, the carbonates of both the Yeli-Liangjiashan Formation and the Fengshan Formatio have been reformed by atmospheric freshwater. The difference in the $^{87}Sr/^{86}Sr$ value between the atmospheric freshwater of the Yeli-Liangjiashan Formation and the Fengsha Formation is reflected by the covariant Sr–O isotope relationship, which indicates that the were modified by atmospheric freshwater in the same period.

When we determined the range of isotopic values of the supergene karstificatio influenced by the Huaiyuan movement, which is characterized as "carbon–oxygen isotop poor, strontium isotope rich", we can preliminarily speculate that the carbonate rocks wit such geochemical characteristics are suitable reservoirs in the Yeli Liangjiashan Formatio and Fengshan Formation. Studying the distribution law of high-quality reservoirs in th future is beneficial.

(3) Evidence of the forming time

In earlier studies on the supergene karstification of the lower Paleozoic strata i the Jiyang Depression, many geologists noted the existence of breccia at the top of th Yeli-Liangjiashan Formation and Fengshan Formation, and some of them have classifie it as a breccia of tectonic origin. Although most of them define it as karst breccia, it i believed that the formation process was closely related to supergene karstification in th Indosinian and Himalayan periods. Many studies have shown that the depth of influenc of supergene karstification caused by atmospheric freshwater on strata is approximatel 150 m [42,48]. In well CB302 in the study area, a set of gray and gray karst breccia tha is approximately 109 m thick developed continuously in the Yeli-Liangjiashan Formatio and Fengshan Formation (Figure 7), of which the karst breccia in the Fengshan Formatio is approximately 90 m thick. The set of karst breccia lies 488 m below the unconformit between the Mesozoic and lower Paleozoic strata (late Caledonian–early Hercynian an Indosinian). In addition, the Majiagou Formation overlying this set of karst breccia contain mostly gypsum salt and gypsum dolomite layers acting as water barriers. Due to the dept of karst processes at the top of the Majiagou Formation and the existence of multiple layer acting as water barriers in the Majiagou Formation, the karst processes occurring at the to of the Majiagou Formation cannot affect the Yeri-Liangjiashan Formation and Fengsha Formation. Therefore, the large sets of karst breccia in the Yeli-Liangjiashan Formation an Fengshan Formation likely formed during the Huaiyuan Movement, and they were likel the products of supergenetic activity in the late Yeli-Liangjiashan Formation or the lat Fengshan Formation.

In the lower Paleozoic strata of the Jiyang Depression, the C and O isotopes of car bonates modified by atmospheric freshwater in the Hercynian and Xishan Stages show positive correlation, and their C and O isotope covariant relationships (the detailed charac teristics of which will be discussed separately) are obviously different from the C and O iso tope covariant relationships formed by paleokarstification in the long Huaiyuan period. I the diagram of the C-O isotope covariant relationship between the Yeli-Liangjiashan Forma tion and Fengshang Formation modified by long-term atmospheric freshwater (Figure 12 two trend lines with different slopes are visible: one trend line with a positive slope an the other trend line with a negative slope (Figure 12). These results indicate that th Yeli-Liangjiashan Formation and Fengshan Formation were subjected to the effects of mod ification and superposition by at least two atmospheric freshwater fluids with differen C and O isotopes. Among them, at the sample point in well CB302 (Table 2), the strat overlying the Ordovician strata are Mesozoic strata, the depth of the boundary is 3520 m and the sampling points are in the Yeli-Liangjiashan Formation (approximately 4003 m according to which the lower Paleozoic–Mesozoic unconformity is approximately 500 m deep. According to previous studies, the depth of influence of atmospheric freshwate leaching is approximately 150 m. The test results show that the sample point in well CB302 plots on the trend line of atmospheric freshwater, with a negative slope (Figure 12). It ca be concluded that the time of atmospheric freshwater occurrence in this period was th Huaiyuan Movement, that is, the early Caledonian Movement. The values of C and C isotopes in this sample were not affected by late supergenetic karstification. On this trend line, C and O isotopes constitute a good negative linear correlation. One end represents the initial rock sample, while the other end represents the long-term atmospheric freshwater end. The C and O isotopes of the initial rock sample are approximately −1‰ and −12‰, respectively. The C and O isotopic values are approximately −10‰ and −9‰, respectively, which are obviously different from the C and O isotopic values of the Hercynian and Xishan Stages. Therefore, the carbonate reservoirs of the Yeli-Liangjiashan Formation and Fengshan Formation are the direct products of the uplift and denudation in the Huaiyuan period (early Caledonian Movement) and the transformation of supergene karstification.

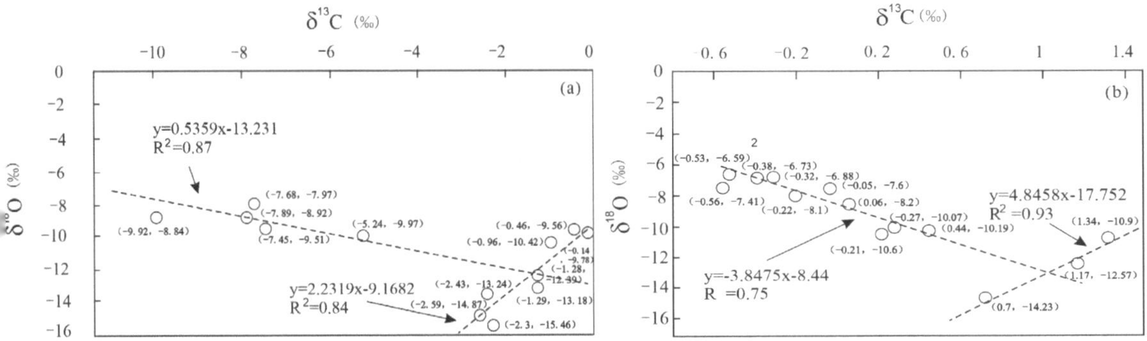

Figure 12. C—O isotope covariant relationship between Yeli—Liangjiashan Formation (**a**) and Fengshan Formation (**b**).

5.2. Influence and Control of Huaiyuan Movement on Reservoir

Through the exploration of peripheral outcrops, core observations, microscopic observations of thin sections, geochemical analysis and analysis of drilling data, the reservoirs of the Yeli-Liangjiashan Formation and Fengshan Formation were determined to be controlled by the original sedimentary facies, dolomitization and supergene karstification.

In the (early Paleozoic) late Cambrian–Early Ordovician sedimentary period, the overall sedimentary environment was an epicontinental sea. In the early sedimentary stages of the Fengshan Formation and Yeli-Liangjiashan Formation, the aquatic environment was relatively stable, the sedimentary facies mainly developed in an open platform environment, and the original lithology was mainly fine-micritic limestone. This study provides a foundation for developing high-quality reservoirs in the study area [52–54]. At the end of the sedimentary period of the Fengshan Formation, the Huaiyuan Movement began and continued until the deposition of the Majiagou Formation. Several seabed uplifting, storm turbulence, and uplift-denudation events affected the entire North China Platform in eastern China [32,33], and diagenesis occurred during this process, including dolomitization and supergene karstification, which actively transformed the reservoir in the studied interval. In general, two critical sedimentary facies changes, dolomitization and supergene karstification, occurred in the late sedimentary period of the Fengshan Formation and the late sedimentary period of the Yeli-Liangjiashan Formation.

In the late sedimentary period of the Fengshan Formation and Yeli-Liangjiashan Formation, tectonic movement caused water turbulence and seabed uplift. Significant changes occurred in the sedimentary facies of the study area: (1) water turbulence and powerful hydrodynamic forces transformed some of the preformed fine-micritic limestones into limestone with a granular structure, and then the deposited sparry granular limestone accumulated to form granular beach facies; (2) the increase in the seabed (sea level decrease) led to the further development of evaporative platform and restricted platform (restricted lagoon) sedimentary facies in the sedimentary environment of the study area based on the open platform facies that formed in the epicontinental sea environment.

When the sea level in the evaporative and restricted platform environments wa relatively high, quasi-simultaneous dolomitization due to evaporation occurred, and th fine micrite changed to fine-micritic dolomite and limy dolomite [55,56]. When the sea lev was low and the evaporative platform or restricted platform environment was expose to the surface, the water salinity increased, and the fine micrite that developed in th platform was dolomitized by backflow osmosis to form limy dolomite and fine micriti dolomite [55]. The granular beach facies formed above the wave base in a high-energ aquatic environment, and the granular limestone that developed in this environment wa dolomitized under the continuous backflow infiltration of high-energy seawater [26,56–59 as well as the formation of granular dolomite, the development of primary pores betwee grains, and better reservoir performance [60–64]. According to the thin-section microscopi identification results, in addition to fine crystalline dolomite, mesocrystalline dolomi also occurs in the granular dolomite of the Yeli-Liangjiashan Formation and Fengsha Formation (Figure 6), and its crystals are well formed. Therefore, burial dolomitizatio occurred in the late study interval [65–67], recrystallization occurred between mud an fine-grained dolomite, and the grains became coarse, forming the mesocrystalline dolomi observed today. After various types of dolomitization occurred in the study interva the developed lithologies included granular dolomite, medium-fine crystalline dolomit mud-microcrystalline dolomite, and limy dolomite. Among them, granular dolomite ha pores, a good intergranular three-dimensional network pore structure, smooth pore flui migration, and good original reservoir performance [68]. The reservoir performance o medium-fine crystalline dolomite is second, and the original reservoir conditions of mu (micro)crystalline dolomite are relatively poor [69,70].

The local surface environment emerged and suffered from leaching by atmospheri freshwater, resulting in supergene karstification, which had a positive effect on the tran formation of carbonate reservoirs [26,50,51,71], wherein the supergene karstification in th late Yeli-Liangjiashan Formation lasted longer and was more intense than that in the la Fengshan Formation. Apparent karst zones (vadose zone and underflow zone) develope vertically on the tops of the two Formations. Because the dolomitization of the Fengsha Formation was less intense than that of the Yeli-Liangjiashan Formation, the litholog of the dolomite of the Fengshan Formation is lower than that of the Yeli-Liangjiasha Formation. For granular beach facies dolomite, when subjected to supergene karstificatio atmospheric freshwater permeated along cracks, intergranular pores formed in the vados zone, and dissolution occurred, forming cavity-type and crack-cavity-type reservoir spac combinations, forming angular karst breccia in the sinkhole and dissolved pores and cave in the areas between sinkholes (Figure 13b). The infiltrated atmospheric freshwater is pron to stratified flow in the underflow zone. Due to the development of the primary pores i the granular dolomite of the granular beach facies, water flows along the existing interna pores, and dissolution occurs, forming a stratified dissolved pore zone or karst cave zon which results in leakage during exploration and drilling in the oilfield. In the underflo zone, high-quality porous reservoirs easily form. Therefore, the high-quality reservoir of the Yeli-Liangjiashan Formation and Fengshan Formation are mainly distributed at th paleokarst boundary (unconformity), and their distributions are planar.

From a regional perspective, micritic dolomite, micrite, and salt-rock are quickl formed in the granular beach environment, forming an impermeable layer between th beach facies; their primary pores are not developed, and atmospheric freshwater ha difficulty penetrating [72–74]. Thus, only a thin surface cavity layer (surface karst zone) i formed in the surface layer. Karstification occurs during the infiltration of cracks, resultin in a lateral or longitudinal channel for local fluid migration [58]. Compared with that o granular beach facies dolomite, its reservoir performance is significantly lower, and it ofte forms a tight reservoir (Figure 13a). Due to the leaching effect of atmospheric freshwate restricted platform limestone and limy dolomite are prone to develop vertically dissolve pores and caves. They are vertically connected with high-angle cracks, which can becom excellent vertical fluid transport channels. When encountering impermeable layers, crack and pores with good reservoir properties can form [73]. Compared with those of granular dolomite, the porosity and permeability of grainy dolomite in an evaporation platform are better, but the order of the grains is relatively disordered [74]. When strata are exposed to denudation and subjected to the leaching of atmospheric freshwater, their developed karst patterns are also different due to the differences in their structural locations. In the higher terrain areas (karst highlands and karst slopes), the thickness of the vadose zone is relatively large. The vertical development of dissolved pores, solution caves, connected cracks, and sinkholes can form an excellent banded reservoir, which is affected by the underflow zone and boundary [75]. A layered dissolved pore zone and an underground river are developed in the lower part of the underflow zone. A good stratified reservoir can be formed. When the grainy dolomite is located in a relatively flat and low-lying area, the vertical flow and dissolution process of atmospheric freshwater occurs in the surface cracks of the grains over a short distance. Then, forward flow occurs in the grains to develop layered dissolved pores, and finally, a high-porosity reservoir can form.

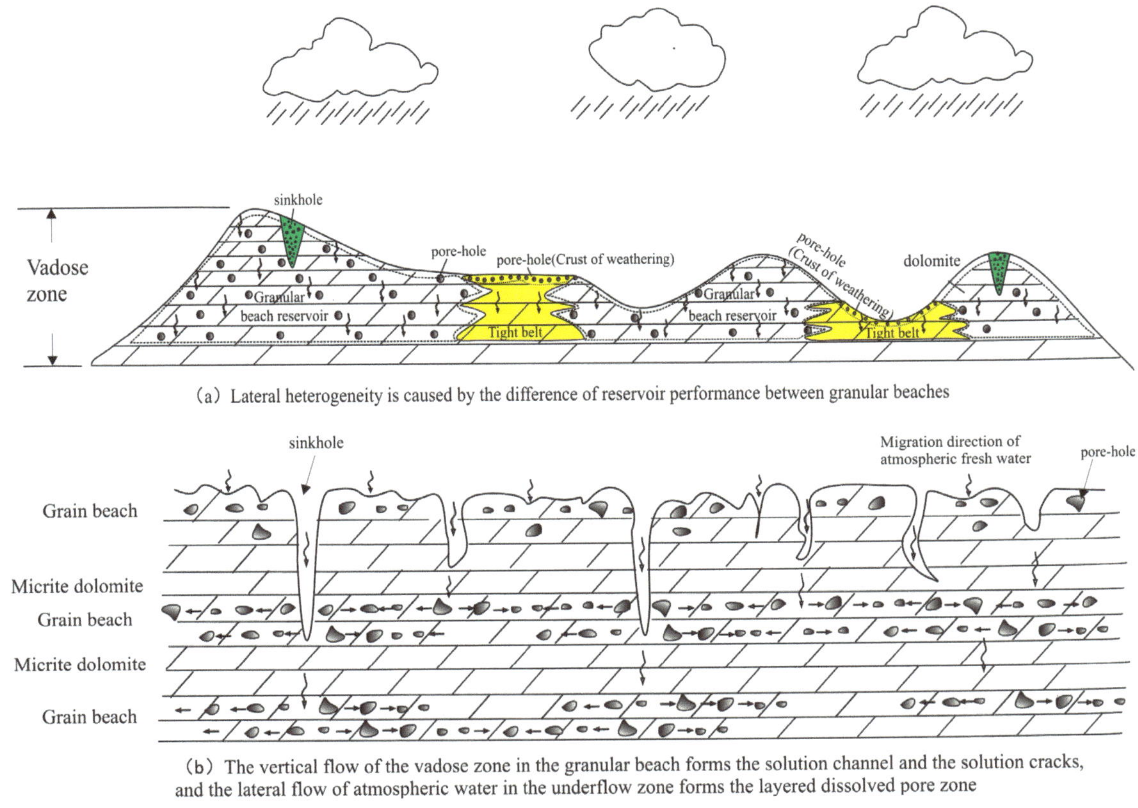

Figure 13. Huaiyuan period reservoir development model map of Yeli-Liangjiashan Formation and Fengshan Formation in Chengdao-Zhuanghai buried hill.

6. Conclusions

This study aimed to reveal the influence of supergene karstification on reservoir formation during the Huaiyuan Movement. The most important conclusions are as follows:

(1) Geological and geochemical data indicate that the Fengshan Formation and Yeli-Liangjiashan Formation were affected and controlled by the Huaiyuan Movement and have apparent vertical karst zonation. From top to bottom, the vadose zone and

the underflow zone can be delimited successively, the karst breccia is present in th two karst zones, and differences are present.

(2) There are apparent differences in the types of reservoir space combinations betwee the vadose and underflow zones. In the vadose zone, the dissolved pores, kars caves, and cracks formed by weathering solutions are present and are connecte longitudinally. It is easy to develop a composite crack-cavity reservoir space. I contrast, in the underflow zone, layered dissolved pores and cracks are present, and i is easy to develop two types of reservoir spaces: cavity type and crack type.

(3) Primitive sedimentary microfacies, dolomitization, and supergene karstification co trolled the Fengshan Formation and Yeli-Liangjiashan Formation reservoirs. Differen sedimentary microfacies have apparent differences in the mode of karst developmen development intensity, reservoir type, and reservoir physical properties. The reservoir are mainly distributed along the paleokarst zone that formed in the Huaiyuan perioc and the dolomite of the granular beach facies has the best reservoir physical propertie which are the features of surface distribution along the upper unconformity in the regio

(4) Limited by the progress of the current work, the data obtained, and the researc methods, the longitudinal and transverse connectivity of the internal reservoir spac of the unconformity structure has not been studied, and it is necessary to increas the logging data and seismic experiments to carry out this research. It is suggeste that there are longitudinal and transverse migration channels in the unconformit affected by the Huaiyuan movement, and combined with faults formed by multi-stag tectonic movement, they form a network of fluid (oil and gas) transport systems. I the strata of the Jiyang Depression, a large number of source rocks are distribute in the Paleogene strata. Under mature hydrocarbon expulsion, the source rock migrate continuously along the Paleogene sand body and reach the buried hill of th Lower Paleozoic. After that, the hydrocarbon-bearing fluid migrates and transport horizontally or vertically along the network transport system, forming hydrocarbo accumulations in the buried hill of the lower Paleozoic in the Jiyang Depression. Thu a new network reservoir formation system is formed, which opens a new avenue fo the exploration of the Jiyang Depression and provides a new research direction fo the study of the area affected by the Huaiyuan movement in Bohai Bay Basin, Chin

Author Contributions: R.L.: literature search, figures, study design, data collection, data analysi data interpretation, writing. G.W.: study design, review, writing. Y.W.: study design, review, writin X.H. and F.Q.: literature search, figures. X.F., W.M. and G.L.: data collection, data analysis. A authors have read and agreed to the published version of the manuscript.

Funding: This study is financially supported through Projects of Sinopec (NO. P22066).

Institutional Review Board Statement: Not applicable.

Informed Consent Statement: Not applicable.

Data Availability Statement: For confidentiality reasons, we are unable to provide raw data report but our experimental test results are highly accurate and we firmly believe in this.

Acknowledgments: We would like to express our thanks for the funding from Shengli Oilfield and th experimental test data from Colleges of Earth Sciences, Chengdu University of Technology laborator

Conflicts of Interest: Authors Ruijuan Liu, Yongshi Wang, Xuefeng Hao, Feng Qin and Wei Men were employed by the company Sinopec ShengliOilfield Company. The remaining authors declar that the research was conducted in the absence of any commercial or financial relationships tha could be construed as a potential conflict of interest.

References

1. Fritz, R.D.; Wilson, J.L.; Yurewicz, D.A. *Paleokarst Related Hydrocarbon Reservoirs*; SEPM Society for Sedimentary Geolog Claremore, OK, USA, 1993.
2. Halbouty, M.T. *Giant Oil and Gas Fields of the 1990s: An Introduction*; American Association of Petroleum Geologists: Tulsa, OK USA, 2003.

Gu, Z.D.; Wang, Z.C.; Hu, S.Y.; Wang, H.; Yin, J.F.; Huang, P.H. Tectonic settings of global marine carbonate giant fields and exploration significance. *Nat. Gas Geosci.* **2012**, *23*, 106–118.

Budd, D.A.; Saller, A.H.; Harris, P.M. Unconformities and porosity in carbonate strata. *AAPG Mem.* **1995**, *79*, 1183–1184.

Clari, P.A.; Pierre, F.D.; Martire, L. Discontinuities in carbonate successions: Identification, interpretation and classification of some Italian examples. *Sediment. Geol.* **1995**, *100*, 97–121. [CrossRef]

Akbar, M.; Vissapragada, B.; Alghamdi, A.H.; Allen, D.; Herron, M.; Carnegie, A.; Dutta, D.; Olesen, J.-R.; Logan, D.; Stief, D.; et al. A snapshot of carbonate reservoir evaluation. *Oilfield Rev.* **2000**, *12*, 20–21.

Wang, B.; Al-Aasm, I.S. Karst-controlled diagenesis and reservoir development: Example from the Ordovician mainreservoir carbonate rocks on the eastern margin of the Ordos basin, China. *AAPG Bull.* **2002**, *86*, 1639–1658.

Sattler, U.T.E.; Immenhauser, A.; Hillgärtner, H.; Esteban, M. Characterization, lateral variability and lateral extent of discontinuity surfaces on a carbonate platform (Barremian to Lower Aptian, Oman). *Sedimentology* **2005**, *52*, 339–361. [CrossRef]

Craig, J.; Thurow, J.; Thusu, B.; Whitham, A.; Abutarruma, Y. *Global Neoproterozoic Petroleum Systems: The Emerging Potential in North Africa*; Geological Society London, Special Publications: London, UK, 2009; Volume 326, pp. 1–25.

10. Su, J.; Zhang, S.C.; Yang, H.J.; Zhu, G.Y.; Chen, J.P.; Zhang, B. The control of carbonate effective reservoir by fault system and its reservoir-forming rule. *Acta Pet. Sin.* **2010**, *31*, 196–203.

11. Chow, N.; Wendte, J. Palaeosols and palaeokarst beneath subaerial unconformities in an Upper Devonian isolated reef complex (Judy Creek), Swan Hills Formation, west-central Alberta, Canada. *Sedimentology* **2011**, *58*, 960–993. [CrossRef]

12. Bhat, G.M.; Craig, J.; Hafiz, M.; Hakhoo, N.; Thurow, J.W.; Thusu, B.; Cozzi, A. *Geology and Hydrocarbon Potential of Neoproterozoic–Cambrian Basins in Asia: An Introduction*; Geological Society, London, Special Publications: London, UK, 2012; Volume 366, pp. 1–17.

13. Bagni, F.L.; Bezerra, F.H.; Balsamo, F.; Maia, R.P.; Dall'Aglio, M. Karst dissolution along fracture corridors in an anticline hinge, Jandaíra Formation, Brazil: Implications for reservoir quality. *Mar. Pet. Geol.* **2020**, *115*, 104249. [CrossRef]

14. Ji, Y.L.; Hu, G.M.; Zhang, S.W.; Zhao, J.Q. Mineralogical and geochemical methods in study of sedimentary sequence boundary. *Tongji Daxue Xuebao/J. Tongji Univ. (Nat. Sci.) (China)* **2004**, *32*, 455–460.

15. Bai, B.; Zou, C.; Zhu, R.; Zhai, W.L.; Liu, L.H.; Dai, Z.C.; Mao, Z.G. Integrated identification of sequence boundaries through outcrop, natural gammaray spectral, rock geochemistry, logging and seismic: A case of upper Triassic Xujiahe formation, Sichuan Basin. *Nat. Gas Geosci.* **2010**, *21*, 78–86.

16. Zhiqian, G.; Tailiang, F. Unconformities and their influence on lower Paleozoic petroleum reservoir development in the Tarim Basin. *J. Pet. Sci. Eng.* **2015**, *133*, 335–351. [CrossRef]

17. Beckert, J.; Vandeginste, V.; John, C.M. Relationship between karstification and burial dolomitization in Permian platform carbonates (Lower Khuff—Oman). *Sediment. Geol.* **2016**, *342*, 165–179. [CrossRef]

18. Vincent, S.J.; Guo, L.; Flecker, R.; BouDagher-Fadel, M.K.; Ellam, R.M.; Kandemir, R. Age constraints on intra-formational unconformities in Upper Jurassic-Lower Cretaceous carbonates in northeast Turkey; geodynamic and hydrocarbon implications. *Mar. Pet. Geol.* **2018**, *91*, 639–657. [CrossRef]

19. Nadeau, O. Sources of fluids in Archean hydrothermal stockwork-disseminated gold deposits of Abitibi, Canada: Insights from Duquesne, Dolodau, Lac Shortt and Canadian Malartic. *Ore Geol. Rev.* **2019**, *111*, 102975. [CrossRef]

20. Lyu, D.; Deng, Y.; Wang, H.; Zhang, F.; Ren, R.; Gao, Z.; Canfield, D.E. Using cyclostratigraphic evidence to define the unconformity caused by the Mesoproterozoic Qinyu Uplift in the North China Craton. *J. Asian Earth Sci.* **2021**, *206*, 104608. [CrossRef]

21. James, N.P.; Choquette, P.W. (Eds.) *Paleokarst*; Springer Science & Business Media: Berlin/Heidelberg, Germany, 2012.

22. Lucia, F.J. *Lower Paleozoic Cavern Development, Collapse, and Dolomitization, Franklin Mountains, El Paso, Texas*; American Association of Petroleum Geologists: Tulsa, OK, USA, 1995.

23. Luczaj, J.A.; Harrison, W.B., III; Smith Williams, N. Fractured hydrothermal dolomite reservoirs in the Devonian Dundee Formation of the central Michigan Basin. *AAPG Bull.* **2006**, *90*, 1787–1801. [CrossRef]

24. Tian, H.; Xiao, X.; Wilkins, R.W.; Tang, Y. New insights into the volume and pressure changes during the thermal cracking of oil to gas in reservoirs: Implications for the in-situ accumulation of gas cracked from oils. *AAPG Bull.* **2008**, *92*, 181–200. [CrossRef]

25. Cao, J.W. *Study on the Distribution and Genetic Mechanism of Paleokarst in the Eastern Part of Lunnan Buried Mountain*; China University of Geosciences: Wuhan, China, 2019.

26. Jiu, B.; Huang, W.; Mu, N.; Hao, R. Petrology, mineralogy and geochemistry of Ordovician rocks in the southwest of Tarim Basin, implications for genetic mechanism and evolution model of the hydrothermal reformed-paleokarst carbonate reservoir. *Mar. Pet. Geol.* **2022**, *140*, 105687. [CrossRef]

27. Li, P.L.; Zhang, S.W.; Wang, Y.S. *Genesis, Accumulation and Exploration of Diverse Buried Hills: A Case Study of Jiyang Depression*; Petroleum Industry Press: Beijing, China, 2003.

28. Song, G.Q.; Zhuo, Q.G.; Sun, L. Migration and accumulation model of Tertiary unconformity hydrocarbon reservoirs in Jiyang Depression. *Oil Gas Geol.* **2008**, *29*, 716–720+732.

29. Sui, F.G.; Song, G.Q.; Zhao, L.Q.; Wang, X.J. Oil and gas transport mode and performance of unconformity in continental faulted basin of Jiyang Depression. *J. China Univ. Pet. (Nat. Sci. Ed.)* **2010**, *34*, 44–48.

30. Ma, L.C. Discussion on genesis of lower Paleozoic insider-type reservoir in Jiyang Depression. *Spec. Reserv.* **2003**, *10*, 13–14+91.

31. Lin, H.X. Karstification of lower Paleozoic buried hill reservoir in Zhuangxicheng Island area, Jiyang Depression. *J. Chengdu Univ. Technol. (Sci. Technol. Ed.)* **2004**, *42*, 490–497.

32. Song, D.N. Reunderstanding of Huaiyuan movement. *Geol. Shandong* **2001**, *17*, 19–23.
33. Li, X.B.; Wang, H.B.; Huang, J.P.; Zhang, C.L.; Zhang, Y.; Wang, Y.T.; Zhang, L.; Wang, J.; Liu, H.Q. Characteristics of Huaiyua movement unconformity in Ordos Basin and its significance for oil and gas exploration. *Oil Gas Geol.* **2021**, *42*, 1043–1055.
34. Zhang, M.; Wu, Z.P.; Huang, Z.; Zhang, B.; Zhang, F.P.; Miao, Y.Z. Development law and genetic type division of buried hill i Jiyang Depression, Bohai Bay Basin. *Earth Sci.* **2023**, *48*, 488–502.
35. Wang, Z.; Zhang, K.; Cheng, Y.; Wu, Q. Identification and evaluation of fault-fracture reservoirs in buried hills of the Low Paleozoic, Chengdao area, China. *Energy Geosci.* **2023**, *4*, 100183. [CrossRef]
36. Jing, A.Y. Fault characteristics and reservoir-controlling effect of Lower Paleozoic buried hill in Chengdao area of Jiyan Depression. *Sci. Technol. Eng.* **2020**, *20*, 6011–6017.
37. Ma, S.; Wang, Y.S.; Wang, X.J.; Jing, A.Y. Development characteristics and genetic mechanism of Lower Paleozoic buried hi insider dolomite reservoir in Jiyang Depression. *Geol. Rev.* **2023**, *69*, 279–280.
38. Zhang, Y.F.; Wang, Q.C. Isotopic characteristics and significance of calcite C and O in Ordovician carbonate rocks and fractur cavity filling in Jiyang Depression. *Chin. J. Geol.* **2007**, *42*, 570–578.
39. Osleger, D.; Read, J.F. Relation of eustasy to stacking patterns of meter-scale carbonate cycles, Late Cambrian, USA. *J. Sedimen Res.* **1991**, *61*, 1225–1252.
40. Liu, B.; Wang, Y.H.; Qian, X.L. Genesis of two Ordovician unconformities and prediction of related regional reservoirs in Nort China. *Acta Sedimentol.* **1997**, *15*, 25–30.
41. Mei, M.X.; Ma, Y.S. Late Cambrian sequence stratigraphy of North China Platform and its correlation with sea level change c North American platform. *Sediment. Tethys Geol.* **2003**, *23*, 14–26.
42. Cross, N.E.; van Veen, L.J.; Al-Enezi, A.; Singh, S.; van Beusekom, G. Seismic geomorphology of karst in Cretaceous to Earl Cenozoic carbonates of North Kuwait. *Mar. Pet. Geol.* **2021**, *128*, 104947. [CrossRef]
43. Li, Y.; Jin, Q.; Zhong, J.H.; Zou, S.Z. Characteristics of Ordovician karst zonation and fissure cavity structure in Tahe Oilfield. *Ac Pet. Sin.* **2016**, *37*, 289–298.
44. Zhang, Q.Y.; Liang, B.; Cao, J.W.; Dan, Y.; Li, J.R.; Chen, L.X. Pre-carboniferous paleo-karst microgeomorphology and kar development model in east Lungu area, Tarim Basin. *Mar. Pet. Geol.* **2017**, *22*, 30–36.
45. Veizer, J.; Ala, D.; Azmy, K.; Bruckschen, P.; Buhl, D.; Bruhn, F.; Strauss, H. $^{87}Sr/^{86}Sr$, $\delta^{13}C$ and $\delta^{18}O$ evolution of Phanerozoi seawater. *Chem. Geol.* **1999**, *161*, 59–88. [CrossRef]
46. Denison, R.E.; Koepnick, R.B.; Burke, W.H.; Hetherington, E.A. Construction of the Cambrian and Ordovician seawater 87Sr/86S curve. *Chem. Geol.* **1998**, *152*, 325–340. [CrossRef]
47. Clayton, R.N.; Degens, E.T. Use of carbon isotope analyses of carbonates for differentiating fresh-water and marine sedimen *AAPG Bull.* **1959**, *43*, 890–897.
48. Meng, Q.Q.; Zhu, D.Y.; Hu, W.X.; Jin, Z.J. Dissolution-filling mechanism of atmospheric precipitation controlled by bot thermodynamics and kinetics. *Sci. China Earth Sci.* **2013**, *56*, 2150–2159. [CrossRef]
49. Jones, B. Diagenetic processes associated with unconformities in carbonate successions on isolated oceanic islands: Case study c the Pliocene to Pleistocene sequence, Little Cayman, British West Indies. *Sediment. Geol.* **2019**, *386*, 9–30. [CrossRef]
50. Bagni, F.L.; Erthal, M.M.; Tonietto, S.N.; Maia, R.P.; Bezerra, F.H.; Balsamo, F.; Fonseca, J.P.T. Karstified layers and caves forme by superposed epigenic dissolution along subaerial unconformities in carbonate rocks–Impact on reservoir-scale permeabilit *Mar. Pet. Geol.* **2022**, *138*, 105523. [CrossRef]
51. Xiong, Y.; Tan, X.; Zhong, S.; Xiao, D.; Wang, B.; Yang, M.; Cao, J. Dynamic paleokarst geochemistry within 130 Myr in the Midd Ordovician Shanganning carbonate platform, North China. *Palaeogeogr. Palaeoclimatol. Palaeoecol.* **2022**, *591*, 110879. [CrossRef]
52. Esrafili-Dizaji, B.; Rahimpour-Bonab, H. Effects of depositional and diagenetic characteristics on carbonate reservoir quality: case study from the South Pars gas field in the Persian Gulf. *Petrol. Geosci.* **2009**, *15*, 325–344. [CrossRef]
53. Mehrabi, H.; Rahimpour-Bonab, H.; Hajikazemi, E.; Jamalian, A. Controls on depositional facies in Upper Cretaceous carbona reservoirs in the Zagros area and the Persian Gulf, Iran. *Facies* **2015**, *61*, 23. [CrossRef]
54. Ma, C.F.; Huang, W.J.; Du, Z.L.; Han, W.Z.; Zhan, Y.; Shi, Z.N.; Zhou, J.; Song, M.Y. Lithofacies classification scheme of continent lacustrine shale and its significance: A case study of the second member of Kongdian Formation in Cangdong Sag. *J. Cent. Sout Univ. Nat. Sci. Ed.* **2022**, *53*, 3287–3300.
55. Vandeginste, V.; Swennen, R.; Reed, M.H.; Ellam, R.M.; Osadetz, K.; Roure, F. Host rock dolomitization and secondary porosit development in the upper Devonian Cairn Formation of the Fairholme carbonate complex (south-west Alberta, Canadia Rockies): Diagenesis and geochemical modelling. *Sedimentology* **2009**, *56*, 2044–2060. [CrossRef]
56. Adams, J.E.; Rhodes, M.L. Dolomitization by seepage refluxion. *AAPG Bull.* **1961**, *44*, 1912–1920.
57. Skall, H. The paleoenvironment of the Pine Point lead-zinc district. *Econ. Geol.* **1975**, *70*, 22–47. [CrossRef]
58. Lucia, F.J.; Major, R.P. Porosity evolution through hypersaline reflux dolomitization. In *Dolomites: A Volume in Honour of Dolomie* John Wiley & Sons, Inc.: Hoboken, NJ, USA, 1994; pp. 325–341.
59. Meng, M.; Ge, H.; Shen, Y.; Ji, W.; Li, Z. Insight into water occurrence and pore size distribution by nuclear magnetic resonance i marine shale reservoirs, southern China. *Energy Fuels* **2023**, *37*, 319–327. [CrossRef]
60. Sullivan, E.C.; Marfurt, K.J.; Lacazette, A.; Ammerman, M. Application of new seismic attributes to collapse chimneys in the For Worth Basin. *Geophysics* **2006**, *71*, B111–B119. [CrossRef]

11. Zhao, W.; Shen, A.; Qiao, Z.; Zheng, J.; Wang, X. Carbonate karst reservoirs of the Tarim Basin, northwest China: Types, features, origins, and implications for hydrocarbon exploration. *Interpretation* **2014**, *2*, SF65–SF90. [CrossRef]
12. Chopra, S.; Marfurt, K.J. Seismic attributes for stratigraphic feature characterization. In *SEG Technical Program Expanded Abstracts 2008*; Society of Exploration Geophysicists: Houston, TX, USA, 2008; pp. 1590–1594.
13. Ma, Y.; Zhang, S.; Guo, T.; Zhu, G.; Cai, X.; Li, M. Petroleum geology of the Puguang sour gas field in the Sichuan Basin, SW China. *Mar. Pet. Geol.* **2008**, *25*, 357–370. [CrossRef]
14. Meng, M.; Ge, H.; Shen, Y.; Ji, W.; Wang, Q. Rock fabric of tight sandstone and its influence on irreducible water saturation in Eastern Ordos Basin. *Energy Fuels* **2023**, *37*, 3685–3696. [CrossRef]
15. Al-Aasm, I. Origin and characterization of hydrothermal dolomite in the Western Canada Sedimentary Basin. *J. Geochem. Explor.* **2003**, *78*, 9–15. [CrossRef]
16. Smith, L.B., Jr. Origin and reservoir characteristics of Upper Ordovician Trenton–Black River hydrothermal dolomite reservoirs in New York. *AAPG Bull.* **2006**, *90*, 1691–1718. [CrossRef]
17. Chen, L.; Zhang, H.; Cai, Z.; Cong, F.; Huang, S.; Tang, P. Characteristics and formation mechanisms of the unconformity-related paleokarst reservoirs in the Upper Sinian, Northwestern Tarim Basin, China. *Mar. Pet. Geol.* **2020**, *120*, 104559. [CrossRef]
18. Flügel, E.; Munnecke, A. *Microfacies of Carbonate Rocks: Analysis, Interpretation and Application*; Springer: Berlin, Germany, 2010; Volume 976, p. 2004.
19. Murray, R.C. Origin of porosity in carbonate rocks. *J. Sediment. Res.* **1960**, *30*, 59–84. [CrossRef]
20. Lucia, F.J.; Kerans, C.; Jennings, J.W., Jr. Carbonate reservoir characterization. *J. Pet. Technol.* **2003**, *55*, 70–72. [CrossRef]
21. Ford, D.; Williams, P.D. *Karst Hydrogeology and Geomorphology*; John Wiley & Sons: Hoboken, NJ, USA, 2007.
22. La Bruna, V.; Bezerra, F.H.; Souza, V.H.; Maia, R.P.; Auler, A.S.; Araujo, R.E.; Sousa, M.O. High-permeability zones in folded and faulted silicified carbonate rocks–implications for karstified carbonate reservoirs. *Mar. Pet. Geol.* **2021**, *128*, 105046. [CrossRef]
23. Scholle, P.A.; Bebout, D.G.; Moore, C.H. *Carbonate Depositional Enviroments*; American Association of Petroleum Geologists: Tulsa, OK, USA, 1983.
24. Meng, M.; Zhang, Y.; Yuan, B.; Li, Z.; Zhang, Y. Imbibition behavior of oil-saturated rock: Implications for enhanced oil recovery in unconventional reservoirs. *Energy Fuels* **2023**, *37*, 13759–13768. [CrossRef]
25. He, J.; Fang, S.; Hou, F.; Yan, R.; Zhao, Z.; Yao, J.; Tang, X.; Wu, G. Vertical zonation of weathered crust ancient karst and reservoir evaluation and prediction—A case study of M55–M51 sub-members of Majiagou Formation in gas fields, central Ordos Basin, NW China. *Pet. Explor. Dev.* **2013**, *40*, 572–581.

Journal of
Marine Science and Engineering

MDPI

Article

Experimental and Numerical Simulation Investigation of Cement Sheath Integrity during Multi-Stage Fracturing in Offshore Tight Oil Reservoir

Yangang Wang and Yongcun Feng *

College of Petroleum Engineering, China University of Petroleum (Beijing), Beijing 102249, China; 2019310110@student.cup.edu.cn
* Correspondence: yfeng@cup.edu.cn

Abstract: The integrity of the cement sheath is susceptible to failure during multi-stage fracturin In this study, the failure mechanisms of cement sheath integrity during multi-stage fracturing i the A offshore tight oil reservoir wells were investigated. The cement samples were subject t triaxial compression test (TCT), triaxial cyclic loading test (TCLT), and permeability test. A ful scale device was constructed for cement sheath integrity experiments. Additionally, a 3-D finit element model was developed to simulate the interface debonding and the subsequent growth o micro-annuli throughout multi-stage fracturing. The results revealed that TCLT induced cumulativ plastic deformation in the cement samples, resulting in a 10.7% decrease in triaxial compressiv strength, an 8.3% decrease in elastic modulus, and a 150% increase in permeability. Despite thes significant variations, no serious damage was caused to the cement sheath matrix. It was observe that gas leakage occurred at the 8th, 10th, and 14th cycles under cyclic loading with upper limits o 70 MPa, 80 MPa, and 90 MPa, respectively. After 15 cycles, the experimentally measured widths o micro-annuli were 117 μm, 178 μm, and 212 μm, which were in good agreement with simulatio results of 130 μm, 165 μm, and 205 μm, respectively. These findings elucidate the causes of cemen sheath integrity failure, providing insights into the failure mechanisms of cement sheath integrit during multi-stage fracturing.

Keywords: offshore tight oil reservoir; multi-stage fracturing; cyclic loading; cement sheath integrit

Citation: Wang, Y.; Feng, Y. Experimental and Numerical Simulation Investigation of Cement Sheath Integrity during Multi-Stage Fracturing in Offshore Tight Oil Reservoir. *J. Mar. Sci. Eng.* **2024**, *12*, 814. https://doi.org/10.3390/jmse12050814

Academic Editor: Hailong Lu

Received: 31 March 2024
Revised: 8 May 2024
Accepted: 11 May 2024
Published: 14 May 2024

1. Introduction

Global demand for oil and gas has witnessed a substantial increase in recent decades [1 As the global development of oil and gas reservoirs with medium to high permeability intensifies, the depletion of high-quality hydrocarbon resources is becoming increasingly evident [2]. Consequently, there is a growing interest in unconventional resources [3]. The exploration of unconventional oil and gas resources, particularly tight and low-permeability reservoirs [4], has steadily expanded and undergone extensive development worldwide [5]

Multi-stage fracturing technology significantly enhances the productivity of oil and gas wells by creating complex fracture networks within reservoirs, making it the most effective method for exploiting unconventional resources, such as tight and low-permeability formations [6]. Offshore unconventional resources, including tight oil reservoirs, constitute a substantial proportion of the world's hydrocarbon potential and necessitate multi-stage fracturing techniques to increase productivity [7].

However, the integrity of the cement sheath is susceptible to failure during multi-stage fracturing, leading to sustained casing pressure (SCP). It poses threats to the safe production of oil and gas wells and potentially causes environmental and safety issues during subsequent drilling operations [8]. Consequently, the sustainable development o unconventional oil and gas reservoirs is compromised [9]. Understanding the key factors affecting cement sheath integrity in multi-stage fracturing is crucial [10]. Therefore, it is imperative to explore the mechanisms behind cement sheath integrity failure during multi-stage fracturing, thus providing a basis for the rationalization of fracturing construction parameters [11].

During multi-stage fracturing, variations in internal pressure within the casing induce alternating stresses in the cement sheath, serving as a primary factor contributing to its integrity failure [12]. In addition, the mechanical environment around the cement sheath becomes more complex during multi-stage fracturing due to abrupt changes in casing pressure and temperature [13]. This complexity makes it challenging to mathematically analyze and physically simulate cement sheath integrity failures under such conditions [14].

The causes of cement sheath integrity failure can be categorized into three types: disking cracks, radial cracks, and debonding. Disking cracks and radial cracks are typically induced by mechanical stresses, whereas debonding is usually caused by the mismatched deformation between the casing and the cement sheath [15].

Goodwin and Crook [16] conducted a laboratory test to investigate the integrity of a casing–cement–casing system. They observed catastrophic shear failure and the formation of radial cracks in the cement sheath under alternating pressure loading conditions. Kuanhai et al. [17] designed a simulated wellbore device for evaluating cement sheath integrity under different pressures. The findings of this study offered valuable insights into the impact of pressure on the mechanical properties of cement sheath interfaces. Su et al. [18] developed a mechanical equivalence method to replicate the wellbore loading on the cement sheath using a self-designed wellbore simulator. Cyclic load tests were conducted, revealing a decrease in the tensile strength of the cement sheath following cyclic loading. Feng et al. [19] and Wang and Taleghani [20] developed a combined numerical model for simulating hydraulic fracturing processes. They observed that fluid pressurization may create cracks in the annular space around the wellbore, specifically between the cement sheath and the formation. This phenomenon subsequently leads to the formation of channels conducive to gas flow.

Li et al. [21] constructed a comprehensive physical simulation experiment to identify the primary determinant in scenarios involving cyclic loading and unloading. The results revealed that plastic strain accumulates throughout the cyclic loading and unloading process, leading to the formation of micro-annuli at the primary interface, which then served as channels for gas migration.

In the previous study investigating the integrity of cement sheath, the material was considered to be elastically brittle with stresses influenced by its properties [22]. Pereira et al. [23] also explored the reduction of shear failure in cement sheaths during hydraulic fracturing, suggesting that lowering the elastic modulus can mitigate this type of failure.

Landry et al. [24] found that in a shale gas field in the USA, SCP always occurs after fracturing and occasionally before fracturing. They attributed the predominance of SCP to issues such as improper cement placement or incomplete consideration of stimulation conditions. Additionally, they emphasized the importance of considering changes in cement sheath stress during stimulation and throughout well operation.

During multi-stage fracturing, the pressure inside the casing frequently rose and fell, subjecting both the casing and the cement sheath to cyclical loading [25]. Earlier research has shown that while the cement sheath might withstand a static load, it is prone to failure under repeated applications of the same load, typical of the pressure fluctuations experienced in multi-stage fracturing [26].

Ramadan et al. [27] developed a wellbore model to simulate gas migration in the cemented annulus. A series of tests were performed using this model to assess the sealability of both neat Class H and Class G cement types. Various additives, including bentonite and latex, were incorporated. The tests demonstrated that the permeability of annulus cement in the wellbore increased as the cement aged.

A strong bond between the casing and cement sheath is essential to mitigate the occurrence of micro-annuli leakage paths at their interface [28]. Various elements, such as the cement type, the surface condition of the casing steel, and the curing environmer significantly affect the bond strength between these materials [29]. Furthermore, tł evaluation of the bond strength between casing and cement sheath commonly involves tł implementation of laboratory pushout tests [30].

Tabatabaei et al. [31] conducted pushout tests to assess the effect of surface-modifie cement additives incorporating graphite nanosheets on bond strength under various sc narios involving oil-based mud residues.

Corina et al. [32] examined the sealing effectiveness of neat and silica–cement systerr in pipes with three varying degrees of surface roughness. In their study, gas leaks we identified in all samples at low differential pressures, demonstrating that the cement–casin interface is the source of these leaks.

The tight oil reservoir A is located in offshore China, and all of its production wel utilize a typical three-casing structure. Among them, the structure of Well H is depicte in Figure 1. After multi-stage fracturing, several wells have exhibited SCP, all of whic occurred in the B annulus (the space between the production casing and intermedia casing). This phenomenon indicates a failure of the cement sheath integrity within the annulus. Thus, investigating the underlying mechanisms responsible for compromising tł integrity of the cement sheath in the B annulus constitutes the primary focus of this study

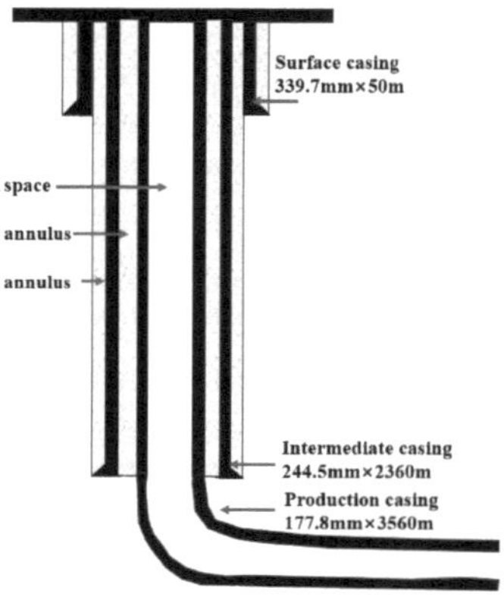

Figure 1. Structure of Well H.

This study aims to investigate the failure mechanisms of cement sheath integrit during multi-stage fracturing in the A offshore tight oil reservoir wells. The cemer samples were subjected to triaxial compression test (TCT), triaxial cyclic loading te (TCLT), and permeability test. It quantifies the cumulative plastic deformation patter of the cement under cyclic loading, assesses the impact of cyclic loads on the mechanica properties and permeability of the cement, and experimentally examines the failure c cement sheath integrity under multi-stage fracturing conditions using a full-scale devic The morphology of the micro-annuli was observed using a scanning electron microscop (SEM), and the widths of the micro-annuli were precisely measured. Based on the actua geological conditions and wellbore structure, a 3-D finite element model was develope to simulate the development of cumulative plastic strain, interface debonding, and th growth of micro-annuli during multi-stage fracturing.

The numerical simulation results are in good agreement with the experimental result providing valuable insights for a more in-depth understanding of the failure mechanism and patterns associated with the cement sheath integrity during multi-stage fracturin This study also provides guidance for the rational design of fracturing parameters.

2. Physical Experiments

2.1. Mechanical and Permeability Testing of Cement

The test cement was formulated using 100% G-grade cement, 23% silica fume, 6% anti-gas channeling agents, 4.5% fluid loss additive, 1.0% retarder, and 0.3% dispersion agent, with a 0.4 water-to-cement ratio and a 1.84 g/cm^3 density. The cement slurry was then cured to obtain cement stone. This cement stone was standardized for cylindrical specimens, measuring 25 mm in diameter and 50 mm in height, as depicted in Figure 2a for samples 1 and 2. These specimens had consistent mechanical and permeability characteristics. Initially, the initial permeability of sample 1 was evaluated using nitrogen gas as the testing medium, as illustrated in Figure 2b, followed by a TCT in Figure 2c. Subsequently, sample 2 underwent a TCLT, which replicated the stress conditions experienced by cement sheaths in multi-stage fracturing. Upon completion of the TCLT, sample 2 was also subjected to a permeability test, followed by a TCT. TCT, TCLT, and permeability tests were conducted under 10 MPa confining pressure. The implementation of TCT and TCLT was achieved with the MTS device, possessing a maximum load of 600 kN and a maximum hydrostatic pressure of 100 MPa.

(a)

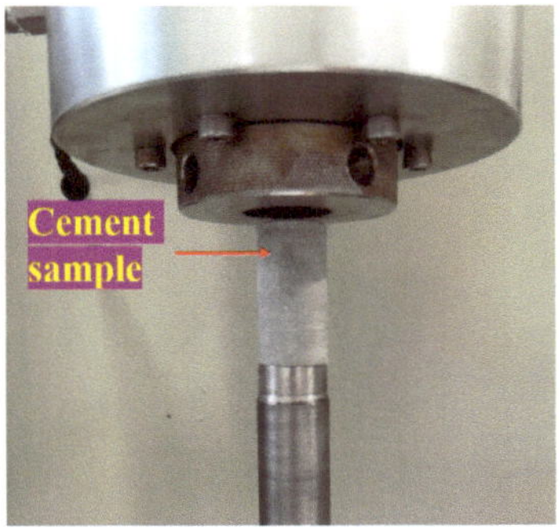

(b)

(c)

Figure 2. TCT, TCLT, and permeability test on cement samples: (**a**) samples 1 and 2, (**b**) permeability test, and (**c**) TCT and TCLT.

TCT procedure: The cement sample was placed within a pressure chamber, the triaxial cell. The employment of oil ensured the uniform application of confining pressure around the sample's lateral surface. On the premise of keeping the confining pressure constant, an axial load was gradually applied to the sample through a loading platen at the top. The applied load was increased until the sample failed.

TCLT procedure: With the triaxial cell, the cement specimen was enclosed in a rubber membrane to isolate it from the confining fluid. During the application of confining pressure, oil was employed to envelope the specimen. After reaching the desired initial confining pressure, cyclic axial loading was applied using a dynamic actuator. The cyclic loading was capped at 80% of the specimen's initial triaxial compressive strength over 15 cycles.

2.2. Experiment on the Cement Sheath Integrity

2.2.1. Experimental System

A comprehensive full-scale device was designed, comprising a foundation, a wellbore configuration, and an integrated control and data acquisition system. This device allows for simulation and evaluation of the cement sheath integrity under multi-stage fracturing. Notably, considering the exclusive detection of SCP in the B annulus under field conditions, the current experiments primarily focused on assessing the cement sheath integrity within this annulus. This device is depicted in Figure 3a, its structural layout is presented in Figure 3b, and the physical images are shown in Figure 3c,d.

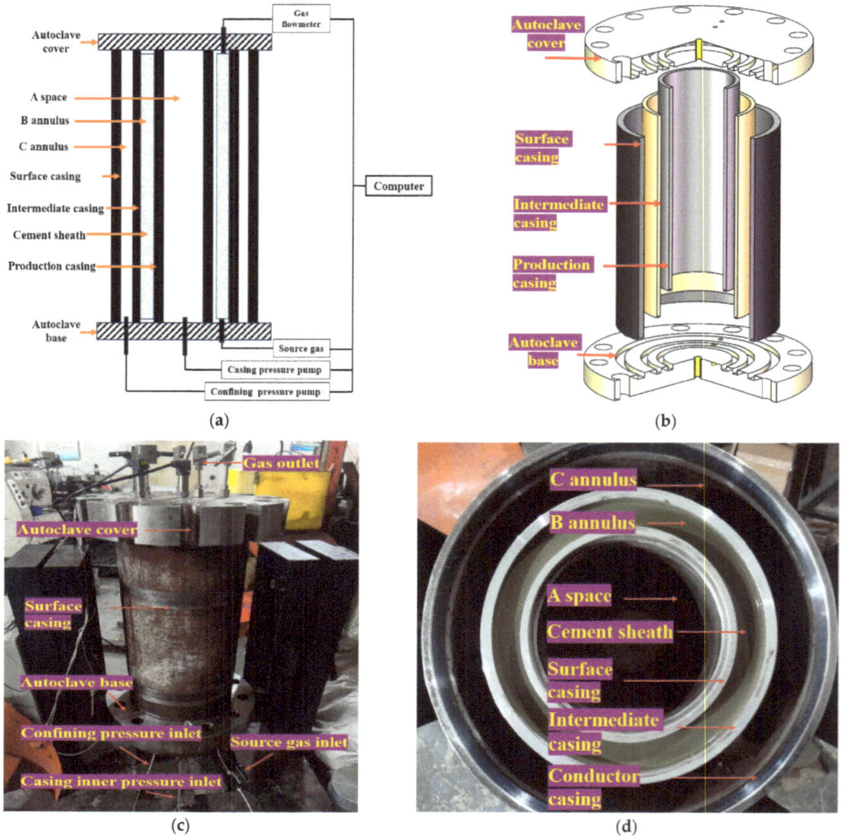

Figure 3. The full-scale device: (**a**) schematic diagram, (**b**) structural layout, (**c**) device assembled and (**d**) casings.

Cement and casing materials for the experiments were obtained from the oil field. This practice ensured that the dimensions and materials of the inner and middle casings were precisely consistent with those used underground. The details of the casing system's specifications are provided in Table 1. Specifically, A space within the production casing was designated to provide cyclic pressure. In contrast, nestled between the production casing and intermediate casing, the B annulus served as cement injection to establish the cement sheath. Additionally, the C annulus, with its position between the surface casing and intermediate casing, exerted confining pressure and simulated the subterranean stress effects. This configuration guarantees that the experimental setup accurately replicates real-world field scenarios.

Table 1. Basic parameters of casings.

Casing	Grade	Diameter (mm)	Thickness (mm)
Surface casing	P110	177.8	9.19
Intermediate casing	P110	244.5	9.65
Conductor casing	N80	339.7	11.99

In this experiment, the adoption of a control system allows for the automated management of various modules, thereby ensuring the automation of the experimental procedure. Briefly, the control system guaranteed the source gas injection from the pump into the cement sheath foundation; additionally, it continuously documented the data from the gas flowmeter at the upper side. Subsequently, the system sent commands to the confining pressure pump to apply the required pressure in the B annulus, thus simulating the geo-stress effects. Lastly, under the instruction of the control system, the casing pressure pump applied the required pressure precisely, creating cyclic pressure within the A space to replicate the multi-stage fracturing process. This automated workflow improved the experiment's safety, convenience, and accuracy.

Two-stage and three-stage cementing jobs are prevalent in cementing operations. Figure 4a,b depict their typical flowcharts. It can be seen that both processes are complex, leading to a prolonged duration. Bottom hole circulating temperature and bottom hole static temperature may impact the performance of cement slurries, potentially delivering adverse effects.

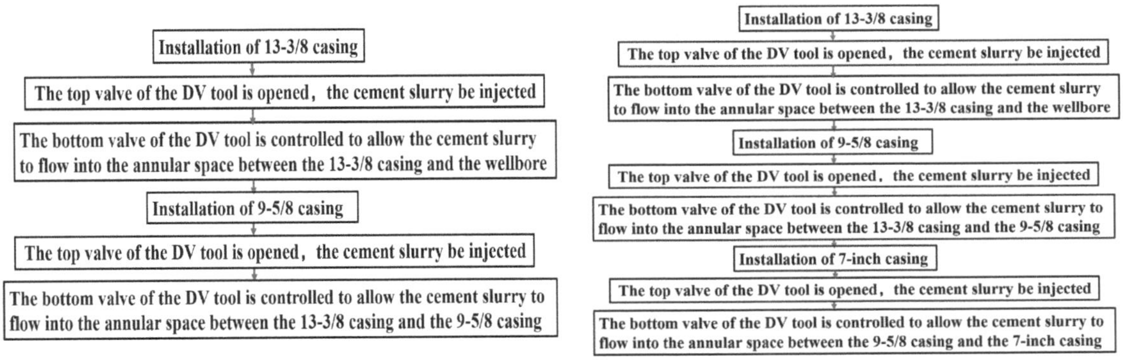

(a) (b)

Figure 4. Diagrams of cementing jobs: (**a**) two-stage and (**b**) three-stage.

During the experiments, we prepared the cement slurry in a single batch and injected it into the annulus within five minutes. This short injection interval ensured consistent properties throughout, including rheology, permeability, and unconfined compressive strength.

2.2.2. Experimental Procedure

Based on the historical fracturing data from the A oil field, three distinct cyclic internal pressure groups featuring upper limit pressures of 70 MPa, 80 MPa, and 90 MPa were selected. The minimum threshold pressure for all cycles was standardized at 20 MPa, consistent with the hydrostatic pressure within the fracturing fluid. The detailed experimental protocol is outlined below:

(a) The casings from the oil field were installed, and the cement slurry was prepared according to the on-site cementing formula. Subsequently, the prepared slurry was injected into Annulus B with a 48 mPa·s plastic viscosity and a 15.3 Pa yield point.

(b) An upper autoclave cover was affixed to the top of the casing strings using bolts to maintain the integrity of the wellbore system. All necessary valves and sensors were connected. Confining pressure was applied to the C annulus via the confining pressure inlet to replicate the compressive force of formation acting on the casing. With the increasing well depth, the temperature of the geological formations rose continuously [33]. The system was heated to 60 °C with a heating jacket to simulate the geothermal temperatures. At this constant temperature, the cement was cured for 168 h, thereby creating a solid cement sheath [34].

(c) Test gas was pressurized to 2 MPa and introduced at the lower end of the B annulus. Cyclic internal pressures were applied using the casing inner pressure inlet under the instruction of the control system to mimic the multi-stage hydraulic fracturing operation. Each cycle was executed for 10 min, summing up to 15 cycles. In the event of a failure in the cement sheath's integrity during the cyclic loading and unloading phases, test gas wa prone to migrate from the annulus's base to its summit and exit through the gas outlet. Th data acquisition system recorded the gas flow rate at the upper B annulus.

The flow chart of the experimental work methodology is shown in Figure 5:

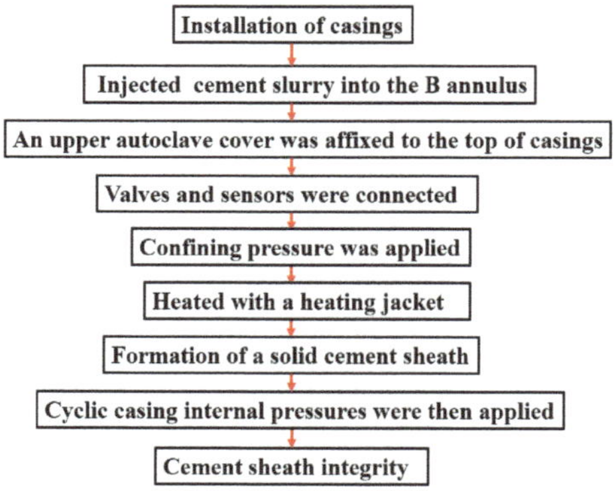

Figure 5. Flow chart of experimental methodology.

3. Numerical Simulation

Full-scale experiments have limitations in accurately simulating non-uniform ge stress and capturing the evolution of micro-annulus widths during fracturing. For thi consideration, a numerical model using ABAQUS 2022 finite element software was deve oped. The developed model helps elucidate the failure mechanism of the cement sheat integrity under multi-stage fracturing.

3.1. Finite Element Model

The simulation focused on the vertical section of Well H (Section A, located at depth of 2000 m), as shown in Figure 6a. A comprehensive 3-D finite element model wa constructed, comprising the production casing, B annulus cement sheath, intermediat casing, C annulus cement sheath, surface casing, and the surrounding formation, consister with the actual dimensions of the wellbore. By employing structured and variable densit meshing techniques, the model measures 5 m × 5 m × 5 m, as demonstrated in Figure 6b,

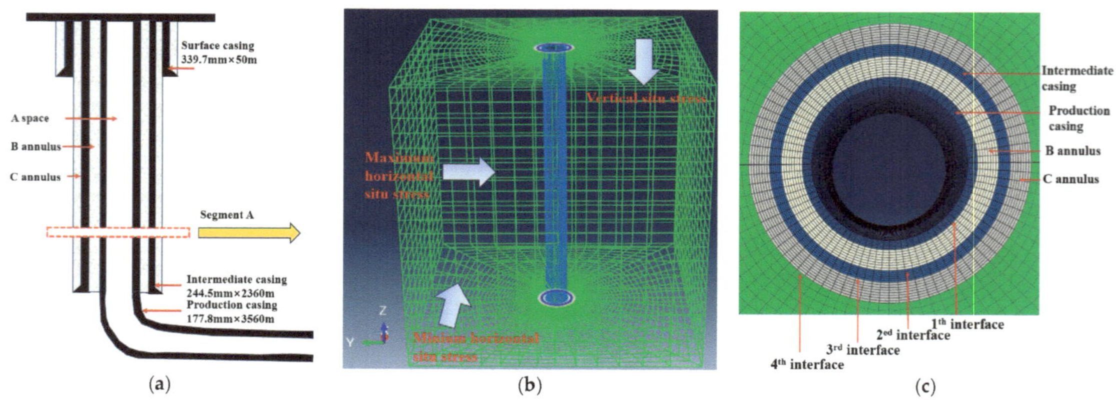

Figure 6. Finite element model based on actual wellbore structure: (**a**) structure diagram of the wel (**b**) finite element model, and (**c**) wellbore structure.

Regarding the simulated geo-stress conditions, the maximum and minimum horizo tal principal stresses, along with vertical stress, were set at 26 MPa, 33 MPa, and 30 MP respectively. The casing was represented through a linear elastic material model. Additionally, the cement sheath and formation were modeled using the Mohr–Coulomb constitutive model, with material specifics outlined in Table 2.

Table 2. Mechanical properties of cement, casing, and formation.

Name	Elastic Modulus (GPa)	Poisson's Ratio	Friction Angle (°)	Cohesive Strength (MPa)
Cement	12	0.23	25	9
Formation	23	0.18	30	6.2
Casing	210	0.28		

To simulate interface debonding, a zero-thickness cohesive element layer was inserted at the 1st, 2nd, 3rd, and 4th interfaces. The mechanical properties of these cohesive elements are detailed in Table 3. The maximum nominal stress damage criterion was used to judge the damage of the cohesive elements in this study, that is:

$$Max\left\{\frac{T_a}{T_a^0}, \frac{T_b}{T_b^0}, \frac{T_c}{T_c^0}\right\} = 1 \quad (1)$$

where T_a^0, T_b^0, and T_c^0 are the cohesive strengths in the normal direction, the first shear direction, and the second shear direction, respectively; T_a, T_b, and T_c are the interface stresses in the normal direction, the first shear direction, and the second shear direction, respectively.

Table 3. Mechanical properties of the cohesive elements.

Name	Normal Stiffness (GPa)	Shear Stiffness (GPa)	Normal Cohesive Strength (MPa)	Shear Cohesive Strength (MPa)	Critical Energy (J/m^2)
The 1st, 2nd, and 3rd interface	0.8	2.2	1.2	2.3	100
The 4th interface	1.5	4.5	2.6	5.6	210

The BK fracture criterion was used in this study as the criterion for the damage evolution, that is:

$$\left\{\begin{array}{l} G^c = G_a^c + G_b^c + G_c^c \\ G_a^c + (G_s^c - G_a^c)\left\{\frac{G_b+G_c}{G_a+G_b+G_c}\right\}^{\beta} = G^c \end{array}\right\} \quad (2)$$

where G_a^c, G_b^c, and G_c^c are the critical energies needed for damage in the normal direction, the first shear direction, and the second shear direction, respectively; G_a, G_b, and G_c are the dissipation energies in the normal direction, the first shear direction, and the second shear direction, respectively; β is an empirical index.

3.2. Boundary Conditions and Simulation Steps

Initially, the finite element model was subjected to far-field geo-stresses to achieve a baseline equilibrium state. Subsequently, cyclic internal pressures were applied to the inner wall of the production casing to capture the actual construction parameters and replicate the fracturing experience. The number of circles was 15, with each involving pressurization and depressurization phases. Moreover, in line with the actual fracturing data from the A oil field, the internal pressure thresholds were set at upper limits of 70 MPa, 80 MPa, and 90 MPa and a lower limit of 20 MPa. The flow chart of the simulation work methodology is illustrated in Figure 7.

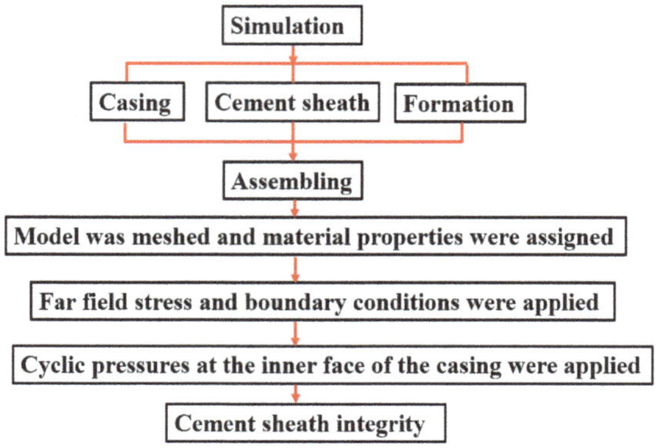

Figure 7. Flow chart of simulation work methodology.

3.3. Mesh Sensitive Analysis

A mesh sensitivity analysis on the model is typically necessary to mitigate the effects of mesh size on computational results [35]. In this process, mesh refinement and the subsequent coarsening around the wellbore when moving away from the well shaft reduce the total number of elements in the model, elevating computational efficiency. We adjusted mesh sizes near the wellbore to obtain different numbers of meshes and varying mesh divisions. On this basis, three configurations were detailed (refer to Table 4): coarse, medium, and fine. The coarse configuration comprises 16,371 elements, the medium configuration contains three times those of the coarse (49,152), and the fine configuration has seven times those of the coarse, totaling 114,600.

Table 4. Comparison of simulation results under different mesh sizes.

Configuration	Mesh Size of the Wellbore (mm)	Total Number of Meshes	Micro-Annuli Width (um)
Coarse	8	16,371	130.33
Medium	4	49,152	130.12
Fine	2	114,600	130

Numerical simulations were performed on the three mesh models under multi-stage hydraulic fracturing conditions at a maximum load of 70 MPa with 15 circles. The micro-annuli widths for the coarse, medium, and fine meshes at the 1st interface (Table 4) were 130.33 μm, 130.12 μm, and 130 μm, respectively. The relative differences between the coarse and medium meshes and between the medium and fine meshes were 0.16% and 0.09%, respectively, indicating the stability of calculations with the fine mesh. Moreover, the grid convergence index (GCI) was calculated [36]. The GCI values of the coarse and medium mesh groups (GCI_{21}) and the medium and fine mesh groups (GCI_{32}) accounted for 1.35% and 0.91%, respectively. These findings indicate that using a fine mesh reduces the computational results' dependence on mesh size, thereby confirming the accuracy and reliability of the numerical simulations with the fine mesh. Based on these analyses, we adopted this mesh without additional refinement, and all subsequent calculations will be derived from this fine mesh model.

3.4. Model Validation

Prior to conducting cement sheath integrity experiments, preliminary tests were conducted to evaluate the stability and safety of the experimental system. We chose 60 MPa as the upper limit for internal cyclic pressure across 16 cycles. As illustrated in Figure 8a, the test revealed a gas leak at 125 mL/min during the 15th cycle. The post-experiment appearance of the cement sheath indicated the micro-annuli generation at the 1st interface, as shown in Figure 8b.

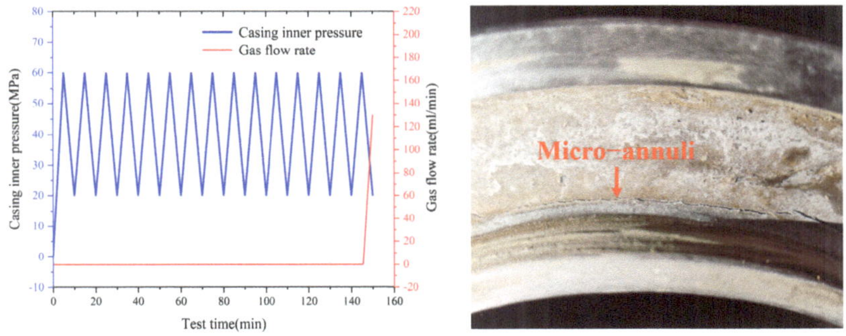

(a) (b)

Figure 8. Results from the cement sheath integrity experiments: (**a**) experiment results and (**b**) morphology of cement sheath.

Using the established finite element model, we performed a numerical simulation of cement sheath integrity with an internal cyclic pressure limit of 60 MPa for 16 cycles. The simulation results showed that debonding occurred at the 1st interface during the 15th cycle, giving rise to micro-annuli formation. As detailed in Figure 9a, the morphology of the micro-annuli was measured at 82 μm in size. The casing was sectioned, and the micro-annulus was examined with a size of 88 μm using SEM, as shown in Figure 9b.

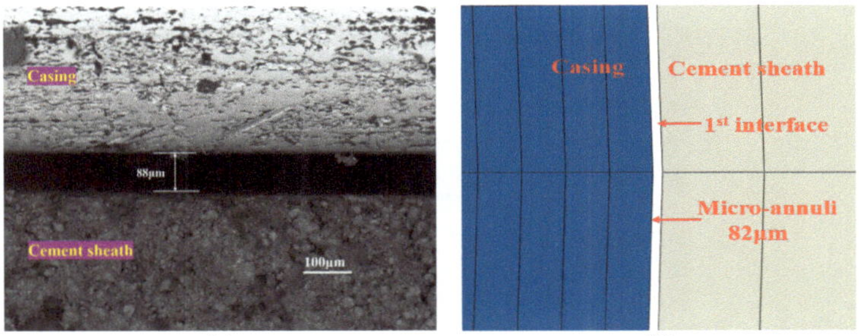

(a) (b)

Figure 9. Microscopic morphology of micro-annuli: (**a**) experiment results and (**b**) simulation outcomes.

In numerical simulations and experiments, micro-annuli appeared at the 1st interface after loaded at the 15th cycle, and their widths were remarkably consistent. This consistency between the experimental results and numerical simulations confirms the effectiveness and reliability of the methods and models employed in this study.

4. Results and Discussion

4.1. Experimental Results

4.1.1. TCT, TCLT, and Permeability Test Results

The post-mechanical testing morphology of the cement samples is illustrated in Figure 10a, and the stress–strain curves before and after the TCT and TCLT are depicted in Figure 10b. The triaxial compressive strength, elastic modulus, and permeability before and after TCLT are detailed in Table 5. Initially, the cement's triaxial compressive strength and elastic modulus were 55 MPa and 7.2 GPa, respectively. The upper limit of cyclic loading was set at 44 MPa. After TCLT, the compressive strength and elastic modulus decreased by 10.7% and 8.3%, respectively, to 49.1 MPa and 6.6 GPa.

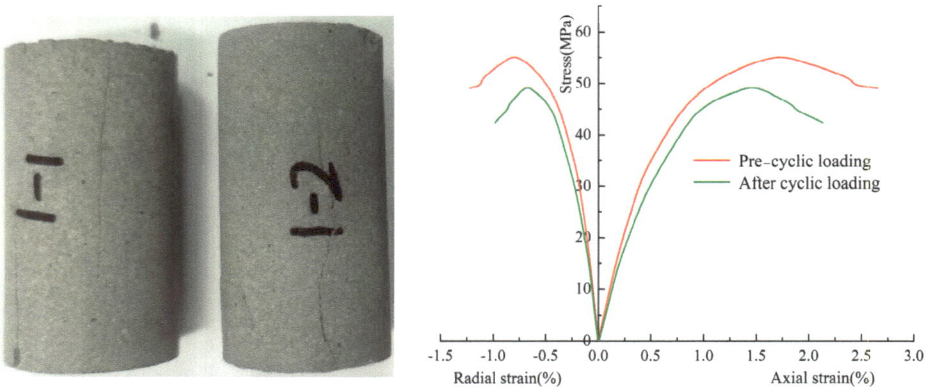

(**a**) (**b**)

Figure 10. Results of mechanical testing on the cement samples: (**a**) morphology of the cemer samples after testing and (**b**) stress–strain curves.

Table 5. Data on cement samples before and after TCLT.

Type	Triaxial Compressive Strength (MPa)	Elastic Modulus (GPa)	Permeability (mD)
Before TCLT	55	7.2	0.022
After TCLT	49.1	6.6	0.053

These changes imply microstructural damage in the cement, including elevated micr crack and microporosity densities, which reduced the matrix's load-bearing capacity, a evidenced by the decreased compressive strength and elastic modulus [37]. Nonetheles these reductions were modest, suggesting that the cement retained its significant resilienc to cyclic loading and its mechanical integrity under downhole fracturing conditions [38]

Following the TCLT, the permeability increased from 0.022 mD to 0.053 mD, reinforcin the observation that cyclic loading facilitated micropore and microcrack developmen thereby enhancing permeability [39]. However, gas leakage through the cement matrix i only possible under permeabilities above 0.1 mD [40]. Consequently, the cement matri maintained its seal integrity even after 15 TCLT cycles, suggesting unlikely bottom-hol gas leakage through the cement matrix, leading to SCP.

4.1.2. Stress–Strain Evolution in Cement under Cyclic Loading

Figure 11a illustrates the stress–strain curve of TCLT. Initially, the curve ascende near-linearly during the loading phase until reaching the predetermined upper-stres limit, triggering the commencement of unloading. The unloading phase featured a linea decrease until the stress diminished to the lower limit, at which point the cycle of reloadin began. Each loading and unloading cycle contributed to the generation of new plasti strain, cumulatively increasing the total plastic strain. Notably, the loading and unloadin curves collectively formed a hysteresis loop, which progressively shifted rightward wit each cycle, indicating a gradual increase in accumulated plastic strain.

The dynamics of accumulated plastic strain throughout the TCLT are further detaile in Figure 11b. From the outset, the irreversible plastic strain (denoted as initial plasti strain) manifested in the 1st cycle at 0.5%. With the progression of cyclic loading an unloading, the magnitude of plastic strain grew steadily to 0.85% by the 15th cycle.

The above analysis reveals that the TCLT induces microstructural damage withi the cement, resulting in irreversible plastic deformation. Moreover, as the TCLT cycle increased, the extent of microstructural damage also intensified, increasing the accumulate plastic strain.

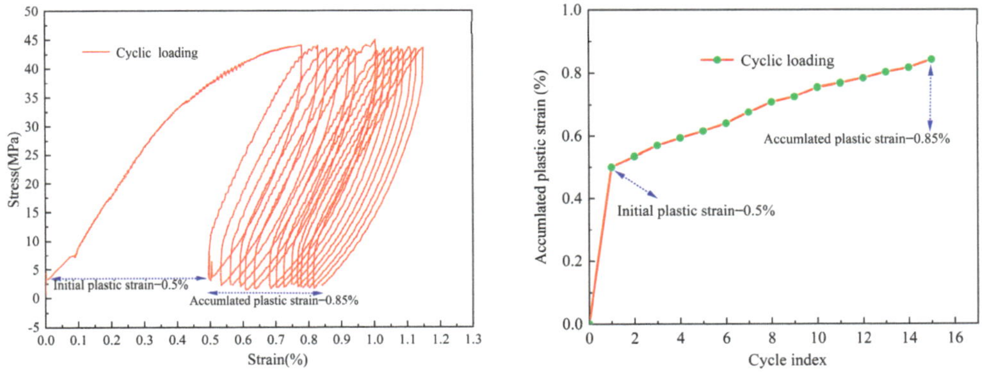

(**a**) (**b**)

Figure 11. Results of cyclic loading: (**a**) cyclic loading curves and (**b**) cumulative plastic strain.

4.1.3. Cement Sheath Integrity Experiment Results

Figure 12 presents the results from three experimental groups, each showing changes in internal casing pressure and the monitoring gas flow rate at the top of the cement sheath over time. Gas leakage was observed at the 14th, 10th, and 8th cycles under upper limit pressures of 70 MPa, 80 MPa, and 90 MPa, respectively, solely during the depressurization phases. The initial leakage rates of the three groups were 151 mL/min, 160 mL/min, and 183 mL/min, respectively. As the cycles increased, the gas leakage rates rose progressively to 160 mL/min, 210 mL/min, and 235 mL/min, respectively, by the 15th cycle. These trends indicated that higher upper limit pressures in cyclic loading and unloading reduced the required cycles for cement sheath integrity failure, with the initial and final cycles resulting in increased gas leakage rates.

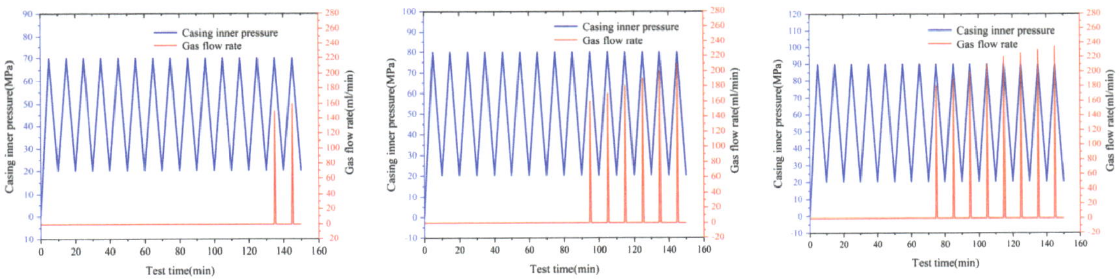

(**a**) (**b**) (**c**)

Figure 12. Results of the cement sheath integrity experiment: (**a**) 70 MPa, (**b**) 80 MPa, and (**c**) 90 MPa.

Figure 13 illustrates the morphology of cement sheaths following the three cyclic loading experiments. While no cracking was observed in any of the cement sheaths across the three experimental groups, distinct micro-annuli appeared at the 1st interface, in contrast to the well-bonded 2nd interface. Thus, cyclic loading induced cumulative plastic strain in the inner walls of the cement sheath at the 1st interface, which progressively increased with each cycle. In the depressurization phase, the inner walls could not return to their original shape, generating tensile stress at the 1st interface. Once surpassing the interface's bonding strength, this stress led to debonding and micro-annuli, serving as gas leakage conduits and causing SCP in the B annulus. Conversely, the 2nd interface showed no signs of debonding, likely due to lower stress levels that produced insufficient (minimal or none) cumulative plastic strain to initiate debonding. Gas leakage occurred solely during depressurization because the micro-annulus formation allowed the compressive forces from the outer wall of the inner casing to close the micro-annuli. Consequently, gas from th bottom of the cement sheath could not breach the 1st interface. However, the micro-annul reopened during depressurization, forming gas leakage pathways.

(**a**) (**b**) (**c**)

Figure 13. Morphology of cement sheaths: (**a**) 70 MPa, (**b**) 80 MPa, and (**c**) 90 MPa.

The casing and cement sheath with micro-annuli were meticulously sectioned, an their morphology was analyzed via SEM (Figure 14). Precise width measurements o the micro-annuli indicated that after 15 loading cycles under upper limit pressures o 70 MPa, 80 MPa, and 90 MPa, the widths reached 117 μm, 178 μm, and 212 μm, respectivel These results demonstrate that the widths of the micro-annuli increase with the uppe limit pressure during cyclic loading. Thus, higher upper limit pressures induce mor significant cumulative plastic strain on the cement sheath's inner walls, resulting in broade micro-annuli and, consequently, higher gas leakage rates.

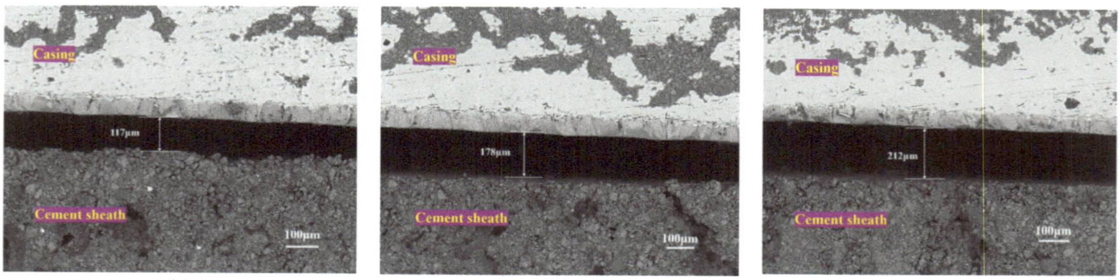

(**a**) (**b**) (**c**)

Figure 14. SEM results of the micro-annulus morphology: (**a**) 70 MPa, (**b**) 80 MPa, and (**c**) 90 MPa.

4.2. Numerical Simulation Results

4.2.1. Equivalent Plastic Strain Evolution in Cement Sheaths

The Plastic Equivalent Strain (PEEQ) quantifies the accumulated plastic strain following cyclic loading. PEEQ = 0 signifies no accumulated plastic strain, whereas PEEQ > (confirms its occurrence, with a larger PEEQ indicating a more significant accumulatior Figure 15 depicts the PEEQ distribution in the cement sheaths of the B and C annuli following 15 loading cycles. A marked plastic strain accumulation is observed in the cemen sheath of the B annulus, whereas the C annulus exhibits no such strain. Despite unequa horizontal principal stresses, the C annulus's cement sheath and the intermediate casing mitigate horizontal geo-stresses, whereas the primary source of accumulated plastic defor mation in the cement sheath is the casing's internal pressure. As a result, PEEQ shows a relatively uniform circumferential distribution. Notably, the highest PEEQ values are a the 1st interface, diminishing progressively toward the 2nd interface, with the 3rd and 4th interfaces exhibiting zero PEEQ values.

Figure 16 illustrates the PEEQ variations at the 1st, 2nd, 3rd, and 4th interfaces o the cement sheath under cyclic loading, with upper limit pressures of 70 MPa, 80 MPa and 90 MPa. The initial PEEQ values at the 1st interface were 0.31%, 0.53%, and 0.66%

respectively, for each pressure setting during the 1st cycle. As the cycles increased, the PEEQ values rose almost linearly to 0.79%, 0.95%, and 1.13% by the 15th cycle. At the 2nd interface, the initial cycle's PEEQ values were 0.11%, 0.18%, and 0.23%, which increased to 0.26%, 0.32%, and 0.47% by the 15th cycle. These results demonstrate that the largest PEEQ increases occurred during the 1st cycle for both the 1st and 2nd interfaces, with the 2nd interface's PEEQ values significantly lower than those at the 1st, indicating lesser cumulative plastic strain. PEEQ values at the 3rd and 4th interfaces remained at zero throughout the whole process.

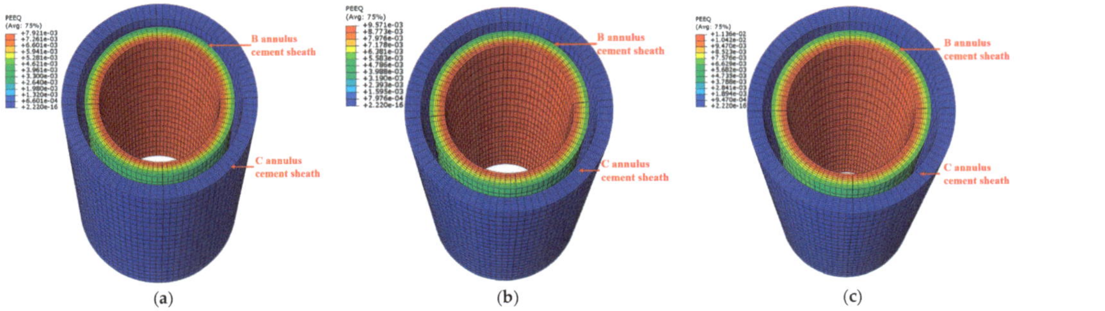

Figure 15. PEEQ distribution in the cement sheaths of the B and C annulus: (**a**) 70 MPa, (**b**) 80 MPa, and (**c**) 90 MPa.

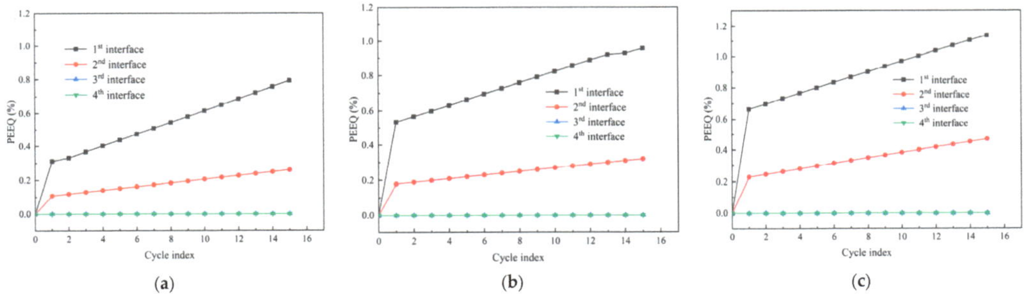

Figure 16. PEEQ evolution: (**a**) 70 MPa, (**b**) 80 MPa, and (**c**) 90 MPa.

4.2.2. Micro-Annulus Evolution at Interfaces

After 15 internal pressure cycles, the interface bonding states are illustrated in Figure 17. Micro-annuli were not observed at the 2nd, 3rd, and 4th interfaces but at the 1st interface.

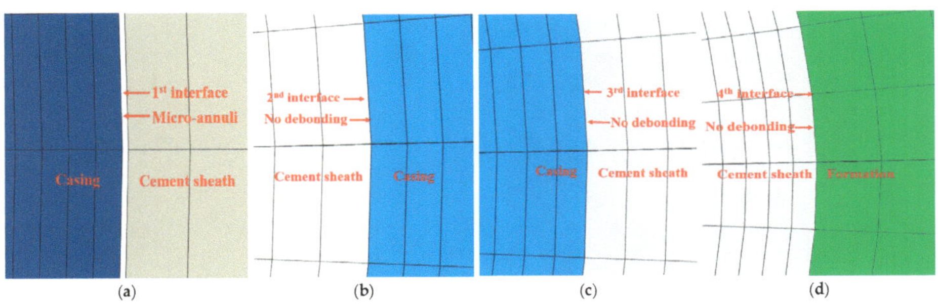

Figure 17. Debonding state at the interfaces: (**a**) the 1st interface, (**b**) the 2nd interface, (**c**) the 3rd interface, and (**d**) the 4th interface.

Post-processing of numerical simulation results yielded the circumferential widths of the micro-annuli at the 1st interface (Figure 18). Micro-annuli first appeared at the 8th, 10th, and 14th cycles under upper limit pressures of 70 MPa, 80 MPa, and 90 MPa, respectivel with widths nearly uniformly distributed around the circumference. The initial widths c these micro-annuli were 83 μm, 95 μm, and 103 μm, respectively. As the cycles increase the widths of the micro-annuli gradually expanded to 130 μm, 165 μm, and 205 μm by th 15th cycle, indicating that the largest width increase occurred during the initial appearanc

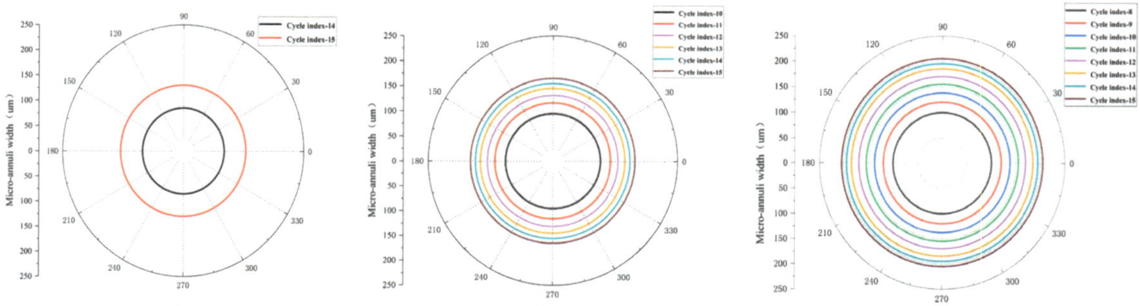

(a) (b) (c)

Figure 18. Evolution of micro-annulus width at the 1st interface: (**a**) 70 MPa, (**b**) 80 MPa, and (**c**) 90 MP

4.3. Correlation between Numerical Simulation and Experimental Results

The results of the three sets of experiments consistently demonstrated the presence c micro-annuli at the 1st interface. This finding was supported by the three sets of numerica simulations that also identified micro-annuli at the same location, confirming alignmen between the numerically simulated and experimentally observed micro-annuli positions

Experimental observations recorded gas leakage during the 8th, 10th, and 14th cycli loading under upper pressure limits of 70 MPa, 80 MPa, and 90 MPa, respectively. Corre sponding numerical simulations displayed micro-annuli at the 1st interface during identica cycles, corroborating the experimental findings. The agreement between the experimen tal cycles with gas leakage and the simulated cycles with micro-annuli substantiates th simulations' accuracy.

After 15 loading cycles, the experimentally measured widths of the micro-annuli wer 117 μm, 178 μm, and 212 μm, respectively, closely mirrored by the simulation results o 130 μm, 165 μm, and 205 μm. This high degree of concordance underscores the consistenc stability, and reliability of this research.

4.4. Sensitivity Analysis

The mechanical properties of cement materials profoundly affect the stress level withi the cement sheath. Choosing a cement formulation with suitable mechanical parameters ca significantly reduce plastic deformation. This study performed a sensitivity analysis on fou key mechanical properties, elastic modulus, Poisson's ratio, cohesive force, and friction angl to guide future cement formulation designs. This analysis utilized a previously establishe finite element model, setting the casing's cyclic upper pressure at 70 MPa over 15 cycles.

Figure 19 depicts the PEEQ curves of cement sheaths under varying elastic modul where the PEEQ values increase progressively with the cycles, and both initial and fina PEEQ values escalate as the elastic modulus increases. These observations suggest tha cement with a lower elastic modulus can effectively reduce the risk of integrity failure i cement sheaths during multi-stage fracturing.

Figure 20 depicts the PEEQ curves under varying Poisson's ratios. Under the Pois son's ratios of 0.13, 0.18, and 0.23, the initial PEEQ values are 0.42%, 0.312%, and 0.09% respectively, while the final PEEQ values are 0.51%, 0.79%, and 0.84%, respectively. Thes results demonstrate that cement with a higher Poisson's ratio helps reduce the stress leve of the cement sheath during multi-stage fracturing.

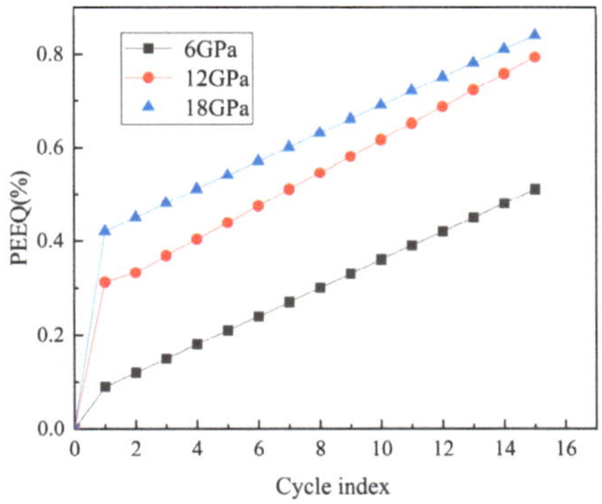

Figure 19. PEEQ values under different elastic moduli.

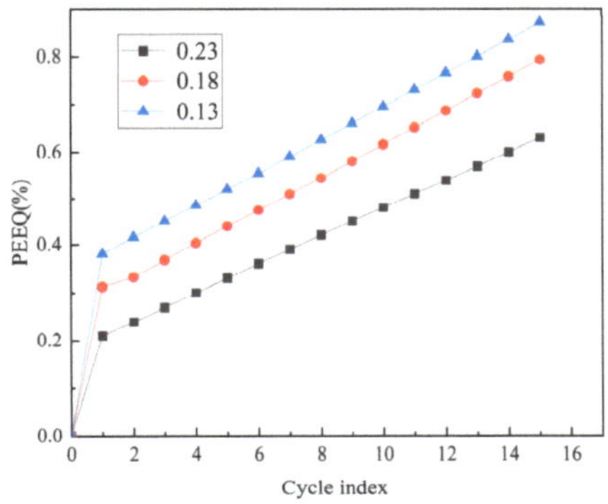

Figure 20. PEEQ values under different Poisson's ratios.

Figure 21 illustrates PEEQ curves under varying friction angles. While both initial and final PEEQ values increase with the friction angle, the variations across different angles are not particularly significant. Nevertheless, a higher friction angle still benefits cement sheath integrity.

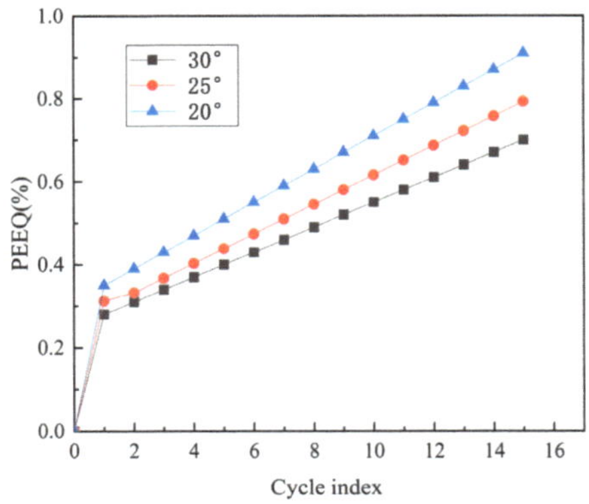

Figure 21. PEEQ values for different friction angles.

Figure 22 depicts the PEEQ curves under various cohesive strengths. The data indicate that at cohesive strengths of 5 MPa, 9 MPa, and 13 MPa, the initial PEEQ values are 0.14, 0.312, and 0.42, respectively, while the final values are 0.56%, 0.79%, and 0.84%. Thus, cement with a higher cohesive strength is advantageous for maintaining the integrity of the cement sheath during multi-stage fracturing.

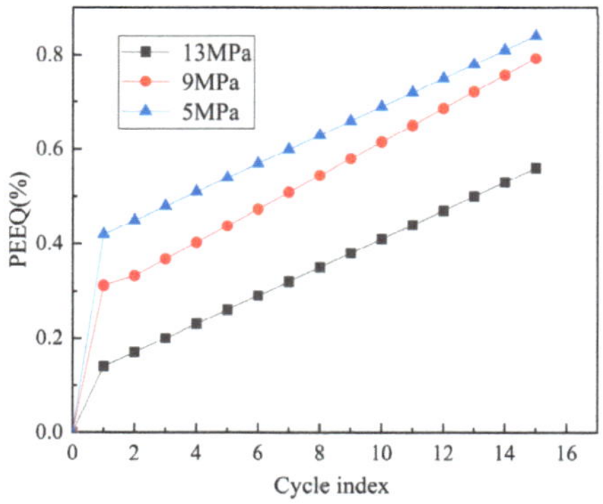

Figure 22. PEEQ values under different cohesive strengths.

4.5. Suggestions for Future Work

Experimental and numerical studies on cement sheath integrity revealed that durin multi-stage fracturing, elevated cyclic pressure and increased cycling promoted micro annulus formation, resulting in SCP. Therefore, it is essential to design the most appropriat multi-stage fracturing pressures and cycles based on the specific conditions of the forma tion and wellbore. Additionally, numerical simulations showed that cement mechanica properties significantly influence the plastic deformation extent of the cement sheatl When selecting cement materials, it is crucial to consider characteristics such as low elasti modulus, high Poisson's ratio, high cohesion, and high friction angle. Optimizing enginee ing parameters and choosing cement materials with superior properties are essential fo enhancing wellbore integrity.

5. Conclusions

This research used the actual parameters from the A oil field in experiments an numerical simulations to investigate the integrity failures of cement sheaths during mult stage fracturing, which uncovered the mechanisms behind these failures. The key finding are as follows:

Cyclic loading and unloading experiments on cement samples demonstrated cumu lative plastic deformation in the cement, with triaxial compressive strength and elasti modulus decreasing by 10.7% and 8.3%, respectively, and permeability increasing by 150% These results confirmed that cyclic loading caused micro-damage in cement. Howeve after 15 cycles, the permeability of the cement remained below the threshold for gas flow indicating that gas would not leak through the cement sheath matrix underground.

Experiments with a full-scale device on cement sheath integrity under multi-stag fracturing confirmed that integrity failure was primarily due to micro-annulus formation a the 1st interface, attributed to cumulative plastic strain in the inner walls of the B annulu cement sheath during cyclic loading, leading to debonding at the 1st interface. While th outer wall of the B annulus cement sheath also experienced plastic strain, it was insufficien to cause debonding at the 2nd interface. No plastic strain was observed in the cemen sheath of the C annulus, and no debonding occurred at the 3rd and 4th interfaces.

Both full-scale experiments and numerical simulations established that higher cycli pressure limits resulted in fewer required cycles for cement sheath integrity failure, wide micro-annuli formation, and increased gas leakage rates through the interfaces.

The number of cycles for cement sheath integrity failure in the experimental group matched perfectly with the cycles at which micro-annuli first appeared at the 1st interface i the numerical simulations. The measured widths of micro-annuli at the 1st interface wer very close to those determined through numerical simulations, showing strong consistenc between experimental and numerical results. These results validated the reliability o both approaches and provided valuable insights into the mechanisms of SCP generatior in offshore A oil field development wells, offering guidance for the design of offshore multi-stage fracturing parameters.

Author Contributions: Y.W.: conceptualization, methodology, software, validation, formal analysis, investigation, resources, and data curation. Y.F.: writing—original draft preparation, writing—review and editing, visualization, and supervision. All authors have read and agreed to the published version of the manuscript.

Funding: This study was supported by the National Natural Science Foundation of China (Grant No. 52004298).

Institutional Review Board Statement: Not applicable.

Informed Consent Statement: Not applicable.

Data Availability Statement: The data presented in this study are available on request from the corresponding author.

Acknowledgments: We extend our heartfelt gratitude to Xiaorong Li for her invaluable assistance in management and technical matters, as well as for the support from the funding projects she has secured.

Conflicts of Interest: The authors declare no conflicts of interest.

References

1. Qiu, K.; Fan, K.; Chen, X.; Lei, G.; Wei, S.; Navik, R.; Li, J. A New Approach for Production Prediction in Onshore and Offshore Tight Oil Reservoir. *J. Mar. Sci. Eng.* **2023**, *11*, 2079. [CrossRef]
2. Meng, M.; Ge, H.; Shen, Y.; Ji, W.; Wang, Q. Rock fabric of tight sandstone and its influence on irreducible water saturation in Eastern Ordos Basin. *Energy Fuels* **2023**, *37*, 3685–3696. [CrossRef]
3. Meng, M.; Zhang, Y.; Yuan, B.; Li, Z.; Zhang, Y. Imbibition Behavior of Oil-Saturated Rock: Implications for Enhanced Oil Recovery in Unconventional Reservoirs. *Energy Fuels* **2023**, *37*, 13759–13768. [CrossRef]
4. Meng, M.; Ge, H.; Shen, Y.; Ji, W.; Li, Z. Insight into water occurrence and pore size distribution by nuclear magnetic resonance in marine shale reservoirs, southern China. *Energy Fuels* **2022**, *37*, 319–327. [CrossRef]
5. Friesen, O.J.; Dashtgard, S.E.; Miller, J.; Schmitt, L.; Baldwin, C. Permeability heterogeneity in bioturbated sediments and im-plications for waterflooding of tight-oil reservoirs, Cardium Formation, Pembina Field, Alberta, Canada. *Mar. Petro-Leum Geol.* **2017**, *82*, 371–387. [CrossRef]
6. Wu, B.; Wu, G.; Wang, L.; Lou, Y.; Liu, S.; Yin, B.; Li, S. Study on fracturing parameters optimization of horizontal wells in low-permeability reservoirs in South China Sea. *Processes* **2023**, *11*, 2999. [CrossRef]
7. Du, F.; Huang, J.; Ru, X.; Ga, Y.; Yu, Z. Status and prospect of offshore horizontal well staged fracturing technology. *Offshore Oil* **2021**, *41*, 22–26. [CrossRef]
8. Liu, K.; Gao, D.; Taleghani, A.D. Analysis on integrity of cement sheath in the vertical section of wells during hydraulic fracturing. *J. Pet. Sci. Eng.* **2018**, *168*, 370–379. [CrossRef]
9. Chu, W.; Shen, J.; Yang, Y.; Li, Y.; Gao, D. Calculation of micro-annulus size in casing-cement sheath-formation system under continuous internal casing pressure change. *Pet. Explor. Dev.* **2015**, *42*, 414–421. [CrossRef]
10. Arshad, W.; Khaqan, K. Understanding the Key Factors Affecting Well Integrity in Horizontal Well Multistage Hydraulic Frac-turing. In Proceedings of the SPE Middle East Oil, Gas and Geosciences Show and Conference, Manama, Bahrain, 19–21 February 2023. [CrossRef]
11. Zeng, Y.; Liu, R.; Li, X.; Zhou, S.; Tao, Q.; Lu, P. Cement sheath sealing integrity evaluation under cyclic loading using large-scale sealing evaluation equipment for complex subsurface settings. *J. Pet. Sci. Eng.* **2019**, *176*, 811–820. [CrossRef]
12. Jiang, H.; Ren, Z.; Xi, Y.; Liu, G.; Li, J. Analysis of dynamic thermal behaviors for multi-stage hydraulic fracturing treatments in horizontal shale oil and shale gas wells. *Appl. Therm. Eng.* **2024**, *240*, 122213. [CrossRef]
13. Arjomand, E.; Bennett, T.; Nguyen, G.D. Evaluation of cement sheath integrity subject to enhanced pressure. *J. Pet. Sci. Eng.* **2018**, *170*, 1–13. [CrossRef]
14. Su, D.; Li, Z.; Huang, S.; Wu, X.; Li, J.; Xue, Y. Experiment and failure mechanism of cement sheath integrity under development and production conditions based on a mechanical equivalent theory. *Energy Sci. Eng.* **2021**, *9*, 2400–2422. [CrossRef]
15. Zhang, W.; Eckert, A. Micro-annulus generation under downhole conditions: Insights from three-dimensional staged finite element analysis of cement hardening and wellbore operations. *J. Rock Mech. Geotech. Eng.* **2020**, *12*, 1185–1200. [CrossRef]
16. Goodwin, K.J.; Crook, R.J. Cement sheath stress failure. *SPE Drill. Eng.* **1992**, *7*, 291–296. [CrossRef]
17. Deng, K.; Yuan, Y.; Hao, Y.; Li, Z.; Lin, Y. Experimental study on the integrity of casing-cement sheath in shale gas wells under pressure and temperature cycle loading. *J. Pet. Sci. Eng.* **2020**, *195*, 107548. [CrossRef]

18. Su, D.; Li, Z.; Wu, X.; Li, J.; Sun, J.; Zheng, G. Cement Sheath Integrity Evaluation Under Multiple Cyclic Loading Usin Mechanical Equivalent Experiment for Gas Storage Wells in Eastern China. In Proceedings of the International Conference o Offshore Mechanics and Arctic Engineering, Hamburg, Germany, 5–10 June 2022. [CrossRef]
19. Feng, Y.; Li, X.; Gray, K.E. Development of a 3D numerical model for quantifying fluid-driven interface debonding of an injecte well. *Int. J. Greenh. Gas Control* **2017**, *62*, 76–90. [CrossRef]
20. Wang, W.; Taleghani, A.D. Impact of hydraulic fracturing on cement sheath integrity; A modelling approach. *J. Nat. Gas Sci. En* **2017**, *44*, 265–277. [CrossRef]
21. Li, J.; Xi, Y.; Tao, Q.; Li, Y.; Qu, G. Experimental investigation and numerical simulation of the emergence and development o micro-annulus in shale gas wells subjected to multistage fracturing. *J. Nat. Gas Sci. Eng.* **2020**, *78*, 103314. [CrossRef]
22. Dahi Taleghani, A.; Li, G.; Moayeri, M. Smart expandable cement additive to achieve better wellbore integrity. *J. Energy Resou Technol.* **2017**, *139*, 062903. [CrossRef]
23. Pereira, F.L.G.; De Simone, M.; Roehl, D. Wellbore integrity assessment considering casing-cement-formation interaction based o a probabilistic approach. In Proceedings of the 51st U.S. Rock Mechanics/Geomechanics Symposium, San Francisco, CA, US 25–28 June 2017.
24. Landry, G.; Welty, R.D.; Thomas, M.; Vaughan, M.L.; Tatum, D. Bridging the gap: An integrated approach to solving sustained casin pressure in the Cana Woodford shale. In Proceedings of the SPE Well Integrity Symposium, Galveston, TX, USA, 2–3 June 201 [CrossRef]
25. Wu, X.; Li, Z.; Hou, Z.; Liu, J.; Huang, S.; Su, D.; Li, J.; Cao, C.; Wu, L.; Song, W. Analytical Perspectives on Cement Sheat Integrity: A Comprehensive Review of Theoretical Research. *ACS Omega* **2024**, *9*, 17741–17759. [CrossRef] [PubMed]
26. Rocha-Valadez, T.; Mentzer, R.A.; Hasan, A.R.; Mannan, M.S. Inherently safer sustained casing pressure testing for well integrit evaluation. *J. Loss Prev. Process Ind.* **2014**, *29*, 209–215. [CrossRef]
27. Al Ramadan, M.; Salehi, S.; Kwatia, G.; Ezeakacha, C.; Teodoriu, C. Experimental investigation of well integrity: Annular ga migration in cement column. *J. Pet. Sci. Eng.* **2019**, *179*, 126–135. [CrossRef]
28. Wolterbeek, T.K.; Cornelissen, E.K.; Nolan, S.; Todea, F.; Stam, W.; Roggeband, S.M.; Keultjes, W.J.G. Restoration of annular zon isolation using localized casing expansion (LCE) technology: A proof of concept based on laboratory studies and field trial result *J. Pet. Sci. Eng.* **2021**, *197*, 108103. [CrossRef]
29. Beltrán-Jiménez, K.; Skadsem, H.J.; Sunde, J.K.; Gardner, D.; Wolterbeek, T.K.; Cornelissen, E.K.; Keultjes, W.J. Restoration c annular zonal isolation using localized casing expansion (LCE) technology: Treatment of near-horizontal test sections containin a free-water channel. *J. Pet. Sci. Eng.* **2022**, *208*, 109792. [CrossRef]
30. Congro, M.; Skadsem, H.J.; Beltrán-Jiménez, K.; Roehl, D. Experimental and numerical study on the pushout shear strength c conventional and expanding cement–casing sections for well integrity. *Geoenergy Sci. Eng.* **2024**, *234*, 212638. [CrossRef]
31. Tabatabaei, M.; Santos, L.; Al Hassan, A.A.; Dahi Taleghani, A. Surface-modified graphite nanoplatelets to limit deteriorativ impacts of oil-based mud residuals on cement bonding. *SPE Drill. Complet.* **2023**, *38*, 235–242. [CrossRef]
32. Corina, A.N.; Opedal, N.; Vrålstad, T.; Skorpa, R.; Sangesland, S. The effect of casing-pipe roughness on cement-plug integrit *SPE Drill. Complet.* **2020**, *35*, 237–251. [CrossRef]
33. Khurshid, I.; Lee, K.J.; Choe, J. Analyses of thermal disturbance in drilling deep and high temperature formations. *Energy Source Part A Recovery Util. Environ. Eff.* **2013**, *35*, 1487–1497. [CrossRef]
34. Xi, Y.; Li, J.; Tao, Q.; Guo, B.; Liu, G. Experimental and numerical investigations of accumulated plastic deformation in cemer sheath during multistage fracturing in shale gas wells. *J. Pet. Sci. Eng.* **2020**, *187*, 106790. [CrossRef]
35. Ali, I.T.; Afgan, I.; Khurshid, I. Stratified Two-Phase Turbulent Pipe Flow Simulations. *Int. J. Adv. Sci. Eng. Inf. Technol.* **202** *12*, 1301–1311. [CrossRef]
36. Aycan, O.; Topuz, A.; Kadem, L. Evaluating uncertainties in CFD simulations of patient-specific aorta models using Gri Convergence Index method. *Mech. Res. Commun.* **2023**, *133*, 104188. [CrossRef]
37. Zhou, S.; Liu, R.; Zeng, H.; Zeng, Y.; Zhang, L.; Zhang, J.; Li, X. Mechanical characteristics of well cement under cyclic loadin and its influence on the integrity of shale gas wellbores. *Fuel* **2019**, *250*, 132–143. [CrossRef]
38. Murphy, B.P.; Prendergast, P.J. Measurement of non-linear microcrack accumulation rates in polymethylmethacrylate bone cemen under cyclic loading. *J. Mater. Sci. Mater. Med.* **1999**, *10*, 779–781. [CrossRef] [PubMed]
39. Rao, F.; Zhang, Z.; Ye, G.; Liu, J. Mechanical behavior and assessment of foamed cement paste under staged cyclic loading. *Mate Struct.* **2021**, *54*, 182. [CrossRef]
40. Thorpe, A.K.; Duren, R.M.; Conley, S.; Prasad, K.R.; Bue, B.D.; Yadav, V.; Miller, C.E. Methane emissions from underground ga storage in California. *Environ. Res. Lett.* **2020**, *15*, 045005. [CrossRef]

Journal of
Marine Science
and Engineering

MDPI

rticle

\ Novel Method for Analyzing Sandbar Distribution in ;helf-Type Tidal Deltas Using Sediment Dynamic Simulation

/ingming Tang [1,2,*], Sichen Xiong [1], Qian Zhang [3], Ruifeng Hong [1], Chenyang Peng [1] and Rong Xie [1]

1 School of Geosciences, China University of Petroleum (East China), Qingdao 266580, China; z21010029@s.upc.edu.cn (S.X.); z22010006@s.upc.edu.cn (R.H.); z22010094@s.upc.edu.cn (C.P.); z22010093@s.upc.edu.cn (R.X.)
2 Key Laboratory of Deep Oil and Gas, China University of Petroleum (East China), Qingdao 266580, China
3 Hainan Branch of China National Offshore Oil Company Ltd., Haikou 570311, China; zhangqian60@cnooc.com.cn
* Correspondence: tangmingming126@126.com

Abstract: Shallow marine shelf sedimentation is a hot and difficult topic in today's reservoir sedimentology research, and it is widely present in the world. The shallow marine shelf sedimentation is not only affected by complex hydrodynamic effects such as tides and waves, but also controlled by bottom tectonic features, forming a complex and varied sedimentation pattern. During the Middle Jurassic period, the northern part of West Siberian Basin was characterized by a shallow marine shelf sedimentary environment. In the central reion of this basin, a typical tectonic uplift zone developed, forming a tectonic background of "one uplift zone between two depressions". Simultaneously, the dominant influence of tides in the shallow marine shelf environment facilitated the formation of a typical shelf-type tidal delta sedimentation system in the Jurassic strata of the northern part of West Siberian Basin. This sedimentation constitutes a significant natural gas reservoir, and it is important to investigate the sedimentary evolution of shelf-type tidal deltas and to clarify the internal structure and distribution of sedimentary sand bodies and interlayers in shelf-type tidal deltas, which is the basis for the fine development of this type of reservoir. This paper takes the Jurassic strata in the Y region of northern part of West Siberian Basin as the research object, and conducts numerical simulation based on sedimentary dynamics for the shelf-type tidal delta sedimentation formed under the tectonic background of "one uplift zone between two depressions". In addition, tidal amplitude and initial water level were selected for different hydrodynamic factors to study the main controlling factors of shelf-type tidal delta sedimentation. The simulation results show that tidal amplitude is positively correlated with three-dimensional configuration characteristic parameters of the sedimentary sand bodies, and the development of tidal bars becomes more and more limited as the initial water level increases. This paper systematically investigates the sedimentary evolution of shelf-type tidal delta under the tectonic background of "one uplift zone between two depressions" by the sedimentary dynamics method, which deepens the understanding of the shelf-type tidal delta sedimentation process and provides a new thinking for the development of this sedimentary reservoir type (School of Geosciences China University of Petroleum (East China)).

Keywords: shelf-type tidal delta; sedimentary dynamics; numerical simulation; Delft3D

:**itation:** Tang, M.; Xiong, S.; Zhang,).; Hong, R.; Peng, C.; Xie, R. A Novel /lethod for Analyzing Sandbar)istribution in Shelf-Type Tidal Deltas Jsing Sediment Dynamic Simulation. *Mar. Sci. Eng.* **2024**, *12*, 1102. ttps://doi.org/10.3390/jmse12071102

\cademic Editors: Wenming Ji, ;uodong Cui and Mianmo Meng

leceived: 3 June 2024
levised: 23 June 2024
ccepted: 25 June 2024
'ublished: 28 June 2024

1. Introduction

The hydrodynamic conditions of shallow marine shelf-type tidal deltas are complex and diverse, including tides, currents, and storm-induced waves [1], and shelf-type tidal delta deposition is a complex depositional system formed under the joint influence of tidal action and paleo-uplift tectonics. The shelf-type tidal delta is a special kind of shallow marine shelf deposition that is mainly developed in the broader continental shelf, shallow sea water in the coastal zone environment, and it is a kind of sedimentary landform formed by the strong influence of tidal action [2]. The sediments of shelf-type tidal deltas ar mainly imported by the fluvial system and interact with periodic tidal currents, forming series of complex morphological sedimentary sand bodies on the shelf, and its sedimer tary sequences often show obvious tidal rhythmic laminations and interbedded lamina structures [3]. The formation process of shelf-type tidal deltas involves a number of link such as fluvial sand transport, tidal current dynamics, sediment resuspension and redep sition [4], and their depositional systems usually include components such as distal bar intermediate plains, and frontal subfacies, where the distal bar portion is often shown to b branching channels and sheet sand bodies due to the action of strong tidal currents, and th frontal subfacies may develop tidal pads, tidal channels, and tidal sand ridges and othe geomorphological units [5,6]. However, in sedimentary geology, there are fewer literatur records on shelf-type tidal delta deposition formed due to tectonics [1], and there is a lac of systematic understanding of the internal structure and distribution of sand bodies an interlayers in shelf-type tidal delta deposition, which is needed to clarify the sedimentar evolution law, to determine the spreading characteristics of the sand bodies, and to provid guiding ideas for the research of the exploration and development of oil and gas in thi kind of sedimentary reservoir. As such, study on shelf-type tidal delta deposition ha become a hot and difficult problem nowadays.

Commonly used methods to study ancient sedimentation include ancient sedimer tary record analysis, modern sedimentary example dissection, flume physics experiment and numerical simulation [7]. Seismic data would be limited by a lack of drilling well resulting in less information available for research [8]. Modern sedimentary exampl dissection depends on the practicalities of outcrops; sometimes rock outcrops can pro vide the two-dimensional structure of sediments but not enough information on interna three-dimensional structure [9]. Physical simulation is mainly carried out through flum experiments for sediments, and it can provide reliable results [10]; however, physical simu lation is limited by many conditions, such as small study size and limited run time [11]. Th utilization of numerical simulation as an efficient tool in sedimentary dynamics primaril relies on the hydrodynamic approaches to investigate sediment transport and geomo phological evolution [12]. Numerical simulation is highly actionable and applicable to wide range of hydrodynamic scenarios, and can provide strong evidence for the study o morphodynamic and stratigraphic patterns arising from sedimentary erosion at differen spatial and temporal scales [13]. To date, numerous numerical simulation methods o sediment dynamic have been developed, which can be classified into various categorie based on different principles and criteria. [14]. This study employs a sediment dynamic numerical simulation model grounded in hydrodynamic equations. It is the numerica simulation of sedimentation–erosion based on the Naiver–Stokes equations, which allow the recovery of the detailed processes of sediment transport and lithofacies distributio and it is appropriate for modelling the evolution of sediment environments over shor timescales measured in millennia [15]. In 2016, Chatzirodou [16] used Delft3D softwar to simulate a deep-water shelf in the Nestor Channel, Scotland, UK, and the model re vealed the interactions between the sand bodies and the complex three-dimensional tida structures; in particular, how channel currents formed during rising tide periods and tida jets formed during ebb tide periods can affect the sediment's movement and morpholog changes. In 2018, Tristan Salles [17] used the numerical simulation method to study th millennial-scale sedimentary evolution of a shallow marine shelf. In the same year, Va de Lageweg [18,19] applied sedimentary dynamics modelling in a land–marine transitio zone environment; the research quantifies the role of factors such as fluvial flow, wave tides, and mud supply in the geometry and sedimentary structure of funnel-shaped basin along the fluvial–marine transition zone region based on a sedimentary dynamics metho In 2020, Nnafie [20] analyzed the evolution of the natural morphodynamics of tidal san ridges on the shelf and the response to human intervention based on numerical simulatio of sedimentary dynamics, and carried out a simulation study of the dynamic process o large sedimentary sand bodies formed under strong tidal influence in a shallow marin shelf environment by Delft3D. These studies exemplify that the Delft3D software is a reliable and widely used tool for the analysis of shelf environments and tidally driven sedimentary dynamics problems.

2. Region Setting

The West Siberian Basin is the largest hydrocarbon-bearing basin in the world, with an area of about 3.5 million square kilometers and a huge potential for hydrocarbon resource development [21]. The study area is the Y area on the uplift zone in the northern part of the West Siberian Basin, as shown in Figure 1a, and its Jurassic stratigraphic system shows a good potential for gas deposits. Shallow marine facies deposits were widely developed in the northern part of West Siberian Basin during the Jurassic period [22], and in the center of the northern part of the basin, a significant paleo-uplift zone was formed due to long-term tectonic movements and crustal uplift [23,24]. This uplift zone serves as an important geological tectonic line, bisecting the vast northern part of Western Siberian Basin into two relatively low-lying depressions, thus constituting a typical tectonic pattern of "one uplift zone between two depressions" [23–26].

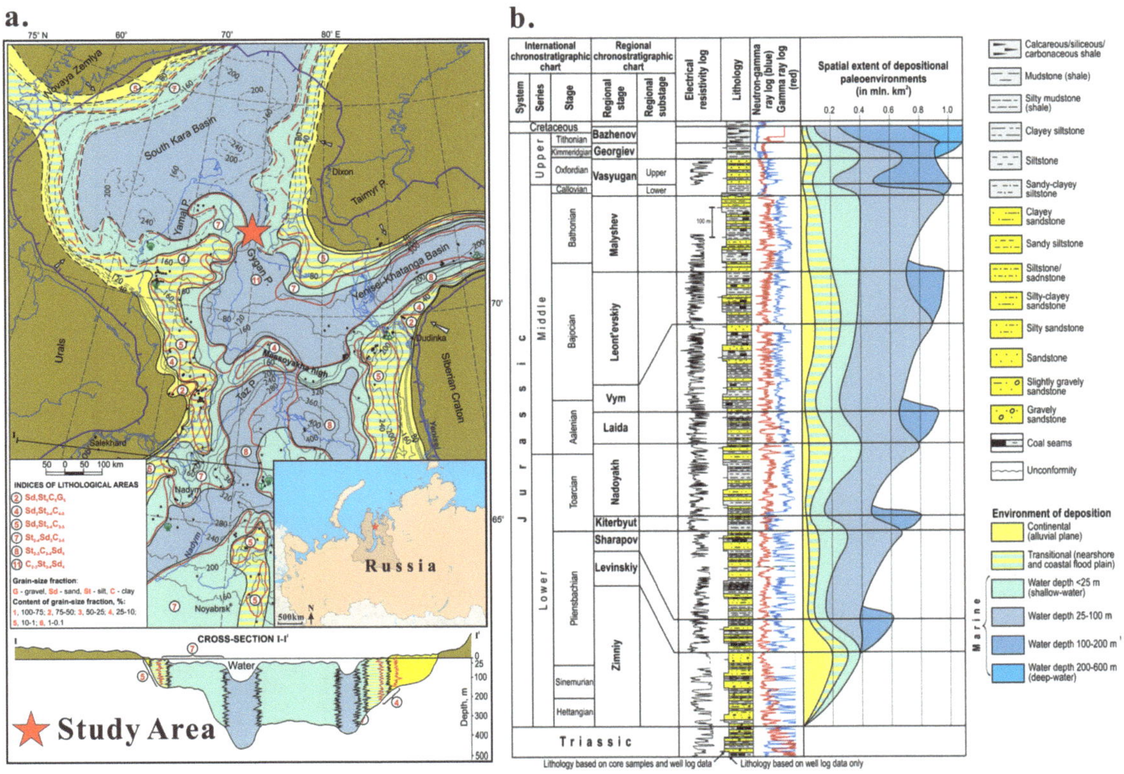

Figure 1. (**a**) Litho–paleogeographic map of the Vym time in the northern part of West Siberian Basin [22]; (**b**) lithostratigraphy and sedimentation conditions of Jurassic deposits in the northern part of the West Siberian basin [22].

Georgiy Shemin [22] in 2019 carried out a detailed Jurassic paleogeography study in the northern part of West Siberian Basin, mapping the litho-paleogeography of the Jurassic strata (Figure 1b). In Vym time, the northern part of West Siberian Basin was in an extensive shallow marine shelf environment, with depths ranging from 25 to 100 m. The study area is located in a region where sandstones and siltstones dominate the sediments, accompanied by the development of some mudstones, in which sandstones generally account for more than 50% of the sediments, and sandy and muddy interbedded sediments with a thicknes of up to 260 m have been deposited. The shelf uplift zone below sea level was under th joint control of tides and paleo-tectonics [22], which resulted in the formation of shel type tidal delta deposition under the tectonic background of "one uplift zone between tw depressions". Under the tectonic background of "one uplift zone between two depressions the gently sloping zone extending to both depressions not only received a large amount c sediments from the Ural Mountains and the Taimyr region, but also deposited mud an sand from the two oceans under the strong influence of tides, which ultimately formed th complex shelf-type tidal delta depositional system [2].

3. Methods

The simulation software used in this study is the open-source software Delft3D ve sion 4.04.01, which is a high-resolution numerical simulation system for dynamic flo of sedimentary sands based on continuous profiles for hydrodynamic–stratigraphic n merical simulations [27]. Its FLOW module is suitable for predicting hydrodynamic an sedimentary geomorphological evolution in semi-enclosed coastal areas such as coasta areas, estuaries, shallow marine shelves, and lagoons [28,29]. Fluid and sediment transpo are discretized on a 3D curvilinear finite-difference grid and solved using the alternat direction implicit format ADI method [29]. In this paper, Delft3D software is used to carr out numerical simulation of the depositional process of shelf-type tidal deltas formed b the Jurassic strata in the study area, so as to determine the main controlling factors for th distribution of sand bodies and their internal configuration in the "one uplift zone betwee two depressions" shelf-type tidal delta.

3.1. Establishment of Basic Model

In this paper, the target strata in the study area correspond to the Jurassic Vym tim strata. The geological conceptual model topography is set up with reference to the lith paleogeographic map of the Vym time in the northern part of West Siberian Basin mappe by Georgiy Shemin. The horizontal surface area of the model corresponds to the area o the paleogeographic map (Figure 1a), which is 1500 km long and 1200 km wide, as show in Figure 2. According to the litho-paleogeographic map, the ocean boundary is set up i the north, east, and northwest of the model to provide tidal action, and the Harmonic tida force type is adopted, that is, the tide with a period of 12 h and amplitude of 6 m. In th Ural Mountains, the Taimyr region, and the Siberian craton, river boundaries are set.

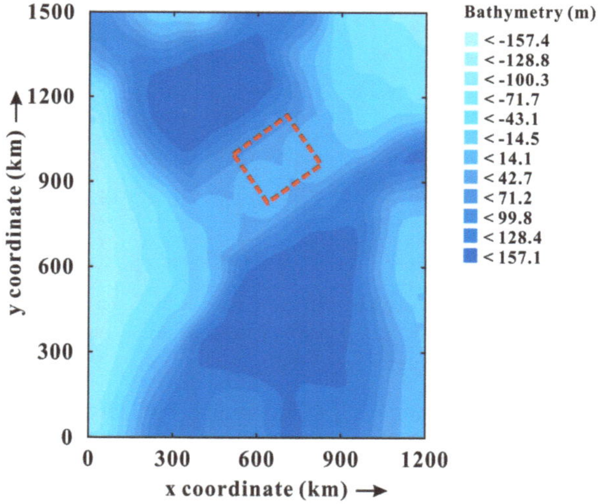

Figure 2. Three−dimensional conceptual model plan of the study area.

3.2. Basic Model Parameter Setting

The research of modern depositional systems, especially the modern continental shelf depositional system, can provide a reference for the hydrodynamic and sediment parameters of the numerical simulation model of the "one uplift zone between two depressions" shelf-type tidal delta [30–37]. The area of the model grid is set to 3 × 3 km^2. During the Jurassic period, the study area was in an extensive shallow marine shelf environment, the shelf uplift zone, lying below sea level, and was concurrently influenced by the synergistic controls of tidal forces and paleotectonic activities [22]. In this context, the height value of the highest point of the uplift zone in the model is defined as 0 m in the Z axis of the model, and the initial water level of the sea in the model is set as 1 m. Semi-diurnal tidal control was adopted, the tidal frequency is 12 h/time, and the tidal height was set to 6 m. The wave frequency is 25 s/time, and the wave height is set to 1 m. The fluvial discharge is set to 3000 m^3/s, and the amount of sediment carried by the river is 0.525 kg/m^3. The sediment types are divided into two types, non-cohesive sandy sediment and cohesive muddy sediment. The non-cohesive sandy sediments are divided into coarse grain size and fine grain size. The content ratio is coarse sand:fine sand:mud = 1:1:1, the simulation time is 150 days, every 10 days is recorded as 1 step, the simulation discrete time step is 1 min, the morphological scale factor H is set to 100, and the results can be obtained for up to 500 months of sedimentary evolution. At the same time, the morphological scale factor does not have any effect on the formation of the sedimentary morphology of the shelf tidal delta [38], as shown in Table 1. In the simulation of Delft3D software, the tidal amplitude and wave height vary automatically with time in a single period, and the input parameters represent the absolute values of the maximum and minimum within a period. Fluvial discharge serves as an auxiliary hydrodynamic force, maintained at a constant rate throughout the simulation. The initial water level height means the water level at the beginning moment of the simulation, and it is subsequently allowed to fluctuate in response to the evolving hydrodynamic conditions as the simulation progresses.

Table 1. Basic model parameter settings.

Parameter Settings	Value
Size of Study Area	1500 × 1200 km^2
Size of Single Grid	3 × 3 km^2
Time Step	1 min
Initial Water Level	1.0 m
Tidal Amplitude	6.0 m
Wave Height	1 m
Fluvial Discharge	3000.0 m^3/s
Sediment Grain Size	130\65\mud μm
Coarse Sand:Fine Sand:Mud	1:1:1
Maximum Water Depth	176 m
Morphological Scale Factor	100

3.3. Main Controlling Factor Analysis Model Parameter Setting

In order to clarify what factors are mainly in control of the sand bodies' development in the shelf-type tidal delta, the tidal amplitude and initial water level were selected for two univariate analyses and the values of the tidal amplitude and the initial water level are referenced by research of modern depositional systems, mentioned in Section 3.2 of this paper. In order to study the influence of the two factors on the sedimentation in depth, the range of values of the two factors in this paper is slightly larger than the reference range of research of modern depositional systems. Two different casaaaaaes are selected for each factor on the basis of the basic model, and a total of five models were simulated. Tidal factors are as follows: high tide, 10 m; medium tide, 6 m (base); and low tide, 2 m. The water level factors are as follows: high water level, 45 m; medium water level, 15 m; and low water level, 1 m (base), as shown in Table 2.

Table 2. Main controlling factor analysis parameters.

Model Name	Factor	Case	Parameter
HIGHWATER		High Water Level	45
MIDDLEWATER	Initial Water Level (m)	Medium Water Level	15
BASE		Low Water Level	1
HIGHTIDE		High Tide	10
BASE	Tidal Amplitude (m)	Medium Tide	6
LOWTIDE		Low Tide	2

4. Results

4.1. Analysis of Basic Model Results

Figure 3 shows the simulated erosion and sedimentation in the northern part of Wes Siberian Basin at simulation time step = 15. Marine boundaries are set in the northwestern northern, and northeastern parts of the study area, and the local high part of the uplif zone and the east and west sides of the target study area are the main sources of sand supply, and in the shallow marine shelf area, the study area is adjacent to the sources and the high-current-shear-stress area, forming sand body deposits. In the semi-deep-wate and deep-water shelf area, the currents are far away from the sources and only receive mud supply from the east and west sides and the northern ancient ocean. The shelf uplift zone in which area Y is situated, as shown in Figure 3b, has a stronger sedimentary erosion in the north than in the south. Sedimentary sand bodies are more developed in the north and sand sheets are widely distributed in the study area, on which some tidal bars are developed, and the sedimentary sand bodies are in the form of sand–mud thin interbedded structures. Tidal channels are distributed on both sides of the uplift zone, dividing the sand bars. The results of the simulation show that the distribution of sand bodies is similar to the conclusions obtained from the paleogeographic study of Kontorovich, A.E. [25,26] which verifies the reliability of this simulation.

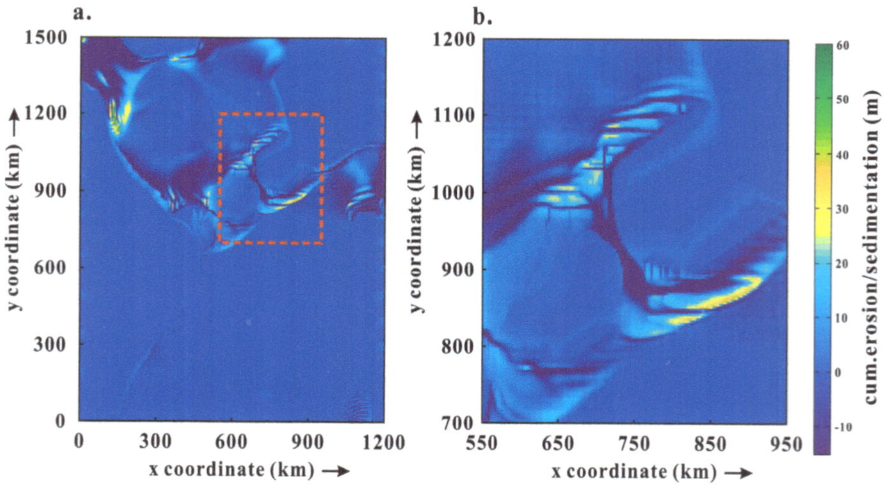

Figure 3. Results of basic model: (**a**) cumulative erosion and sedimentation of basic model; (**b**) cumulative erosion and sedimentation of target study area.

Figure 4 shows the simulation results of the erosion and sedimentation distribution in the target study area during the simulation time. Due to tidal action, sediment accumulation occurs, and as the simulation time grows, the sheet sand bodies gradually increase in length and width, and tidal channels appear in some areas; as the bars continue to be deposited they begin to move towards the center of the basin, and the rate of deposition slows down

At the marine boundary in the northwestern part of the study area, various bars and tidal channels are developed. In the early period of the simulation, when step = 1, there is sediment accumulation in the study area, although tidal channels are not obviously developed, but there is a trend; in the middle period of the simulation, when step = 8, the basic morphology of the tidal bars and tidal channels in the study area is initially visible, and in the late period of the simulation, when step = 15, the development of the tidal bars and tidal channels becomes particularly significant.

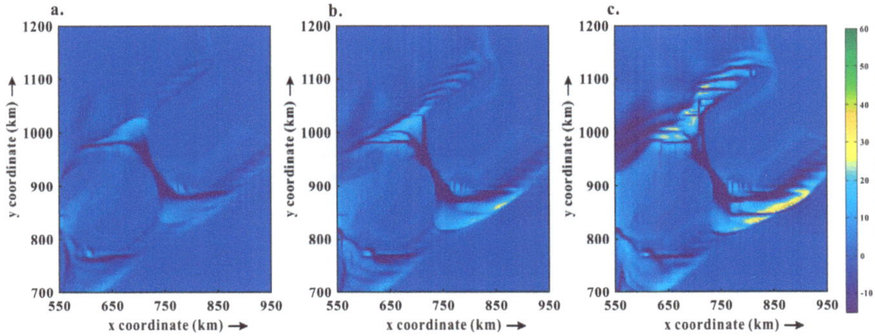

Figure 4. Process of basic model study area simulation: (**a**) cumulative erosion and sedimentation of basic model at step 1; (**b**) cumulative erosion and sedimentation of basic model at step 8; (**c**) cumulative erosion and sedimentation of basic model at step 15.

Especially in the study area, it can be observed that under the strong influence of the tidal channels, strip-shaped bar structure with separation characteristics is gradually formed. The type of sand bar is mostly dominated by northwest-trending bars, and statistical results show that the average length of the tidal bar is 11.83 km, the average width is 5.11 km, and the average thickness is 8.61 m. A large number of thin sand sheets are developed on both sides of the uplift zone, with the simulation going on, where sand sheets become more developed and expanded, and statistical results show that the average length of the sand sheets is 14.83 km, the average width is 12.60 km, and the average thickness is 0.77 m.

4.2. Analysis of Tide Model Results

Tidal amplitude is one of the important factors affecting the development of bars in the shelf depositional system. The rest of the parameters are fixed to carry out simulations with different tidal amplitudes, which are set to be a low tide with 2 m amplitude, a medium tide with 6 m (basic model) amplitude, and a high tide with 10 m amplitude.

4.2.1. Analysis of Sediment Distribution

As shown in Figure 5, comparison of the models with different tidal amplitudes shows that, compared to the basic model for the medium tide, the bars develop slowly at low tide, but the sediment thickness is greater, little sediment is reworked, and tidal channels are developed along both sides of the uplift zone. At high tide, the tidal modification is stronger, the bars and tidal channels are more developed, and the bar morphology is heavily eroded and redeposited. As tidal amplitude increases, tidal bars grow faster and tidal channels are more developed; the greater the tidal amplitude, the wider the range of tidal influence, the greater the number of bars, and the more complex the morphology.

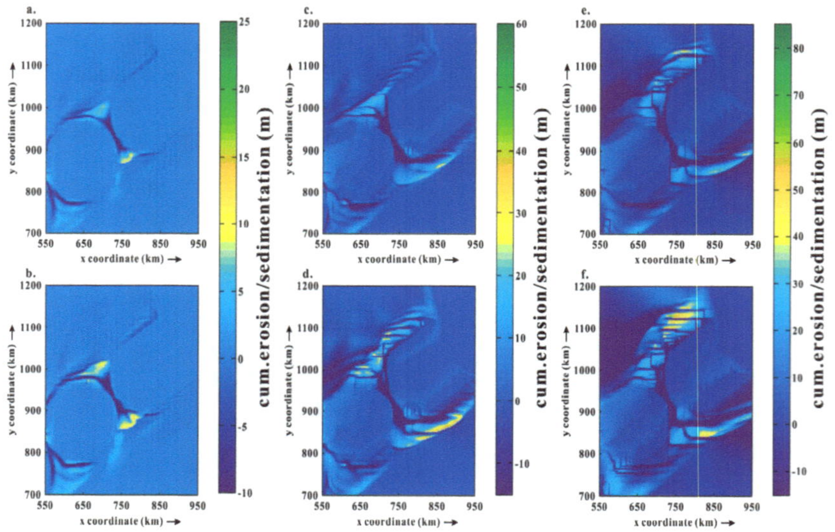

Figure 5. Changes in cumulative erosion and sedimentation of shelf-type tidal delta with differer tidal amplitude: (**a**) low tide model at step 8; (**b**) low tide model at step 15; (**c**) medium tide model a step 15; (**d**) medium tide model at step 8; (**e**) high tide model at step 15; (**f**) high tide model at step 1!

4.2.2. Analysis of Flow Velocity Distribution

As shown in Figure 6, at low tide, deposition occurs only at the ends of the channe of the uplift zone, with less tidal action, and the water flows in the channel during risin tide, resulting in the formation of a bar at that location. At medium tide, the oscillation c water flow caused by tidal action is significantly enhanced, and multiple tidal channel develop on both sides of the northern side of the uplift zone under the impact of rising tida currents. At high tide, the tidal action is the strongest, and the water flow in the channe meets with the tidal current to form a vortex current, resulting in the rapid formation c bars on both sides of the uplift zone.

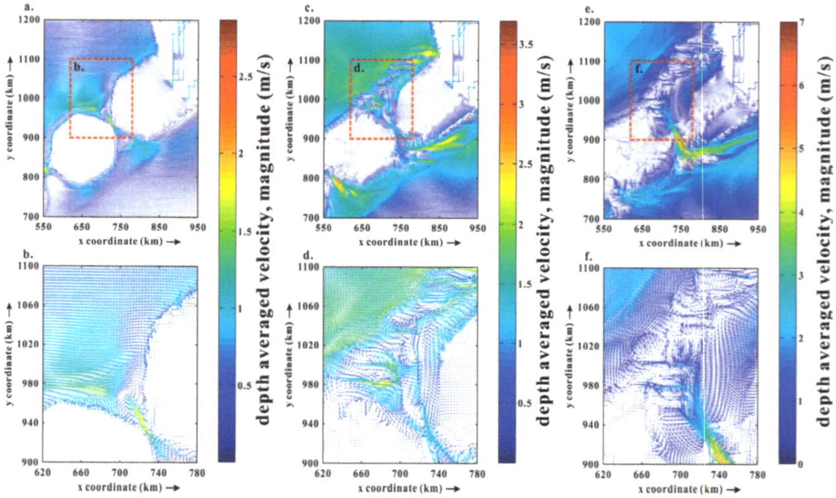

Figure 6. Changes in velocity of shelf-type tidal delta with different tidal amplitude: (**a**) study area a low tide; (**b**) zoom in on study area at low tide; (**c**) study area at medium tide; (**d**) zoom in on stud area at medium tide; (**e**) study area at high tide; (**f**) zoom in on study area at high tide.

4.2.3. Analysis of Sand Bodies

According to the simulation results, as shown in Table 3, the three-dimensional configuration characteristic parameters of the sand bodies for each tidal amplitude model were counted. At low tide, the average length, average width, and average thickness of the sand bodies decreased to different degrees compared to the medium tidal amplitude model used as the basic model. The average length of the tidal bars decreased by 21.13%, the average width by 9.78%, and the average thickness by 10.57%, while the average length of the sand sheet decreased by 7.42%, the average width by 16.19%, and the average thickness by 29.87%. On the other hand, at high tide, the three-dimensional configuration characteristic parameters of the sand bodies increased significantly compared to the basic model. The average length of the tidal bars increased by 26.63%, the average width by 23.09%, and the average thickness by 19.28%, while the average length of the sand sheet increased by 34.05%, the average width by 7.30%, and the average thickness by 27.27%. The simulation results show that the tidal energy effect has the greatest influence on the length of the deposited sand body, and with the increase in tidal amplitude, there is a trend of advancing towards the northern ocean of the study area, indicating that the tidal amplitude is one of the main factors controlling the sedimentation in the study area.

Table 3. Three-dimensional configuration characteristic parameters of the sand bodies with different tidal amplitude.

Tidal Amplitude	Average Length/km		Average Width/km		Average Thickness/m	
	Tidal Bar	Sand Sheet	Tidal Bar	Sand Sheet	Tidal Bar	Sand Sheet
Low tide (2 m)	10.33	13.73	4.61	10.56	7.70	0.54
Medium tide (6 m)	11.83	14.83	5.11	12.60	8.61	0.77
High tide (10 m)	13.48	19.88	6.49	13.52	10.27	0.98

Analyzing the relationships of the three-dimensional configuration characteristic parameters of tidal bars, as shown in Figure 7, the values of the length-to-width ratio of the tidal bars ranged from 1.15 to 5, and the values of the length-to-thickness ratio ranged from 0.62 to 2.33. Tidal amplitude is positively correlated with the length, width, thickness, length-to-width ratio, and length-to-thickness ratio of tidal bars.

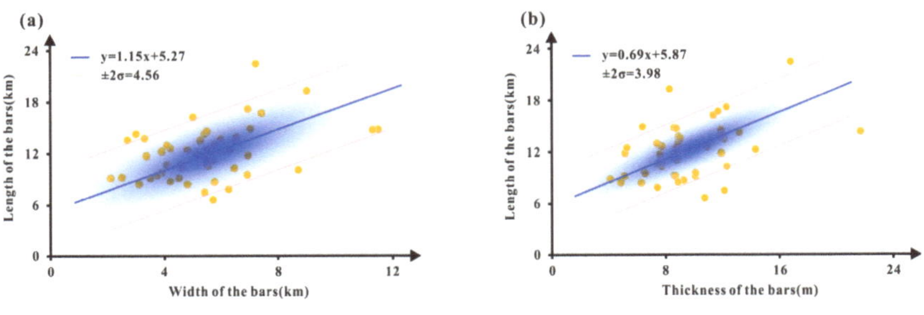

Figure 7. (**a**) Tidal bars' length-to-width ratio distribution map with different tidal amplitude; (**b**) tidal bars' length-to-thickness ratio distribution map with different tidal amplitude.

4.2.4. Analysis of Interlayers

As shown in Figure 8, at low tide, the length of the interlayers ranges from 8 to 25 km, the thickness of the interlayers ranges from 1.0 to 2.5 m, and the length of the interlayers is concentrated at 18.73 km; at medium tide, the length of the interlayers ranges from 4 to 18 km, the thickness of the interlayers ranges from 0.3 to 0.6 m, and the length of the interlayers is concentrated at 9.16 km; at high tide, the length of the interlayers ranges from 2 to 12 km, the thickness of the interlayers ranges from 0.1 to 0.3 m, and the length of the interlayers is concentrated at 5.94 km. The statistical result shows that the interlayers are widely developed at low tide; at medium and high tide, the development of the interlayer is limited and the thickness is thinner. With the increase in tidal amplitude, the length of the interlayers tends to become shorter, and the thickness of the interlayers gradually becomes smaller.

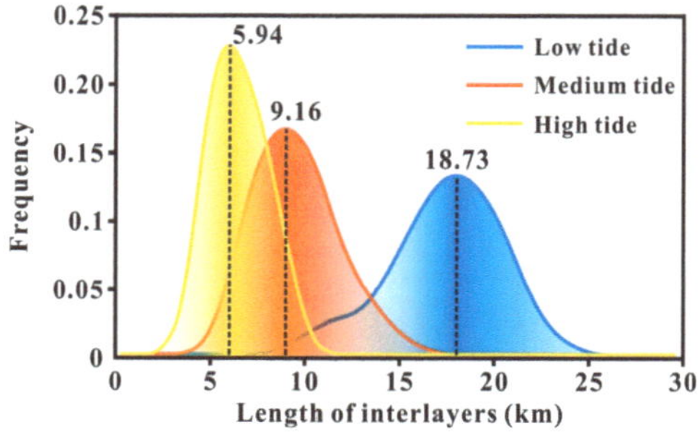

Figure 8. Distribution of interlayer length with different tidal amplitude.

After the above analysis, we know that three-dimensional configuration characteristic parameters of the sand bodies all increase with the increase in the tidal amplitude, and the larger the tidal amplitude, the more bars are eroded and redeposited. The larger the tidal amplitude, the larger the average length and average width, the thicker average thickness of the tidal bars, and the larger the length-to-width and length-to-thickness ratios of the bars; the larger the tidal amplitude, the larger the length and width, the thicker the thickness of the sand sheets; the tidal factor is positively correlated with the above parameters. The interlayers are widely developed at low tide with a large number, and the length and the thickness of the interlayers are large. The development of interlayers is limited at medium and high tide with a decreasing number, while the length and thickness of the interlayers decrease with the increase in tidal amplitude.

4.3. *Analysis of Initial Water Level Model Results*

The initial water level is also one of the important factors affecting the shelf tidal delta depositional system, and different initial water levels produce very different results in the sedimentary numerical simulation. Simulations in different initial water level conditions are carried out with remaining parameters fixed; the initial water level is set as 1 m for low water level (basic model), 15 m for medium water level, and 45 m for high water level, with the bottom of the channel in the uplift zone as the 0 m water level surface. A simulation is carried out in different initial water level conditions.

4.3.1. Analysis of Sediment Distribution

As shown in Figure 9, the low water level model (basic model) has formed tidal bars and tidal channels at the study area in the early stage of the simulation, but the segmentation is not clear enough, and the development of tidal bars and tidal channels is obvious in the later stage of the simulation, with the development of distinctly segmented bars under the tidal action in the study area. The medium water level model has formed a large area of sand sheets in the early stage of the simulation, and there is a slight development of the tidal channels but not the development of tidal bars. In the later stage of the simulation, large sand sheets are further developed in the study area, and a small number of tidal bars and tidal channels are developed near the area. In the high water level model, the tidal bar and tidal channel are not developed in the whole simulation process, and only part of the sand sheets are developed. Comparison of the different water level models shows that from the tidal delta to the shallow-water shelf to the deep-water shelf, the cutting of the channel becomes weaker and weaker, and the concentration of the sand bodies becomes less and less, and the thickness becomes thinner and thinner.

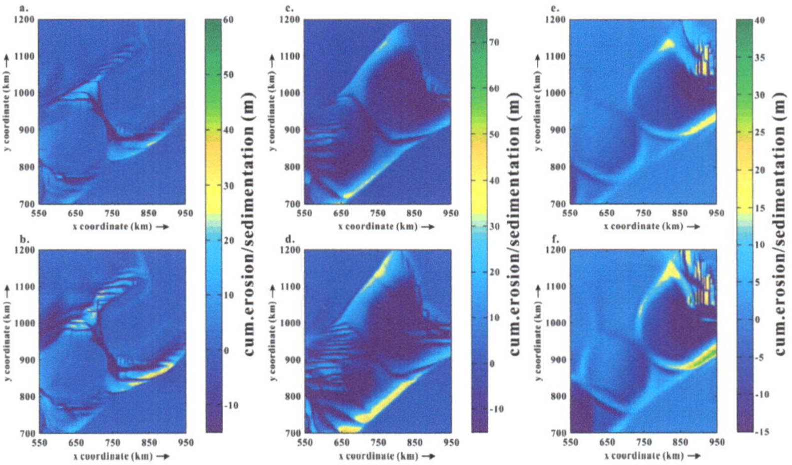

Figure 9. Changes in cumulative erosion and sedimentation of shelf-type tidal delta with different water level: (**a**) low water level model at step 8; (**b**) low water level model at step 15; (**c**) medium water level model at step 15; (**d**) medium water level model at step 8; (**e**) high water level model at step 15; (**f**) high water level model at step 15.

4.3.2. Analysis of Flow Velocity Distribution

As shown in Figure 10, at low water level, the current oscillation caused by tidal action is obvious, and in the north side of the uplift zone, many tidal channels and tidal bars are developed on the east and west sides of the study area under the impact of rising tidal currents. At medium water level, the flow velocity in the study area does not change significantly under the same tidal action, and only a small number of tidal bars and tidal channels are developed on the west side. At high water level, the uplift zone is in deep water and less affected by tidal action, the flow oscillation in the study area is not obvious, and neither tidal bar nor tidal channel is developed.

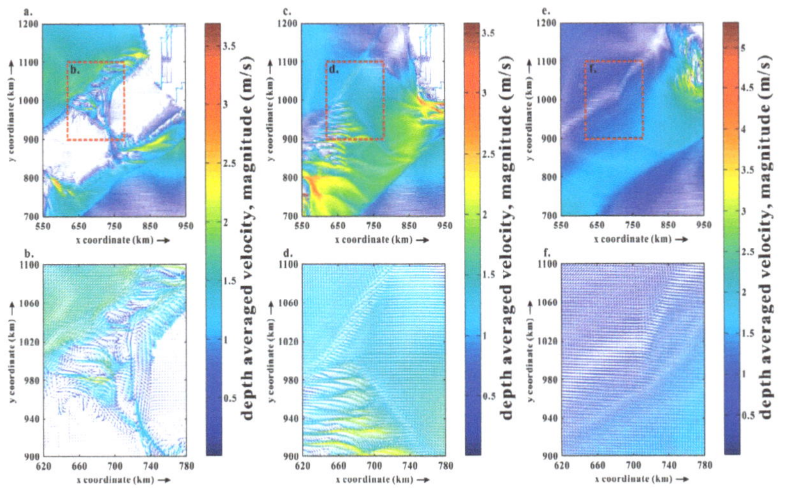

Figure 10. Changes in velocity of shelf-type tidal delta with different water levels: (**a**) study area at low water level; (**b**) zoom in on study area at low water level; (**c**) study area at medium water level; (**d**) zoom in on study area at medium water level; (**e**) study area at high water level; (**f**) zoom in on study area at high water level.

4.3.3. Analysis of Sand Bodies

According to the simulation results, as shown in Table 4, the three-dimensional con figuration characteristic parameters of the sand bodies for each water level model wer counted, and the low water level model (basic model) can be used as a control grou for comparison. When the initial water level increases to 15 m, the length and widt of the deposited sand bodies show a decreasing trend, while the thickness of the tida bars increased and the thickness of the sand sheet decreased. The average length an average width of the tidal bars decreased by 5.07% and 12.13%, respectively, and th average thickness increased by 5.23%. The depositional condition in the study area change drastically when the water level increases significantly up to 45 m. At high water, the stud area is at a lower depth below the water's surface, and tidal bar is not developed. Only larg thin sand sheets are developed, and the length of the sand sheet increases substantiall The simulation results show that the development of sedimentary sand bodies is more an more limited as the water level rises and the whole study area is gradually submerge deeper and deeper below the water's surface. The simulation results under different wate levels with other parameters kept constant are quite different, indicating that the wate level is also one of the main controlling factors of sedimentation in the study area.

Table 4. Three-dimensional configuration characteristic parameters of the sand bodies with differer water levels.

Water Level	Average Length/km		Average Width/km		Average Thickness/m	
	Tidal Bar	Sand Sheet	Tidal Bar	Sand Sheet	Tidal Bar	Sand Sheet
Low water level (1 m)	11.83	14.83	5.11	12.60	8.61	0.77
Medium water level (15 m)	11.23	13.55	4.49	11.43	9.06	0.68
High water level (45 m)	-	49.21	-	13.29	-	0.64

Analyze the relationships of the three-dimensional configuration characteristic pa rameters of tidal bars, as shown in Figure 11, the values of the length to width ratio of th tidal bars ranged from 1.5 to 5, and the values of the length to thickness ratio ranged fron 0.6 to 3.36. Initial water level is positively correlated with the length, width, and length t thickness ratio of tidal bars and it is negatively correlated with the thickness and length t width ratio of tidal bars.

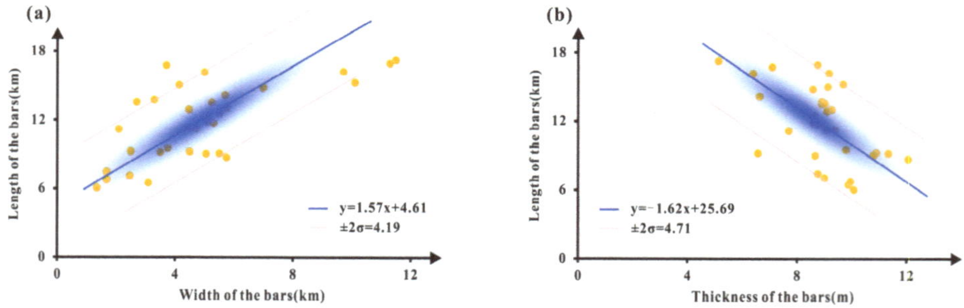

Figure 11. (**a**) Tidal bars' length-to-width ratio distribution map with different water levels; (**b**) tida bars' length-to-thickness ratio distribution map with different water levels.

4.3.4. Analysis of Interlayers

As shown in Figure 12, at low water level, the length of the interlayers ranges fron 4 to 18 km, the thickness of the interlayers ranges from 0.3 to 0.6 m, and the length of th interlayers is concentrated at 9.16 km; at medium water level, the length of the interlayer ranges from 6 to 40 km, the thickness of the interlayers ranges from 0.5 to 1.8 m, and th length of the interlayers is concentrated at 23.88 km; at high water level, the length of th interlayers ranges from 12 to 50 km, the thickness of the interlayers ranges from 1.0 to 3.0 m, and the length of the interlayers is concentrated at 31.43 km. From the statistical results, it shows that at low water level, the interlayers are developed with thin thickness; at medium and high water level, the interlayers are more developed and become thicker.

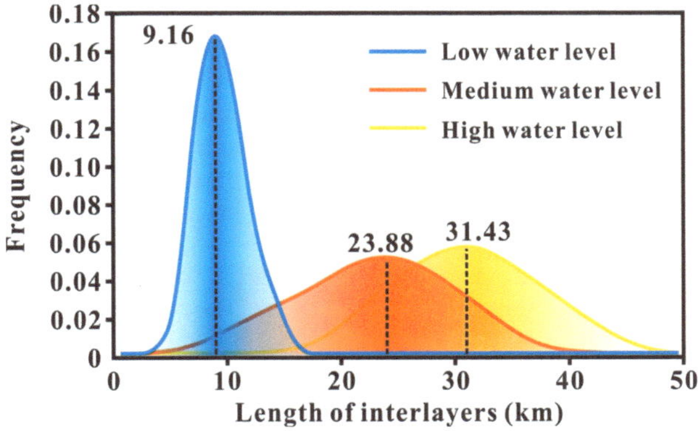

Figure 12. Distribution of interlayer length with different water levels.

After the above analysis, we know that at low water level, sedimentary sand bodies on the uplift zone are more developed, forming shelf-type tidal delta deposition. The development of tidal channels is obvious, as well as the development of strip-shaped bars with clear divisibility under its action. At medium water level, the development of sand bars is limited and it is easier to form dispersed and spreading thin sand sheets, with only a few tidal bars and tidal channels developed. At high water level, a deep-water shelf is formed, so that sand deposits do not easily to develop. Sedimentation is dominated by remotely suspended and slowly settling mud, with no tidal bar or tidal channel developed, and only a large area of thin sand sheets formed. With the increase in water level height, the length of the interlayers tends to become longer, and the thickness of the interlayers increases.

5. Discussion

This study has conducted extensive research on parameter selection during the preliminary model construction phase and has carried out multiple simulation experiments to adjust the modeling parameters. To validate the rationality of the models built in this research, this section compares the simulation results of the series of models with the parameters of the Jurassic strata sand bodies in the study area, as shown in Table 5.

Table 5. Comparison of study area's actual conditions and numerical simulation.

Model	Hydrodynamic Condition				Statistical Result of Tidal Bars		
	Tide/m	Water Level/m	Wave/m	Discharge/($m^3 \cdot s^{-1}$)	Average Length/km	Average Width/km	Average Thickness/m
Study area	-	-	-	-	12.02	5.38	8.93
Base 1	6	1	1	3000	11.83	5.11	8.61
Base 2	4.5	1	2	3000	10.71	4.69	7.98
Base 3	3	5	1	4500	9.66	6.05	11.06
Base 4	7.5	5	2	4500	13.24	6.67	10.62

Extensive simulation experiments were performed to establish a basic model that best represents the characteristics of the study area. Table 5 presents a comparison of typical simulation outcomes with the actual parameters of the sand bodies in the study area, and the three-dimensional configuration characteristic parameters of simulated sand bodies of the Base 1 model show the highest degree of congruence with the real condition Subsequent random fine-tuning of the Base 1 model's parameters led to a decline in th degree of congruence between the simulation results and the actual conditions of the stud area. In this study, based on existing data from the study area, we present the sedimentar facies map of the Jurassic Vym time in the study area. As shown in Figure 13, a comparativ examination of the map of cumulative erosion and sedimentation from the Base 1 mode and the sedimentary facies map of the Jurassic Vym time in the northern study area reveal obvious similarities. Both illustrations show the prominent development of tidal bar atop sand sheets. Notably, these tidal sand bars, under the partitioning influence of tida channels, have evolved into a diverse array of morphologies, thereby testifying to th model's capability in replicating complex depositional patterns akin to those observed i natural settings. This visual correlation reinforces the notion that the model simulates th tide-dominated dynamic processes in a shallow marine shelf environment during Jurassi Vym time, including the intricate interplay between sedimentation and erosion forces givin rise to the varied forms of tidal bars. The comparison shows the optimality of the Base model's parameter settings and further validates the Base 1 model's similarity with th depositional characteristics of the Vym time strata. Therefore, the Base 1 model is adopte as the basic model in this paper, serving as a platform for subsequent univariate analyse

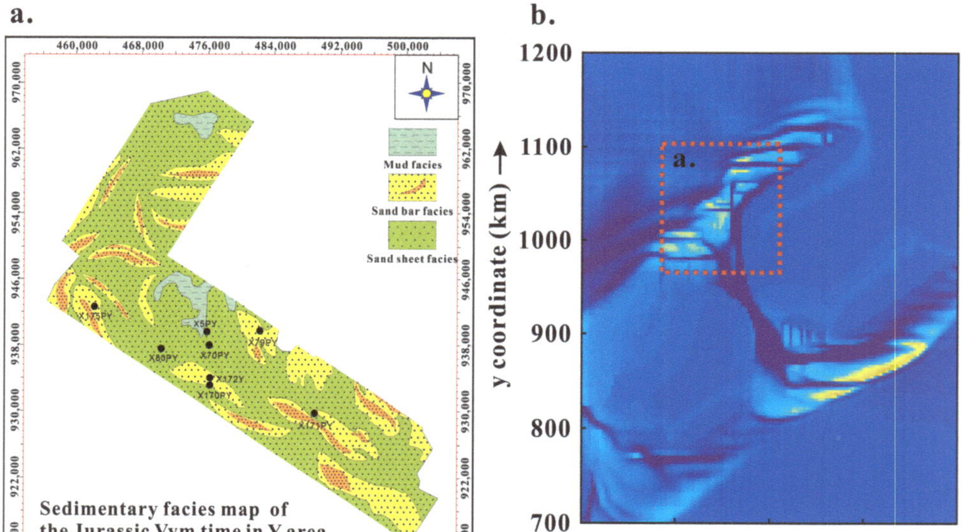

Figure 13. Comparison between the actual conditions and numerical simulation in study area (**a**) sedimentary facies map of the Jurassic Vym time in the study area; (**b**) cumulative erosion and sedimentation simulation result of basic model.

6. Conclusions

The present study utilizes numerical simulation of sedimentary dynamics to investigate the distribution of sand bodies and interlayers in the target strata in the study area and to provide prediction and guidance for the development of oil and gas reservoirs In addition, we carry out numerical simulation with two factors, namely model tide and water level, and establish a numerical evolution model of sedimentary dynamics under the control of the two factors. By comparing the morphology and three-dimensional configuration characteristic parameters of sand bodies under different single-factor conditions the shelf-type tidal delta depositional system with mixed fluvial and tide is quantitatively studied. The sedimentary numerical simulation can obtain the following insights about the main controlling factors of the development of the sedimentary sand bodies and interlayers.

Based on numerical simulation of sedimentary dynamics, models adopt conceptual paleogeographic topography in this study, with tidal hydrodynamics as the main role. The analysis of the results shows that the tide and water level as the main controlling factors all have a significant influence on the development of tidal bars, sand sheets, and the length and width of tidal channels and interlayers in the location of the study area.

1. The tidal amplitude effect is the factor that produces the greatest impact on the shelf tidal delta; the average length, width, and thickness of tidal bars and sand sheets increase with tidal amplitude, and the three-dimensional configuration characteristic parameters are positively correlated with tidal amplitude.
2. The effect of initial water level height on the development of shelf-type tidal delta sand bodies shows that the development of tidal bars, sand sheets, and tidal channels is limited with the increase in water level from a low water level to medium water level, and only a large area of thin sand sheets is formed under a high water level, indicating that the development of sediment is more appropriate in the low water level, namely the shallow-water shelf condition. Sand bodies do not easily form in a deep-water shelf.
3. Quantitative characterization of the distribution of interlayers in the bars shows that the tidal amplitude and initial water level have a strong influence on the morphology of the interlayers. In this case, the tidal amplitude is negatively correlated with both the length of the interlayers and the thickness of the interlayers, and the initial water level is positively correlated with both the length and the thickness of the interlayers.

Author Contributions: Conceptualization, M.T. and S.X.; methodology, M.T.; software, M.T.; validation, M.T. and S.X.; formal analysis, S.X.; investigation, S.X. and Q.Z.; resources, S.X. and Q.Z.; data curation, S.X.; writing—original draft preparation, S.X. and Q.Z.; writing—review and editing, M.T., S.X. and Q.Z.; visualization, R.H. and C.P.; supervision, R.X.; project administration, M.T.; funding acquisition, M.T. All authors have read and agreed to the published version of the manuscript.

Funding: This research was funded by the National Natural Science Foundation (grant numbers. 42072163, 41972250), and the Foundation of Shandong Province (grant number ZR2019MD006), and the Foundation of CNPC (grant number 2021DJ3302), and the National Natural Science Foundation of China (62305196), and the National Postdoctoral Researcher Program (GZC20231498).

Institutional Review Board Statement: Not applicable.

Informed Consent Statement: Not applicable.

Data Availability Statement: The original data contributions presented in the study are included in the article; further inquiries can be directed to the corresponding author.

Conflicts of Interest: The authors declare no conflicts of interest.

References

1. Snedden, J.W.; Bergman, K.M. Isolated shallow marine sand bodies: Deposits for all interpretations. In *Isolated Shallow Marine Sand Bodies: Sequence Stratigraphic Analysis and Sedimentologic Interpretation*; Bergman, K.M., Snedden, J.W., Eds.; SEPM: Claremore, OK, USA, 1999; pp. 1–11.
2. Bourget, J.; Ainsworth, R.B.; Thompson, S. Seismic stratigraphy and geomorphology of a tide or wave dominated shelf-edge delta (NW Australia): Process-based classification from 3D seismic attributes and implications for the prediction of deep-water sands. *Mar. Pet. Geol.* **2014**, *57*, 359–384. [CrossRef]
3. Gil-Ortiz, M.; McDougall, N.D.; Cabello, P.; Marzo, M.; Ramos, E. Sedimentary architecture of a Middle Ordovician embayment in the Murzuq Basin (Libya). *Mar. Pet. Geol.* **2022**, *135*, 105339. [CrossRef]
4. Collins, D.S.; Avdis, A.; Wells, M.R.; Dean, C.D.; Mitchell, A.J.; Allison, P.A.; Johnson, H.D.; Hampson, G.J.; Hill, J.; Piggott, M.D. Prediction of shoreline–shelf depositional process regime guided by palaeotidal modelling. *Earth-Sci. Rev.* **2021**, *223*, 103827. [CrossRef]
5. Desjardins, P.R.; Buatois, L.A.; Praat, B.R.; Mángano, M.G. Sedimentological–ichnological model for tide-dominated shelf sandbodies: Lower Cambrian Gog Group of western Canada. *Sedimentology* **2012**, *59*, 1452–1477. [CrossRef]

6. Desjardins, P.R.; Buatois, L.A.; Mángano, M.G. Chapter 18—Tidal Flats and Subtidal Sand Bodies. In *Developments in Sedimentolog* Knaust, D., Bromley, R.G., Eds.; Elsevier: Amsterdam, The Netherlands, 2012; Volume 64, pp. 529–561.
7. Liu, X.; Lu, S.; Tang, M.; Sun, D.; Tang, J.; Zhang, K.; He, T.; Qi, N.; Lu, M. Numerical Simulation of Sedimentary Dynamics Estuarine Bar under the Coupled Fluvial-Tidal Control. *Earth Sci.* **2021**, *46*, 2944–2957. [CrossRef]
8. Qian, W.; Yin, T.; Hou, G. A new method for clastic reservoir prediction based on numerical simulation of diagenesis: A cas study of Ed1 sandstones in Bozhong depression, Bohai Bay Basin, China. *Adv. Geo-Energy Res.* **2019**, *3*, 82–93. [CrossRef]
9. Tang, M.; Zhang, K.; Huang, J.; Lu, S. Facies and the Architecture of Estuarine Tidal Bar in the Lower Cretaceous Mcmurra Formation, Central Athabasca Oil Sands, Alberta, Canada. *Energies* **2019**, *12*, 1769. [CrossRef]
10. Hu, C.; Yu, M.; Wei, H.; Liu, C. The mechanisms of energy transformation in sharp open-channel bends: Analysis based o experiments in a laboratory flume. *J. Hydrol.* **2019**, *571*, 723–739. [CrossRef]
11. Zhang, X.; Wu, C.; Zhou, Q.; Weng, Z.; Yuan, L.; Zhu, F.; Li, Z.; Zhang, Z.; Yang, B.; Zhao, Y. Multi-Scale 3D Modeling an Visualization of Super Large Manganese Ore Gathering Area in Guizhou China. *Earth Sci.* **2020**, *45*, 634–644. [CrossRef]
12. Liechoscki de Paula Faria, D.; Tadeu dos Reis, A.; Gomes de Souza, O. Three-dimensional stratigraphic-sedimentological forwar modeling of an Aptian carbonate reservoir deposited during the sag stage in the Santos basin, Brazil. *Mar. Pet. Geol.* **2017**, *8* 676–695. [CrossRef]
13. Dalman, R.; Weltje, G.J.; Karamitopoulos, P. High-resolution sequence stratigraphy of fluvio–deltaic systems: Prospects c system-wide chronostratigraphic correlation. *Earth Planet. Sci. Lett.* **2015**, *412*, 10–17. [CrossRef]
14. Lin, C.; Chen, B.; Ren, L.; Dong, C.; Zhang, X. A review of depositional numerical simulation and a case study. *Acta Geol. Si* **2023**, *97*, 2756–2773.
15. van der Vegt, H.; Storms, J.E.A.; Walstra, D.J.R.; Howes, N.C. Can bed load transport drive varying depositional behaviour i river delta environments? *Sediment. Geol.* **2016**, *345*, 19–32. [CrossRef]
16. Chatzirodou, A.; Karunarathna, H.; Reeve, D.E. Investigation of deep sea shelf sandbank dynamics driven by highly energeti tidal flows. *Mar. Geol.* **2016**, *380*, 245–263. [CrossRef]
17. Salles, T.; Ding, X.; Webster, J.M.; Vila-Concejo, A.; Brocard, G.; Pall, J. A unified framework for modelling sediment fate fror source to sink and its interactions with reef systems over geological times. *Sci. Rep.* **2018**, *8*, 5252. [CrossRef] [PubMed]
18. van de Lageweg, W.I.; Feldman, H. Process-based modelling of morphodynamics and bar architecture in confined basins wit fluvial and tidal currents. *Mar. Geol.* **2018**, *398*, 35–47. [CrossRef]
19. van de Lageweg, W.I.; Braat, L.; Parsons, D.R.; Kleinhans, M.G. Controls on mud distribution and architecture along th fluvial-to-marine transition. *Geology* **2018**, *46*, 971–974. [CrossRef]
20. Nnafie, A.; Wolf, T.B.J.; de Swart, H.E. Tidal sand ridges on the shelf: A numerical study of their natural morphodynami evolution and response to interventions. *Cont. Shelf Res.* **2020**, *205*, 104195. [CrossRef]
21. Ulmishek, G.F. Petroleum Geology and Resources of the West Siberian Basin, Russia. *US Geol. Surv. Bull.* **2003**, *2201G*, 49.
22. Shemin, G.; Deev, E.; Vernikovsky, V.A.; Drachev, S.S.; Moskvin, V.; Vakulenko, L.; Pervukhina, N.; Sapyanik, V. Jurassi paleogeography and sedimentation in the northern West Siberia and South Kara Sea, Russian Arctic and Subarctic. *Mar. Pet. Gec* **2019**, *104*, 286–312. [CrossRef]
23. Kontorovich, V.A.; Ayunova, D.V.; Gubin, I.A.; Kalinin, A.Y.; Kalinina, L.M.; Kontorovich, A.E.; Malyshev, N.A.; Skvortso M.B.; Solov'ev, M.V.; Surikova, E.S. Tectonic evolution of the Arctic onshore and offshore regions of the West Siberian petroleur province. *Russ. Geol. Geophys.* **2017**, *58*, 343–361. [CrossRef]
24. Sobornov, K.; Afanasenkov, A.; Gogonenkov, G. Strike-slip faulting in the northern part of the West Siberian Basin and Enisey Khatanga Trough: Structural expression, development and implication for petroleum exploration. In Proceedings of the 3P Arcti Conference, Stavanger, Norway, 29 September–2 October 2015.
25. Kontorovich, A.E.; Kontorovich, V.A.; Ryzhkova, S.V.; Shurygin, B.N.; Vakulenko, L.G.; Gaideburova, E.A.; Danilova, V.F Kazanenkov, V.A.; Kim, N.S.; Kostyreva, E.A.; et al. Jurassic paleogeography of the West Siberian sedimentary basin. *Russ. Gec Geophys.* **2013**, *54*, 747–779. [CrossRef]
26. Kontorovich, A.E.; Burshtein, L.M.; Malyshev, N.A.; Safronov, P.I.; Gus'kov, S.A.; Ershov, S.V.; Kazanenkov, V.A.; Kim, N.S Kontorovich, V.A.; Kostyreva, E.A.; et al. Historical-geological modeling of hydrocarbon generation in the Mesozoic–Cenozoi sedimentary basin of the Kara Sea (basin modeling). *Russ. Geol. Geophys.* **2013**, *54*, 917–957. [CrossRef]
27. Lesser, G.R.; Roelvink, J.A.; van Kester, J.A.T.M.; Stelling, G.S. Development and validation of a three-dimensional morphologica model. *Coast. Eng.* **2004**, *51*, 883–915. [CrossRef]
28. Carballo, R.; Iglesias, G.; Castro, A. Numerical model evaluation of tidal stream energy resources in the Ría de Muros (NW Spain *Renew. Energy* **2009**, *34*, 1517–1524. [CrossRef]
29. Zheng, X.; Mayerle, R.; Wang, Y.; Zhang, H. Study of the wind drag coefficient during the storm Xaver in the German Bight usin data assimilation. *Dyn. Atmos. Ocean.* **2018**, *83*, 64–74. [CrossRef]
30. dos Santos, V.H.M.; da Silva Dias, F.J.; Torres, A.R.; Soares, R.A.; Terto, L.C.; de Castro, A.C.L.; Santos, R.L.; Cutrim, M.V. Hydrodynamics and suspended particulate matter retention in macrotidal estuaries located in Amazonia-semiarid interfac (Northeastern-Brazil). *Int. J. Sediment Res.* **2020**, *35*, 417–429. [CrossRef]
31. Tan, L.S.; Ge, Z.M.; Fei, B.L.; Xie, L.N.; Li, Y.L.; Li, S.H.; Li, X.Z.; Ysebaert, T. The roles of vegetation, tide and sediment in th variability of carbon in the salt marsh dominated tidal creeks. *Estuar. Coast. Shelf Sci.* **2020**, *239*, 106752. [CrossRef]

2. Geleynse, N.; Storms, J.E.A.; Walstra, D.-J.R.; Jagers, H.R.A.; Wang, Z.B.; Stive, M.J.F. Controls on river delta formation; insights from numerical modelling. *Earth Planet. Sci. Lett.* **2011**, *302*, 217–226. [CrossRef]
3. Dmitrenko, I.; Volkov, D.; Stadnyk, T.; Tefs, A.; Babb, D.; Kirillov, S.; Crawford, A.; Sydor, K.; Barber, D. Atmospherically forced sea-level variability in western Hudson Bay, Canada. *Ocean Sci.* **2021**, *17*, 1367–1384. [CrossRef]
4. Pinet, N.; Lavoie, D.; Dietrich, J.; Hu, K.; Keating, P. Architecture and subsidence history of the intracratonic Hudson Bay Basin, northern Canada. *Earth-Sci. Rev.* **2013**, *125*, 1–23. [CrossRef]
5. Ottesen, D.; Dowdeswell, J.A.; Bugge, T. Morphology, sedimentary infill and depositional environments of the Early Quaternary North Sea Basin (56°–62°N). *Mar. Pet. Geol.* **2014**, *56*, 123–146. [CrossRef]
6. Sundal, A.; Nystuen, J.P.; Rørvik, K.-L.; Dypvik, H.; Aagaard, P. The Lower Jurassic Johansen Formation, northern North Sea—Depositional model and reservoir characterization for CO_2 storage. *Mar. Pet. Geol.* **2016**, *77*, 1376–1401. [CrossRef]
7. Meng, M.; Ge, H.; Shen, Y.; Ji, W.; Li, Z. Insight into water occurrence and pore size distribution by nuclear magnetic resonance in marine shale reservoirs, southern China. *Energy Fuels* **2023**, *37*, 319–327. [CrossRef]
8. Schuurman, F.; Marra, W.; Kleinhans, M. Physics-based modeling of large braided sand-bed rivers: Bar pattern formation, dynamics, and sensitivity. *J. Geophys. Res. Earth Surf.* **2013**, *118*, 2509–2527. [CrossRef]

Journal of
Marine Science and Engineering

MDPI

Article

Intelligent Prediction of Sampling Time for Offshore Formatior Testing Based on Hybrid-Driven Methods

Yiying Nie [1], Caoxiong Li [2], Yanmin Zhou [3], Qiang Yu [3], Youxiang Zuo [3], Yuexin Meng [3] and Chenggang Xian [2,*]

1 College of Artificial Intelligence, China University of Petroleum, Beijing 102200, China; 2022216670@student.cup.edu.cn
2 Unconventional Petroleum Science and Technology Institute, China University of Petroleum, Beijing 102200, China; licaoxiong@cup.edu.cn
3 China Oilfield Services Limited, Langfang 065201, China; zhouym3@cosl.com.cn (Y.Z.); yuqiang3@cosl.com.cn (Q.Y.); zuoyx@cosl.com.cn (Y.Z.); mengyx@cosl.com.cn (Y.M.)
* Correspondence: xianchenggang@cup.edu.cn

Abstract: Formation testing is widely used in offshore oil and gas development, and predicting th sampling time of pure fluids during this process is very important. However, existing formatio testing methods have problems such as long duration and low efficiency. To address these issues, thi paper proposes a hybrid-driven method based on physical models and machine learning models t predict fluid sampling time in formation testing. In this hybrid-driven model, we establish a digita twin model to simulate a large amount of experimental data (6000 cases, totaling over 1 millio data points) and significantly enhance the correlation between features using physical formulas. B applying advanced machine learning algorithms, we achieve real-time predictions of fluid samplin time with an accuracy of up to 92%. Additionally, we use optimizers to improve the model's accurac by 3%, ultimately reaching 95%. This model provides a novel approach for optimizing formatio testing that is significant for the efficient development of offshore oil and gas.

Keywords: formation testing; probe sampling; neural network; MLP; hybrid-driven

Citation: Nie, Y.; Li, C.; Zhou, Y.; Yu, Q.; Zuo, Y.; Meng, Y.; Xian, C. Intelligent Prediction of Sampling Time for Offshore Formation Testing Based on Hybrid-Driven Methods. *J. Mar. Sci. Eng.* **2024**, *12*, 1348. https://doi.org/10.3390/jmse12081348

Academic Editor: Atilla Incecik

Received: 18 June 2024
Revised: 3 August 2024
Accepted: 5 August 2024
Published: 8 August 2024

1. Introduction

As a vital component of petroleum resources, offshore oil and gas have always been a focus of attention in oil and gas development [1]. Formation testing is crucial in offshore oil development, especially given offshore oil extraction's high costs and risks. The overal development costs of offshore oil extraction are significant, including those for drilling platform construction, and maintenance, making minimizing trial-and-error costs essentia Formation testing provides critical subsurface reservoir information by directly measuring formation pressure, temperature, and fluid properties, helping to determine the presence scale, and production potential of oil and gas reservoirs [2]. Figure 1 shows the applicatio of formation testing in offshore oil development. Formation testing is essential in offshore oil development to detect subsurface information. These formation test data are crucia for optimizing extraction plans, maximizing recovery rates, and reducing uncertainties thereby effectively lowering development risks and costs [3].

Formation testing plays an indispensable role at each stage. During the exploratior stage, formation testing provides preliminary data about the reservoir, confirming the presence and scale of oil and gas reservoirs, thus avoiding the risks and costs associatec with blind drilling. Formation testing optimizes well placement and pattern layout in the development stage, ensuring maximum recovery rates and economic benefits. During the production phase, continuous formation testing can monitor changes in reservoir pressure and fluid properties, helping adjust production strategies to extend the oil field's life anc ensure sustainable and economic production [4].

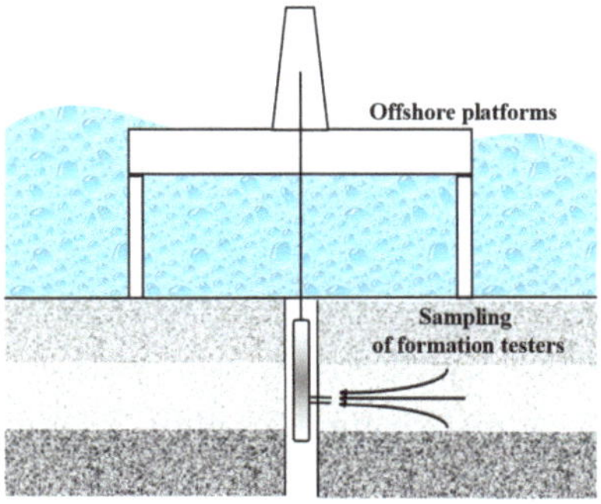

Figure 1. Application of formation testing in offshore oil development.

Additionally, formation testing can detect and assess potential leakage risks early, allowing for preventive measures to protect the marine environment. Therefore, formation testing is an essential part of offshore oil development, ensuring a more efficient, safe, and economical oil extraction process.

Predicting pure fluid sampling time is crucial in formation testing, especially when oil-based mud (OBM) or synthetic-based mud (SBM) filtrates mix with crude oil and are hard to separate. When crude oil samples are contaminated by OBM filtrates beyond a certain level, it becomes difficult to determine the properties of the original crude oil [5]. Figure 2 shows a profile of contaminated crude oil samples. The pollution zones represent the part contaminated by mud filtrates, while the non-pollution zones represent the formation fluid, which is the desired sample for analysis. During sampling, the probe first collects contaminated samples. As sampling time increases, the contamination level decreases, eventually obtaining pure fluid. Accurately determining the contamination level to infer the properties of the original crude oil from contaminated samples is a critical part of formation testing.

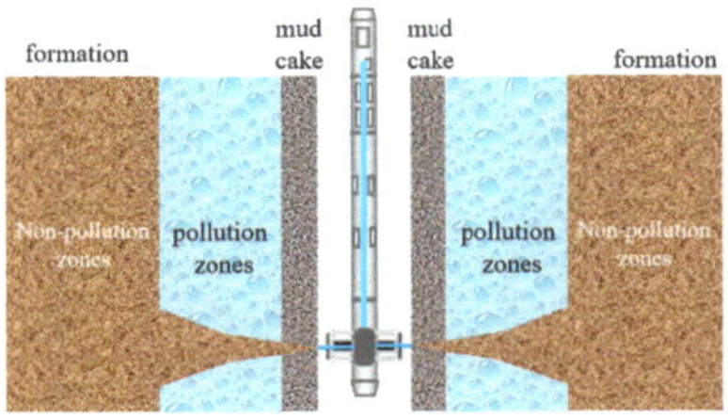

Figure 2. Diagram of crude oil sample contamination by invasion.

Current methods for predicting and determining pure fluid sampling time and fluid contamination levels rely mainly on optical fluid identification, sensor fluid property measurements, and resistivity. In 2000, Mullins and Schroer introduced an optical fluid identification module for real-time monitoring using optical fluid analysis (OFA) data to assess OBM filtrate contamination during MDT sampling [6]. By 2006, C. Del Campo and colleagues developed a new "focused sampling" device with an innovative formation testing probe that effectively separated drilling fluid filtrate contamination, allowing for faster acquisition of clean reservoir fluid samples [7]. In 2008, Hsu and others created a new model to calculate contamination using multi-wavelength OD measurement data, recognizing that traditional fluid cleanup simulations were overly optimistic because of incomplete mud cake formation during LWD measurements [8]. In 2009, Abdolhamid's study used a 3D multiphase, multicomponent reservoir simulator to understand mud filtrate invasion, considering gravity and capillary pressure, and assessed the impact of sampling time on fluid sample quality [9]. With advances in real-time downhole fluid measurement technology, Zuo and colleagues in 2015 developed a contamination monitoring workflow using multi-sensor fluid property measurements, improving the accuracy an robustness of quantifying hydrocarbon contamination mixed with OBM. Gisolf and other proposed a method for quantifying water sample contamination using on-site fluid densit and resistivity measurements [10]. In 2016, Ryan Lee and colleagues introduced a new focused sampling parameter estimation algorithm using direct sensor measurements fc more accurate and reliable estimation of formation fluid properties [11]. Overall, currer methods predict or observe other indicators to indirectly determine the timing for pur fluid sampling, which is time-consuming and affects timely decision-making. Achievin real-time downhole determination of water sample filtrate contamination and direct pr diction of sampling time will significantly reduce offshore oil development costs and risk Table 1 presents the main methods for determining formation test sampling times and th various explorations conducted by previous researchers.

Table 1. Review of formation test sampling times.

Author	Year	Input	Description	Output
O.C. Mullins [6]	2000	OFA data	Developed an optical fluid-recognition module for real-time monitoring using OFA data to determine the percentage of OBM filtrate contamination during sampling	Filtrate contamination
M.E. Chenevert [12]	2001	Filtration measurement data of 100 water-based muds	Developed a theory to predict mud cake buildup and filtrate invasion, measured various filtration characteristics of water-based muds, and developed a corresponding numerical simulator	Prediction of mud cake buildup and filtrate invasion
J. Wu [9]	2004	Formation parameters	Developed an effective time-dependent flow rate function that captures the effects of mud cake buildup to simulate complex mud filtrate invasion scenarios	Simulation of mud filtrate invasion
K. Hsu [8]	2006	Parameters of OBM filtrate contamination	Studied the physical mechanisms of OBM filtrate contamination cleanup and constructed and validated a numerical model capable of handling multicomponent fluid flow and the thermodynamics of phase behavior.	Numerical model output
Del Campo [7]	2006	Drilling mud filtrate and reservoir fluid samples	This paper introduced a new "focused sampling" device that rapidly separates drilling mud filtrate, improving sample quality, and presented real-time fluid characterization techniques to optimize the sampling process	Faster acquisition of clean fluid samples and real-time fluid property information
Bon Johannes [13]	2007	Methods for collecting reservoir fluid samples and downhole conditions	Explores the impact of downhole conditions and fluid characteristics on sample quality as well as the application of single-phase and isokinetic sampling methods	Quality and accuracy of representative fluid samples
A. Hadibeik [14]	2009	Formation parameters	Developed a 3D multiphase, multicomponent reservoir simulator considering the gravity and capillary pressure, studied the impact of the pollution function, and evaluated the impact of sampling time on fluid sample quality	Sampling time
F.O. Alpak [15]	2015	Probe shape	The shape and layout of the sampling probe are crucial for obtaining low-contamination samples in a short time	Sampling time
J.Y. Zuo [10]	2015	Sensor fluid characteristic measurement data	Developed a pollution monitoring workflow based on multiple sensor fluid characteristic measurements, improving the accuracy and robustness of pollution quantification	Contamination degree
R. Lee [11]	2016	Fluid density and resistivity measurement data	Proposed a new water-sampling contamination quantification method applicable to all fluid combinations, demonstrating its effectiveness and robustness through multiple case studies	Contamination degree

In recent years, the developments of digital twin technology [16], artificial intelligence (AI) [17], and big data [18] have provided new approaches for predicting pure fluid sampling time. Digital twin technology combines physical systems with digital simulations, enabling virtual modeling of real operations and making sampling time predictions more accurate and reliable. Through digital twin models, the sampling process can be simulated under various conditions and continuously adjusted and optimized based on real-time monitoring data, achieving precise predictions of pure fluid sampling time. Additionally, the use of AI and big data technology allows for a more comprehensive and in-depth analysis of various variables in offshore oil development, providing more data support and an analytical basis for predicting pure fluid sampling time. These new technologies offer powerful tools and methods for optimizing the sampling process and improving sample quality, which is expected to further enhance the efficiency and reliability of offshore oil development.

This study addresses the prediction of pure fluid sampling time in offshore oil development. By combining physical models and machine learning with large-scale simulated data generated by surrogate models, we establish a hybrid-driven model. This model significantly enhances the correlation between features and target variables, enabling real-time prediction of pure fluid sampling time with an accuracy exceeding 95%. In field development, using this model to predict sampling time allows for better estimation of the pump-out duration and reduces the number of sampling analyses required. This has significant implications for the efficient development of offshore oil resources.

2. Materials and Methods

This paper presents a hybrid-driven model that combines physical methods and machine learning, significantly enhancing the accuracy and speed of predicting pure fluid sampling time in offshore oil development while adhering to physical principles. The detailed process is illustrated in Figure 3. Initially, this paper develops a proxy model to simulate the probe sampling seepage. Next, this paper processes the simulated data to construct a comprehensive database for the digital twin. Finally, this paper integrates a machine learning model with physical relationships to create a hybrid-driven model for predicting probe sampling time.

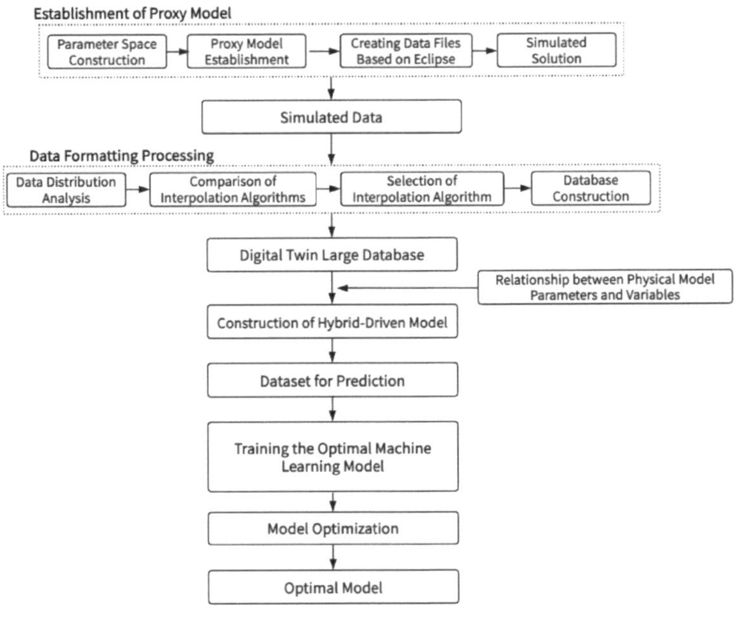

Figure 3. Workflow of the hybrid-driven model for predicting pure fluid sampling time.

2.1. Proxy Model Development

This paper establishes a proxy model based on Eclipse for the pumping process c downhole fluid sampling during formation testing with 3D probes, standard probes, larg type inlet probes, large-type plate probes, elliptical probes, and dual packers. The mode introduced a new OBM-contamination-monitoring (OCM) algorithm using the inversio of downhole-fuid-analysis(DFA) data,references the new OCM algorithm proposed b Morten Kristensen in 2019, which uses a full 3D numerical flow model to invert DFA dat and reverses sensor data in real time to provide contamination predictions [19].

In this proxy model, six independent parameters affect the cleanup behavior: pe meability anisotropy (kv/kh), the radius of filtrate invasions (doi), Wellbore diamete (rw), formation thickness (H), relative tool distance from formation top (h), and formatio fluid/mud-filtrate viscosity ratio (vrat) [20]. The proxy model constructs input parameter according to Equation (1). By running this model, this paper simulates the entire cleanu process and examines various parameter indicators, mainly including cleanup time (time cleanup volume (wvpt), pump rate (wvpr), and pressure (wbhp). Only one of these fou outputs is needed to fully describe the cleanup behavior. For example, a tool operating a a constant pump rate can be described by cleanup volume and pressure drop; similarl a tool operating at a constant pressure drop can be described by cleanup volume an rate. Equations (2)–(5) are derived from the paper by Kristensen, M. [19], illustrating th relationships among the four observed outputs. In the equations, t_j, V_j, Q_j, and P_j represen the cleanup time, cleanup volume, cleanup rate, and drawdown pressure at contaminatio level j, respectively. $\tilde{t}$, $\tilde{V}$, $\tilde{Q}$, and $\tilde{P}$ are the corresponding scaled quantities simulated by th proxy model.

$$\mathbf{x} = \left(\ln\frac{k_v}{k_h}, \ln\frac{\mu_o}{\mu_{mf}}, \ln R_{inv}, \ln D_w, \ln\frac{H}{\sqrt{k_v/k_h}}, z\right)^T \quad (1$$

$$\tilde{t} = [\tilde{t}_1, \cdots, \tilde{t}_l]^T, \quad \tilde{t}_j = \ln\left(\frac{t_j \cdot M \cdot P}{\phi}\right) \quad (2$$

$$\tilde{V} = [\tilde{V}_1, \cdots, \tilde{V}_l]^T, \quad \tilde{V}_j = \ln\left(\frac{V_j}{\phi}\right) \quad (3$$

$$\tilde{Q} = [\tilde{Q}_1, \cdots, \tilde{Q}_l]^T, \quad \tilde{Q}_j = \ln\left(\frac{Q_j}{M \cdot P}\right) \quad (4$$

$$\tilde{P} = [\tilde{P}_1, \cdots, \tilde{P}_l]^T, \quad \tilde{P}_j = \ln\left(\frac{P_j \cdot M}{Q}\right) \quad (5$$

2.2. Constructing Training Data

Based on the proxy model, this paper constructed the input parameter space, ensurin parameters are within typical maximum and minimum values, as shown in Table 2.

Table 2. Typical range of parameters in proxy model.

Parameter	Min Value	Max Value	Mean	Median	Standard Deviation	Skewness	Unit
Wellbore Diameter	2.9375	6.125	4.53	4.53	0.93	0	inch
Radius of Filtrate Invasions	2	30	16	16	8.17	0	inch
Permeability Anisotropy	0.01	100	0	0	1.167	0	-
Formation Thickness	0.5	100	50	50	30	0	ft
Fluid Viscosity Ratio	0.01	100	0	0	1.167	0	-
Relative Tool Distance	0	0.5	0.25	0.25	0.15	0	-

To ensure diversity and comprehensiveness, each parameter was randomly distribute within its range, resulting in 6000 different cases for each probe type. The range of maximu and minimum values for the selected parameters was based primarily on the following sources: first, a statistical analysis of actual experimental data provided the typical range for each parameter; second, a review of numerous research papers in related fields helped determine reasonable values for parameters in different contexts; finally, industry standards and regulations were referenced to ensure that the parameter settings meet practical application requirements. The data within this parameter range included almost all on-site conditions. Therefore, as a large database, it was reasonable to include as much data as possible on all on-site situations, and the models trained with these data had a certain representativeness for the site. Table 2 shows the specific maximum and minimum values for each parameter. For example, the well diameter range was set as [5.27, 12.25] inches, and the filtrate invasion depth range was [2, 30] inches. These ranges ensure the parameter space is broad and representative.

To generate randomly distributed parameters, this paper used Python's random number generation library, numpy. The specific implementation is as follows: define the range for each parameter and then use the numpy.random.uniform function [21] to generate uniformly distributed random numbers within these ranges. Each parameter generated 6000 cases, ensuring data diversity and representativeness. Based on the above parameter space, this paper used commercial simulation software Eclipse to build a seepage model, including basic settings like geological structure, fluid properties, and boundary conditions. This paper then created numerous input data files from the randomly generated parameter space to serve as inputs for Eclipse simulations. Using Eclipse's batch processing feature, this paper performed batch simulations for all the generated input data files, with each case corresponding to an independent seepage simulation. Finally, this paper collected and organized the raw data generated by Eclipse simulations to form the initial dataset. After extensive and long-term multiple simulations, this paper generated a large amount of raw data. This data formed the initial dataset, providing a solid foundation for subsequent analysis and model optimization.

2.3. Data Processing and Construction of a Large Digital Twin Database

In the raw data generated by the proxy model, the sampling times vary for each parameter, and each data length is different. To examine the sampling timing on the same scale, we needed to process the data to obtain rate (wvpr), volume (wvpt), time (time), and pressure (wbhp) at the same contamination level (wspc) intervals. Before data interpolation, we had to examine the data distribution to choose the appropriate preprocessing method and the optimal interpolation function [22]. Taking the standard probe data as an example, this paper selected data with wspc in the range of [0, 1] from the raw data and plotted the relationships between wspc and wvpr, wspc and wvpt, and wspc and time, as shown in Figure 4.

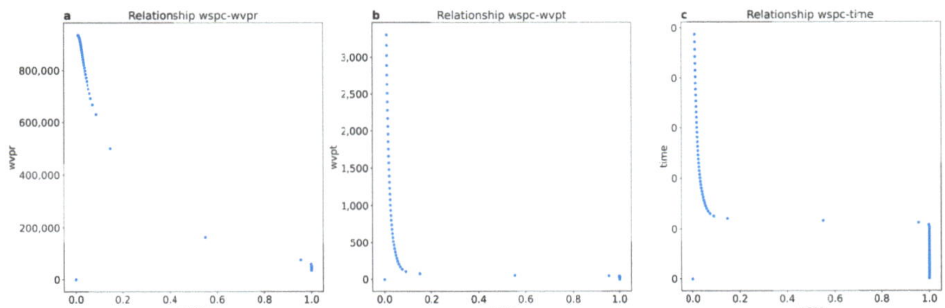

Figure 4. Relationships between wspc and wvpr (**a**), wspc and wvpt (**b**) and wspc and time (**c**).

We find that wspc (horizontal axis) follows a logarithmic distribution. Therefore, this paper interpolated wspc at interval points in the logarithmic space.

After analyzing the data distribution, we needed to choose the appropriate interpo lation algorithm. We tested various interpolation methods, considering the nonlinearit of parameter distribution, and experimented with several nonlinear interpolation meth ods. Using the first dataset of the standard probe as an example, this paper tested th effectiveness of each interpolation function, as shown in Figure 5.

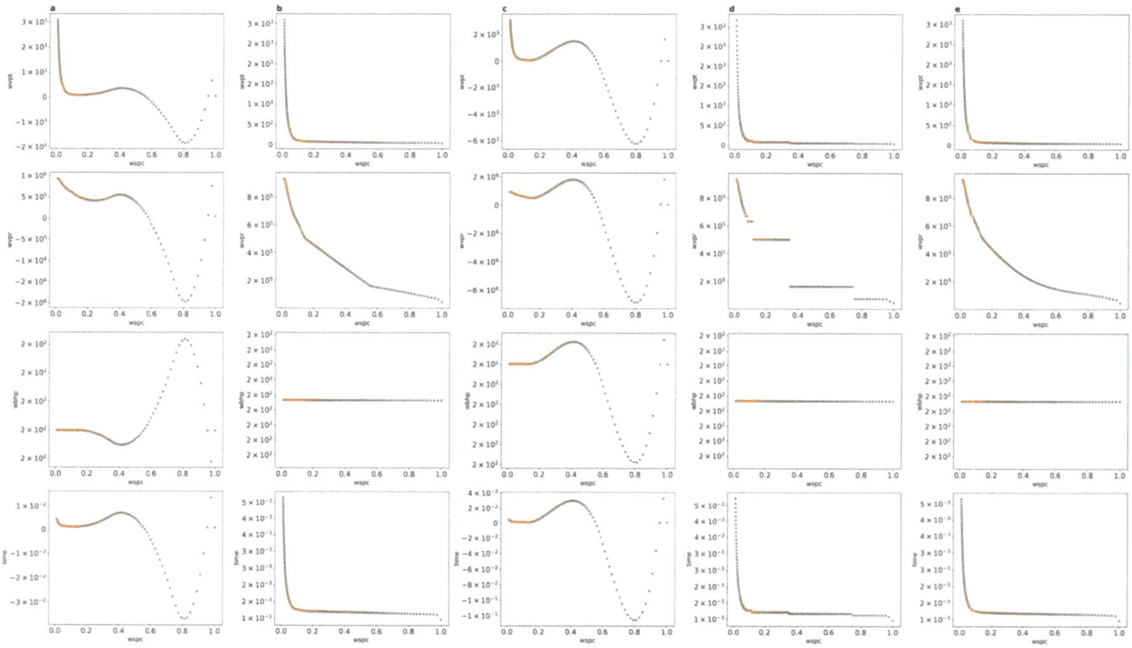

Figure 5. Interpolation results (from left to right, (**a**) quadratic spline interpolation, (**b**) linear interpo lation, (**c**) cubic spline interpolation, (**d**) nearest interpolation, and (**e**) Pchip spline interpolation).

These tests helped us determine the most suitable interpolation method to accuratel reflect the probe measurements. From the analysis of the interpolation results, it is eviden that the nearest interpolation does not conform to the original data pattern on wvpr. Both quadratic and cubic spline interpolations exhibit the Runge phenomenon [23] becaus of their high degrees and the uneven distribution of the data. Among the remainin interpolation methods, Pchip spline interpolation [24] handles data points more smoothl and aligns better with the original data trend. Therefore, this study ultimately selecte Pchip spline interpolation for subsequent interpolation processing.

Regarding the interpolation range, based on experience and the range and distributio of most data, this paper first interpolated to generate 200 points for wspc within the interva $[-2, 0]$ in the logarithmic space. Then, this paper interpolated the remaining columns base on wspc. Each case resulted in a target parameter matrix with 200 rows and 5 column (5 evaluation parameters). After completing the interpolation process described abov a digital twin database containing 6000×200 observation target values under differen stratigraphic parameters was formed for each probe.

2.4. Correlation Analysis

After constructing the database using proxy models and data interpolation, this pape used stratigraphic parameters as input features and fluid sampling time as the target vari able. The goal was to predict sampling time using six input features (log(kv/kh), doi, H, h log(vrat)). Improving the correlation between features and the target variable could signif cantly enhance the model's predictive accuracy and performance. This approach reduce training time and resource consumption, lowers the risk of overfitting, increases model interpretability, and optimizes the feature engineering and selection process. Ultimately, it makes the model more effective and reliable in practical applications [25]. For example, using the XLProbe, this paper employed Python's heatmap function [26] to visualize the correlation between formation parameters (input features) and the target variable (fluid sampling time) (see Figure 6).

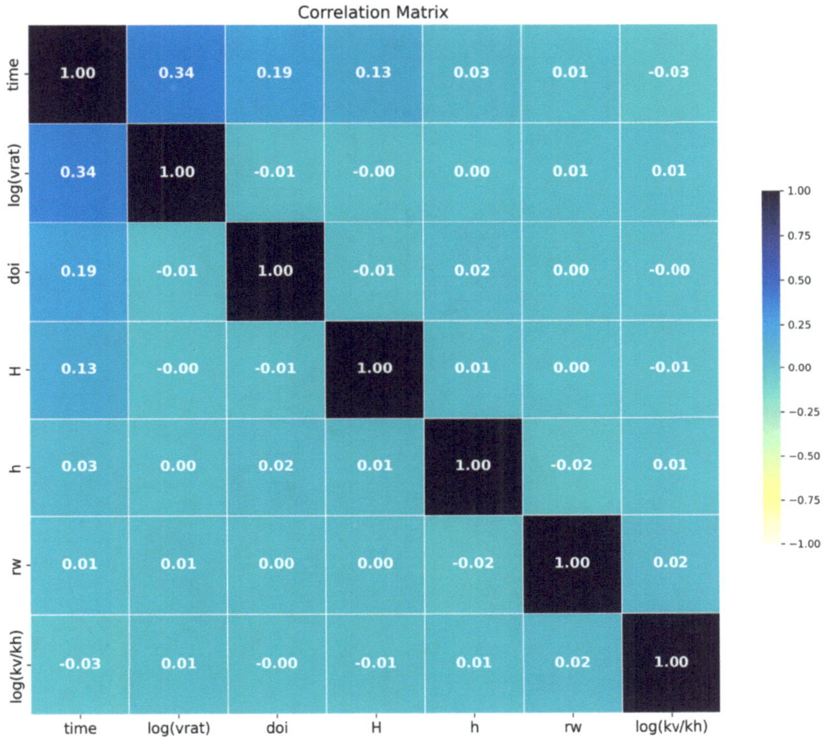

Figure 6. Feature–target correlation before adding physical correlations.

The results show that the correlation between the features and the target variable is generally low, with only log(vrat) and time showing a correlation of 0.34, while the rest are below 0.2 with an average of just 0.11.

$$\tilde{t_j} = \ln\left(\frac{t_j \cdot M \cdot P}{\phi}\right) \tag{6}$$

The low correlation between features and the target variable directly affects the training and prediction of subsequent machine learning models. To improve this correlation, this paper incorporated physical relationships (Equations (1)–(5)) to enhance the input features' relevance to the target variable. The physical model used was the same as that employed during the proxy model construction. Time satisfies Equation (6), where $M = k_h/\mu_o$, P is the drawdown pressure and ϕ is the porosity. The input features adhere to the relationships defined in Equation (1).

After incorporating the physical model, the changes in features are shown in Table 3.

Table 3. Comparison of features and targets before and after processing.

Before Processing	After Processing
t	$\ln\left(\frac{tj \cdot M \cdot P}{\phi}\right)$
h	h
doi	$\ln(doi)$
rw	$\ln(rw)$
H	$\ln\left(\frac{H}{\sqrt{kv/kh}}\right)$
$\log 10(kv/kh)$	$\ln(kv/kh)$
$\log 10(vrat)$	$\ln(vrat)$

The correlation between features and the target variable improved significantly. The heatmap of the correlation after adding the hybrid-driven features is shown in Figure 7.

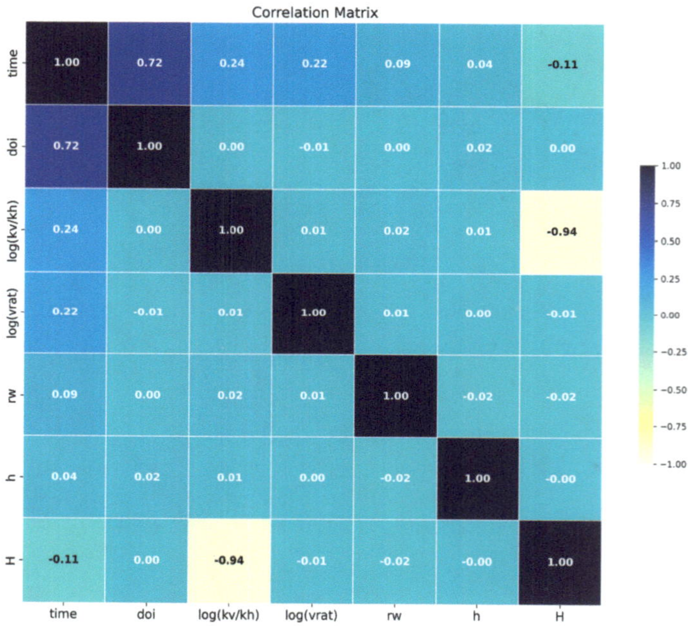

Figure 7. Feature–target correlation after adding physical correlations.

The highest correlation increased from 0.34 to 0.72, and the overall correlation also improved, with the average correlation rising to 0.2. This enhancement in feature–target correlation, driven by the physical model, lays a solid foundation for the subsequent model development.

2.5. Construction of Hybrid-Driven Model

After data interpolation and enhancing correlation, this paper transformed the data generated by the proxy models into high-quality big data suitable for machine learning training. Building upon this foundation, this paper introduced machine learning models integrating traditional physics simulations with artificial intelligence techniques. This integration resulted in a hybrid-driven model that adheres to physical principles while leveraging machine learning for real-time and accurate predictions.

Within the context of data-driven simulation, this paper employed machine learning and artificial intelligence methods, utilizing different models for training to obtain the optimal predictive model. The primary goal of these models is to accurately predict the variation in the target purity of formation fluids obtained by various sealing mechanisms such as probes and packers, during the sampling process. Through training these models, this paper aimed to understand and capture the dynamic evolution of formation fluid purity. This enables us to better predict the target purity of formation fluids collected by different sealing mechanisms over time during sampling operations. Such predictive models are expected to play a crucial role in the oil and gas exploration and production field, enhancing sampling efficiency and accuracy and providing more reliable support for geological and engineering decisions. The machine learning model training process [27] is shown in Figure 8, where the input features are the following processed features: $\ln(kv/kh)$, $\ln(vrat)$, $\ln(doi)$, $\ln(rw)$, $\ln(H/\sqrt{kv/kh})$, and h. The target variable is $\ln(tj \cdot M \cdot P/\phi)$.

In selecting the research model, this paper experimented with various machine learning models, including linear models [28], support vector machines [29], XGBoost [30], decision trees, random forests [31], and multilayer perceptrons [32]. Additionally, selecting an appropriate optimizer is a critical step in the neural network training process. The optimizer determines how to update the model's weights to minimize the loss function, thereby improving model performance and accuracy. Testing and selecting the best optimizer during model optimization helps enhance model performance. For instance, comparing the performance of Adam, Nadam, and RMSprop allowed us to choose the most suitable optimizer for model training, achieving higher accuracy and faster convergence speeds.

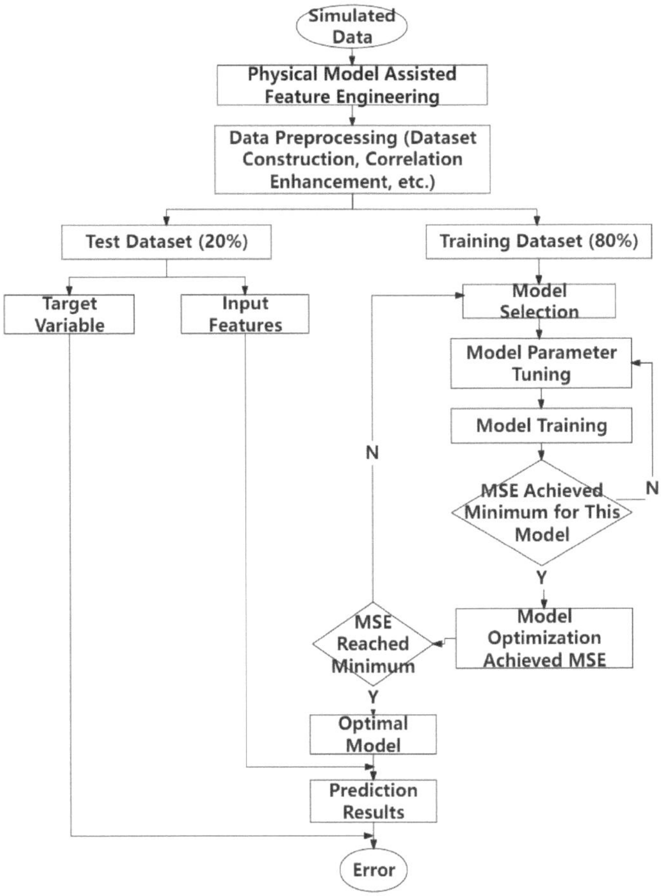

Figure 8. Hybrid-driven machine learning model training process.

3. Results and Discussion

In this section, this paper used simulated data generated by proxy models for machin learning training and prediction. The training and testing datasets were split in an 80:2 ratio [33]. This paper evaluated each model using R^2 score, MSE, and loss. The coefficier of determination R^2 (R-squared score) is a statistical metric used in regression analysis t assess the predictive performance of the model [34]. It indicates the proportion of varianc in the target variable that is explained by the model. R^2 score is a unitless measure rangin between zero and one, where values closer to one indicate better model performance. Th coefficient of determination (R^2) is given by $R^2 = 1 - \frac{SS_{\text{res}}}{SS_{\text{tot}}}$, where $SS_{\text{res}} = \sum_{i=1}^{n}(y_i - \hat{y}_i)$ and $SS_{\text{tot}} = \sum_{i=1}^{n}(y_i - \bar{y})^2$, with $\bar{y} = \frac{1}{n}\sum_{i=1}^{n} y_i$. Here, y_i is the actual value of the i-th sampl (i.e., the true sampling time), $\hat{y}_i$ is the predicted value of the i-th sample (i.e., the model' predicted sampling time), and $\bar{y}$ is the mean of the target variable's sampling times. Th mean squared error (MSE) [35–37] is a commonly used metric to evaluate the performanc of regression models [38]. It measures the average of the squares of the differences betwee the predicted and actual values: $\text{MSE} = \frac{1}{n}\sum_{i=1}^{n}(y_i - \hat{y}_i)^2$. The loss function of a model i very similar to MSE [39]. The loss function is the metric used to optimize the model durin training. Different models may use different loss functions, depending on the model typ and the nature of the problem. In neural networks, a weighted MSE is typically used as th loss function: $\text{Loss} = \frac{1}{n}\sum_{i=1}^{n}(y_i - \hat{y}_i)^2 + \lambda\sum_{j=1}^{m} w_j^2$, where λ is the regularization paramete and w_j is the j-th weight of the model.

In the model evaluation, this study integrated multiple indicators mentioned above t comprehensively assess the model's performance from various dimensions and ultimatel select the optimal model.

3.1. Correlation Enhancement

After incorporating physical relationships, the model showed significant improvemer in target–feature correlations. Compared to the model without these enhancements, th model saw a 74.21% reduction in error percentage.

As shown in Figure 9, before incorporating physical correlations, the model's perfo mance was poor.

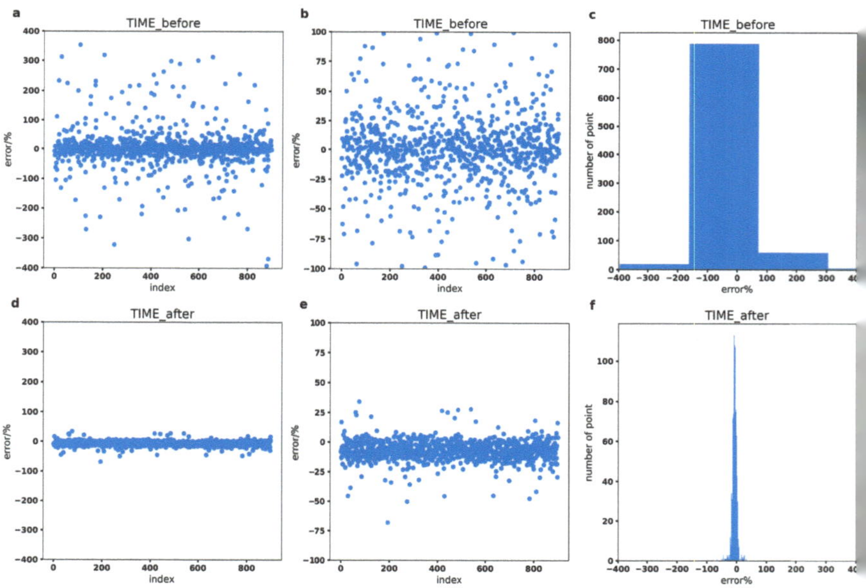

Figure 9. Before (**a**–**c**) and after (**d**–**f**) incorporating physical relationships, scatter plots of mode error percentages are shown overall (**left**), locally (**middle**), and as a distribution histogram (**right**

The error scatter plot indicates that most points have large errors, reflecting poor model fitting, making prediction nearly impossible. The error distribution histogram also shows that about half of the points have errors exceeding 100%. Overall, the model's performance was suboptimal and unsuitable for practical predictions, with an average error percentage calculated at 82.31%. However, after integrating physical correlations for auxiliary processing, the model's performance improved significantly. As illustrated in Figure 9, most of the absolute error percentages fall within 15% or even 10%. The average error percentage dropped to 8.10%, demonstrating that combining physical correlations with machine learning models, or the hybrid-driven approach, greatly enhanced the accuracy of the machine learning model.

3.2. Optimal Model Selection

To select the optimal model, this study employed experimental testing. The models tested included linear regression, support vector machines, XGBoost, decision trees, random forests, and multilayer perceptrons. Using the same training data, different models were evaluated based on R^2 score, MSE, and mean percentage error. After testing various models, the performance metrics were obtained, as shown in Table 4. From Table 4, it can be seen that the MLP model, trained using forward and backward propagation algorithms to adjust network parameters gradually and minimize the loss function, performed exceptionally well in handling the large-scale and complex dataset generated by the proxy model in this study. It achieved the highest R^2 score and the lowest MSE among all models. Linear regression showed the poorest performance, while MLP had the best performance, with support vector machine (SVM) following closely.

Table 4. Comparison of model performance.

Model	R^2	MSE	Mean Absolute Percentage Error (%)
MLP	0.9969	0.0107	8.0997
Support Vector Machine	0.9872	0.0322	11.3467
Xgboost	0.9691	0.1064	30.9935
Decision Tree	0.7721	0.7857	258.1168
Linear Regression	0.6622	1.1647	1820.6289

Figure 10 shows the comparison of the normal distribution curves of errors for each model. It visually illustrates that the error distribution is most compact for the MLP model, followed by SVM. The XGBoost model also exhibited decent error performance, outperforming the remaining models but still falling short compared to MLP and SVM.

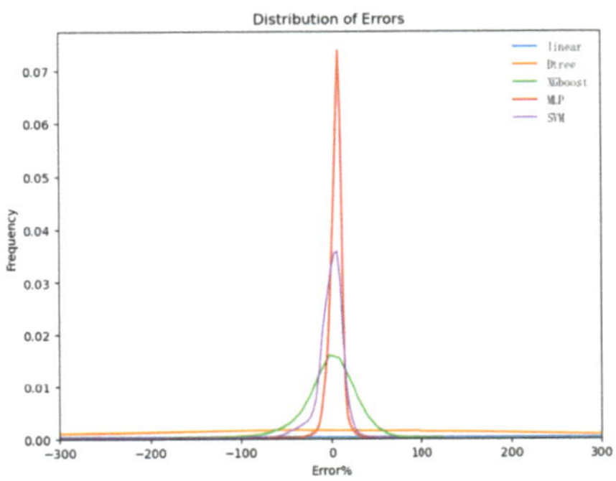

Figure 10. Performance of various machine learning models.

Overall, the MLP model significantly outperformed all other models, with an average error percentage 3.25% lower than that of the SVM model. Considering multiple evaluation metrics, the MLP model was ultimately selected as the machine learning prediction model.

3.3. Model Parameter Optimization

The optimizer plays a crucial role in MLP optimization. MLP is a neural networ comprising input, hidden, and output layers. During training, the optimizer continuousl adjusts model parameters to minimize the error between predicted and actual values, which is a critical step in model learning and optimization. Testing and selecting the best optimize during MLP optimization helps enhance model performance and accuracy. In this stud seven common MLP optimizers were tested, including Adam [40], Adamax, RMSprop Adagrad, Nadam, Adadelta, and FTRL. Evaluating these optimizers aimed to identify th most effective one for improving MLP model performance and prediction accuracy.

Table 5 presents the performance test results of the selected seven optimizers. From the table, it is evident that Adam and Adamax demonstrated the best performance. Adam performed the highest, while Adamax closely followed. On the other hand, FTRL, which is more suitable for sparse and high-dimensional data, showed the poorest performance which was not well-aligned with the characteristics of the dataset in this study. Overall, th Adam optimizer performs the best.

Table 5. Comparison of optimizer performance.

Optimizer	R^2	Loss
Adam	0.9976	0.0080
Adamax	0.9974	0.0085
RMSprop	0.9894	0.0356
Adagrad	0.9849	0.0507
Nadam	0.9837	0.0548
Adadelta	0.7881	0.7140
FTRL	0.6639	1.1324

Figure 11 provides a clearer visualization of the optimizers' performance. Adam optimizer showed higher R^2 scores and lower loss compared to the other optimizers. Based on these results, Adam optimizer was selected for subsequent model development to achieve higher precision and faster convergence speed.

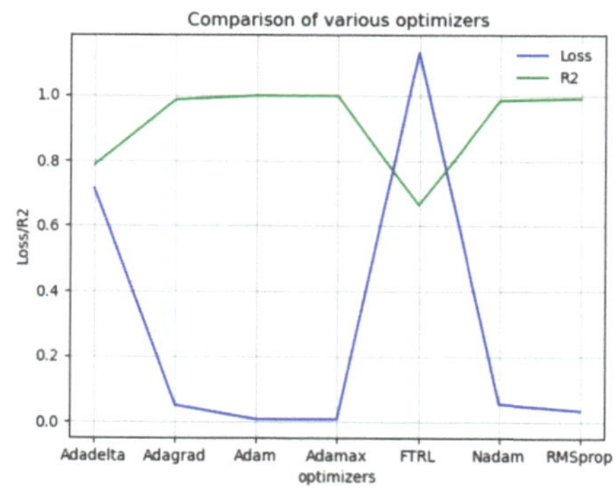

Figure 11. Comparison of performance of multiple optimizers.

Parameter tuning is also a significant optimization direction for neural networks. This study performed parameter optimization for the neural network, conducting simple tuning for epochs, batch size, and neurons using cross-validation. The test results are shown in Figure 12. As can be clearly seen from the figure, the model performed best when the epochs were set to 1000, the batch size to 32, and the neurons to 128.

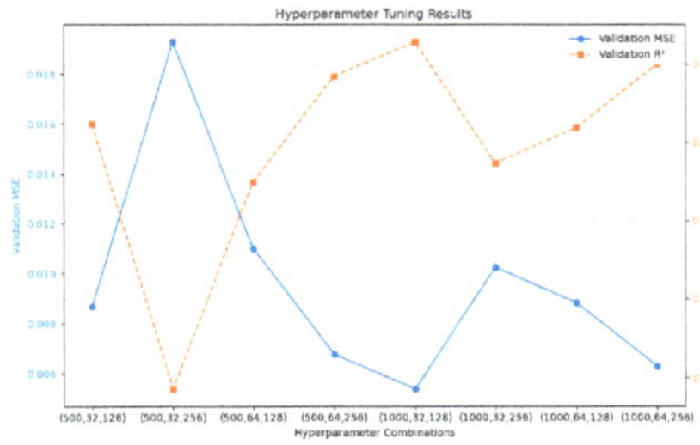

Figure 12. Neural network parameter tuning.

Figure 13 illustrates the performance comparison between the optimized and unoptimized models. It clearly shows that using the superior optimizer improved the model's performance. With Adam's optimizer, the scatter plot of errors converged better, and overall errors decreased. After computation, the error percentage of the MLP model using Adam optimizer reduced to 4.88%, which represents a 3.22% improvement compared to previous results.

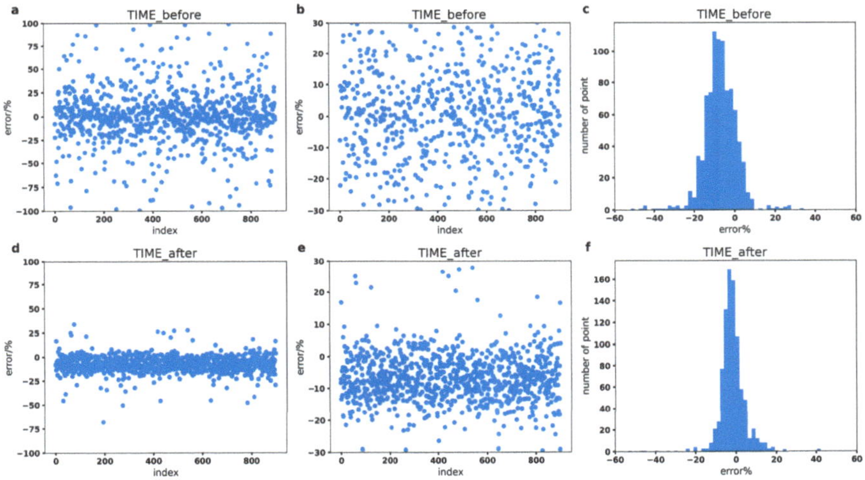

Figure 13. Before (**a–c**) and after (**d–f**) using Adam optimizer, scatter plots of model error percentages are shown overall (**left**), locally (**middle**), and as a distribution histogram (**right**).

In conclusion, optimizing MLP with Adam optimizer significantly enhanced model performance across various metrics, underscoring its suitability for achieving higher accuracy and faster convergence in this study.

3.4. Optimal Model Performance

After optimizing the model, the optimal MLP prediction model was obtained. Figure 14 displays the scatter plot of actual vs. predicted values for the optimized model. This plot shows the relationship between model predictions and actual values. The horizontal axis represents actual values, and the vertical axis represents predicted values, typically used to assess prediction accuracy. Ideally, all points should lie on the 45-degree diagonal line if the model predicts perfectly. From Figure 14, it is evident that most points are closely aligned along the diagonal line, indicating excellent model performance and accurate predictions.

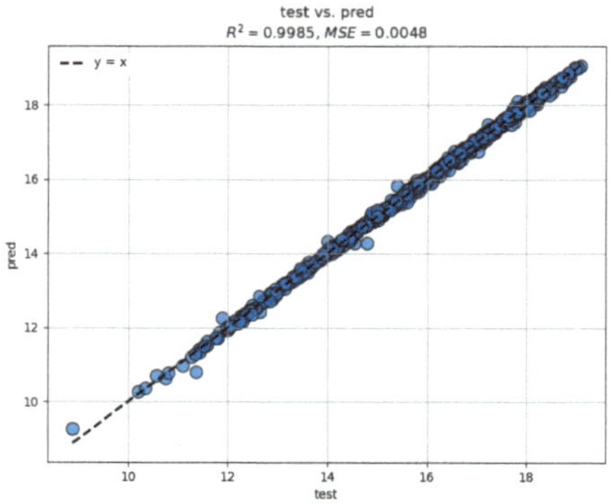

Figure 14. Scatter plot of actual vs. predicted values.

A residual plot displays the scatter of predicted values against residuals (the diffe ences between actual and predicted values). The horizontal axis represents predicte values, and the vertical axis represents residuals, which are used to check for patterns i prediction errors. Ideally, residuals should randomly scatter evenly around the horizonta zero line without noticeable patterns or trends [41]. Figure 15 shows the residual plot o the optimized model.

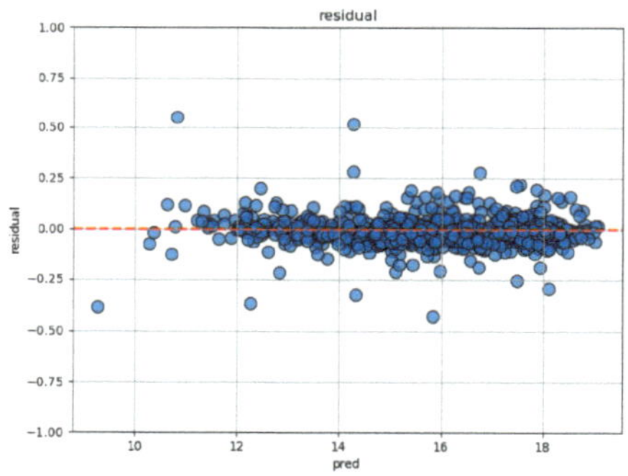

Figure 15. Model residuals.

From Figure 15, it can be observed that residuals are tightly distributed around th horizontal zero line without any obvious nonlinear patterns or spreading trends, indicatin ideal model performance.

Figure 16 presents the line plot of predicted vs. actual values for the optimized mode

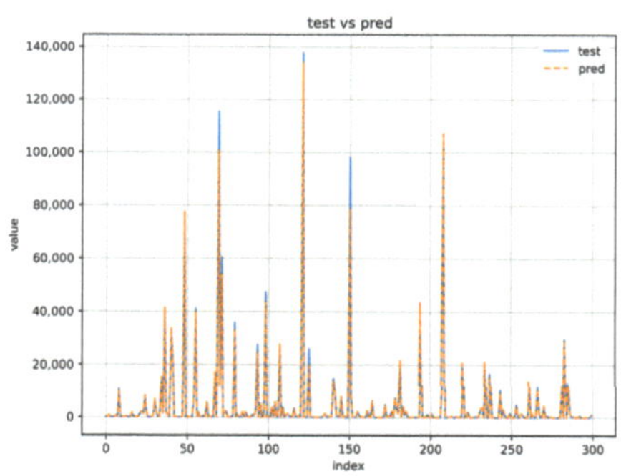

Figure 16. Line plot of predicted vs. actual values.

This plot provides a visual comparison of predicted values against actual values in terms of numerical differences. The x-axis represents point IDs, and the y-axis represents values. The orange dashed line represents predicted values, while the blue solid line represents actual values. From Figure 16, it is clear that very few predicted values deviate slightly from actual values, with the majority closely matching, demonstrating the excellent performance of the optimized MLP model.

3.5. Comprehensive Discussion and Practical Implications

This study integrated physical methods with machine learning to develop and validate a high-precision model for predicting the sampling time of pure fluids during offshore oil development. Compared to traditional field methods, this model significantly improves the accuracy of sampling time predictions, thereby optimizing formation testing plans and reducing the required number of sampling analyses. Traditional methods often rely on empirical judgments, which can lead to prolonged sampling procedures due to uncertainties and inaccuracies. In contrast, our model combines physical principles with machine learning to quickly and accurately predict the optimal sampling time, resulting in a more streamlined and efficient drilling process that reduces downtime and operational delays caused by fluid sampling.

Our model outperforms existing advanced technologies by offering higher predictive accuracy and better generalizability. Compared to traditional empirical methods and some of the most advanced machine learning models, our model demonstrates superior performance, highlighting its effectiveness in improving prediction accuracy. We compared the sampling times predicted by our model with the actual times recorded by formation testers. The results showed that the predicted sampling times closely matched the actual times, indicating that our model can reliably predict the optimal moments for fluid sampling. This not only improves the accuracy of the sampling process but also reduces the overall time spent on drilling operations.

In terms of time and cost efficiency, the methodology presented in this study offers significant advantages. By utilizing this method, it is possible to estimate the sampling time early in the logging process, eliminating the need for multiple downhole contaminated sample collections and analyses. This reduces both the total sampling duration and the number of sampling attempts. By shortening the time required for fluid sampling, the model helps decrease the overall duration of drilling operations, thereby directly lowering operational costs. Furthermore, reducing the frequency of equipment usage saves additional costs by decreasing wear and tear on sampling equipment, thus extending its operational lifespan. Overall, the application of this predictive model enhances both the time efficiency and cost-effectiveness of drilling operations, making it a valuable tool for the oil and gas industry.

The methodology developed in this study significantly enhances the automation and digitization of drilling operations. By integrating physical principles with machine learning algorithms, the model enables real-time, data-driven decision-making, reducing reliance on manual judgments. The predictive model can be embedded into automated control systems to monitor and adjust drilling parameters, ensuring optimal performance and minimizing downtime. Additionally, the digitized process allows for comprehensive data logging and analysis, facilitating continuous improvement and operational insights. This approach aligns with the industry's shift toward smarter, more sustainable practices.

Despite significant progress, the model has certain limitations. The current dataset may not encompass all possible geological conditions, which can affect the model's generalizability. Additionally, the model's performance may be influenced by the quality and diversity of the input data. Future research should focus on expanding the dataset sources and dimensions to enhance the model's robustness and applicability, optimizing algorithms to improve performance, and increasing the model's stability and reliability. In complex geological conditions or situations where real-time prediction results are required, the method proposed in this study can be attempted.

By reducing the frequency of equipment usage and optimizing the sampling process, this study also contributes to safety and environmental protection, aiding in energy efficiency and emission reduction. This aligns with the decarbonization goals of the oil and gas industry. Furthermore, the developed methodology is not limited to offshore oil development but can be extended to other drilling contexts, such as onshore oil exploration and geothermal energy extraction, providing valuable tools for various subsurface fluid sampling operations.

This study addresses a critical gap in predicting fluid sampling times by offering a quantitative, data-driven approach, filling the void left by current practices that largely rely on empirical judgment. The model provides a more rigorous prediction tool, enhancing both prediction accuracy and operational efficiency.

4. Conclusions

This article proposes a hybrid-driven model that combines big data and artificial intelligence data to achieve high-precision prediction of formation testing sampling time. Our main contributions include the following:

- This study established a digital twin model for downhole formation testers to simulate the process of obtaining pure fluid samples, forming a large database of sampling simulations.
- In the prediction of pure fluid sampling time, this research model improved the data feature correlation through physical formulas and combined machine learning to establish a hybrid-driven model; the accuracy of the model improved by 74.21%. Moreover, on high-quality processed data, the optimal selected model outperformed others by 3.25% in accuracy and post-parameter optimization; it improved accuracy by 3.22% compared to before optimization. The final accuracy of the model is 95.12%.
- Based on simulated cleaning process data, this study devised an intelligent prediction method, enabling rapid forecasting of the onset time for pure formation fluid extraction without the need for modeling on offshore platforms. It has the advantages of accuracy, speed, and real-time feedback. Subsequently, it will play a crucial role in determining the timing of downhole fluid sampling.
- In future research, efforts can be directed toward expanding the data sources and dimensions to enhance the model's generalizability. Additionally, optimizing algorithms through model integration and adaptive learning can improve performance robustness. Strengthening interpretability and stability also presents opportunities for further refinement.

Author Contributions: Conceptualization: Q.Y., Y.M. and C.X.; methodology: Y.N. and Y.Z. (Youxiang Zuo); software: C.L. and Y.N.; Validation: Y.N. and Y.Z. (Youxiang Zuo); writing—original draft preparation: Y.N.; writing—review and editing: C.L. and Y.N.; project administration: Y.Z. (Yanmin Zhou); formal analysis, Y.Z. (Yanmin Zhou); data curation: Y.Z. (Yanmin Zhou) and Y.Z. (Youxiang Zuo); resources: Y.Z. (Yanmin Zhou) and Y.Z. (Youxiang Zuo). All authors have read and agreed to the published version of the manuscript.

Funding: This research was supported by the project "Research on Formation Testing Pressure Interpretation and Sampling Methods Using Mini-DST" (Project No. G2317A-0414T063). We gratefully acknowledge the funding and support provided.

Institutional Review Board Statement: Not applicable.

Informed Consent Statement: Not applicable.

Data Availability Statement: The datasets presented in this article are not readily available due to commercial restrictions.

Conflicts of Interest: The authors declare no conflict of interest.

Nomenclature

Abbreviation	Full Term
OBM	Oil-based mud
SBM	Synthetic-based mud
MLP	Multilayer perceptron
AI	Artificial intelligence
DFA	Downhole fluid analysis
OCM	OBM-contamination-monitoring
Pchip	Piecewise cubic hermite interpolating polynomial
MSE	Mean squared error
R^2	R-squared (coefficient of determination)
Adam	Adaptive moment estimation
Nadam	Nesterov-accelerated adaptive moment estimation
RMSprop	Root mean square propagation
Adagrad	Adaptive gradient algorithm
Adadelta	Adaptive learning rate method
FTRL	Follow-the-regularized-leader
kv/kh	Permeability anisotropy
rw	Wellbore diameter
H	Formation thickness
h	Relative tool distance from formation top
vrat	Formation-fluid/mud-filtrate viscosity ratio
wspc	Wellbore sampling contamination
wvpr	Wellbore volume pump rate
wvpt	Wellbore volume pump time
wbhp	Wellbore bottom hole pressure
SVM	Support vector machine
XGBoost	Extreme gradient boosting

eferences

Tian, G.; Han, P. Research on the Application of Offshore Smart Oilfield Construction Based on Computer Big Data and Internet of Things Technology. *J. Phys. Conf. Ser.* **2021**, *1992*, 032002. [CrossRef]

Proett, M.; Walker, M.; Welshans, D.; Gray, C. Formation Testing While Drilling, a New Era in Formation Testing. In Proceedings of the SPE Annual Technical Conference and Exhibition, Denver, CO, USA, 5–8 October 2003; p. SPE–84087–MS. [CrossRef]

Proett, M.; Welshans, D.; Sherrill, K.; Wilson, J.; House, J.; Shokeir, R.; Solbakk, T. Formation Testing Goes Back To The Future. In Proceedings of the SPWLA 51st Annual Logging Symposium, Perth, Australia, 13–23 June 2010; p. SPWLA–2010–95856. Available online: https://onepetro.org/SPWLAALS/proceedings-pdf/SPWLA10/All-SPWLA10/SPWLA-2010-95856/1756838/spwla-2010-95856.pdf (accessed on 4 August 2024).

Golovko, J.; Jones, C.; Dai, B.; Pelletier, M.; Gascooke, D.; Olapade, P.; Van Zuilekom, A. Formation Fluid Microsampling While Drilling: A New PVT and Geochemical Formation Evaluation Technique. In Proceedings of the SPE Annual Technical Conference and Exhibition, Calgary, AB, Canada, 30 September–2 October 2019; p. D032S098R001. [CrossRef]

Partouche, A.; Yang, B.; Tao, C.; Sawaf, T.; Xu, L.; Nelson, K.; Chen, H.; Dindial, D.; Edmundson, S.; Pfeiffer, T. Applications of Wireline Formation Testing: A Technology Update. In Proceedings of the OTC Offshore Technology Conference, Houston, TX, USA, 4–7 May 2020; p. D031S038R001. [CrossRef]

Mullins, O.C.; Schroer, J. Real-Time Determination of Filtrate Contamination during Openhole Wireline Sampling by Optical Spectroscopy. In Proceedings of the SPE Annual Technical Conference and Exhibition, Dallas, TX, USA, 1–4 October 2000; p. SPE–63071–MS. [CrossRef]

Del Campo, C.; Dong, C.; Vasques, R.; Hegeman, P.; Yamate, T. Advances in Fluid Sampling with Formation Testers for Offshore Exploration. In Proceedings of the OTC Offshore Technology Conference, Houston, TX, USA, 1–4 May 2006; p. OTC–18201–MS. [CrossRef]

Hsu, K.; Hegeman, P.; Dong, C.; Vasques, R.R.; O'Keefe, M.; Ardila, M. Multichannel Oil-Base Mud Contamination Monitoring Using Downhole Optical Spectrometer. In Proceedings of the SPWLA Annual Logging Symposium, Austin, TX, USA, 25–28 May 2008; p. SPWLA–2008–QQQQ. Available online: https://onepetro.org/SPWLAALS/proceedings-abstract/SPWLA08/All-SPWLA08/SPWLA-2008-QQQQ/27872 (accessed on 4 August 2024).

Wu, J.; Torres-Verdín, C.; Sepehrnoori, K.; Delshad, M. Numerical Simulation of Mud-Filtrate Invasion in Deviated Wells. *SPE Reserv. Eval. Eng.* **2004**, *7*, 143–154. [CrossRef]

10. Zuo, J.Y.; Gisolf, A.; Dumont, H.; Dubost, F.; Pfeiffer, T.; Wang, K.; Mishra, V.K.; Chen, L.; Mullins, O.C.; Biagi, M.; et al. Breakthrough in Accurate Downhole Fluid Sample Contamination Prediction in Real Time . *Petrophys.-Spwla J. Form. Eval. Reser Descr.* **2015**, *56*, 251–265. Available online: https://onepetro.org/petrophysics/article-pdf/56/03/251/2202504/spwla-2015-v 6n3a2.pdf (accessed on 4 August 2024).
11. Lee, R.; Chen, L.; Gisolf, A.; Zuo, J.Y.; Meyer, J.C.; Campbell, T. Real-Time Formation Testing Focused-Sampling Contaminatio Estimation. In Proceedings of the SPWLA Annual Logging Symposium, Reykjavik, Iceland, 25–29 June 2016; p. SPWLA–2016 LLLL.
12. Chenevert, M.; Dewan, J. A Model For Filtration Of Water-base Mud During Drilling: Determination of Mudcake Param ters. *Petrophys.-Spwla J. Form. Eval. Reserv. Descr.* **2001**, *42*, SPWLA-2001-v42n3a4. Available online: https://onepetro.org petrophysics/article-pdf/2201111/spwla-2001-v42n3a4.pdf (accessed on 4 August 2024).
13. Bon, J.; Sarma, H.; Rodrigues, T.; Bon, J. Reservoir-Fluid Sampling Revisited—A Practical Perspective. *SPE Reserv. Eval. En* **2007**, *10*, 589–596. [CrossRef]
14. Hadibeik, A.; Proett, M.; Torres-Verdin, C.; Sepehrnoori, K.; Angeles, R. Wireline and While-Drilling Formation-Tester Samplin with Oval, Focused, and Conventional Probe Types in the Presence of Water- and Oil-Base Mud-Filtrate Invasion in Deviated Wel In Proceedings of the SPWLA Annual Logging Symposium, The Woodlands, TX, USA, 21–24 June 2009; p. SPWLA–2009–8680 Available online: https://onepetro.org/SPWLAALS/proceedings-pdf/SPWLA09/All-SPWLA09/SPWLA-2009-86800/1799 42/spwla-2009-86800.pdf (accessed on 4 August 2024).
15. Alpak, F.O.; Elshahawi, H.; Hashem, M.; Mullins, O. Compositional Modeling of Oil-Based Mud-Filtrate Cleanup Durin Wireline Formation Tester Sampling. In Proceedings of the SPE Annual Technical Conference and Exhibition, San Antonio, T USA, 24–27 September 2006; p. SPE–100393–MS. [CrossRef]
16. Strielkowski, W.; Rausser, G.; Kuzmin, E. Digital Revolution in the Energy Sector: Effects of Using Digital Twin Technolog In *Proceedings of the Digital Transformation in Industry*; Kumar, V., Leng, J., Akberdina, V., Kuzmin, E., Eds.; Springer: Chan Switzerland, 2022; pp. 43–55.
17. Minsky, M. Steps toward Artificial Intelligence. *Proc. IRE* **1961**, *49*, 8–30. [CrossRef]
18. Al-Jarrah, O.Y.; Yoo, P.D.; Muhaidat, S.; Karagiannidis, G.K.; Taha, K. Efficient Machine Learning for Big Data: A Review. *B Data Res.* **2015**, *2*, 87–93. [CrossRef]
19. Kristensen, M.; Chugunov, N.; Gisolf, A.; Biagi, M.; Dubost, F. Real-Time Formation Evaluation and Contamination Predictio Through Inversion of Downhole Fluid-Sampling Measurements. *SPE Reserv. Eval. Eng.* **2018**, *22*, 531–547. [CrossRef]
20. Kristensen, M.; Ayan, C.; Chang, Y.; Lee, R.; Gisolf, A.; Leonard, J.; Corre, P.Y.; Dumont, H. Flow Modeling and Comparativ Analysis for a New Generation of Wireline Formation Tester Modules. In Proceedings of the SPE Latin America and Caribbea Petroleum Engineering Conference, Maracaibo, Venezuela, 21–23 May 2014; p. D031S028R001. [CrossRef]
21. Meng, X. Scalable Simple Random Sampling and Stratified Sampling. In Proceedings of the 30th International Conference on M chine Learning, Atlanta, GA, USA, 17–19 June 2013; Proceedings of Machine Learning Research; Dasgupta, S., McAllester, D., Eds PMLR: Sacramento, CA, USA, 2013; Volume 28, pp. 531–539.
22. Lam, N.S.N. Spatial Interpolation Methods: A Review. *Am. Cartogr.* **1983**, *10*, 129–150. [CrossRef]
23. Ye, C.; Feng, S.; Xue, Z.; Guo, C.; Zhang, Y. Defeating Runge Problem by Coefficients and Order Determination Method wit Various Approximation Polynomials. In Proceedings of the 2018 37th Chinese Control Conference (CCC), Wuhan, Chin 25–27 July 2018; pp. 8622–8627. [CrossRef]
24. Aràndiga, F.; Donat, R.; Santágueda, M. The PCHIP subdivision scheme. *Appl. Math. Comput.* **2016**, *272*, 28–40. [CrossRef]
25. Peng, H.; Long, F.; Ding, C. Feature selection based on mutual information criteria of max-dependency, max-relevance, an min-redundancy. *IEEE Trans. Pattern Anal. Mach. Intell.* **2005**, *27*, 1226–1238. [CrossRef] [PubMed]
26. Gu, Z. Complex heatmap visualization. *iMeta* **2022**, *1*, e43. Available online: https://onlinelibrary.wiley.com/doi/pdf/10.1002 imt2.43 (accessed on 4 August 2024). [CrossRef] [PubMed]
27. Jordan, M.I.; Mitchell, T.M. Machine learning: Trends, perspectives, and prospects. *Science* **2015**, *349*, 255–260. [CrossRe [PubMed]
28. Pavlyshenko, B. Machine learning, linear and Bayesian models for logistic regression in failure detection problems. In Proceeding of the 2016 IEEE International Conference on Big Data (Big Data), Washington, DC, USA, 5–8 December 2016; pp. 2046–205 [CrossRef]
29. Hearst, M.; Dumais, S.; Osuna, E.; Platt, J.; Scholkopf, B. Support vector machines. *IEEE Intell. Syst. Appl.* **1998**, *13*, 18–2 [CrossRef]
30. Chen, T.; Guestrin, C. XGBoost: A Scalable Tree Boosting System. In Proceedings of the 22nd ACM SIGKDD Internation Conference on Knowledge Discovery and Data Mining, KDD '16, New York, NY, USA, 13–17 August 2016; pp. 785–794. [CrossRe
31. Ho, T.K. Random decision forests. In Proceedings of the 3rd International Conference on Document Analysis and Recognitio Montreal, QC, Canada, 14–16 August 1995; Volume 1, pp. 278–282. [CrossRef]
32. Kruse, R.; Mostaghim, S.; Borgelt, C.; Braune, C.; Steinbrecher, M. Multi-layer Perceptrons. In *Computational Intelligence: Methodological Introduction*; Springer International Publishing: Cham, Switzerland, 2022; pp. 53–124. [CrossRef]
33. van der Goot, R. We Need to Talk About train-dev-test Splits. In Proceedings of the 2021 Conference on Empirical Methods i Natural Language Processing, Punta Cana, Dominican Republic, 7–11 November 2021; Moens, M.F., Huang, X., Specia, L., Yi S.W.t., Eds.; Association for Computational Linguistics: Stroudsburg, PA, USA, 2021; pp. 4485–4494. [CrossRef]

4. Lewis-Beck, M.S.; Skalaban, A. The R-Squared: Some Straight Talk. *Political Anal.* **1990**, *2*, 153–171. [CrossRef]
5. Malvić, T.; Ivšinović, J.; Velić, J.; Rajić, R. Interpolation of Small Datasets in the Sandstone Hydrocarbon Reservoirs, Case Study of the Sava Depression, Croatia. *Geosciences* **2019**, *9*, 201. [CrossRef]
6. Barudžija, U.; Ivšinović, J.; Malvić, T. Selection of the Value of the Power Distance Exponent for Mapping with the Inverse Distance Weighting Method—Application in Subsurface Porosity Mapping, Northern Croatia Neogene. *Geosciences* **2024**, *14*, 155. [CrossRef]
7. Ivšinović, J.; Malvić, T. Comparison of mapping efficiency for small datasets using inverse distance weighting vs. moving average, Northern Croatia Miocene hydrocarbon reservoir. *Geologija* **2022**, *65*, 47–57. [CrossRef]
8. Sara, U.; Akter, M.; Uddin, M. Image Quality Assessment through FSIM, SSIM, MSE and PSNR—A Comparative Study. *J. Comput. Commun.* **2019**, *7*, 8–18. [CrossRef]
9. Zhu, X.; Suk, H.I.; Shen, D. A novel matrix-similarity based loss function for joint regression and classification in AD diagnosis. *NeuroImage* **2014**, *100*, 91–105. [CrossRef]
0. Zhang, Z. Improved Adam Optimizer for Deep Neural Networks. In Proceedings of the 2018 IEEE/ACM 26th International Symposium on Quality of Service (IWQoS), Banff, AB, Canada, 4–6 June 2018; pp. 1–2. [CrossRef]
1. Larsen, W.A.; McCleary, S.J. The Use of Partial Residual Plots in Regression Analysis. *Technometrics* **1972**, *14*, 781–790. [CrossRef]

Journal of
Marine Science and Engineering

MDPI

Article

Occurrence Mechanism and Controlling Factors of Shale Oil from the Paleogene Kongdian Formation in Cangdong Sag, Bohai Bay Basin, East China

Binyu Ma [1,2,*], Qinhong Hu [3,4,*], Xiugang Pu [5], Shengyu Yang [3], Xuyang Wang [3], Wenzhong Han [5] and Jiacheng Wen [3]

1 Key Laboratory of Exploration Technologies for Oil and Gas Resources, Ministry of Education, Yangtze University, Wuhan 430100, China
2 School of Geosciences, Yangtze University, Wuhan 430100, China
3 National Key Laboratory of Deep Oil and Gas, China University of Petroleum (East China), Qingdao 266580, China; s.yang@upc.edu.cn (S.Y.)
4 Laboratory for Marine Mineral Resource, Qingdao Marine Science and Technology Center, Qingdao 266237, China
5 Exploration and Development Research Institute, Dagang Oilfield Company, PetroChina, Tianjin 300280, China; puxgang@petrochina.com.cn (X.P.); hanwzhong@petrochina.com.cn (W.H.)
* Correspondence: mabinyu@yangtzeu.edu.cn (B.M.); huqinhong@upc.edu.cn (Q.H.)

Abstract: Free oil, rather than adsorbed oil, is the main contributor to shale oil production wit current development technologies, and assessing oil contents in different occurrence states (adsorbe oil vs. free oil) is a critical component in evaluating the economics of shale wells and plays. Althoug various methodologies have been developed, there are still some fundamental issues in assessin the oil contents in different occurrence states in shale. In this study, a new method was develope to estimate the adsorbed and free oil contents in the Second Member of the Eocene Kongdia Formation (Ek_2) shales in Cangdong Sag, Bohai Bay Basin. This method combines the results c standard Rock-Eval pyrolysis and multi-step Rock-Eval pyrolysis with thin section petrography, X-ra diffraction for mineralogy, total organic carbon analyses, field emission scanning electron microscop for pore morphology, and pore structure analyses by nitrogen physisorption and mercury intrusio porosimetry. Nine lithofacies were identified in a total of 50 shale samples, and the results show tha the adsorbed and free oil are mainly contained in pores with diameters > 20 nm, and their content are mainly controlled by organic matter abundance and thermal maturity of shales. While por space volume influences the storage of shale oil, it is not a major determinant. Models of shale o occurrence and its evolution are proposed, suggesting that the high S_1 contents of organic-rich an -fair shales, which the latter resulted from oil migration, are the most favorable exploration targets c Ek_2 shales. The findings of this study will help prioritize shale oil exploration targets in Ek_2 shales

Keywords: lacustrine shale; Kongdian Formation; oil distribution; free oil; adsorbed oil

Citation: Ma, B.; Hu, Q.; Pu, X.; Yang, S.; Wang, X.; Han, W.; Wen, J. Occurrence Mechanism and Controlling Factors of Shale Oil from the Paleogene Kongdian Formation in Cangdong Sag, Bohai Bay Basin, East China. *J. Mar. Sci. Eng.* **2024**, *12*, 1557. https://doi.org/10.3390/jmse12091557

Academic Editor: Anabela Oliveira

Received: 8 July 2024
Revised: 24 August 2024
Accepted: 27 August 2024
Published: 5 September 2024

1. Introduction

Although large reserves of oil have been discovered in lacustrine shales in Chin economic exploration and development of shale oil are still at an early stage due to th complex properties of lacustrine shales (e.g., high heterogeneity, low thermal maturity and the poor mobility of shale oil [1–6]. In recent years, understanding the accumulatio mechanism of shale oil and its controlling factors has attracted considerable attentio since they are essential for identifying the "sweet spots" for the occurrence of poter tially productive shale oil reservoirs [7–10]. In addition, the occurrence characteristic o shale oil is a critical factor influencing its sustainable development [11,12]. Generall shale oil occurs primarily as free oil and adsorbed oil, with free oil being the primar contributor to oil production using current technology [13–15]. Consequently, the quantification of free oil content plays a crucial role in the evaluation of shale oil resources and production potential.

The current techniques to quantify the contents of free and adsorbed oil primarily involve the multi-step solvent extraction method, swelling method, and multi-step Rock-Eval pyrolysis method [15–17]. For example, using a multi-step solvent extraction method, Zhang et al. (2019) investigated the free and adsorbed oil contents of shales in the Shahejie Formation in Dongying Depression [17]. In their study, core chips (8–10 mm in size) were extracted with a mixed organic solvent (chloromethane: methanol = 9:1 by volume) for 48 h to obtain the extracts as free oil. The extracted core chips were crushed to 2–5 mm in size for another extraction under the same conditions to obtain the adsorbed oil [17]. This method is not only complicated in operation but is also influenced by various factors affecting the amount of extracts, such as the types and ratios of organic solvents, as well as the extraction time [18]. On the other hand, the swelling method has been utilized mainly to study the oil adsorbed by kerogen [19–21]. In recent years, some researchers have used the swelling method to study the adsorption of oil on minerals and have found that the adsorbed oil content of clay minerals is higher than that of siliceous and calcareous minerals [16,22,23]. However, it is likely that the adsorbed capacities of minerals extracted from shales or pure minerals do not represent the actual contents of oil adsorbed of minerals in the bulk shale since oil is not adsorbed on all pore surfaces due to variable pore connectivity, wettability, and oil saturation, as shown by scanning electron microscopy (SEM) observations [9,10,24,25]. Consequently, the adsorbed oil obtained using the swelling method is obviously overestimated, occasionally resulting in higher adsorbed oil content than the total oil content [16].

The multi-step Rock-Eval pyrolysis method has been widely used to analyze shale oil contents in adsorbed versus free oil state since free and adsorbed oils volatilize at different temperatures and can be detected and quantified instrumentally [11]. Free oil mainly includes small-sized hydrocarbon molecules, which are released at low temperatures, while adsorbed oil mainly consists of macromolecular hydrocarbons (HC) that are typically released at high temperatures [14]. Consequently, a multi-step Rock-Eval pyrolysis can quantitatively evaluate the contents of free and adsorbed oil using the appropriate temperature ranges. For example, Romero-Sarmiento (2019) proposed a scheme where volatiles at pyrolysis temperatures of <200 °C, 200–350 °C, and 350–650 °C correspond to free, adsorbed, and cracked oil, respectively [15]. In contrast, Li et al. (2020) defined the volatiles released at heating temperatures of <350 °C, 350–450 °C, and >450 °C as free, adsorbed, and cracked oil, respectively [26]. However, it has been confirmed that alkenes crack from kerogen at temperatures above 350 °C, while some residual oils remain in shales at temperatures above 350 °C and even above 450 °C [7,13]. As a result, effectively distinguishing the temperature ranges of free and adsorbed oils by only using the multi-step Rock-Eval pyrolysis method presents challenges.

In this study, a novel method is proposed to quantify the contents of free and adsorbed oils that combines standard Rock-Eval pyrolysis with multi-step Rock-Eval pyrolysis to quantify the contents of free and adsorbed oils in shales. This method not only offers ease of operation but also addresses the inaccuracies associated with solely using multi-step Rock-Eval pyrolysis. Shales from the Second Member of the Kongdian Formation (Ek_2) shales in Cangdong Sag, Bohai Bay Basin, were studied to determine the contents of total oil, free oil, and adsorbed oil. In addition, the samples were analyzed to determine the factors controlling their free versus adsorbed oil contents, including TOC contents, thermal maturity, and mineralogy and lithofacies. Models of oil occurrence in Ek_2 shales are proposed that can help prioritize favorable shale oil exploration targets in lacustrine-sourced Ek_2.

2. Samples and Methods

2.1. Samples

In this study, a total of 50 samples of Ek_2 shales were selected from three well (G108-8, GD12, and GD14) from depths of 2900–4200 m. The locations of the well are shown in Figure 1, and detailed information about each sample is listed in Table Each sample was prepared in three different forms for the following analyses: (1) cube (~1 cm × 1 cm × 1 cm) for thin sections and field emission scanning electron microscop (FE-SEM); (2) grains with a size of 500–841 μm (#20–35 mesh) for low-temperature nitroge adsorption (LNA) and mercury intrusion porosimetry (MIP) analyses; and (3) powder with a particle size <75 μm (<#200 mesh) for X-ray diffraction (XRD), total organic carbo (TOC), and standard and multi-step Rock-Eval pyrolysis analyses.

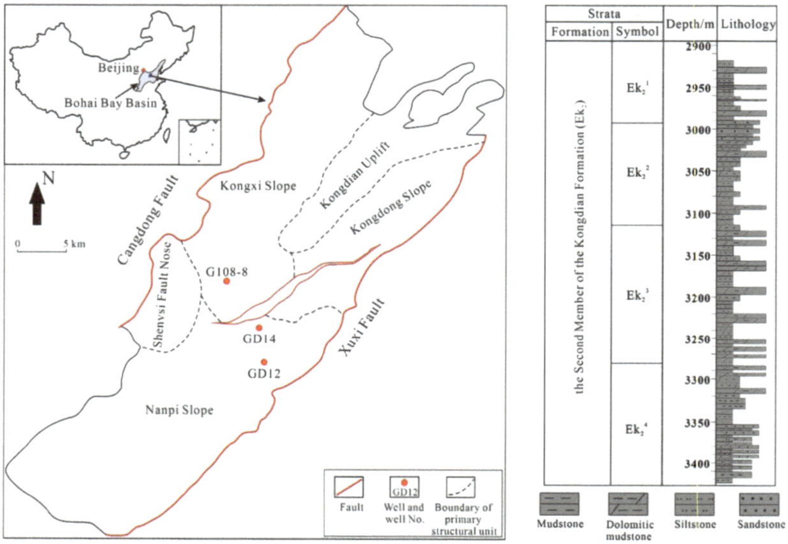

Figure 1. Location of Cangdong Sag and three sampling wells and stratigraphic column of Ek_2 i Cangdong Sag (modified from [27]).

Table 1. Sample list, mineral compositions, and TOC contents of Ek_2 shales used in this study.

Sample ID	Well Name	Depth (m)	Quartz (wt.%)	Feldspar (wt.%)	Calcite (wt.%)	Dolomite (wt.%)	Clays (wt.%)	Analcime (wt.%)	Others (wt.%)	TOC (%)	Lithofacies
G108-8 2928	G108-8	2928.56	13	15	17	24	21	10	0	1.78	OFMM
G108-8 2944	G108-8	2944.63	12	8	12	36	15	17	1	3.98	ORLC
G108-8 2945	G108-8	2945.49	12	13	13	21	19	21	1	4.68	ORLM
G108-8 2946	G108-8	2946.83	13	5	21	26	21	13	1	1.21	OFLC
G108-8 2947	G108-8	2947.97	19	13	8	11	16	31	1	5.00	ORLM
G108-8 2949	G108-8	2949.65	12	14	13	28	16	16	1	4.55	ORLC
G108-8 2958	G108-8	2958.90	17	11	13	10	17	30	3	2.59	ORLM
G108-8 2964	G108-8	2964.34	7	6	9	63	10	5	0	2.63	ORLC
G108-8 2971	G108-8	2971.88	13	14	11	20	19	20	4	3.47	ORLM
G108-8 2973	G108-8	2973.07	9	6	9	48	20	8	1	2.18	ORMC
G108-8 3027	G108-8	3027.85	16	14	12	24	22	12	1	1.07	OFMM
G108-8 3050	G108-8	3050.26	16	18	15	23	17	11	1	1.36	OFLM
G108-8 3052	G108-8	3052.22	16	16	6	4	15	41	2	3.23	ORLS
G108-8 3059	G108-8	3059.18	14	13	12	19	15	24	3	4.19	ORLM
G108-8 3088	G108-8	3088.09	10	9	6	38	24	13	0	1.15	OFMC
G108-8 3104	G108-8	3104.93	12	25	7	9	26	20	3	5.58	ORLM
G108-8 3114	G108-8	3114.88	14	16	7	38	16	10	0	1.81	OFLM
G108-8 3127	G108-8	3127.43	14	25	5	13	20	23	1	5.53	ORLS

Table 1. *Cont.*

Sample ID	Well Name	Depth (m)	Quartz (wt.%)	Feldspar (wt.%)	Calcite (wt.%)	Dolomite (wt.%)	Clays (wt.%)	Analcime (wt.%)	Others (wt.%)	TOC (%)	Lithofacies
G108-8 3135	G108-8	3135.72	11	15	0	50	13	11	0	4.46	ORLC
G108-8 3143	G108-8	3143.87	16	10	2	26	24	21	2	2.49	ORLM
G108-8 3150	G108-8	3150.23	19	15	8	7	15	36	1	1.66	OFLS
G108-8 3161	G108-8	3161.47	16	12	6	32	18	17	0	1.56	OFMM
G108-8 3183	G108-8	3183.08	20	20	7	24	17	10	2	4.58	ORLM
G108-8 3204	G108-8	3204.05	20	11	4	33	29	0	4	3.56	ORLM
G108-8 3212	G108-8	3212.16	19	37	6	1	29	4	4	6.74	ORLS
G108-8 3235	G108-8	3235.44	11	23	14	26	18	6	2	4.74	ORLM
G108-8 3268	G108-8	3268.09	17	14	3	8	33	24	1	6.44	ORLM
GD12 3823	GD12	3823.87	20	36	1	27	10	7	0	1.15	OFLS
GD12 3826	GD12	3826.02	17	14	3	8	33	24	1	0.39	OFMM
GD12 3831	GD12	3831.92	20	12	6	5	36	20	1	0.48	OFMM
GD12 3833	GD12	3833.48	8	10	8	34	23	16	2	1.28	OFMC
GD12 3834	GD12	3834.15	15	9	2	8	29	36	1	0.44	OFMM
GD12 3847	GD12	3847.25	18	42	6	12	19	3	0	2.79	ORLS
GD12 3855	GD12	3855.72	9	60	1	24	6	0	0	3.84	ORLS
GD12 3859	GD12	3859.39	10	64	3	9	11	0	3	4.72	ORLS
GD12 3894	GD12	3894.93	18	25	4	31	22	0	0	3.37	ORLM
GD14 4078	GD14	4078.07	7	11	21	38	15	7	1	0.51	OFLC
GD14 4081	GD14	4081.55	14	47	18	5	13	0	3	2.60	ORLS
GD14 4082	GD14	4082.09	10	44	14	17	14	0	2	4.38	ORLS
GD14 4084	GD14	4084.38	11	68	11	6	4	0	0	2.70	ORLS
GD14 4095	GD14	4095.27	12	34	6	25	20	0	3	1.79	OFLM
GD14 4096	GD14	4096.66	14	44	9	15	19	0	0	2.02	ORLS
GD14 4103	GD14	4103.11	9	52	10	5	22	0	2	3.26	ORLS
GD14 4113	GD14	4113.26	15	41	8	24	13	0	0	2.20	ORLS
GD14 4115	GD14	4115.10	8	28	14	39	10	0	1	0.92	OFLC
GD14 4116	GD14	4116.29	19	20	10	44	7	0	0	3.37	ORMC
GD14 4117	GD14	4117.69	6	16	15	54	10	0	0	1.07	OFMC
GD14 4126	GD14	4126.07	14	23	10	43	10	0	0	2.40	ORMC
GD14 4134	GD14	4134.52	15	34	15	19	16	0	1	2.77	ORLM
GD14 4136	GD14	4136.21	11	44	5	32	6	0	2	4.13	ORLS

2.2. Mineralogical and Geochemical Analyses

An X'Pert Pro X-ray diffractometer from Panalytical Company (Almelo, Holland) was utilized to determine the mineral compositions of shales at a working voltage of 40 kV and a current of 40 mA. The powdered samples were scanned from 3° to 65° with a step of 0.02°, and the mineralogical compositions were semi-quantified by using the K-value method following the Chinese Oil and Gas Industry Standard (SY/T 5163-2010). Thin sections of the Ek_2 lacustrine shales were studied using an optical microscope to analyze the laminar structures. The TOC content of the shale was measured using a LECO CS230 instrument (LECO, St. Joseph Charter Township, MI, USA) after undergoing decarbonation treatment, as described in previous studies [28].

Standard Rock-Eval pyrolysis analyses were performed using the Rock-Eval 7 Analyzer manufactured by Vinci Technologies (Nanterre, France). The powdered samples were initially heated at 300 °C for 3 min and then heated to 650 °C with a heating rate of 25 °C/min. Free hydrocarbon content (S_1), thermal cracking hydrocarbon content (S_2), and pyrolysis peak temperature for S_2 (T_{max}) were obtained [29]. Then, a second series of pyrolysis analyses using the same procedures and equipment were performed on the samples after they had been extracted using a mixture of dichloromethane and methanol (volumetric ratio of 93:7) for seven days to remove residual oil [30].

The multi-step Rock-Eval pyrolysis experiments were also carried out using the Rock-Eval 7 Analyzer. Using the pyrolysis procedures outlined by Romero-Sarmiento (2019) [15], powdered shales (~100 mg) were initially heated at 200 °C for 5 min to obtain *Sh*0, then heated to 350 °C with a heating rate of 25 °C/min and held for 5 min to obtain *Sh*1, and followed by heating to 650 °C at a rate of 25 °C/min to obtain *Sh*2.

2.3. Field Emission Scanning Electron Microscopy Imaging

A FEI Quanta FEG 650 scanning electron microscope (FEI, Lexington, KY, USA equipped with an energy-dispersive spectrometer was used to analyze the pore mo phology of shales. A surface of a cubic sample oriented perpendicular to the laminae wa polished using a GATAN Ilion II 697 Argon Beam Milling System (GATAN, San Diego, CA USA) and then coated with a 10-nm-thick carbon film to enhance the electrical conductivit and SEM image resolution.

2.4. Low-Temperature Nitrogen Adsorption Analyses

The LNA analyses were performed using a Micromeritics ASAP 2460 instrumen (Micromeritics, Norcross, GA, USA) to analyze the pore structure of shales. Granular shal samples were first extracted with a mixed organic solvent (dichloromethane and methanc at a volumetric ratio of 93:7) for seven days and then oven-dried at 110 °C for 24 h t eliminate the residual oil and moisture. The dried and organic solvent-extracted shale were degassed at 110 °C for 24 h and then adsorbed with nitrogen incrementally at 77.3 K The relative pressure P/P_0 (the ratio of absolute pressure to saturation pressure) range from 0.002 to 0.998. The specific surface area, pore volume, and pore size distribution wer determined using the BET (Brunauer-Emmette-Teller) and BJH (Barrette-Joynere-Halenda methods, respectively [31,32]. To investigate the distribution of residual oil in shales, second series of LNA analyses were carried out by using as-received (non-extracted) shale following the same procedures outlined above.

2.5. Mercury Intrusion Porosimetry Analyses

Similar to LNA analyses, two sets of granular shale samples were analyzed: a se of organic solvent-extracted shales for pore structure evaluation and a set of as-receive shales for residual oil distribution. The samples were prepared for MIP analyses usin a Micromeritics AutoPore IV 9520 instrument (Micromeritics, USA) at an incrementa intrusion pressure of mercury from 0.2 psi (0.001 MPa) to 60,000 psi (414 MPa). To elim inate the conformance effect caused by voids between shale grains, which artificiall increases the total pore volume, the effective pressure range of MIP tests for data analyse was set at 20–60,000 psi (0.138–414 MPa), corresponding to a pore-throat size range o 2.8 nm–12 μm, based on the Washburn Equation and the nanopore confinement corre tion [33,34], as described in previous studies [35].

3. Results

3.1. Petrological and Geochemical Properties

The mineral composition of Ek_2 shales is dominated by quartz, feldspar, calcit dolomite, analcime, and clays, as well as minor minerals such as pyrite and siderit (Table 1). The quartz content ranges from 6.0 to 20.0%, with an average of 13.8%, while th feldspar content ranges from 5.0 to 68.0%, with an average of 23.5%. The average content of calcite and dolomite are 8.9 and 23.7%, respectively. The content of clay minerals range from 4.0 to 36.0%, with an average of 17.9%, while analcime content ranges from 0 to 41.0% with an average of 11.3%. The average content of other minerals is only 1.3%. Accordin to the relative contents of siliceous minerals (quartz and feldspar), carbonates (calcit dolomite, and siderite), and clay minerals, the Ek_2 shales can be categorized into siliceou shale, mixed shale, and calcareous shale (Figure 2) [36]. The lithology is mainly compose of mixed and calcareous shales in Well G108-8, siliceous and mixed shales for Well GD1 and mainly siliceous and calcareous shales in Well GD14.

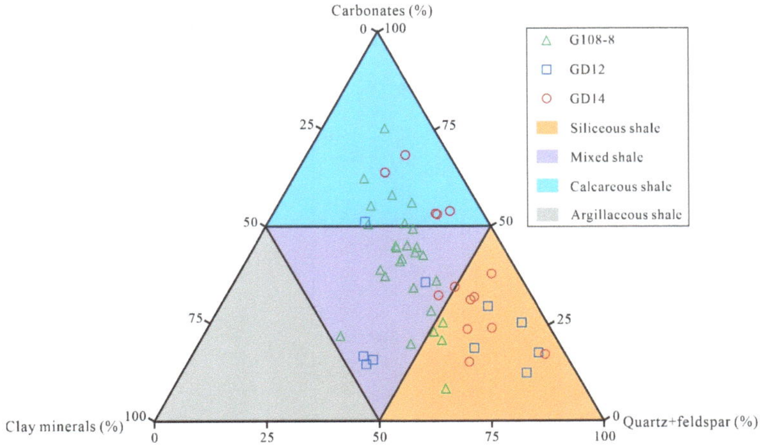

Figure 2. Ternary mineralogy classification of Ek_2 shales from three wells.

The TOC contents of Ek_2 shales range from 0.39 to 6.74%, with a mean value of 2.90% (Table 1). The shales are categorized into organic-rich shale (TOC > 2%) and organic-fair shale (TOC < 2%) [37]. The S_1 and S_2 values of as-received shales vary from 0.11–6.71 to 0.48–49.71 mg HC/g rock, with averages of 1.92 and 16.78 mg HC/g rock, respectively, indicating good source rock potential (Table 2). The free hydrocarbon content (S_{1E}) and thermal cracking hydrocarbon content (S_{2E}) of the organic solvent-extracted shales have decreased to 0.02–0.33 and 0.05–40.22 mg HC/g rock, respectively (Table 2).

Table 2. Results of standard Rock-Eval pyrolysis of as-received and organic solvent-extracted shales and oil contents.

Sample ID	As-Received Shales		Organic Solvent-Extracted Shales		T_{max} (°C)	Residual Oil Content (mg HC/g Rock)	Total Oil Content (mg HC/g Rock)	Heavy Oil Content (mg HC/g Rock)
	S_1 (mg HC/g Rock)	S_2 (mg HC/g Rock)	S_{1E} (mg HC/g Rock)	S_{2E} (mg HC/g Rock)				
G108-8 2928	0.58	7.09	0.11	3.98	444	3.69	4.33	3.11
G108-8 2944	0.94	27.17	0.09	20.27	444	7.84	8.88	6.90
G108-8 2945	0.95	35.86	0.15	27.31	445	9.50	10.54	8.55
G108-8 2946	0.19	2.09	0.06	0.18	442	2.10	2.31	1.91
G108-8 2947	1.52	49.56	0.03	38.43	447	12.65	14.33	11.13
G108-8 2949	0.86	27.25	0.08	20.33	446	7.78	8.73	6.92
G108-8 2958	1.33	25.05	0.04	18.55	440	7.83	9.30	6.50
G108-8 2964	0.87	12.73	0.07	8.55	440	5.05	6.00	4.18
G108-8 2971	0.90	18.53	0.11	13.26	440	6.17	7.16	5.27
G108-8 2973	2.20	9.82	0.06	6.19	438	5.83	8.25	3.63
G108-8 3027	0.17	1.03	0.11	0.18	438	1.02	1.21	0.85
G108-8 3050	0.11	3.44	0.09	1.02	446	2.53	2.65	2.42
G108-8 3052	1.03	23.48	0.15	17.55	442	6.96	8.09	5.93
G108-8 3059	0.95	24.93	0.06	18.45	446	7.43	8.48	6.48
G108-8 3088	0.13	0.90	0.03	0.18	440	0.85	0.99	0.72
G108-8 3104	1.80	46.49	0.08	35.94	439	12.35	14.33	10.55
G108-8 3114	0.25	6.94	0.04	3.86	444	3.33	3.61	3.08
G108-8 3127	0.99	41.97	0.07	32.27	438	10.69	11.78	9.70
G108-8 3135	1.60	29.51	0.11	22.16	438	8.95	10.71	7.35
G108-8 3143	1.28	18.08	0.25	12.13	443	7.23	8.64	5.95
G108-8 3150	0.81	9.68	0.19	6.79	442	3.70	4.59	2.89
G108-8 3161	0.54	9.62	0.16	7.02	444	3.14	3.73	2.60
G108-8 3183	2.00	36.54	0.17	23.98	444	14.56	16.76	12.56
G108-8 3204	3.21	23.82	0.06	17.55	442	9.48	13.01	6.27
G108-8 3212	5.60	49.71	0.11	40.22	439	15.09	21.25	9.49
G108-8 3235	2.12	35.85	0.20	26.82	448	11.15	13.48	9.03
G108-8 3268	2.61	43.83	0.33	36.44	446	10.00	12.87	7.39

Table 2. *Cont.*

Sample ID	As-Received Shales		Organic Solvent-Extracted Shales		T_{max} (°C)	Residual Oil Content (mg HC/g Rock)	Total Oil Content (mg HC/g Rock)	Heavy Oil Content (mg HC/g Rock)
	S_1 (mg HC/g Rock)	S_2 (mg HC/g Rock)	S_{1E} (mg HC/g Rock)	S_{2E} (mg HC/g Rock)				
GD12 3823	0.70	2.32	0.06	1.06	445	1.96	2.73	1.26
GD12 3826	0.96	1.44	0.08	0.22	442	2.18	3.24	1.22
GD12 3831	0.29	0.59	0.07	0.18	494	0.70	1.02	0.41
GD12 3833	0.28	7.10	0.03	3.64	444	3.74	4.05	3.46
GD12 3834	0.23	0.48	0.03	0.05	450	0.66	0.91	0.43
GD12 3847	3.93	10.81	0.11	7.62	446	7.12	11.44	3.19
GD12 3855	2.67	22.32	0.05	17.85	442	7.14	10.08	4.47
GD12 3859	3.87	22.73	0.07	18.70	446	7.90	12.16	4.03
GD12 3894	4.45	10.97	0.02	5.03	445	10.39	15.29	5.94
GD14 4078	0.32	0.77	0.04	0.37	439	0.72	1.07	0.40
GD14 4081	3.39	14.55	0.11	11.06	450	6.88	10.61	3.49
GD14 4082	3.25	16.35	0.12	11.02	444	8.58	12.16	5.33
GD14 4084	6.71	12.35	0.13	9.35	447	9.71	17.09	3.00
GD14 4095	4.52	6.67	0.11	2.76	442	8.43	13.40	3.91
GD14 4096	4.02	8.46	0.09	4.69	443	7.79	12.21	3.77
GD14 4103	2.84	17.50	0.15	16.36	447	3.98	7.10	1.14
GD14 4113	1.60	9.33	0.06	8.45	447	2.48	4.24	0.88
GD14 4115	2.52	2.50	0.03	0.34	435	4.68	7.45	2.16
GD14 4116	3.03	8.01	0.08	4.21	443	6.83	10.16	3.80
GD14 4117	1.49	3.44	0.04	1.32	444	3.61	5.25	2.12
GD14 4126	2.47	9.11	0.07	4.56	445	7.02	9.74	4.55
GD14 4134	2.34	12.39	0.11	11.00	448	3.73	6.30	1.39
GD14 4136	4.73	17.99	0.06	10.44	446	12.28	17.48	7.55

In Figure 3, thin section petrography images of the shales show a range of lamin types, including siliceous lamina (SIL), calcareous lamina (CAL), clay lamina (CLL), organi matter lamina (OML), analcime lamina (ANL), and mixed organic matter-clay lamin (OM-CLL) (Figure 3a–f). The Ek_2 shales are mainly classified as laminated shales an massive shales depending on the extent of the development of laminae (Figure 3).

Considering the mineral compositions, TOC contents, and the development of lamina of shales, nine lithofacies are identified, including organic-rich laminated siliceous sha (ORLS), organic-rich laminated mixed shale (ORLM), organic-rich massive calcareous sha (ORMC), organic-rich laminated calcareous shale (ORLC), organic-fair laminated siliceou shale (OFLS), organic-fair massive mixed shale (OFMM), organic-fair laminated mixe shale (OFLM), organic-fair massive calcareous shale (OFMC), and organic-fair laminate calcareous shale (OFLC), among which ORLS, ORLM and ORMM are the main lithofacie (Table 1). In the organic-rich shales, the ORLM and ORLC shales have higher TOC content but lower S_1 contents compared to ORLS and ORMC shales (Figure 4). This is possibl influenced by the variation in thermal maturities (expressed as the vitrinite reflectance R of the shale core samples at different depths. Zhao et al. (2020) found that the Ro values o shales from the shallow Well G108-8 ranged from 0.66 to 0.91%, with an average of 0.76 while shales from the deep Wells GD12 and GD14 have similar Ro values, ranging fro 0.80 to 1.21%, with a mean value of 1.02% [27].

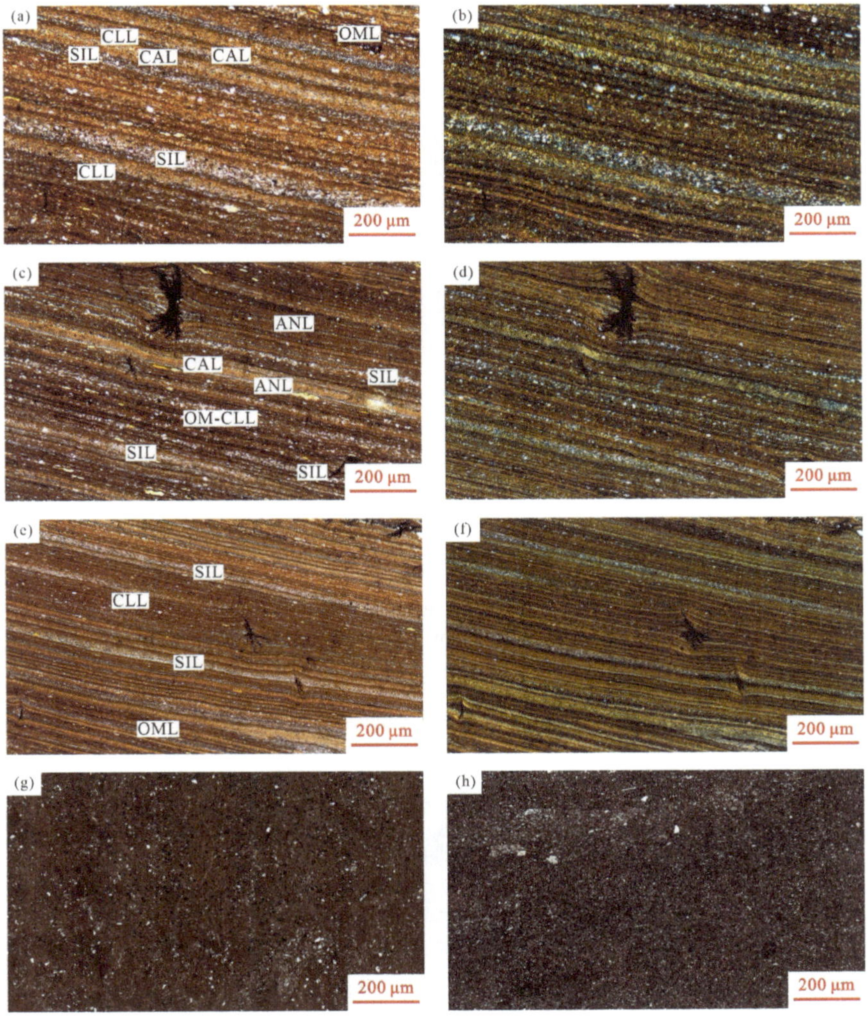

Figure 3. Thin section petrography of typical Ek_2 shales: (**a**) Laminated structure, G108-8 well, 3183.08 m, plane polarized light; (**b**) Laminated structure, G108-8 well, 3183.08 m, crossed polarized light; (**c**) Laminated structure, G108-8 well, 3235.44 m, plane polarized light; (**d**) Laminated structure, G108-8 well, 3235.44 m, crossed polarized light; (**e**) Laminated structure, GD14 well, 4103.11 m, plane polarized light; (**f**) Laminated structure, GD14 well, 4103.11 m, crossed polarized light; (**g**) Massive structure, GD12 well, 3831.92 m, plane polarized light; (**h**) Massive structure, GD14 well, 4126.07 m, plane polarized light.

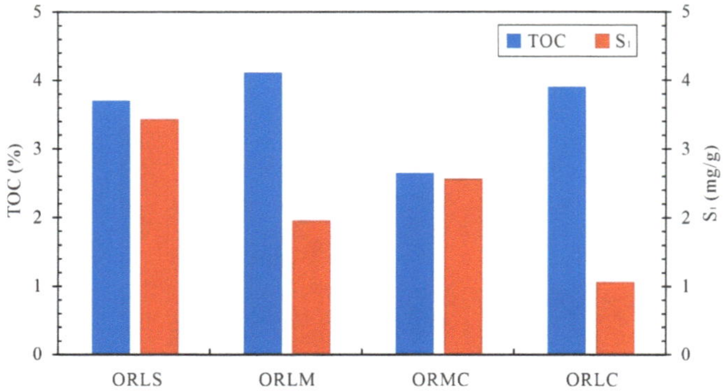

Figure 4. Organic matter abundance of organic-rich shales related to lithofacies.

3.2. Pore Properties

3.2.1. Qualitative Description of Pore Morphology from FE-SEM Imaging

According to the classification scheme proposed by Loucks et al. (2012), the matrix-related pores in shales can be categorized into interparticle, intraparticle, and OM-hosted pores [38]. All of these pore types have been observed in this study. Examples of interparticle pores are shown in Figure 5a–c: these pores mainly occur between quartz, feldspar, and dolomite grains, and typically have angular or rounded shapes. They are usually coated with an OM film (bitumen) varying from tens to hundreds of nm in thickness. This film is thought to have been formed by migration of mobile bitumen, suggesting good connectivity of pre-existing pores [39,40]. Additionally, there are some primary and secondary interparticle pores without OM films in Ek_2 shales, mainly developed between brittle minerals or between brittle minerals and clay minerals, with sizes ranging up to a dozen microns (Figure 5d–f). The secondary interparticle pores are mainly developed at the edges of feldspar, probably resulting from the dissolution of minerals by organic acids released during kerogen maturation (Figure 5e,f) [41].

Intraparticle pores include both dissolved intraparticle and intercrystalline pores. The former mainly develops in calcite and dolomite, have rounded shapes and range from tens of nm to several microns in size (Figure 5g,h). Intercrystalline pores occur mainly between clay aggregates and occasionally in dolomite framboids. They have angular or rounded shapes and also range from tens of nm to several microns in size (Figure 5i–l). The OM films also occur in intraparticle pores, suggesting good connectivity of these pores.

Examples of OM-hosted pores are shown in Figure 5l–o, and can be classified as inherited OM pores and hydrocarbon-generating OM pores. The inherited OM pores are densely distributed with rounded shapes and range in size from hundreds of nm to tens of microns (Figure 5m,n). Their genesis is probably related to the OM types rather than thermal maturity [40]. The hydrocarbon-generating OM pores are developed mainly within OM and between OM and minerals, exhibiting rounded and crescent shapes with sizes up to several microns (Figure 5l,o).

In addition, microfractures are widely observed in Ek_2 shales, mainly at the edges of OM and minerals (clay and calcite), and can be tens of microns in length, which are speculated to be related to hydrocarbon generation and mineral transformations, respectively (Figure 5a,i,p). Some microfractures occur within or cut through mineral particles, possibly resulting from overpressure induced by compaction and/or hydrocarbon generation (Figure 5q). Shale oil can be observed in microfractures, indicating the enhancement of shale oil mobility by microfracturing (Figure 5r).

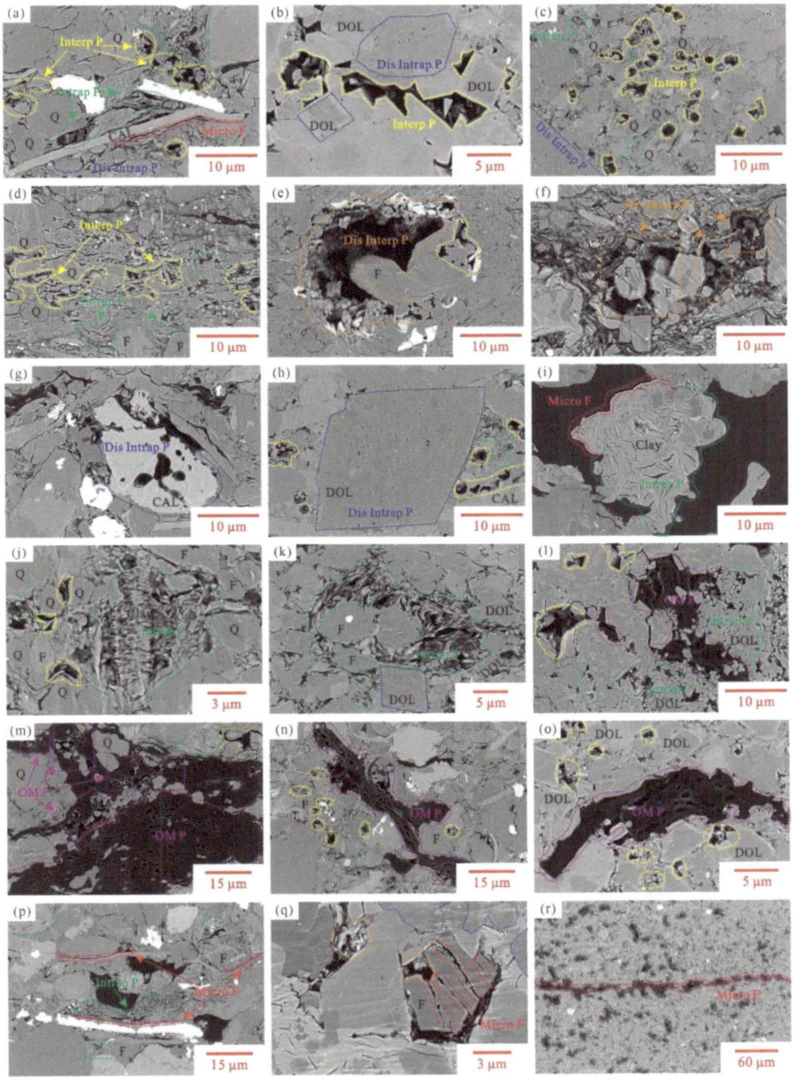

Figure 5. SEM images of pore morphologies in Ek_2 shales: (**a**) G108-8 well, 3212.16 m; (**b**) GD12 well, 3894.93 m; (**c**) GD14 well, 4082.09 m; (**d**) G108-8 well, 2949.65 m; (**e**) G108-8 well, 3204.05 m; (**f**) G108-8 well, 3212.16 m; (**g**) G108-8 well, 3212.16 m; (**h**) GD14 well, 4082.09 m; (**i**) G108-8 well, 3050.26 m; (**j**) GD14 well, 4082.09 m; (**k**) GD14 well, 4136.21 m; (**l**) G108-8 well, 3204.05 m; (**m**) G108-8 well, 3204.05 m; (**n**) G108-8 well, 3212.16 m; (**o**) G108-8 well, 3212.16 m; (**p**) G108-8 well, 3212.16 m; (**q**) GD12 well, 3833.48 m; (**r**) GD14 well, 4095.27 m. Interp P: interparticle pore; Intrap P: intraparticle pore; Dis Intrap P: dissolved intraparticle pore; Dis Interp P: dissolved interparticle pore; OM P: organic matter pore; Micro F: micro-fracture; Q: quartz; F: feldspar; CAL: calcite; DOL: dolomite; Clay: clay minerals.

3.2.2. Quantitative Analyses of Pore Structure from LNA Analyses

Typical N_2 adsorption/desorption curves of organic solvent-extracted shale samples and their corresponding pore size distributions, ranging from ~2 to 200 nm in diameters, are shown in Figure 6. The pore size distribution curves gradually change from unimodal to bimodal patterns with increasing maximum adsorption volumes (Figure 6). Zou et al.

(2015) found that the oil seepage through shale pores exhibited lower and upper threshol at 20 nm and 200 nm, respectively [42]. It was observed that oil did not penetrate por with diameters smaller than 20 nm, while pores with diameters exceeding 200 nm we easily permeable to oil regardless of the chemical or physical conditions. So that pores i shales can be classified into three types: non-seepage-pores (<20 nm in diameter), potenti seepage-pores (20–200 nm in diameter) and seepage-pores (>200 nm in diameter). The LN mainly analyzes the characteristics of non-seepage-pores and potential seepage-pores.

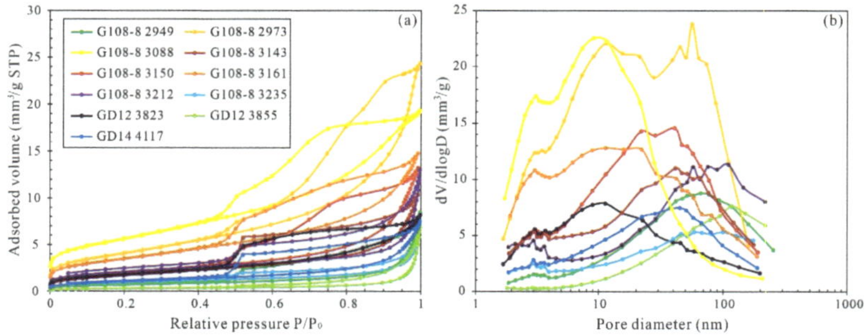

Figure 6. (**a**) Nitrogen adsorption-desorption curves; and (**b**) Pore size distributions of typical organ solvent-extracted Ek_2 shales.

The specific surface areas and pore volumes for the organic solvent-extracted sha samples are shown in Table 3. The specific surface area ranges from 0.85 to 19.63 $m^2/$ with an average of 6.58 m^2/g. The pore volume ranges from 3.41 to 35.77 mm^3/g, wit an average of 14.34 mm^3/g. In general, the specific surface areas and pore volumes c organic-rich shales are lower than those of organic-fair shales (Figure 7a), possibly du to pore filling by bitumen. Among the organic-rich shales, the ORMC lithofacies exhib the highest specific surface area, pore volume, non-seepage-pore and potential seepag pore volume. Other organic-rich shales and organic-fair shales have similar potenti seepage-pore volumes but differ significantly in non-seepage-pore volumes. This differenc is probably the cause of the high specific surface areas and pore volumes of organic-fa shales (Figure 7b).

Table 3. Pore parameters of as-received and organic solvent-extracted shales from LNA analyses.

Sample ID	As-Received Shales		Organic Solvent-Extracted Shales		Increase in Pore Volume (mm^3/g)	Increase in Non-Seepage-Pore Volume (mm^3/g)	Increase in Potential Seepage-Pore Volume (mm^3/g)
	Specific Surface Area (m^2/g)	Pore Volume (mm^3/g)	Specific Surface Area (m^2/g)	Pore Volume (mm^3/g)			
G108-8 2928	13.10	22.96	18.32	25.78	2.83	2.28	0.55
G108-8 2944	1.05	8.82	2.32	11.27	2.45	1.39	1.06
G108-8 2945	0.54	3.69	1.19	5.09	1.40	0.49	0.91
G108-8 2946	14.56	24.37	17.42	25.88	1.51	1.49	0.02
G108-8 2947	0.62	5.09	1.68	8.57	3.48	1.25	2.23
G108-8 2949	0.96	7.44	2.34	11.21	3.76	1.26	2.50
G108-8 2958	0.75	4.41	2.62	7.67	3.26	1.94	1.32
G108-8 2964	1.99	13.61	6.50	19.39	5.78	3.82	1.96
G108-8 2971	1.00	5.73	7.19	17.75	12.03	5.44	6.58
G108-8 2973	5.02	23.75	14.56	35.77	12.02	9.03	2.99
G108-8 3027	18.10	28.33	19.14	29.80	1.47	1.08	0.39
G108-8 3050	9.37	18.64	11.57	19.66	1.02	0.92	0.10
G108-8 3052	2.33	11.55	5.15	14.25	2.71	1.86	0.85
G108-8 3059	0.49	2.27	1.83	4.76	2.49	1.32	1.17
G108-8 3088	18.62	25.16	19.63	26.26	1.10	0.40	0.70
G108-8 3104	0.18	1.25	0.85	3.59	2.34	0.46	1.89
G108-8 3114	12.26	20.92	14.61	22.30	1.38	0.84	0.55
G108-8 3127	0.20	1.11	2.33	3.41	2.30	1.07	1.10

Table 3. *Cont.*

Sample ID	As-Received Shales Specific Surface Area (m^2/g)	As-Received Shales Pore Volume (mm^3/g)	Organic Solvent-Extracted Shales Specific Surface Area (m^2/g)	Organic Solvent-Extracted Shales Pore Volume (mm^3/g)	Increase in Pore Volume (mm^3/g)	Increase in Non-Seepage-Pore Volume (mm^3/g)	Increase in Potential Seepage-Pore Volume (mm^3/g)
G108-8 3135	0.13	0.89	3.10	7.25	6.36	2.41	3.96
G108-8 3143	1.04	6.32	6.16	14.93	8.61	4.71	3.90
G108-8 3150	3.77	16.50	6.97	19.34	2.85	1.98	0.87
G108-8 3161	6.69	17.96	12.34	20.50	2.54	2.38	0.17
G108-8 3183	1.46	9.05	12.48	22.53	13.48	7.13	6.36
G108-8 3204	1.31	6.99	18.60	24.65	17.65	11.47	6.19
G108-8 3212	0.20	1.59	7.35	14.62	13.03	4.13	8.90
G108-8 3235	0.77	5.25	3.12	7.46	2.21	1.48	0.72
G108-8 3268	0.54	3.47	1.83	5.58	2.11	0.77	1.35
GD12 3823	3.75	9.24	6.37	10.87	1.64	1.46	0.17
GD12 3826	16.73	23.20	17.99	23.70	0.49	0.36	0.13
GD12 3831	15.41	21.78	15.86	22.11	0.33	0.03	0.30
GD12 3833	13.91	27.47	16.53	30.02	2.55	1.42	1.13
GD12 3834	14.23	21.67	14.91	21.91	0.24	0.17	0.07
GD12 3847	0.31	3.04	1.68	10.95	7.91	2.04	5.87
GD12 3855	0.26	2.44	0.93	7.46	5.02	0.80	4.22
GD12 3859	0.40	3.29	1.13	7.03	3.74	0.97	2.77
GD12 3894	0.86	6.14	3.60	13.83	7.70	3.42	4.27
GD14 4078	4.85	12.25	5.98	14.03	1.78	1.12	0.66
GD14 4081	0.63	5.21	1.31	8.13	2.92	1.06	1.86
GD14 4082	0.24	2.10	1.34	9.52	7.42	1.70	5.71
GD14 4084	0.46	3.45	1.60	10.16	6.71	1.86	4.84
GD14 4095	0.38	3.88	1.60	9.76	5.87	2.03	3.84
GD14 4096	0.48	4.01	1.38	8.69	4.68	1.49	3.18
GD14 4103	0.39	3.35	0.97	7.35	4.00	0.91	3.09
GD14 4113	0.56	4.39	1.13	7.61	3.22	0.83	2.39
GD14 4115	0.77	4.67	3.25	10.40	5.74	3.11	2.63
GD14 4116	0.52	5.14	1.62	11.43	6.30	1.70	4.59
GD14 4117	0.42	2.58	3.69	9.71	7.13	3.72	3.41
GD14 4126	0.48	4.26	2.03	12.49	8.23	2.38	5.84
GD14 4134	0.49	4.33	1.43	8.58	4.25	1.48	2.77
GD14 4136	0.48	4.50	1.67	11.96	7.45	1.73	5.72

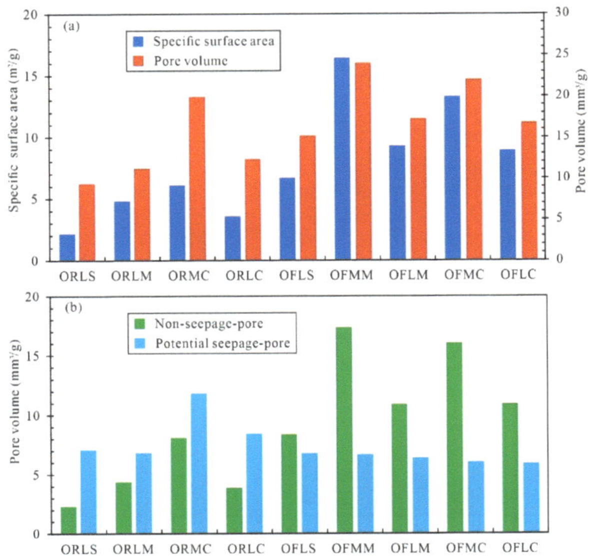

Figure 7. Pore structure parameters of organic solvent-extracted shales from LNA analyses; (**a**) Specific surface area and pore volume; (**b**) Volume of different types of pores.

3.2.3. Quantitative Analyses of Pore Structure from MIP Analyses

Table 4 shows the porosity and total pore volume of organic solvent-extracted shale obtained from MIP analyses. The porosity and total pore volume of Ek_2 shales range from 1.80–12.95% and 7.16–43.87 mm^3/g, with averages of 5.18% and 22.49 mm^3/g, respectively. The volumes of non-seepage-pores, potential seepage-pores and seepage-pores are 8.56, 4.57, and 9.36 mm^3/g, with relative proportions of 38%, 20%, and 42%, respectively, indicating that a significant proportion of pore space is contained in the non-seepage-pores. The volume of non-seepage-pores is higher than potential seepage-pores, which is different from the results from LNA analyses. This discrepancy is probably due to the different measurement methodologies, since the measured pore volume using MIP represents the volume of throats and their controlling pore bodies, while the latter are measured by LNA [33,43,44].

Table 4. Pore parameters of as-received and organic solvent-extracted shales from MIP analyses.

Sample ID	As-Received	Organic Solvent-Extracted		Increase in Pore Volume (mm^3/g)
	Pore Volume (mm^3/g)	Porosity (%)	Pore Volume (mm^3/g)	
G108-8 2928	18.22	3.70	21.26	3.04
G108-8 2944	20.94	5.50	26.94	6.00
G108-8 2945	13.31	6.29	26.47	13.16
G108-8 2946	22.08	5.44	23.42	1.34
G108-8 2947	5.86	4.25	20.43	14.57
G108-8 2949	16.92	5.84	25.36	8.44
G108-8 2958	11.12	3.11	14.61	3.49
G108-8 2964	13.50	5.08	22.50	9.00
G108-8 2971	19.40	5.28	24.56	5.17
G108-8 2973	22.12	7.54	34.50	12.38
G108-8 3027	25.54	5.88	26.34	0.79
G108-8 3050	11.27	3.73	16.15	4.88
G108-8 3052	26.33	7.68	36.20	9.87
G108-8 3059	7.41	2.35	12.36	4.95
G108-8 3088	21.34	5.05	22.82	1.47
G108-8 3104	9.54	5.49	22.43	12.88
G108-8 3114	11.94	3.87	17.55	5.61
G108-8 3127	2.38	2.78	13.76	11.38
G108-8 3135	5.50	3.23	12.11	6.60
G108-8 3143	9.82	5.19	20.12	10.30
G108-8 3150	14.09	4.31	18.73	4.64
G108-8 3161	15.01	4.14	18.03	3.02
G108-8 3183	9.30	7.18	28.65	19.35
G108-8 3204	18.67	7.86	34.00	15.32
G108-8 3212	29.36	12.95	43.87	14.51
G108-8 3235	15.17	7.11	28.76	13.59
G108-8 3268	17.92	8.13	30.84	12.92
GD12 3823	0.47	1.80	7.16	6.69
GD12 3826	13.94	3.16	17.78	3.84
GD12 3831	17.32	3.94	18.25	0.93
GD12 3833	27.18	6.18	27.99	0.81
GD12 3834	18.80	4.73	19.91	1.11
GD12 3847	18.92	6.81	29.83	10.91
GD12 3855	14.74	5.09	22.80	8.06
GD12 3859	13.73	5.88	24.91	11.18
GD12 3894	7.22	4.85	20.90	13.68

Table 4. *Cont.*

Sample ID	As-Received	Organic Solvent-Extracted		Increase in Pore Volume (mm^3/g)
	Pore Volume (mm^3/g)	Porosity (%)	Pore Volume (mm^3/g)	
GD14 4078	10.77	2.77	12.27	1.50
GD14 4081	14.03	5.18	22.29	8.26
GD14 4082	15.69	6.46	27.64	11.96
GD14 4084	13.32	5.46	25.82	12.50
GD14 4095	6.15	3.69	15.86	9.71
GD14 4096	18.26	6.84	27.50	9.24
GD14 4103	17.12	5.02	21.86	4.74
GD14 4113	19.81	5.18	22.29	2.48
GD14 4115	10.99	3.80	15.70	4.71
GD14 4116	14.93	4.97	22.33	7.40
GD14 4117	6.10	2.08	7.94	1.84
GD14 4126	13.13	4.15	20.53	7.40
GD14 4134	17.08	4.92	21.13	4.05
GD14 4136	19.15	6.99	28.87	9.72

The pore structures of organic solvent-extracted shales of different lithofacies are shown in Figure 8. The porosities and total pore volumes of the extracted organic-rich shales are higher than those of organic-fair shales, and the organic-rich shales also have higher volumes of potential seepage-pores and seepage-pores volumes compared with organic-fair shales. This indicates that the promotion of OM to larger pores, which is consistent with the findings from FE-SEM observations (Figure 5i–o). In the organic-rich shales, the ORLS and ORLM lithofacies have the higher volumes of potential seepage-pores and seepage-pores, followed by ORLC lithofacies, showing a higher potential for shale oil storage (Figure 8b).

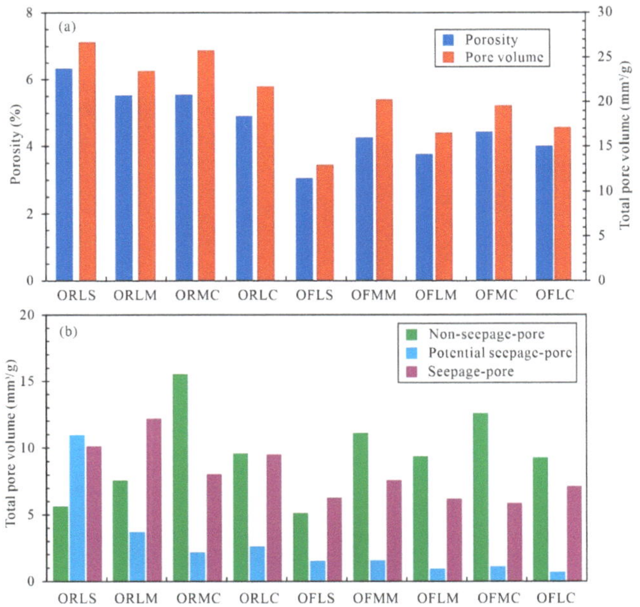

Figure 8. Pore structure parameters of organic solvent-extracted shales from MIP analyses; (**a**) Porosity and total pore volume; (**b**) Volume of different types of pores.

3.3. Estimation of Total Oil Content

According to the method proposed by Jarvie (2012) [45], the residual oil content c shale can be determined by Equation (1):

$$\textbf{\textit{Residual oil}} = (S_1 + S_2) - (S_{1E} + S_{2E}) \quad ($$

where S_1 and S_2 are the free and thermal cracking hydrocarbon contents of as-receive shale samples, while S_{1E} and S_{2E} are the free and thermal cracking hydrocarbon contents c organic solvent-extracted shales. The S_{1E} of organic solvent-extracted shales is possibly th residual oil in isolated pores that are inaccessible by organic solvent [46]. Thus, the residu oil content can be calculated by Equation (2):

$$\textbf{\textit{Residual oil}} = S_1 + S_2 - S_{2E} \quad (2$$

According to Equation (2), the oil content contains the light-medium oil (S_1) an heavy oil ($S_2 - S_{2E}$), but still does not account for the light hydrocarbons lost due t volatilization during storage, so that the contents of total oil and movable oil are clearl underestimated [47]. Zhao et al. (2021) compared the S_1 contents of Ek_2 shales that ha been frozen by liquid nitrogen and those exposed to air for several days and determine the correction factor for light hydrocarbon loss is a value of 2.1 [48]. Although it is strictl inappropriate to apply the same correction factor to all shales, the lack of sealed core dat is evident. The estimated total oil content can be obtained using Equation (3):

$$\textbf{\textit{Total oil}} = 2.1 \times S_1 + S_2 - S_{2E} \quad (3$$

The total oil contents of Ek_2 shales range from 0.91 to 21.25 mg HC/g rock, with a average of 8.62 mg HC/g rock, while the contents of heavy oil range from 0.40 to 12.56 m HC/g rock (Table 2), with a mean value of 4.59 mg HC/g rock, accounting for 53.2% of th total oil content, which is unfavorable for shale oil mobility. The total oil contents of shale from different lithofacies are shown in Figure 9, in which the total oil contents of organi rich shales are generally higher than those of organic-fair shales, and the ORLS and ORLM shales have slightly higher total oil contents compared with ORMC and ORLC shales. I addition, there are some organic-fair shales, such as samples GD14 4095 (OFLM), GD1 4115 (OFLC), and GD14 4117 (OFMC) that have higher total oil contents than some organi rich shales (Table 2); probably due to the migration of fluids from adjacent organic-ric intervals [49–51].

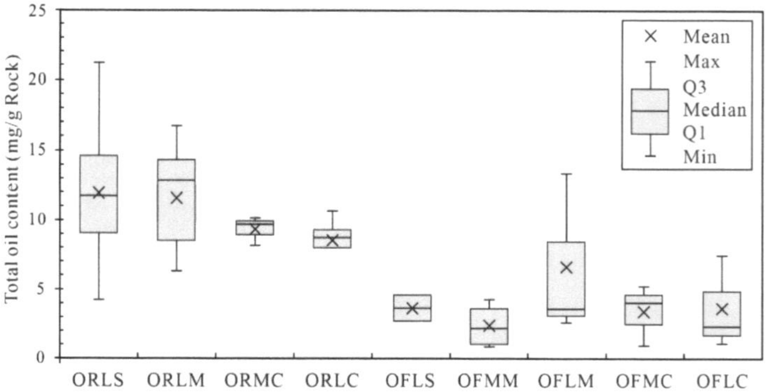

Figure 9. Comparison of the total oil content of shales of different lithofacies.

3.4. Estimation of Oil Contents in Different Occurrence States

As described above, the contents of shale oil in different occurrence states cannot b accurately determined solely by multi-step Rock-Eval pyrolysis analyses. In this study, b combining the standard Rock-Eval pyrolysis results of both as-received and organic solvent-extracted shales with the data for multi-step Rock-Eval pyrolysis of as-received shales, the shale oil contents in different occurrence states can be obtained. As illustrated in Figure 10, *Sh*0 means light hydrocarbons in 0–200 °C; *Sh*1-1 means light-medium hydrocarbons in 200–300 °C and is the difference between S_1 and ***Sh*0** of as-received shale; *Sh*1-2 are medium hydrocarbons in the 300–350 °C range and are determined by the difference between *Sh*1 and *Sh*1-1 of shale; *Sh*2-1 means heavy hydrocarbons, determined by the difference between *Sh*2 of as-received shale and S_{2E} of organic solvent-extracted shale; *Sh*2-2 means hydrocarbons generated by the cracking of kerogen, equivalent to S_{2E} of shales. Consequently, free oil contains *Sh*0, *Sh*1-1, and *Sh*1-2, while the adsorbed oil is related to *Sh*2-1. Since the sum of *Sh*0, *Sh*1-1, and *Sh*1-2 is the residual free oil content, the total free oil content can be determined by the difference between total oil content and adsorbed oil content.

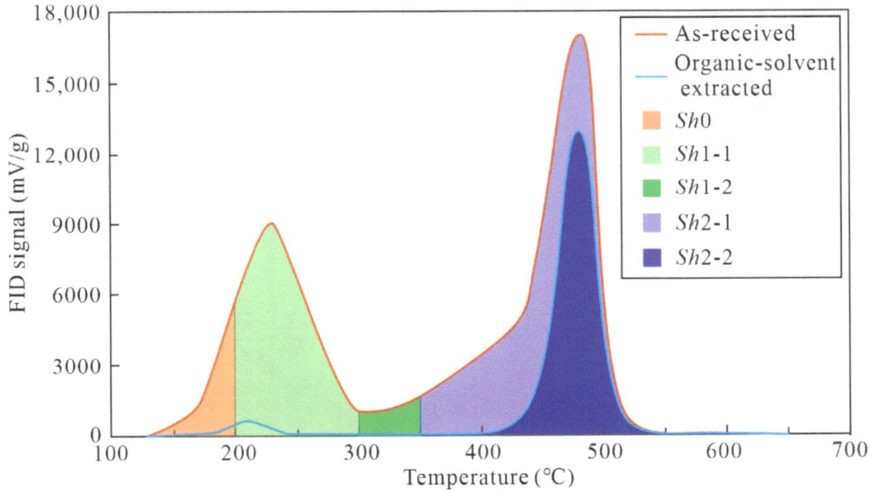

Figure 10. Schematic diagram of pyrolysis spectra of as-received and solvent-extracted shales and division of shale oil into different occurrence states (FID: flame ionization detector).

Table 5 shows the contents of adsorbed and free oil, which range from 0.06–11.90 mg HC/g rock and 0.23–15.79 mg HC/g rock, with averages of 4.04 and 4.59 mg HC/g rock, showing a higher proportion of free oil (53.2%) than adsorbed oil (46.8%). Figure 11a presents the adsorbed oil contents of shales of different lithofacies. The ORLS and ORMC shales contain similar adsorbed oil contents that are slightly higher than organic-fair shales but are clearly lower than ORLM and ORLC shales. The free oil contents of shales of different lithofacies are shown in Figure 11b, in which some organic-fair shales (GD14 4095 of OFLM, GD14 4115 of OFLC, GD14 4117 of OFMC) contain higher free oil contents that are higher than some organic-rich shales (Table 5), indicating that the migrated oil increases the content of free oil but not adsorbed oil. The ORLS shales have the highest amount of free oil content, followed by ORMC and ORLM shales, and so have a higher potential for shale oil exploitation (Figure 11b).

Table 5. Adsorbed and free oil contents of Ek_2 shales.

Sample ID	Adsorbed Oil Content (mg HC/g Rock)	Free Oil Content (mg HC/g Rock)	Sample ID	Adsorbed Oil Content (mg HC/g Rock)	Free Oil Content (mg HC/g Rock)
G108-8 2928	3.17	1.16	G108-8 3235	7.98	5.51
G108-8 2944	6.59	2.29	G108-8 3268	6.09	6.78
G108-8 2945	8.25	2.29	GD12 3823	0.97	1.76
G108-8 2946	1.89	0.42	GD12 3826	0.82	2.42
G108-8 2947	10.85	3.47	GD12 3831	0.31	0.71
G108-8 2949	6.62	2.11	GD12 3833	3.35	0.70
G108-8 2958	6.00	3.30	GD12 3834	0.67	0.24
G108-8 2964	4.10	1.90	GD12 3847	1.35	10.09
G108-8 2971	4.97	2.19	GD12 3855	4.04	6.03
G108-8 2973	2.68	5.56	GD12 3859	3.28	8.88
G108-8 3027	0.87	0.33	GD12 3894	5.08	10.20
G108-8 3050	2.42	0.23	GD14 4078	0.19	0.89
G108-8 3052	5.87	2.22	GD14 4081	3.00	7.61
G108-8 3059	6.22	2.26	GD14 4082	3.54	8.62
G108-8 3088	0.72	0.27	GD14 4084	1.30	15.79
G108-8 3104	10.07	4.26	GD14 4095	1.76	11.64
G108-8 3114	3.13	0.48	GD14 4096	3.00	9.21
G108-8 3127	9.44	2.34	GD14 4103	0.64	6.47
G108-8 3135	6.92	3.78	GD14 4113	0.06	4.18
G108-8 3143	5.57	3.07	GD14 4115	1.71	5.74
G108-8 3150	2.70	1.90	GD14 4116	3.21	6.95
G108-8 3161	2.53	1.20	GD14 4117	2.01	3.24
G108-8 3183	11.90	4.86	GD14 4126	3.40	6.34
G108-8 3204	5.42	7.59	GD14 4134	0.94	5.36
G108-8 3212	7.73	13.52	GD14 4136	6.44	11.04

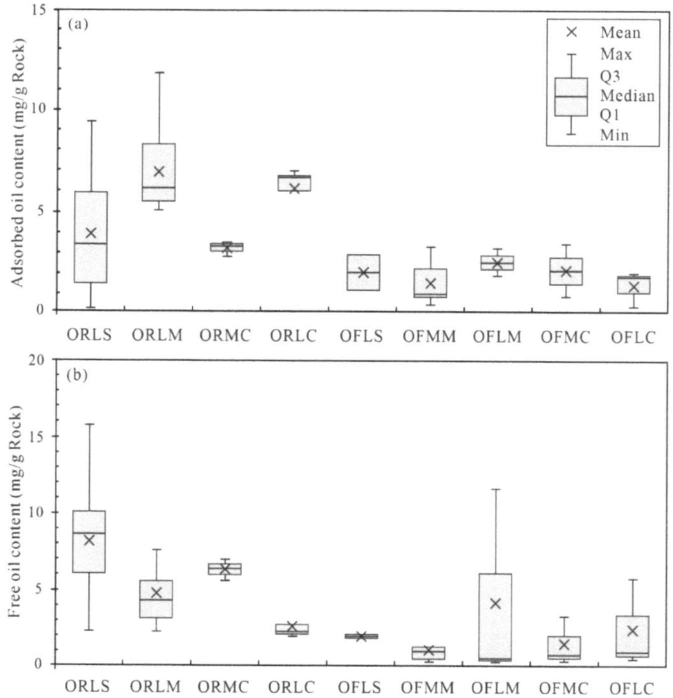

Figure 11. Comparison of the (**a**) adsorbed oil content and (**b**) free oil content of shales of different lithofacie

4. Discussion

4.1. Occurrence Space of Shale Oil at Ek_2

4.1.1. Occurrence Space of Shale Oil Based on LNA Analyses

In previous studies, the shale oil (actually residual oil) distribution was usually determined by the difference in pore size distributions between as-received and organic solvent-extracted shales [10,52]. In this study, the pore volume of as-received shales obtained from LNA analyses and the increase volumes of pores with different types after extraction are shown in Table 3. The increase in pore volume of organic-rich shales (mean value of 5.97 mm^3/g) is higher than that of organic-fair shales (a mean value of 2.38 mm^3/g), and their ratios to pore volume (diameters < 200 nm) of organic solvent-extracted shales are 53.6% and 11.8%, indicating that more than half of the volume with diameters <200 nm of organic-rich shales is filled with residual oil. The proportions of pore volumes filled by residual oil of ORLS, ORLM, ORMC, and ORLC shales are 58.7%, 55.9%, 44.5%, and 37.4%, respectively. The relative proportions of residual oil in non-seepage-pore and potential seepage-pore are 28.8% and 71.2% (ORLS), 51.0% and 49.0% (ORLM), 49.4% and 50.6% (ORMC), 48.4% and 51.6% (ORLC), in which the ORLS shale shows different residual oil distribution from other organic-rich shales. In addition, there is only a slightly positive relationship between the increase in pore volume and residual oil content of organic-rich shales (Figure 12a), suggesting that amounts of oil being stored in pores with diameters >200 nm.

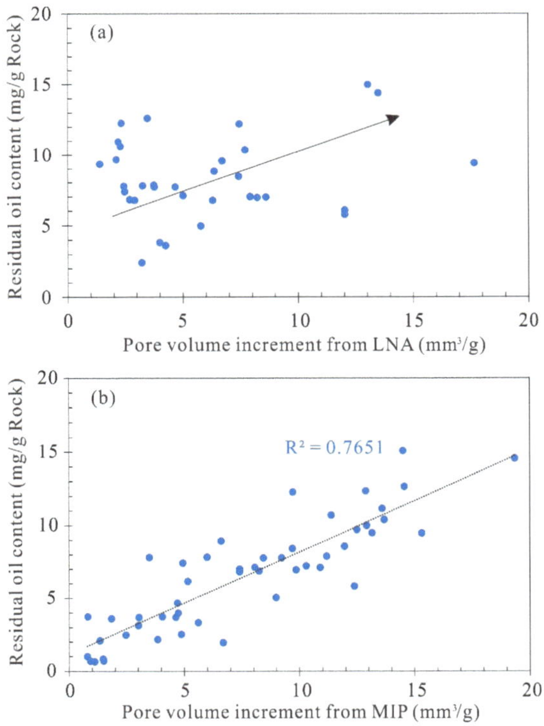

Figure 12. Correlations between pore volume increment and residual oil content: (**a**) Pore volume increment from LNA vs. residual oil content of organic-rich shales; (**b**) Pore volume increment from MIP vs. residual oil content of all shale samples.

4.1.2. Occurrence Space of Shale Oil Based on MIP Analyses

The pore-throat size range obtained from MIP analyses is 2.8 nm–12 μm in diameter, which includes almost all the storage space of residual oil. A positive relationship between the increase in pore volume and residual oil content is shown in Figure 12b, indicating th effectiveness of MIP analyses of residual oil distribution. The increases in pore volum are shown in Table 4. The pore volume increases in the organic-rich shale (mean value c 9.86 mm^3/g) are obviously higher than that of organic-fair shale (mean value c 3.29 mm^3/g), and their ratios to total pore volume of organic solvent-extracted shale are 39.8% and 18.2%, indicating that ~40% of the pore volume is filled by the residual oil i organic-rich shales. The proportions of pore volumes filled by residual oil of ORLS, ORLM ORMC, and ORLC shales are 35.9%, 47.0%, 35.1%, and 34.6%, respectively, and the relativ proportions of residual oil in the three different types of pores are shown in Figure 13. Th sums of relative proportions of residual oil in potential seepage-pore and seepage-por are 80% (ORLS), 70% (ORLM), 35% (ORMC), and 56% (ORLC). This indicates that th laminated ORLS, ORLM, and ORLC shales have higher potential for oil exploitation tha the massive ORMC shales.

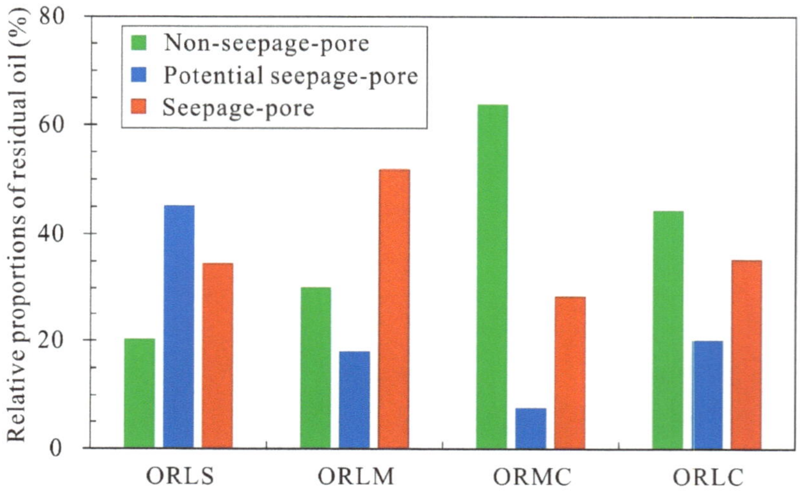

Figure 13. Relative proportions of residual oil in pores with different types.

4.2. *Factors Controlling the Total Oil Content*

4.2.1. Effects of OM Abundance on Total Oil Content

OM is the source of hydrocarbons and the concentration of OM is a key determinant c the hydrocarbon-generating potential of source rock [53]. In this study, the total oil conten of organic-rich and organic-fair shales (except for three shales in which oil migration ha occurred) are 11.2 and 2.60 mg HC/g rock (Table 2), showing a significant differenc between the two. The TOC and S_1 contents both show good relationships with total o content, indicating that they are effective parameters for total oil evaluation, especially th S_1 (Figure 14a,b).

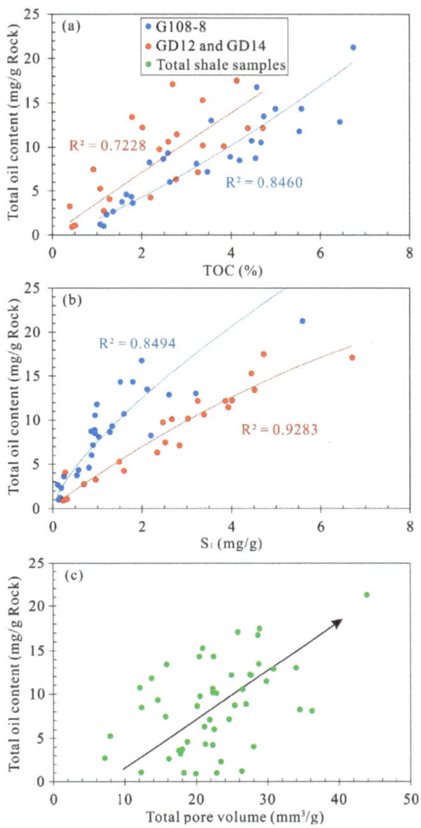

Figure 14. Correlations between total oil content and (**a**) TOC, (**b**) S_1, and (**c**) total pore volume.

4.2.2. Effect of Thermal Maturity on Total Oil Content

Previous studies found that Eocene samples from Well G108-8 have lower thermal maturities (Ro values of 0.66 to 0.91%) compared with those from Wells GD12 and GD14 (Ro values 0.80 to 1.21%) [27]. To eliminate the influence of TOC on oil content, TOC normalization is first carried out for total oil content. The TOC-normalized total oil content of organic-rich shales from Well G108-8 ranges from 1.92 to 3.79 mg/g rock with a mean value of 2.70 mg/g rock, which is lower than that of Wells GD12 and GD14, in which the range is 1.93 to 6.33 mg/g rock (mean of 3.63 mg/g rock), indicating shales with higher thermal maturity have higher oil content. In addition, the proportion of heavy oil to total oil of organic-rich shales from Well G108-8 (67.6%) is obviously higher than that of Wells GD12 and GD14 (33.7%), which suggests that the differences in the proportion of heavy oil to light oil are related to differences in thermal maturities between shales from these wells.

4.2.3. Effect of Pore Space on Total Oil Content

A shale oil reservoir is a self-generating and self-storing petroleum system, and generally, the larger the pore volume of shales is, the more oil that can be stored [54]. This is the reason for a positive relationship shown in Figure 14c. However, non-hydrocarbon fluids also occur in pores of shale, such as the formation water [55], and Zhao et al. (2019) found that the average oil saturation of Ek_2 shales from Well G108-8 is ~40% [56]. The presence of non-hydrocarbon fluids results in only a slight positive correlation between pore volume and total oil content (Figure 14c).

4.2.4. Effect of Mineral Compositions on Total Oil Content

Statistically, the mean TOC-normalized total oil contents of organic-rich siliceou organic-rich calcareous, and organic-rich mixed shales are 3.44, 2.91, and 2.81 mg HC/ rock, respectively. This suggests that siliceous minerals are more favorable for the reservoi ing of shale oil, possibly due to the pore space being better-preserved in siliceous mineral and shale oil migration from adjacent OM-clays mixed lamina. This concept is supporte by the optical microscopy and SEM observations in the literature [48,57].

4.3. Factors Controlling the Adsorbed and Free Oil

4.3.1. Effects of OM Abundance on Adsorbed and Free Oil

Compared with inorganic minerals, OM has stronger adsorbed capacity for shal oil [16]. Figure 15a,c show the relationships between TOC, S_1, and S_2 and adsorbed oi content, in which positive relationships are observed between TOC, S_2 and adsorbed oi content, especially for shales from Well G108-8, indicating the control of TOC on adsorbe oil, and S_2 can be used to estimate the amount of adsorbed oil, especially for shales wit lower maturity. There is a clearly positive relationship between S_1 and free oil content, an there are slight positive relationships between TOC, S_2 and free oil content, which suggest that the controlling of TOC on free oil and S_1 can be used to estimate the amount of free o (Figure 15d–f). As the material source of hydrocarbons, the relationships between TOC an adsorbed oil and free oil are predictable, which has also been reported in the literature [58

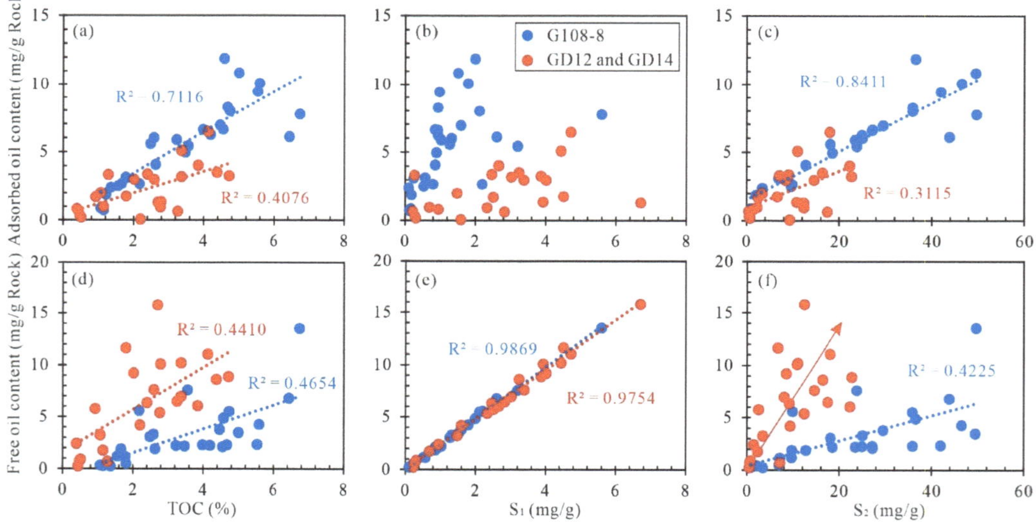

Figure 15. Relationships between TOC, S_1, S_2 and adsorbed oil and free oil contents. (**a–c**) Th relationships between TOC, S_1, S_2 and adsorbed oil content, respectively; (**d–f**) the relationship between TOC, S_1, S_2 and free oil content, respectively.

4.3.2. Effect of Thermal Maturity on Adsorbed and Free Oil

In Figure 15a,d, the shales from Well G108-8 have higher adsorbed oil contents an lower free oil contents, compared with those of Wells GD12 and GD14 shales with simila TOC contents, probably resulting from the thermal maturities difference between shale from Wells G108-8, GD12, and GD14. Generally, higher thermal maturity means more ligh and medium oil, as well as more free oil, enhancing the overall mobility of shale oil.

4.3.3. Effects of Mineral Compositions on Adsorbed and Free Oil

Figure 16 shows the relationships between the siliceous, calcareous, and clay mine als contents and the TOC-normalized adsorbed and free oil contents. As can be seen i

Figure 16, slightly negative and positive relationships are observed between TOC-normalized adsorbed oil content and free oil content and siliceous minerals, respectively, and there are no relationships with calcareous and clay minerals. The pore space preserved by brittle siliceous minerals and shale oil migrated from adjacent OM-clays mixed lamina, and the lower adsorption capacity of siliceous minerals [16], results in a slightly negative relationship with adsorbed oil (Figure 16a) and positive relationship with free oil (Figure 16b). Although there are abundant pores within and between/in calcareous mineral particles, the organic-rich calcareous shales have lower total oil contents compared with organic-rich siliceous and mixed shales (Figure 9), which is possibly the main reason for the lack of relationships between calcareous minerals content and adsorbed and free oil contents.

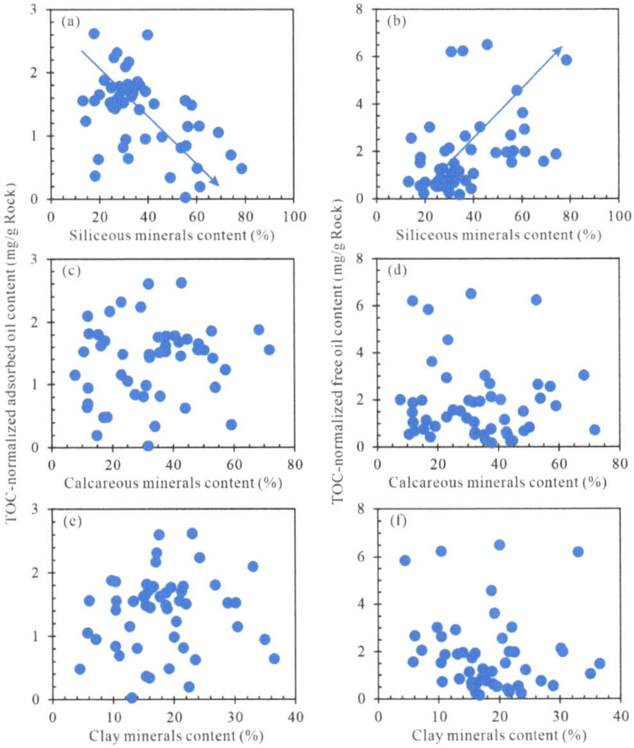

Figure 16. Relationships between TOC-normalized adsorbed oil content, TOC-normalized free oil content, and siliceous, calcareous, and clay minerals. (**a**,**c**,**e**) The relationships between siliceous, calcareous, clay minerals and TOC-normalized adsorbed oil content, respectively; (**b**,**d**,**f**) the relationships between siliceous, calcareous, clay minerals and TOC-normalized free oil content, respectively.

4.3.4. Effects of Pore Structure on Adsorbed and Free Oil

It has been generally believed that adsorbed oil occurs on the pore surface, while free oil mainly occurs in the center of pores. Consequently, the volumes of adsorbed and free oil may be related to the pore surface area and pore volume, respectively [9]. However, there is no relationship between adsorbed oil and pore surface area, while a positive relationship is shown between adsorbed oil content and the seepage-pores volume (Figure 17). In addition, there are positive relationships between porosity, potential seepage-pores, seepage-pores volume, and free oil content (Figure 18), suggesting that porosity and abundance of larger pores significantly influence the volume of free oil. This suggests that both the adsorbed and free oil are mainly stored in larger pores (potential seepage-pores and seepage-pores), which is also implied by the occurrence of ~70% residual oil stored in potential seepage-pores and seepage-pores of the organic-rich shales (Figure 13). Liang et al. (2018) propose that the shale oil preferentially migrates along microfractures and large pores with goo connectivity and low resistance, and as a result, the adsorbed and free oil are mainly i large pores [41].

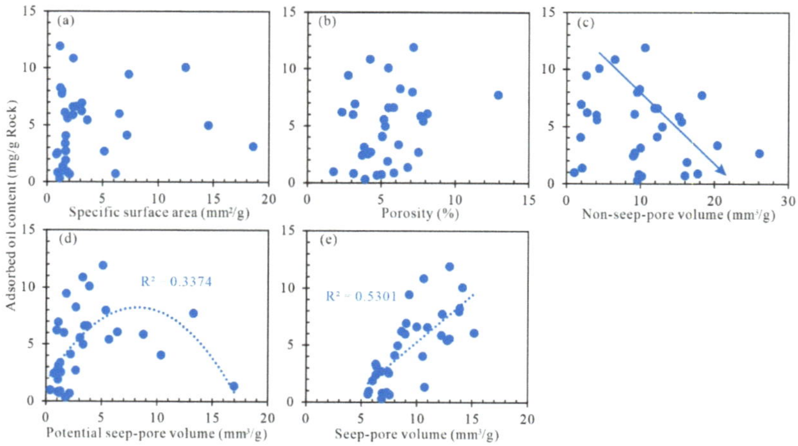

Figure 17. Correlation between adsorbed oil content and parameters of pore structure. (**a**,**b**) The relatio ships between specific surface area, porosity and adsorbed oil content, respectively; (**c**–**e**) the relationshi between non-seep-pore, potential seep-pore, seep-pore volumes and adsorbed oil content, respectively

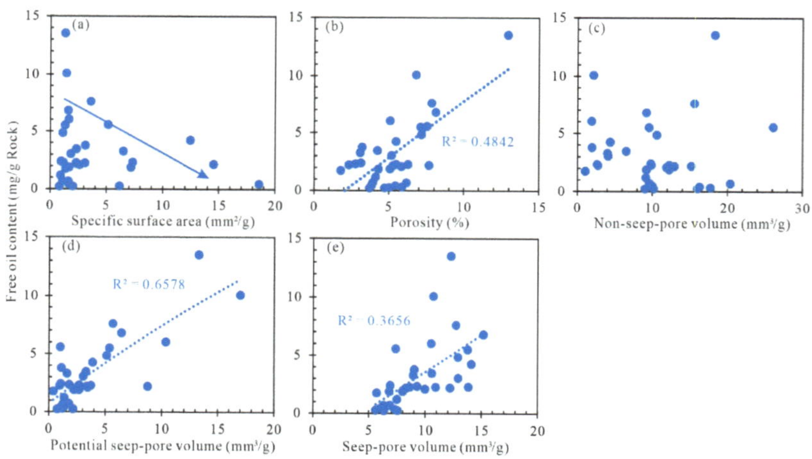

Figure 18. Correlation between free oil content and parameters of pore structure. (**a**,**b**) The relatio ships between specific surface area, porosity and free oil content, respectively; (**c**–**e**) the relationshi between non-seep-pore, potential seep-pore, seep-pore volumes and free oil content, respectively.

4.4. Models for Shale Oil Occurrence

Generally, the total oil content of Ek_2 shales is primarily controlled by TOC conten and thermal maturity, in which variation of the TOC content is the principal factor causin significant differences in total oil content between organic-rich and organic-fair shales. Th variations in TOC are combined with differences in thermal maturity to influence tot oil volumes. For example, the ORLS shales have lower TOC contents but similar total o contents compared to the ORLM shales (Figures 4 and 9), which is due to the influence c thermal maturity. The ORLS shales, which predominantly occur in Wells GD12 and GD1 exhibit a higher thermal maturity, while the ORLM shales, mainly found in Well G108-

show a lower thermal maturity (Table 1). Pore space influences the storage of shale oil, but the variation in oil saturation in shales diminishes the influence of pore space on total oil content. Additionally, the migration of oil causes some organic-fair shales, such as GD14 4095 (OFLM), GD14 4115 (OFLC), and GD14 4117 (OFMC), to have high total oil contents.

The adsorbed oil contents of Ek_2 shales are mainly controlled by thermal maturity and TOC contents. Thermal maturity directly controls generated oil components and determines the adsorbed and free oil contents. In addition, OM not only generates oil during the thermal evolution but also has high adsorption capacity for oil. Consequently, low thermal maturity and high TOC contents are the main reasons for high adsorbed oil contents of ORLM and ORLC shales (Figure 4). The free oil contents are also mainly controlled by TOC contents and thermal maturity, which are the main reasons for high free oil content of organic-rich shales, especially the ORLS and ORMC shales. Similar to the factors controlling the total oil content, pore space is a necessary condition for free oil storage but has a smaller influence due to the variation of oil saturation. The organic-fair shales with oil migration (GD14 4095 in OFLM, GD14 4115 in OFLC, GD14 4117 in OFMC) contain high values of free oil contents, some of which are higher than some organic-rich shales (Table 5), indicating that the immigrated oil increases the contents of free oil, not adsorbed oil.

Based on the results and analyses of this study as well as previously published studies of Ek_2 shales [24,56–58], models of shale oil with different occurrence states in organic-rich laminated shale and organic-fair massive shale are proposed (Figure 19). In shales with a low maturity (such as shales in shallow Well G108-8), the generated oil from OM was first stored in OM pores and then migrated to adjacent large mineral pores, showing a higher proportion of adsorbed oil (heavy hydrocarbons) than of free oil (Figure 19a). With the increase of thermal maturity (such as shales in deep Wells GD12 and GD14), a higher proportion of lighter petroleum and a lower proportion of heavy hydrocarbons were generated, promoting the migration of oil to large pores in mineral laminae and organic-fair shales (Figure 19b). This suggests that the organic-rich shales, especially the ORLS, ORMC, and ORLM shales, and organic-fair shales in which oil migration has occurred are the most favorable exploration targets in the Ek_2 shales.

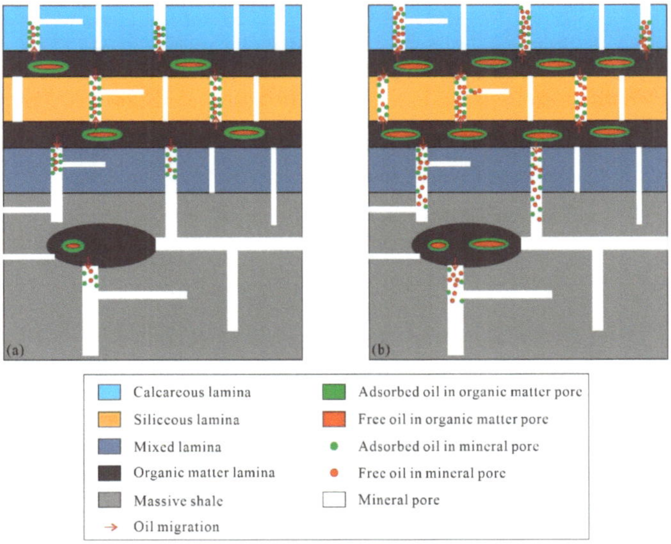

Figure 19. Models of shale oil occurrence for (**a**) adsorbed and free oil distribution in shales with lower maturity and (**b**) adsorbed and free oil distribution in shales with higher maturity.

5. Conclusions

In this study, the total oil contents of Ek_2 shales of different lithofacies were determine and the factors controlling the total oil content were found to include OM abundance (TO contents), thermal maturity, pore space, and mineral compositions. OM abundance an thermal maturity are the primary factors controlling the total shale oil content. Organic-ric shales and organic-fair shales with oil migration show high total oil contents. In organi rich shales, ~40% of the pore volume is filled with residual oil, and ~70% of this residual o is stored in pores with diameters >20 nm.

The adsorbed oil contents of Ek_2 shales are mainly controlled by thermal maturity an TOC contents: lower thermal maturity and higher TOC content resulted in high adsorbe oil contents. The free oil contents are mainly controlled by TOC and thermal maturit higher TOC contents and thermal maturity resulted in high free oil content. In addition, S can be used to estimate the free oil contents in Ek_2 shales.

Oil migration within shale reservoirs can lead to readjustments in both the content and occurrence states of shale oil. With increasing thermal maturity, a higher proportion c lighter oil and a lower proportion of heavy hydrocarbons were generated, which enhance the migration of oil to larger pores in mineral laminae and organic-fair shales. The organi rich shales, especially the ORLS, ORMC, and ORLM shales, and organic-fair shales wit oil migration have high free oil contents and are the most favorable exploration target of Ek_2 shales. In addition, in other lacustrine shale reservoirs, organic-fair shales with o migration may demonstrate a higher proportion of free oil and benefit oil developmen Quantifying the content of migration oil in shale reservoirs is a crucial challenge fo further research.

Author Contributions: B.M.: writing—original draft preparation, methodology, investigation, dat curation. Q.H.: writing—review and editing, conceptualization, funding acquisition, project a ministration. X.P.: funding acquisition, resources, supervision. S.Y.: writing—review and editin investigation, formal analysis. X.W.: validation, methodology, data curation. W.H.: funding acquis tion, resources, supervision. J.W.: methodology, data curation. All authors have read and agreed t the published version of the manuscript.

Funding: This research was funded by the National Natural Science Foundation of China (N 42302175), Provincial Major Type Grant for Research and Development from the Department c Science & Technology of Shandong Province (No. 2020ZLYS08), the China Postdoctoral Scienc Foundation (No. 2024M752744), and Open Fund of Key Laboratory of Exploration Technologies fc Oil and Gas Resources (Yangtze University), Ministry of Education (No. K2023-05).

Institutional Review Board Statement: Not applicable.

Informed Consent Statement: Not applicable.

Data Availability Statement: The original data contributions presented in the study are included i the article; further inquiries can be directed to the corresponding author.

Conflicts of Interest: Binyu Ma has received research grants from the National Natural Scienc Foundation of China and China Postdoctoral Science Foundation, and Key Laboratory of Exploratio Technologies for Oil and Gas Resources (Yangtze University), Ministry of Education. Shengyu Yan has received research grants from the Department of Science & Technology of Shandong Province.

References

1. He, Z.; Han, Y.; Luo, Q.; Wang, F.; Wang, F.; Wu, S. New insights on the retention and migration of shale oil within the hypersalin Qianjiang Formation in the Jianghan Basin, China. *AAPG Bull.* **2024**, 108, 479–493. [CrossRef]
2. Hu, S.; Zhao, W.; Hou, L.; Yang, Z.; Zhu, R.; Wu, S.; Bai, B.; Jin, X. Development potential and technical strategy of continent shale oil in China. *Petrol. Explor. Dev.* **2020**, 47, 819–828. [CrossRef]
3. Jiao, F.; Zou, C.; Yang, Z. Geological theory and exploration & development practice of hydrocarbon accumulation insid continental source kitchens. *Petrol. Explor. Dev.* **2020**, 47, 1067–1078.
4. Jin, Z.; Zhu, R.; Liang, X.; Shen, Y. Several issues worthy of attention in current lacustrine shale oil exploration and developme *Petrol. Explor. Dev.* **2021**, 48, 1471–1484. [CrossRef]

5. Sun, L.; Liu, H.; He, W.; Li, G.; Zhang, S.; Zhu, R.; Jin, X.; Meng, S.; Jiang, H. An analysis of major scientific problems and research paths of Gulong shale oil in Daqing Oilfield, NE China. *Petrol. Explor. Dev.* **2021**, *48*, 453–463. [CrossRef]
6. Zhao, W.; Hu, S.; Hou, L. Connotation and strategic role of in-situ conversion processing of shale oil underground in the onshore China. *Petrol. Explor. Dev.* **2018**, *45*, 537–545. [CrossRef]
7. Gorynski, K.E.; Tobey, M.H.; Enriquez, D.A.; Smagala, T.M.; Dreger, J.L.; Newhart, R.E. Quantification and characterization of hydrocarbon-filled porosity in oil-rich shales using integrated thermal extraction and pyrolysis and solvent extraction. *AAPG Bull.* **2019**, *103*, 723–744. [CrossRef]
8. Guan, M.; Liu, X.; Jin, Z.; Lai, J.; Liu, J.; Sun, B.; Liu, T.; Hua, Z.; Xu, W.; Shu, H.; et al. Quantitative characterization of various oil contents and spatial distribution in lacustrine shales: Insight from petroleum compositional characteristics derived from programed pyrolysis. *Mar. Petrol. Geol.* **2022**, *138*, 105522. [CrossRef]
9. Li, J.; Wang, M.; Jiang, C.; Lu, S.; Li, Z. Sorption model of lacustrine shale oil: Insights from the contribution of organic matter and clay minerals. *Energy* **2022**, *260*, 125011. [CrossRef]
10. Wang, M.; Ma, R.; Li, J.; Lu, S.; Li, C.; Guo, Z.; Li, Z. Occurrence mechanism of lacustrine shale oil in the Paleogene Shahejie Formation of Jiyang Depression, Bohai Bay Basin, China. *Petrol. Explor. Dev.* **2019**, *46*, 789–802. [CrossRef]
11. Hu, T.; Pang, X.; Jiang, F.; Wang, Q.; Liu, X.; Wang, Z.; Jiang, S.; Wu, G.; Li, C.; Xu, T.; et al. Movable oil content evaluation of lacustrine organic-rich shales: Methods and a novel quantitative evaluation model. *Earth-Sci. Rev.* **2021**, *214*, 103545. [CrossRef]
12. Meng, M.; Hu, Q.; Wang, Q.; Hong, Z.; Zhang, L. Effect of initial water saturation and water film on imbibition behavior in tight reservoirs using nuclear magnetic resonance technique. *Phys. Fluids* **2024**, *36*, 056603. [CrossRef]
13. Abrams, M.A.; Gong, C.; Garnier, C.; Sephton, M.A. A new thermal extraction protocol to evaluate liquid rich unconventional oil in place and in-situ fluid chemistry. *Mar. Petrol. Geol.* **2017**, *88*, 659–675. [CrossRef]
14. Jiang, Q.; Li, M.; Qian, M.; Li, Z.; Li, Z.; Huang, Z.; Zhang, C.; Ma, Y. Quantitative characterization of shale oil in different occurrence states and its application. *Petrol. Geol. Exp.* **2016**, *38*, 842–849, (In Chinese with English abstract).
15. Romero-Sarmiento, M.-F. A quick analytical approach to estimate both free versus sorbed hydrocarbon contents in liquid-rich source rocks. *AAPG Bull.* **2019**, *103*, 2031–2043. [CrossRef]
16. Li, Z.; Zou, Y.; Xu, X.; Sun, J.; Li, M.; Peng, P. Adsorption of mudstone source rock for shale oil-Experiments, model and a case study. *Org. Geochem.* **2016**, *92*, 55–62. [CrossRef]
17. Zhang, H.; Huang, H.; Li, Z.; Liu, M. Oil physical status in lacustrine shale reservoirs: A case study on Eocene Shahejie Formation shales, Dongying Depression. *East China Fuel* **2019**, *257*, 116027. [CrossRef]
18. Qian, M.; Jiang, Q.; Li, M.; Li, Z.; Liu, P.; Ma, Y.; Cao, T. Quantitative characterization of extractable organic matter in lacustrine shale with different occurrences. *Petrol. Geol. Exp.* **2017**, *39*, 278–286. (In Chinese with English abstract)
19. Ertas, D.; Kelemen, S.R.; Halsey, T.C. Petroleum expulsion Part 1. Theory of kerogen swelling in multicomponent solvents. *Energy Fuels* **2006**, *20*, 295–300. [CrossRef]
20. Wei, Z.; Zou, Y.; Cai, Y.; Wang, L.; Luo, X.; Peng, P. Kinetics of oil group-type generation and expulsion: An integrated application to Dongying Depression, Bohai Bay Basin, China. *Org. Geochem.* **2012**, *52*, 1–12. [CrossRef]
21. Cao, H.; Shi, J.; Zhan, Z.; Wu, H.; Wang, X.; Cheng, X.; Li, H.; Zou, Y.; Peng, P. Shale oil potential and mobility in low- to medium-maturity lacustrine shales: A case study of the Yanchang Formation shale in southeast Ordos Basin, China. *Int. J. Coal Geol.* **2024**, *282*, 104421. [CrossRef]
22. Cao, H.; Zou, Y.; Lei, Y.; Xi, D.; Wan, X.; Peng, P. Shale oil assessment for the Songliao Basin, northeastern China, using oil generation-sorption method. *Energy Fuels* **2017**, *31*, 4826–4842. [CrossRef]
23. Ribeiro, R.C.; Correia, J.C.G.; Seidl, P.R. The influence of different minerals on the mechanical resistance of asphalt mixtures. *J. Petrol. Sci. Eng.* **2009**, *65*, 171–174. [CrossRef]
24. Xin, B.; Zhao, X.; Hao, F.; Jin, F.; Pu, X.; Han, W.; Xu, Q.; Guo, P.; Tian, J. Laminae characteristics of lacustrine shales from the Paleogene Kongdian Formation in the Cangdong Sag, Bohai Bay Basin, China: Why do laminated shales have better reservoir physical properties? *Int. J. Coal Geol.* **2022**, *260*, 104056. [CrossRef]
25. Zhang, P.; Yin, Y.; Lu, S.; Wang, J.; Zhang, J.; Zhi, Q.; Huang, H. Key factors controlling oil contents in different lithofacies shales from the Funing Formation, Subei Basin: Evidence from scanning electron microscopy. *Geoenergy Sci. Eng.* **2023**, *229*, 212115. [CrossRef]
26. Li, J.; Jiang, C.; Wang, M.; Lu, S.; Chen, Z.; Chen, G.; Li, J.; Li, Z.; Lu, S. Adsorbed and free hydrocarbons in unconventional shale reservoir: A new insight from NMR T1-T2 maps. *Mar. Petrol. Geol.* **2020**, *116*, 104311. [CrossRef]
27. Zhao, X.; Pu, X.; Zhou, L.; Jin, F.; Shi, Z.; Han, W.; Jiang, W.; Zhang, W. Typical geological characteristics and exploration practices of lacustrine shale oil: A case study of the Kong-2 member strata of the Cangdong Sag in the Bohai Bay Basin. *Mar. Petrol. Geol.* **2020**, *113*, 103999.
28. Ma, B.; Hu, Q.; Yang, S.; Zhang, T.; Qiao, H.; Meng, M.; Zhu, X.; Sun, X. Pore structure typing and fractal characteristics of lacustrine shale from Kongdian Formation in East China. *J. Nat. Gas Sci. Eng.* **2021**, *85*, 103709. [CrossRef]
29. Peters, K.E. Guidelines for evaluating petroleum source rock using programmed pyrolysis. *AAPG Bull.* **1986**, *70*, 318–329.
30. Yang, S.; Qiao, H.; Cheng, B.; Hu, Q. Solvent extraction efficiency of an Eocene-aged organic-rich lacustrine shale. *Mar. Petrol. Geol.* **2021**, *126*, 104941. [CrossRef]
31. Barrett, E.P.; Joyner, L.G.; Halenda, P.P. The determination of pore volume and area distributions in porous substances. I. Computations from nitrogen isotherms. *J. Am. Chem. Soc.* **1951**, *73*, 373–380. [CrossRef]

32. Gregg, S.; Sing, K. *Adsorption, Surface Area and Porosity*, 2nd ed.; Academic Press: London, UK, 1982.
33. Washburn, E.W. The dynamics of capillary flow. *Phys. Rev.* **1921**, *17*, 273–283. [CrossRef]
34. Wang, S.; Javadpour, F.; Feng, Q. Confinement correction to mercury intrusion capillary pressure of shale nanopores. *Sci. Re* **2016**, *6*, 20160. [CrossRef]
35. Hu, Q.; Zhang, Y.; Meng, X.; Li, Z.; Xie, Z.; Li, M. Characterization of multiple micro-nano pore networks in shale oil reservoirs c Paleogene Shahejie Formation in Dongying Sag of Bohai Bay Basin, East China. *Petrol. Explor. Dev.* **2017**, *44*, 720–730. [CrossRe
36. Lazar, O.R.; Bohacs, K.M.; Macquaker, J.H.S.; Schieber, J.; Demko, T.M. Capturing key attributes of fine-grained sedimentary rock in outcrops, cores, and thin sections: Nomenclature and description guidelines. *J. Sediment. Res.* **2015**, *85*, 230–246. [CrossRef]
37. Lu, S.; Huang, W.; Chen, F.; Li, J.; Wang, M.; Xue, H.; Wang, W.; Cai, X. Classification and evaluation criteria of shale oil and ga resources: Discussion and application. *Petrol. Explor. Dev.* **2012**, *39*, 249–256. [CrossRef]
38. Loucks, R.G.; Reed, R.M.; Ruppel, S.C.; Hammes, U. Spectrum of pore types and networks in mudrocks and a descriptiv classification for matrix-related mudrock pores. *AAPG Bull.* **2012**, *96*, 1071–1098. [CrossRef]
39. Ko, L.T.; Loucks, R.G.; Zhang, T.; Ruppel, S.C.; Shao, D. Pore and pore network evolution of Upper Cretaceous Boquillas (Eag Ford-equivalent) mudrocks: Results from gold tube pyrolysis experiments. *AAPG Bull.* **2016**, *100*, 1693–1722. [CrossRef]
40. Reed, R.M.; Loucks, R.; Ko, L. Scanning electron microscope petrographic differentiation among different types of pores associate with organic matter in mudrocks. *GCAGS J.* **2020**, *9*, 17–27.
41. Liang, C.; Cao, Y.; Liu, K.; Jiang, Z.; Wu, J.; Hao, F. Diagenetic variation at the lamina scale in lacustrine organic-rich shale Implications for hydrocarbon migration and accumulation. *Geochim. Cosmochim. Acta* **2018**, *229*, 112–128. [CrossRef]
42. Zou, C.; Jin, X.; Zhu, R.; Gong, G.; Sun, L.; Dai, J.; Meng, D.; Wang, X.; Li, J.; Wu, S.; et al. Do shale pore throats have a threshol diameter for oil storage? *Sci. Rep.* **2015**, *5*, 13619. [CrossRef]
43. Giesche, H. Mercury porosimetry: A general (practical) overview. *Part. Part. Syst. Char.* **2006**, *23*, 9–19. [CrossRef]
44. Jarvie, D.M. Shale resource systems for oil and gas: Part 2-Shale-oil resource systems. *AAPG Mem.* **2012**, *97*, 87–119.
45. Li, M.; Chen, Z.; Ma, X.; Cao, T.; Li, Z.; Jiang, Q. A numerical method for calculating total oil yield using a single routine Rock-Ev program: A case study of the Eocene Shahejie formation in Dongying depression, Bohai Bay Basin, China. *Int. J. Coal Geol.* **201** *191*, 49–65. [CrossRef]
46. Ma, W.; Li, J.; Wang, M. Determination of in situ hydrocarbon contents in shale oil plays: Part 3: Quantification of ligh hydrocarbon evaporative loss in old cores based on preserved shales. *Mar. Petrol. Geol.* **2024**, *160*, 106574. [CrossRef]
47. Zhao, X.; Pu, X.; Zhou, L.; Jin, F.; Han, G.; Shi, Z.; Han, W.; Ding, Y.; Zhang, W.; Wang, G.; et al. Enrichment theory, exploratio technology and prospects of shale oil in lacustrine facies zone of deep basin: A case study of the Paleogene in Huanghu depression, Bohai Bay Basin. *Acta Pet. Sin.* **2021**, *42*, 143–162. (In Chinese with English abstract)
48. Guo, Q.; Yao, Y.; Hou, L.; Tang, S.; Pan, S.; Yang, F. Oil migration, retention, and differential accumulation in "sandwiche lacustrine shale oil systems from the Chang 7 member of the Upper Triassic Yanchang Formation, Ordos Basin, China. *Int. J. Co Geol.* **2022**, *261*, 104077. [CrossRef]
49. Hu, T.; Jiang, F.; Pang, X.; Liu, Y.; Wu, G.; Zhou, K.; Xiao, H.; Jiang, Z.; Li, M.; Jiang, S.; et al. Identification and evaluation of sha oil micro-migration and its petroleum geological significance. *Petrol. Explor. Dev.* **2024**, *51*, 114–126. [CrossRef]
50. Jin, Z.; Wang, G.; Liu, G.; Gao, B.; Liu, Q.; Wang, H.; Liang, X.; Wang, R. Research progress and key scientific issues of continent shale oil in China. *Acta Pet. Sin.* **2021**, *42*, 821–835. (In Chinese with English abstract)
51. Liu, C.; Xu, X.; Liu, K.; Bai, J.; Liu, W.; Chen, S. Pore-scale oil distribution in shales of the Qingshankou formation in the Changlin Sag, Songliao Basin, NE China. *Mar. Petrol. Geol.* **2020**, *120*, 104553. [CrossRef]
52. Tissot, B.P.; Welte, D.H. *Petroleum Formation and Occurrence*, 2nd ed.; Springer: Berlin, Germany, 1984.
53. Zou, C.; Yang, Z.; Tao, S.; Li, W.; Wu, S.; Hou, L.; Zhu, R.; Yuan, X.; Wang, L.; Gao, X.; et al. Nano-hydrocarbon and th accumulation in coexisting source and reservoir. *Petrol. Explor. Dev.* **2012**, *39*, 13–26. [CrossRef]
54. Zhao, X.; Zhou, L.; Pu, X.; Han, W.; Jin, F.; Xiao, D.; Shi, Z.; Deng, Y.; Zhang, W.; Jiang, W. Exploration breakthroughs an geological characteristics of continental shale oil: A case study of the Kongdian Formation in the Cangdong Sag, China. *Ma Petrol. Geol.* **2019**, *102*, 544–556. [CrossRef]
55. Meng, M.; Ge, H.; Shen, Y.; Ji, W.; Li, Z. Insight into water occurrence and pore size distribution by nuclear magnetic resonance i marine shale reservoirs, southern China. *Energy Fuels* **2023**, *37*, 319–327. [CrossRef]
56. Han, W.; Zhao, X.; Jin, F. Sweet spots evaluation and exploration of lacustrine shale oil of the 2nd member of Paleogene Kongdia Formation in Cangdong Sag, Dagang Oilfield, China. *Petrol. Explor. Dev.* **2021**, *48*, 1–10. [CrossRef]
57. Bai, L.; Liu, B.; Fu, X.; Hu, Q.; Huo, Q.; Wang, L.; Wang, Q.; Fu, L.; Ostadhassan, M. A new method for evaluating the oil mobili based on the relationship between pore structure and state of oil. *Geosci. Front.* **2023**, *14*, 101684. [CrossRef]
58. Deng, Y.; Chen, S.; Pu, X.; Yan, J.; Chen, J. Formation mechanism and environmental evolution of fine-grained sedimentary rock from the second member of Kongdian Formation in the Cangdong Sag, Bohai Bay Basin. *Petrol. Geol. Exp.* **2020**, *41*, 811–823+89 (In Chinese with English abstract)

Journal of
*Marine Science
and Engineering*

MDPI

Article

Salmon Salar Optimization: A Novel Natural Inspired Metaheuristic Method for Deep-Sea Probe Design for Unconventional Subsea Oil Wells

ia Guo [1,2,3,4], Zhou Yan [2], Yuji Sato [4] and Qiankun Zuo [1,2,3,*]

1 Hubei Key Laboratory of Digital Finance Innovation, Hubei University of Economics, Wuhan 430205, China; guojia@hbue.edu.cn
2 School of Information Engineering, Hubei University of Economics, Wuhan 430205, China
3 Hubei Internet Finance Information Engineering Technology Research Center, Hubei University of Economics, Wuhan 430205, China
4 Faculty of Computer and Information Sciences, Hosei University, Tokyo 102-8160, Japan; yuji@hosei.ac.jp
* Correspondence: qiankun_zuo@163.com

Abstract: As global energy demands continue to rise, the development of unconventional oil resources has become a critical priority. However, the complexity and high dimensionality of these problems often cause existing optimization methods to get trapped in local optima when designing key tools, such as deep-sea probes. To address this challenge, this study proposes a novel metaheuristic approach—the Salmon Salar Optimization algorithm, which simulates the social structure and collective behavior of salmon to perform high-precision searches in high-dimensional spaces. The Salmon Salar Optimization algorithm demonstrated superior performance across two benchmark function sets and successfully solved the constrained optimization problem in deep-sea probe design. These results indicate that the proposed method is highly effective in meeting the optimization needs of complex engineering systems, particularly in the design optimization of deep-sea probes for unconventional oil exploration.

Keywords: salmon salar optimization; high-dimensional optimization; collective behavior; deep-sea probe design

:itation: Guo, J.; Yan, Z.; Sato, Y.; Zuo,). Salmon Salar Optimization: A Iovel Natural Inspired Metaheuristic Iethod for Deep-Sea Probe Design or Unconventional Subsea Oil Wells. *Mar. Sci. Eng.* **2024**, *12*, 1802. ttps://doi.org/10.3390/ nse12101802

cademic Editor: Marco Cococcioni

eceived: 12 August 2024
evised: 20 September 2024
ccepted: 28 September 2024
ublished: 10 October 2024

1. Introduction

The development of unconventional oil resources is increasingly recognized as a critical component in addressing global energy demands. As conventional oil reserves dwindle and geopolitical factors introduce uncertainties in supply, the ability to access unconventional resources, such as shale oil, oil sands, and ultra-deepwater reserves, has become imperative [1,2]. Although these resources are abundant, they require advanced extraction technologies and sophisticated engineering solutions due to their complex geological formations and challenging extraction environments [3,4].

Deep-sea probes play an indispensable role in the exploration and extraction of unconventional oil resources, particularly in ultra-deepwater environments. These probes must navigate and analyze extreme underwater conditions, characterized by high geological complexity and a hostile operational environment. The design and deployment of deep-sea probes present high-dimensional optimization challenges, where multiple factors—such as structural integrity, sensor accuracy, energy efficiency, and environmental resilience—must be considered simultaneously [5]. The complexity of this task is further amplified by the need to ensure reliable long-term operation under conditions of extreme pressure, temperature fluctuations, and corrosive seawater.

While metaheuristic methods have demonstrated success in solving many complex optimization problems, several challenges remain. These include the risk of getting trapped in local optima, particularly in high-dimensional optimization problems where the search space is vast, making it difficult for algorithms to escape local optima and find the globa solution. Additionally, the convergence speed of these algorithms can be slow, especiall in high-dimensional settings, where computational resource consumption is substantia leading to reduced efficiency. In applications, tasks like deep-sea probe design for ur conventional oil exploration require highly precise, adaptable, and robust optimizatio methods capable of handling complex, multi-dimensional environments under extrem conditions. To address these limitations, this study proposes the Salmon Salar Optimizatio (SSO) algorithm, inspired by the social and cooperative behavior of salmon, as a nove approach to achieving more effective and scalable solutions in high-dimensional optimiz tion tasks.

2. Related Works

Evolutionary algorithms, such as particle swarm optimization (PSO) [6], firefly alg rithms [7], differential evolution [8], and the wolf pack algorithm [9], are effective method for solving high-dimensional optimization problems. PSO is inspired by the social behavic of animals like birds and bees and is used to find global optimal solutions to optimizatio problems. In PSO, a group of particles represents potential solutions moving through th search space. These particles are initialized with random positions and velocities and ar guided by two factors: their personal best position (the best solution each particle ha found) and the global best position (the best solution found by any particle in the swarm At each iteration, particles update their velocities and positions based on these two factor aiming to move toward more promising regions of the search space. The movement of th particles is controlled by acceleration constants, which determine how much influence th personal and global best positions have on the particle's trajectory.

In recent years, convex optimization, which deals with optimization problems wher the objective function is convex, has gained increasing importance. Other active researcl areas in optimization include global optimization [10] and multi-objective optimization [11

Optimization problems are widely applied across various fields, such as supply chai management [12], transportation [13], colored TSP problems [14,15], networks [16], pe manent magnets [17], sound classification [18], feature selection [19], risk prediction [20 route design [21], and financial management [22]. As the Internet continues to grow rapidl the complexity and scale of optimization problems have also increased, making traditiona optimization algorithms less suited to modern needs. As a result, researchers continue t develop and improve optimization algorithms to meet these evolving challenges.

In 2015, Gao [23] proposed a selectively informed PSO (SIPSO). In SIPSO, densel connected particles receive information from all their neighbors, while sparsely connecte particles follow only the best-performing neighbor. Liang [24] introduced an adaptive PS(based on clustering (APSO-C), which involves two main steps: first, a K-means clusterin method is used to divide the swarm into several sub-groups, and second, the inerti weights of individuals are adjusted based on the evaluation of clusters and swarm state Li [25] proposed a composite PSO with historical memory (HMPSO), in which each particl considers three candidate positions: its historical memory, personal best position, and th global best of the swarm.

In 2016, Pornsing [26] proposed techniques for self-adaptive inertia weight and tim varying adaptive swarm topology to enhance the performance of particle swarm optimiz tion (PSO). Guo [27] introduced several variants of the Bare Bones PSO (BBPSO) algorithn including pair-wise bare bones PSO (PBBPSO), dynamic local search bare bones PSO (DL BBPSO) [28], and hierarchical bare bones PSO (HBBPSO) [29], all aimed at improving th search capabilities of the original BBPSO [30].

In 2017, Xu [31] proposed a chaotic PSO (CPSO) for combinatorial optimization prol lems, which combines a chaos method with the traditional PSO. In 2018, Tian [32] i troduced a modified PSO that incorporates chaos-based initialization and robust updat mechanisms. That same year, Guo [28] developed a dynamic allocation bare bones PS(

(DABBPSO) and a dynamic reconstruction bare bones PSO (DRBBPSO) to further improve the original BBPSO.

In 2019, Ghasemi [33] proposed the Phasor PSO (PPSO), while Guo [34] introduced a fission-fusion hybrid BBPSO (FBBPSO) for single-objective optimization problems. Xu [35] presented a dimensional learning strategy for PSO.

In 2020, Xu [36] proposed a reinforcement learning-based communication topology for PSO. In 2021, Yamanaka [37] developed a gravitational PSO (SPSO) for multi-modal optimization problems, and Liu [38] introduced a Sigmoid-Function-Based adaptive weighted PSO (SAWPSO). Wang [39] proposed an adaptive granularity learning distributed PSO (AGLDPSO). In 2022, Li [40] introduced a multi-population PSO with neighborhood learning (MPSO-NL), and Tian [41] proposed an electronic transition-based BBPSO for high-dimensional problems. Guo [42] later introduced a twinning strategy for BBPSO, and Zhou [43] proposed an atomic retrospective learning BBPSO (ARBBPSO).

Many researchers have also drawn inspiration from the natural world and developed various metaheuristic algorithms, such as the Dung Beetle Optimizer (DBO) [44], Whale Optimization Algorithm (WOA) [45], Grey Wolf Optimizer (GWO) [46], Harris Hawks Optimization (HHO) [47], African Vultures Optimization Algorithm (AVOA) [48], and Gorilla Troops Optimizer [49]. These algorithms have demonstrated strong performance in solving both benchmark and engineering problems.

With the advancement of artificial intelligence technology, optimization problems have grown in both scale and dimensionality. Traditional optimization algorithms are no longer sufficient to address the wide range of complex optimization challenges. To overcome these issues, this work proposes a new metaheuristic method, Salmon Salar Optimization (SSO). The strength of the SSO method lies in its ability to simulate the social structure and collective behavior of salmon, effectively balancing exploration and exploitation in high-dimensional search spaces. This approach helps the algorithm escape local optima and improves the convergence rate, even in highly complex and constrained environments. These features make SSO particularly well-suited for solving complex engineering optimization problems, such as the design of deep-sea probes for unconventional oil exploration. The major contributions of this work are summarized as follows:

(1) Biologically-Inspired Design: The Salmon Salar Optimization algorithm is a novel high-dimensional optimization method inspired by the collective behavior of salmon in nature. Unlike traditional algorithms, it emulates the migration and cooperative strategies of salmon, making it more effective in tackling high-dimensional optimization problems.

(2) High-Precision Solutions: The SSO algorithm excels in high-dimensional spaces, demonstrating a unique ability to discover highly accurate solutions. It is particularly effective at addressing complex multi-dimensional problems, providing robust solutions across engineering, scientific, and commercial applications.

(3) Adaptiveness: The SSO algorithm exhibits a high degree of adaptiveness, allowing it to dynamically adjust based on the nature and dimensionality of the problem. This adaptability enables it to handle various types of high-dimensional optimization problems without requiring extensive parameter tuning beforehand.

The rest of this paper is organized as follows: Section 2 introduces the design and structure of the Salmon Salar Optimization algorithm; Section 3 presents the experimental details and results and Section 4 provides the conclusion of this work.

3. Materials and Methods

3.1. Salmon Salar Optimization

Drawing inspiration from the structural composition and social behaviors observed in salmon salar populations, the Salmon Salar Optimization (SSO) Algorithm is proposed. The SSO Algorithm derives its conceptual foundation from three principal search objectives: food procurement, breeding habitat identification, and hazard avoidance. Primarily, the food procurement objective corresponds to the imperative for sustenance, enabling the algorithm to pursue optimal solutions within the problem domain to ensure individual health and viability. Subsequently, the breeding habitat identification objective reflects the algorithm's capacity to discern suitable problem domain solutions in support of reproduction and ensuing population expansion. Lastly, the hazard avoidance objective equips the algorithm with adaptability to volatile problem conditions, enabling it to circumvent potential threats or adverse factors, thereby ensuring search process stability.

Consequently, the SSO amalgamates these three fundamental search strategies to achieve efficient, diversified, and adaptive search processes. This approach, when applied to the resolution of intricate high-dimensional optimization problems, evinces significant potential by not only furnishing robust solutions but also introducing an innovative optimization instrument applicable across diverse domains, encompassing engineering, scientific inquiry, and commercial enterprises. By emulating the efficacious survival strategies exhibited by salmon salar populations, the SSO introduces biologically inspired paradigms into the realm of optimization, promising noteworthy advancements in the solution of complex problem sets.

3.2. The Multi-Purpose Fusion Search Strategy

Considering the complexity of the high-dimensional optimization problem, the Gaussian distribution is used to select the position of the particle in the next iteration. The next position of a salmon is calculated by Equation (1).

$$\begin{aligned} \alpha &= (Salmon + LeaderSalmon)/2 \\ \beta &= |Salmon - LeaderSalmon| \\ Salmon\_candi &= Gausi(\alpha,\ \beta) \end{aligned} \quad (1)$$

where *Salmon* is the initial position of a normal salmon; *LeaderSalmon* contains the food memory, breeding memory, and danger memory. $Gausi(\alpha, \beta)$ is a Gaussian distribution with a mean α and a standard deviation β. Since *LeaderSalmon* has three levels, *Salmon_candi* will contain three candidate positions. After the candidate position selection, the next position of a normal *Salmon* is calculated by Equation (2).

$$\begin{aligned} candidatelist^{t+1} &= [Salmon^{t}, Salmon\_candi^{t+1}] \\ Salmont^{t+1} &= FindBest(candidatelist^{t}, 1) \end{aligned} \quad (2)$$

where $candidatelist^{t+1}$ is a list containing the position of a *Salmon* in the tth generation and three candidate positions in the $(t+1)$th generation, $FindBest(X, 1)$ is a function used to find the best position from X. After every *Salmon* finds a new position in the $(t+1)$th generation, the next position of the *LeaderSalmon* is calculated by Equation (3).

$$\begin{aligned} LeaderSalmon\_candi^{t+1} &= FindBest(Salmont^{t+1}, 1) \\ leader_{c}andidatelist^{t+1} &= [LeaderSalmon^{t}, LeaderSalmon\_candi^{t+1}] \\ LeaderSalmont^{t+1} &= FindBest(candidatelist^{t}, 3) \end{aligned} \quad (3)$$

where $leader_{c}andidatelist^{t+1}$ is a list containing the position of the *LeaderSalmon* in the generation and the best position of all Salmons in the $(t+1)$th generation. $FindBest(X, 3)$ is a function used to find the best three positions from X.

To better describe the operation mode of SSO, this section details the pseudo-code and flowchart of SSO. The flowchart is shown in Figure 1; pseudo-code of SSO is shown in Algorithm 1.

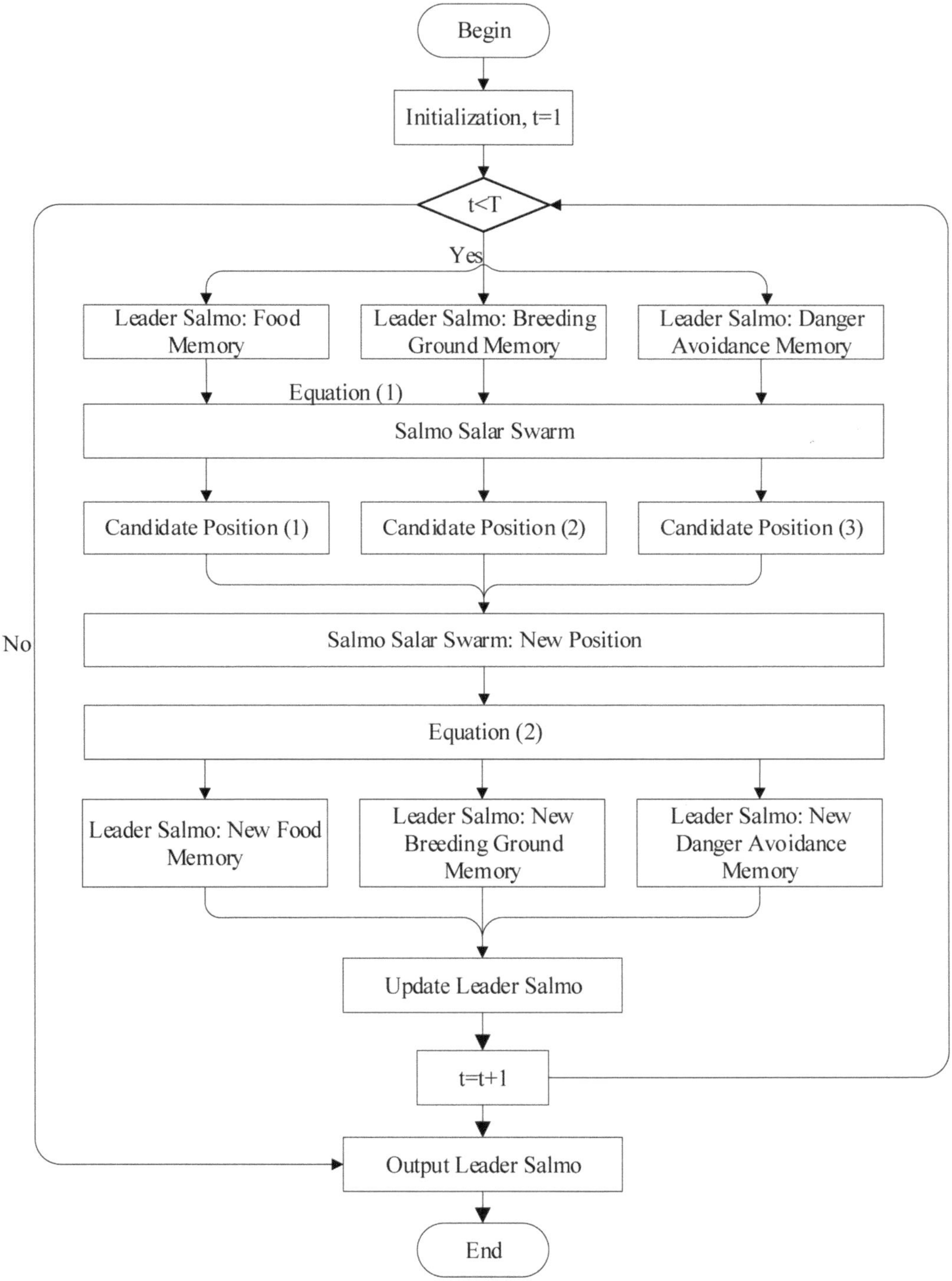

Figure 1. Flowchart of SSO.

Algorithm 1 Pseudo-code of SSO.

Require: IT, max iteration time
Require: $func$, fitness function,
Require: Ran, searching Range
Require: Max, Max generation time
Require: n, number of Salmon
for gen=1 to Max **do**
  Randomly generate the initial position of $Salmons$
  Find the $LeaderSalmon$ from all $Salmons$
  Update the position of $Salmons$, using Equations (1) and (2)
  Update the position of $LeaderSalmon$, using Equation (3)
  gen=gen+1
end for
Output: $LeaderSalmon$

4. Results

4.1. Numerical Experiments with CEC2017

To explore the optimization ability of SSO, the CEC2017 benchmark functions ar selected in the simulation test. The CEC2017 contains four different groups of types:

1. Unimodal Functions, $f_1 - f_2$;
2. Simple Multimodal Functions, $f_3 - f_9$;
3. Hybrid Functions, $f_{10} - f_{19}$;
4. Composition Functions, $f_{20} - f_{29}$.

Eight state-of-the-art optimization algorithms, DBO, WOA, GWO, HHO, AVOA, GTC ARBBPSO, and ETBBPSO are selected in the control group. To ensure the fairness of th experiment, all algorithms use the same number of particles with the same number o iterations. To reduce the error caused by chance, all calculations are repeated thirty-seve times. The mean and variance of these thirty-seven times are recorded, and these result are used to evaluate the performance of the algorithms. The parameters of experiments ar listed below:

1. Population size for all algorithms, 100;
2. Max iteration times, 1.000×10^5;
3. Dimension, 100;
4. Individual runs, 37.
5. search range, [−100, +100]

To easily compare the optimization ability of each algorithm, the FE (final error) i used in the results analysis. The FE is defined in Equation (4).

$$FE = |finalgbest - TOV| \quad (4$$

where the $finalgbest$ is the gbest value of an algorithm after the last evolution, TOV is th theoretical optimal value of the test function. Obviously, when comparing the two method the method with a smaller FE has better optimization performance. Numerical and ranke results of CEC2017 are shown in Tables 1–4. The convergence curves of the experiments ar shown in Figures 2–7.

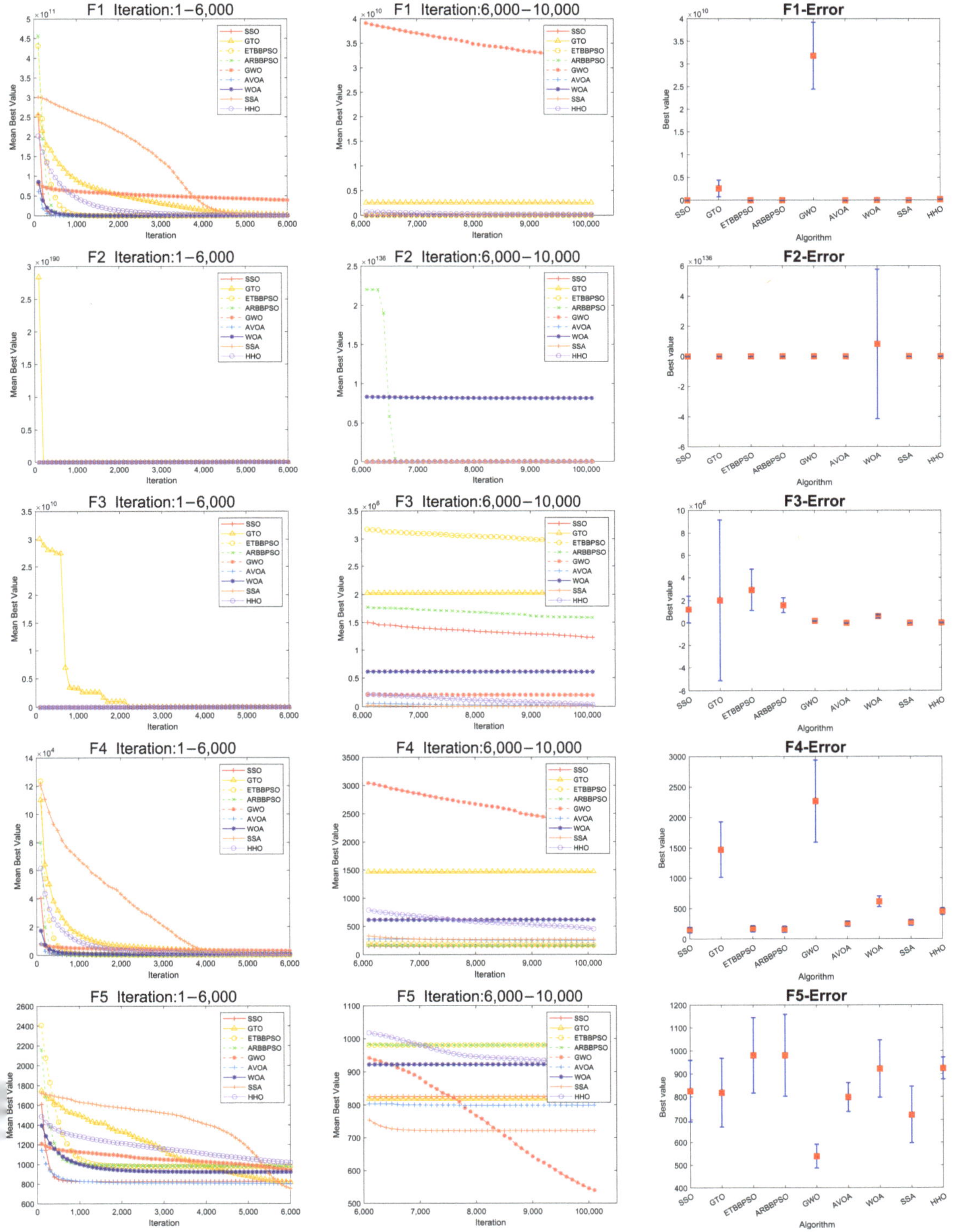

Figure 2. Convergence graph and standard deviation of F1 to F5. In the Error figure, the red square stands for the mean value, and the blue line stands for the error.

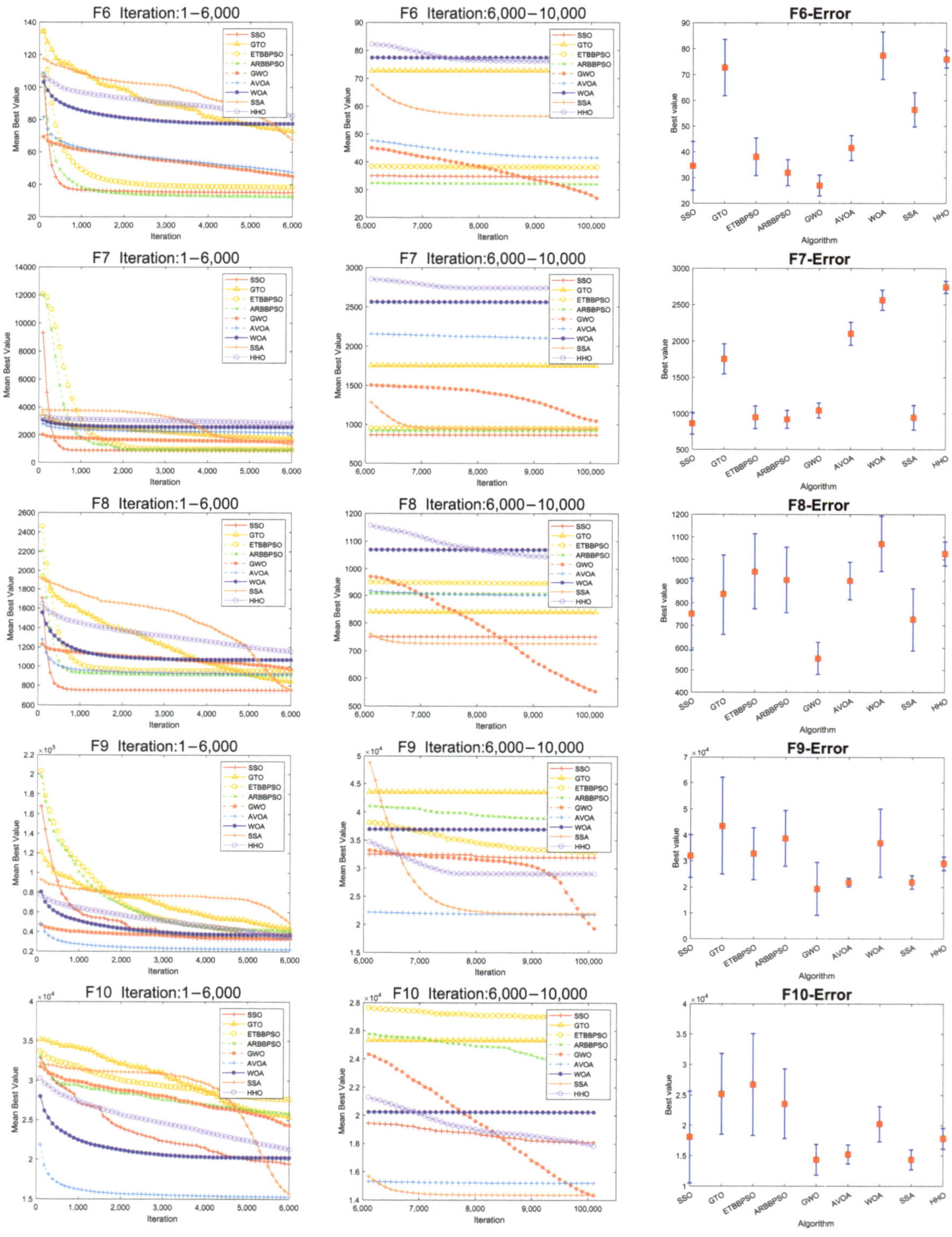

Figure 3. Convergence graph and standard deviation of F6 to F10. In the Error figure, the red square stands for the mean value, and the blue line stands for the error.

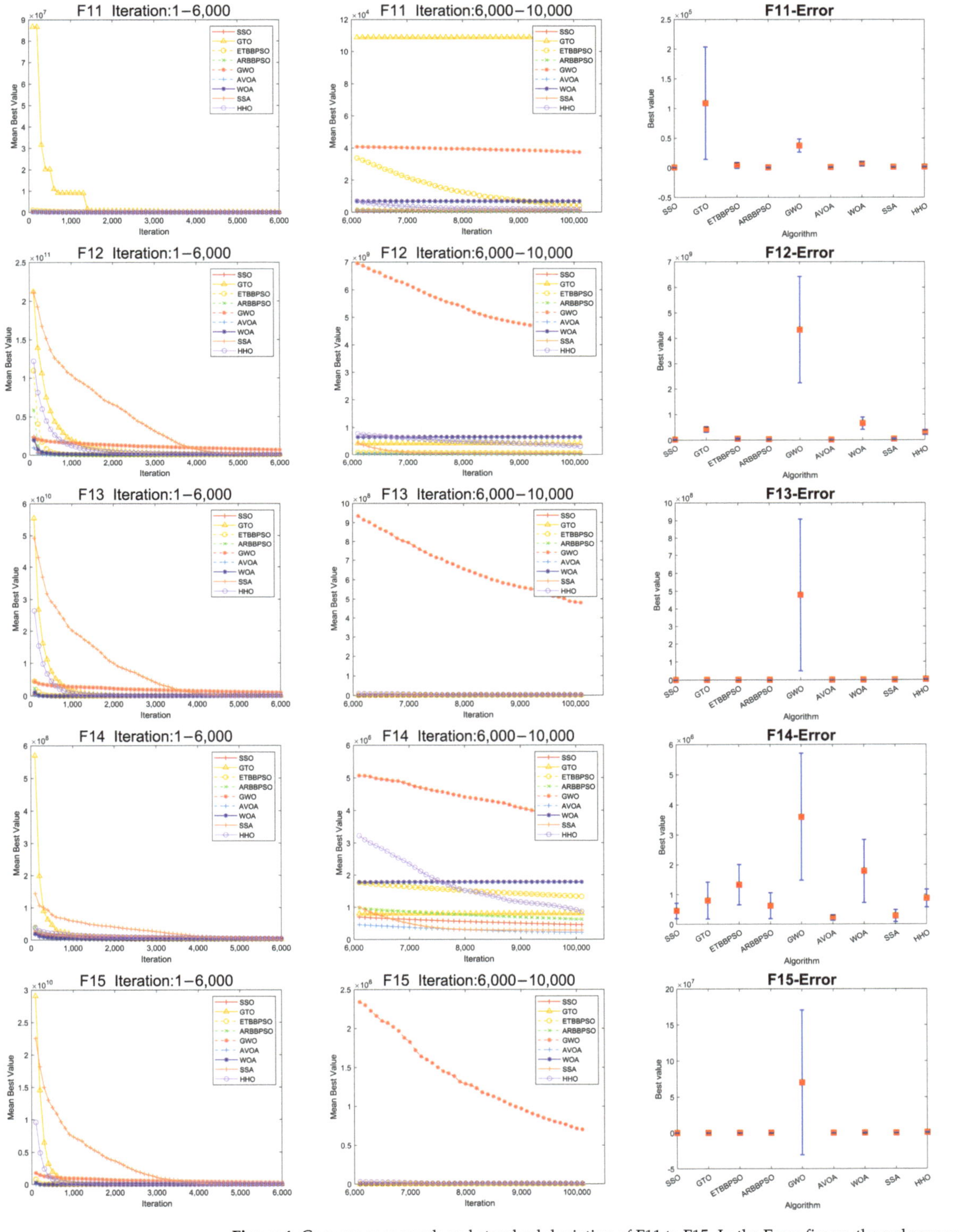

Figure 4. Convergence graph and standard deviation of F11 to F15. In the Error figure, the red square stands for the mean value, and the blue line stands for the error.

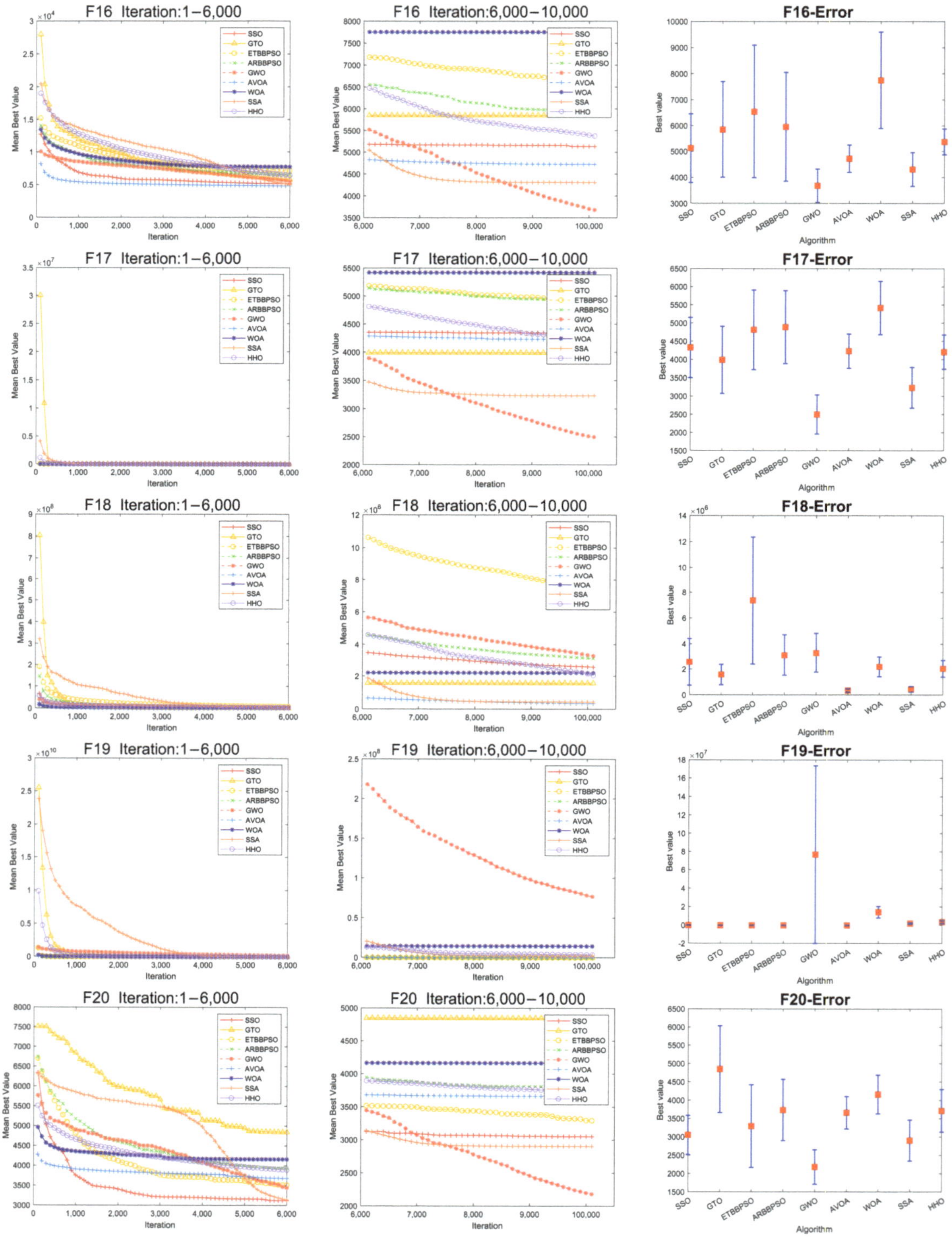

Figure 5. Convergence graph and standard deviation of F16 to F20. In the Error figure, the red square stands for the mean value, and the blue line stands for the error.

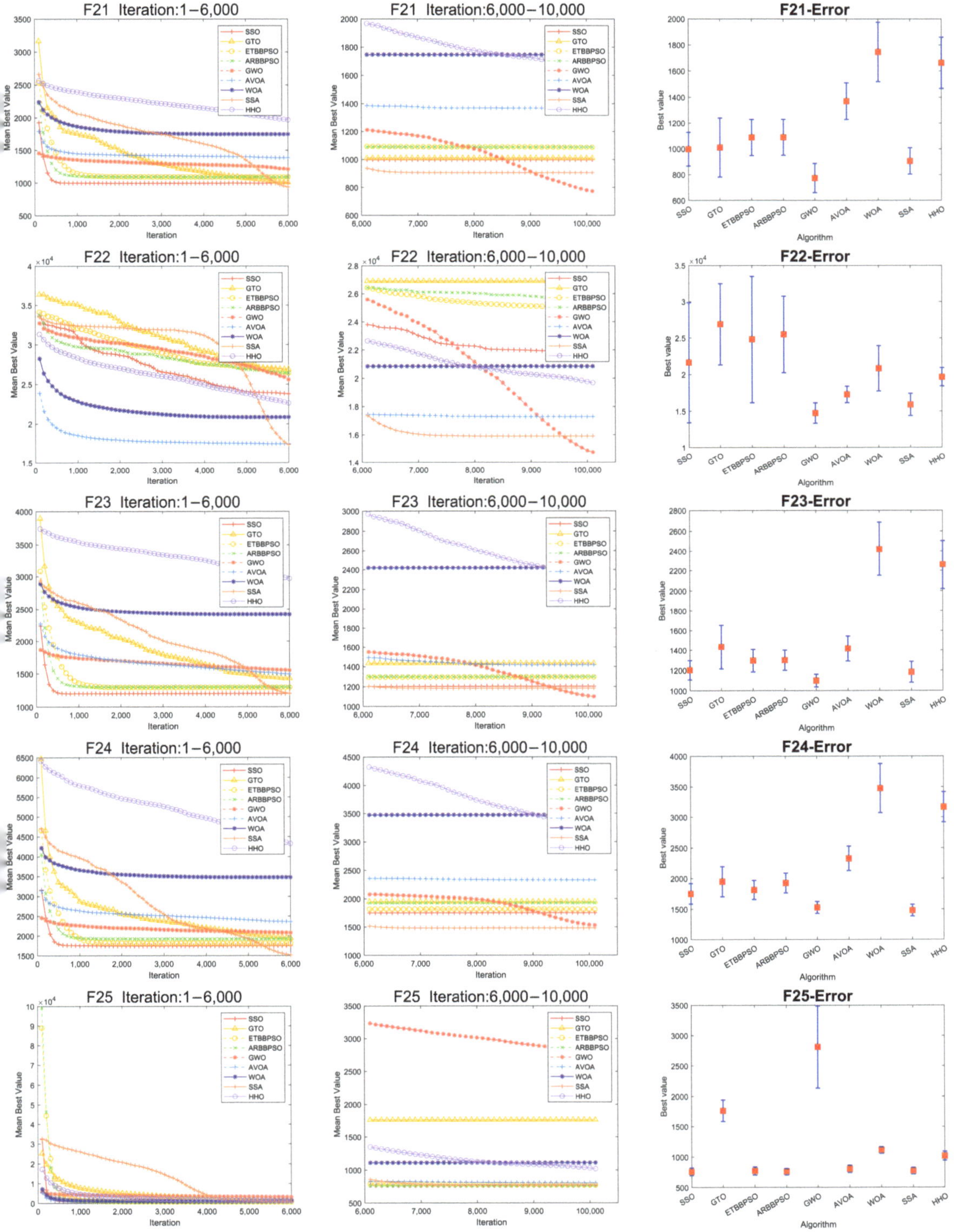

Figure 6. Convergence graph and standard deviation of F21 to F25. In the Error figure, the red square stands for the mean value, and the blue line stands for the error.

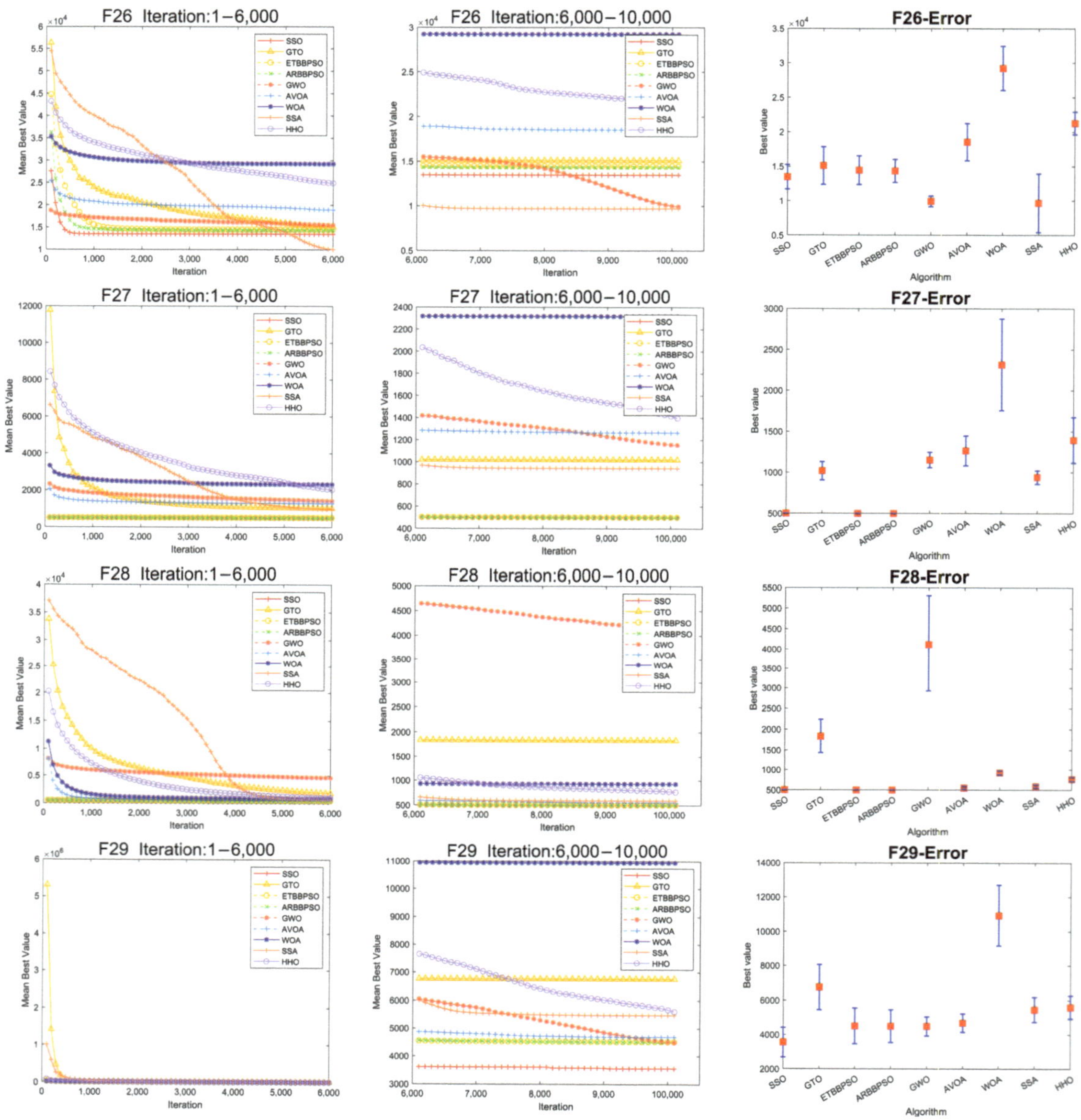

Figure 7. Convergence graph and standard deviation of F26 to F29. In the Error figure, the red square stands for the mean value, and the blue line stands for the error.

Table 1. The simulation results of SSO, DBO, WOA, GWO, HHO, AVOA, GTO, ARBBPSO and ETBBPSO, f1–f8, best results are shwon in bold.

F	Type	SSO	DBO	WOA	GWO	HHO	AVOA	GTO	ARBBPSO	ETBBPSO
1	Mean	2.808×10^{4}	5.748×10^{7}	3.515×10^{7}	3.184×10^{10}	2.097×10^{8}	**6.690×10^{3}**	2.604×10^{9}	2.912×10^{4}	1.656×10^{4}
	Std	2.760×10^{4}	3.705×10^{7}	1.173×10^{7}	7.385×10^{9}	2.181×10^{7}	9.224×10^{3}	1.822×10^{9}	3.091×10^{4}	2.186×10^{4}
	Best	1.851×10^{1}	1.490×10^{6}	1.560×10^{7}	2.009×10^{10}	1.621×10^{8}	9.860×10^{0}	7.170×10^{8}	1.369×10^{2}	4.990×10^{0}
	Worst	8.870×10^{4}	1.431×10^{8}	7.225×10^{7}	5.204×10^{10}	2.517×10^{8}	3.450×10^{4}	8.877×10^{9}	1.281×10^{5}	1.120×10^{5}
	Rank	3	6	5	9	7	1	8	4	2
2	Mean	1.909×10^{89}	1.845×10^{134}	8.159×10^{135}	6.547×10^{125}	1.117×10^{73}	**1.326×10^{30}**	3.553×10^{119}	9.078×10^{125}	7.782×10^{114}
	Std	7.879×10^{89}	1.122×10^{135}	4.955×10^{136}	3.976×10^{126}	5.615×10^{73}	5.600×10^{30}	1.423×10^{120}	5.522×10^{126}	3.073×10^{115}
	Best	1.679×10^{68}	2.434×10^{98}	2.676×10^{110}	5.418×10^{97}	6.231×10^{59}	3.731×10^{13}	9.303×10^{99}	6.574×10^{84}	2.233×10^{88}
	Worst	4.579×10^{90}	6.822×10^{135}	3.014×10^{137}	2.418×10^{127}	3.375×10^{74}	2.681×10^{31}	7.773×10^{120}	3.359×10^{127}	1.718×10^{116}
	Rank	3	8	9	6	2	1	5	7	4
3	Mean	1.227×10^{6}	3.241×10^{5}	6.068×10^{5}	2.008×10^{5}	3.406×10^{4}	**2.034×10^{4}**	2.030×10^{6}	1.589×10^{6}	2.934×10^{6}
	Std	1.186×10^{6}	1.856×10^{4}	1.497×10^{5}	1.767×10^{4}	8.364×10^{3}	6.828×10^{3}	7.150×10^{6}	6.647×10^{5}	1.808×10^{6}
	Best	4.535×10^{5}	2.578×10^{5}	3.368×10^{5}	1.638×10^{5}	1.796×10^{4}	8.612×10^{3}	2.438×10^{5}	6.560×10^{5}	1.059×10^{6}
	Worst	4.888×10^{6}	3.516×10^{5}	8.400×10^{5}	2.421×10^{5}	5.648×10^{4}	3.839×10^{4}	4.276×10^{7}	3.261×10^{6}	8.476×10^{6}
	Rank	6	4	5	3	2	1	8	7	9
4	Mean	**1.493×10^{2}**	4.300×10^{2}	6.164×10^{2}	2.268×10^{3}	4.530×10^{2}	2.495×10^{2}	1.473×10^{3}	1.578×10^{2}	1.683×10^{2}
	Std	4.601×10^{1}	7.513×10^{1}	8.827×10^{1}	6.743×10^{2}	5.750×10^{1}	4.472×10^{1}	4.581×10^{2}	5.156×10^{1}	5.145×10^{1}
	Best	8.963×10^{1}	2.765×10^{2}	4.498×10^{2}	1.096×10^{3}	3.622×10^{2}	1.930×10^{2}	7.983×10^{2}	7.623×10^{1}	7.982×10^{1}
	Worst	2.934×10^{2}	5.789×10^{2}	8.222×10^{2}	4.021×10^{3}	5.927×10^{2}	3.663×10^{2}	2.757×10^{3}	3.117×10^{2}	2.893×10^{2}
	Rank	1	5	7	9	6	4	8	2	3
5	Mean	8.255×10^{2}	1.089×10^{3}	9.223×10^{2}	**5.395×10^{2}**	9.242×10^{2}	7.989×10^{2}	8.184×10^{2}	9.809×10^{2}	9.808×10^{2}
	Std	1.336×10^{2}	1.545×10^{2}	1.244×10^{2}	5.188×10^{1}	4.758×10^{1}	6.290×10^{1}	1.499×10^{2}	1.781×10^{2}	1.639×10^{2}
	Best	5.462×10^{2}	8.141×10^{2}	7.312×10^{2}	4.092×10^{2}	8.221×10^{2}	6.398×10^{2}	6.021×10^{2}	6.248×10^{2}	6.706×10^{2}
	Worst	1.081×10^{3}	1.361×10^{3}	1.250×10^{3}	6.650×10^{2}	1.038×10^{3}	9.522×10^{2}	1.393×10^{3}	1.369×10^{3}	1.270×10^{3}
	Rank	4	9	5	1	6	2	3	8	7
6	Mean	3.455×10^{1}	7.188×10^{1}	7.737×10^{1}	**2.697×10^{1}**	7.590×10^{1}	4.148×10^{1}	7.272×10^{1}	3.194×10^{1}	3.806×10^{1}
	Std	9.455×10^{0}	9.560×10^{0}	9.166×10^{0}	4.047×10^{0}	3.312×10^{0}	4.854×10^{0}	1.090×10^{1}	5.052×10^{0}	7.284×10^{0}
	Best	1.108×10^{1}	5.496×10^{1}	6.168×10^{1}	1.967×10^{1}	6.720×10^{1}	3.192×10^{1}	4.958×10^{1}	1.972×10^{1}	2.120×10^{1}
	Worst	5.381×10^{1}	9.361×10^{1}	1.048×10^{2}	3.501×10^{1}	8.179×10^{1}	5.283×10^{1}	9.305×10^{1}	4.202×10^{1}	5.119×10^{1}
	Rank	3	6	9	1	8	5	7	2	4
7	Mean	**8.635×10^{2}**	1.507×10^{3}	2.561×10^{3}	1.044×10^{3}	2.740×10^{3}	2.101×10^{3}	1.754×10^{3}	9.204×10^{2}	9.485×10^{2}
	Std	1.510×10^{2}	3.905×10^{2}	1.393×10^{2}	1.037×10^{2}	8.270×10^{1}	1.582×10^{2}	2.081×10^{2}	1.256×10^{2}	1.575×10^{2}
	Best	6.102×10^{2}	1.036×10^{3}	2.212×10^{3}	8.211×10^{2}	2.560×10^{3}	1.773×10^{3}	1.337×10^{3}	6.931×10^{2}	6.999×10^{2}
	Worst	1.147×10^{3}	3.145×10^{3}	2.894×10^{3}	1.309×10^{3}	2.932×10^{3}	2.442×10^{3}	2.239×10^{3}	1.212×10^{3}	1.326×10^{3}
	Rank	1	5	8	4	9	7	6	2	3
8	Mean	7.509×10^{2}	1.143×10^{3}	1.069×10^{3}	**5.528×10^{2}**	1.026×10^{3}	9.013×10^{2}	8.395×10^{2}	9.049×10^{2}	9.439×10^{2}
	Std	1.620×10^{2}	1.525×10^{2}	1.231×10^{2}	7.252×10^{1}	5.340×10^{1}	8.687×10^{1}	1.798×10^{2}	1.495×10^{2}	1.705×10^{2}
	Best	5.174×10^{2}	7.631×10^{2}	8.041×10^{2}	4.036×10^{2}	8.413×10^{2}	7.323×10^{2}	5.891×10^{2}	5.801×10^{2}	5.612×10^{2}
	Worst	1.132×10^{3}	1.405×10^{3}	1.476×10^{3}	7.012×10^{2}	1.108×10^{3}	1.102×10^{3}	1.533×10^{3}	1.296×10^{3}	1.223×10^{3}
	Rank	2	9	8	1	7	4	3	5	6

Table 2. The simulation results of SSO, DBO, WOA, GWO, HHO, AVOA, GTO, ARBBPSO and ETBBPSO, f9–f16, best results are shwon in bold.

F	Type	SSO	DBO	WOA	GWO	HHO	AVOA	GTO	ARBBPSO	ETBBPSO
9	Mean	3.189×10^4	3.533×10^4	3.694×10^4	$\mathbf{1.929 \times 10^4}$	2.902×10^4	2.176×10^4	4.350×10^4	3.872×10^4	3.285×10^4
	Std	8.257×10^3	9.423×10^3	1.313×10^4	1.023×10^4	2.590×10^3	1.672×10^3	1.855×10^4	1.076×10^4	1.002×10^4
	Best	1.398×10^4	1.751×10^4	2.303×10^4	9.616×10^3	2.394×10^4	1.901×10^4	2.294×10^4	1.777×10^4	1.762×10^4
	Worst	4.255×10^4	4.880×10^4	8.197×10^4	4.220×10^4	3.367×10^4	2.774×10^4	7.876×10^4	8.490×10^4	4.980×10^4
	Rank	4	6	7	1	3	2	9	8	5
10	Mean	1.808×10^4	1.735×10^4	2.023×10^4	$\mathbf{1.434 \times 10^4}$	1.783×10^4	1.521×10^4	2.516×10^4	2.355×10^4	2.671×10^4
	Std	7.565×10^3	1.543×10^3	2.902×10^3	2.535×10^3	1.718×10^3	1.573×10^3	6.647×10^3	5.728×10^3	8.382×10^3
	Best	1.069×10^4	1.446×10^4	1.545×10^4	1.057×10^4	1.334×10^4	1.213×10^4	1.270×10^4	1.233×10^4	9.839×10^3
	Worst	3.297×10^4	2.025×10^4	2.601×10^4	2.693×10^4	2.084×10^4	1.979×10^4	3.375×10^4	2.947×10^4	3.364×10^4
	Rank	5	3	6	1	4	2	8	7	9
11	Mean	$\mathbf{4.957 \times 10^2}$	1.867×10^4	6.803×10^3	3.741×10^4	1.861×10^3	1.202×10^3	1.087×10^5	5.786×10^2	3.974×10^3
	Std	1.357×10^2	3.081×10^4	3.448×10^3	1.107×10^4	1.889×10^2	2.255×10^2	9.458×10^4	2.244×10^2	5.112×10^3
	Best	2.573×10^2	2.700×10^3	4.322×10^3	1.755×10^4	1.512×10^3	7.107×10^2	2.676×10^4	2.266×10^2	3.565×10^2
	Worst	7.958×10^2	1.483×10^5	2.567×10^4	6.831×10^4	2.247×10^3	1.602×10^3	5.289×10^5	1.058×10^3	2.318×10^4
	Rank	1	7	6	8	4	3	9	2	5
12	Mean	1.830×10^7	4.235×10^8	6.533×10^8	4.330×10^9	3.029×10^8	$\mathbf{1.190 \times 10^7}$	4.165×10^8	2.783×10^7	4.958×10^7
	Std	1.196×10^7	2.298×10^8	2.408×10^8	2.095×10^9	9.060×10^7	6.478×10^6	1.147×10^8	1.477×10^7	2.681×10^7
	Best	6.615×10^6	7.719×10^7	2.185×10^8	1.473×10^9	1.635×10^8	3.719×10^6	2.026×10^8	7.105×10^6	5.747×10^6
	Worst	5.860×10^7	9.782×10^8	1.079×10^9	9.595×10^9	5.550×10^8	2.737×10^7	7.120×10^8	8.331×10^7	1.144×10^8
	Rank	2	7	8	9	5	1	6	3	4
13	Mean	9.419×10^3	9.405×10^6	9.002×10^4	4.787×10^8	3.105×10^6	3.811×10^4	3.234×10^4	9.378×10^3	$\mathbf{7.642 \times 10^3}$
	Std	1.186×10^4	1.835×10^7	3.417×10^4	4.293×10^8	5.306×10^5	1.059×10^4	1.306×10^4	1.530×10^4	1.233×10^4
	Best	6.528×10^2	1.196×10^5	4.051×10^4	9.201×10^4	2.014×10^6	2.153×10^4	1.419×10^4	2.912×10^2	2.868×10^2
	Worst	3.599×10^4	9.361×10^7	1.596×10^5	1.853×10^9	4.130×10^6	6.888×10^4	6.511×10^4	8.090×10^4	4.816×10^4
	Rank	3	8	6	9	7	5	4	2	1
14	Mean	4.505×10^5	3.598×10^6	1.782×10^6	3.591×10^6	8.700×10^5	$\mathbf{2.184 \times 10^5}$	7.977×10^5	6.176×10^5	1.325×10^6
	Std	2.583×10^5	3.441×10^6	1.059×10^6	2.115×10^6	3.002×10^5	9.406×10^4	6.201×10^5	4.363×10^5	6.751×10^5
	Best	1.425×10^5	1.266×10^5	4.800×10^5	6.837×10^5	3.137×10^5	5.758×10^4	9.775×10^4	1.427×10^5	4.959×10^5
	Worst	1.131×10^6	1.302×10^7	4.792×10^6	9.880×10^6	1.634×10^6	4.261×10^5	3.816×10^6	2.290×10^6	2.803×10^6
	Rank	2	9	7	8	5	1	4	3	6
15	Mean	$\mathbf{6.377 \times 10^3}$	7.555×10^5	1.308×10^5	7.000×10^7	7.655×10^5	2.200×10^4	7.456×10^3	7.748×10^3	7.439×10^3
	Std	5.773×10^3	1.481×10^6	2.144×10^5	1.005×10^8	3.090×10^5	7.872×10^3	4.781×10^3	1.019×10^4	1.312×10^4
	Best	1.849×10^2	4.700×10^4	2.620×10^4	4.352×10^5	1.394×10^5	6.418×10^3	3.032×10^3	1.612×10^2	1.635×10^2
	Worst	2.138×10^4	7.265×10^6	1.054×10^6	3.816×10^8	1.849×10^6	4.234×10^4	2.806×10^4	4.044×10^4	7.599×10^4
	Rank	1	7	6	9	8	5	3	4	2
16	Mean	5.132×10^3	5.921×10^3	7.751×10^3	$\mathbf{3.684 \times 10^3}$	5.377×10^3	4.724×10^3	5.854×10^3	5.957×10^3	6.546×10^3
	Std	1.326×10^3	9.500×10^2	1.853×10^3	6.423×10^2	5.029×10^2	5.282×10^2	1.846×10^3	2.099×10^3	2.552×10^3
	Best	3.655×10^3	3.863×10^3	4.193×10^3	2.205×10^3	4.483×10^3	3.733×10^3	3.794×10^3	3.357×10^3	3.885×10^3
	Worst	9.912×10^3	7.955×10^3	1.340×10^4	4.936×10^3	6.998×10^3	6.212×10^3	9.924×10^3	1.031×10^4	1.209×10^4
	Rank	3	6	9	1	4	2	5	7	8

Table 3. The simulation results of SSO, DBO, WOA, GWO, HHO, AVOA, GTO, ARBBPSO and ETBBPSO, f17–f24, best results are shwon in bold.

F	Type	SSO	DBO	WOA	GWO	HHO	AVOA	GTO	ARBBPSO	ETBBPSO
17	Mean	4.331×10^{3}	5.316×10^{3}	5.418×10^{3}	$\mathbf{2.500 \times 10^{3}}$	4.207×10^{3}	4.230×10^{3}	3.994×10^{3}	4.892×10^{3}	4.818×10^{3}
	Std	8.247×10^{2}	1.054×10^{3}	7.340×10^{2}	5.376×10^{2}	4.684×10^{2}	4.697×10^{2}	9.162×10^{2}	1.001×10^{3}	1.093×10^{3}
	Best	2.686×10^{3}	3.639×10^{3}	4.328×10^{3}	1.546×10^{3}	3.531×10^{3}	3.003×10^{3}	2.678×10^{3}	2.963×10^{3}	3.252×10^{3}
	Worst	6.121×10^{3}	8.138×10^{3}	7.694×10^{3}	4.048×10^{3}	5.525×10^{3}	4.990×10^{3}	6.177×10^{3}	7.715×10^{3}	7.388×10^{3}
	Rank	5	8	9	1	3	4	2	7	6
18	Mean	2.582×10^{6}	5.795×10^{6}	2.207×10^{6}	3.302×10^{6}	2.056×10^{6}	$\mathbf{3.554 \times 10^{5}}$	1.581×10^{6}	3.123×10^{6}	7.391×10^{6}
	Std	1.838×10^{6}	5.146×10^{6}	7.891×10^{5}	1.528×10^{6}	6.659×10^{5}	1.237×10^{5}	8.044×10^{5}	1.601×10^{6}	4.976×10^{6}
	Best	7.702×10^{5}	4.827×10^{5}	7.334×10^{5}	7.652×10^{5}	8.964×10^{5}	1.826×10^{5}	4.847×10^{5}	7.828×10^{5}	1.886×10^{6}
	Worst	7.819×10^{6}	2.227×10^{7}	4.676×10^{6}	6.225×10^{6}	3.805×10^{6}	6.416×10^{5}	4.410×10^{6}	8.323×10^{6}	2.448×10^{7}
	Rank	5	8	4	7	3	1	2	6	9
19	Mean	$\mathbf{5.543 \times 10^{3}}$	1.876×10^{6}	1.434×10^{7}	7.688×10^{7}	3.721×10^{6}	1.077×10^{4}	2.169×10^{4}	7.613×10^{3}	1.652×10^{4}
	Std	7.741×10^{3}	2.501×10^{6}	6.244×10^{6}	9.678×10^{7}	1.493×10^{6}	8.269×10^{3}	2.103×10^{4}	1.145×10^{4}	1.577×10^{4}
	Best	1.186×10^{2}	4.900×10^{4}	2.480×10^{6}	5.702×10^{6}	1.125×10^{6}	2.071×10^{3}	1.116×10^{3}	1.396×10^{2}	1.491×10^{2}
	Worst	4.335×10^{4}	1.205×10^{7}	2.536×10^{7}	4.576×10^{8}	6.880×10^{6}	4.351×10^{4}	7.993×10^{4}	4.363×10^{4}	5.512×10^{4}
	Rank	1	6	8	9	7	3	5	2	4
20	Mean	3.053×10^{3}	3.737×10^{3}	4.162×10^{3}	$\mathbf{2.184 \times 10^{3}}$	3.717×10^{3}	3.664×10^{3}	4.850×10^{3}	3.737×10^{3}	3.296×10^{3}
	Std	5.343×10^{2}	6.889×10^{2}	5.279×10^{2}	4.699×10^{2}	5.790×10^{2}	4.433×10^{2}	1.186×10^{3}	8.359×10^{2}	1.128×10^{3}
	Best	1.318×10^{3}	2.061×10^{3}	3.318×10^{3}	1.209×10^{3}	2.461×10^{3}	2.420×10^{3}	2.189×10^{3}	2.255×10^{3}	1.451×10^{3}
	Worst	3.761×10^{3}	4.976×10^{3}	5.401×10^{3}	3.215×10^{3}	4.876×10^{3}	4.418×10^{3}	7.470×10^{3}	5.557×10^{3}	6.233×10^{3}
	Rank	2	7	8	1	5	4	9	6	3
21	Mean	9.983×10^{2}	1.432×10^{3}	1.747×10^{3}	$\mathbf{7.743 \times 10^{2}}$	1.664×10^{3}	1.370×10^{3}	1.011×10^{3}	1.090×10^{3}	1.088×10^{3}
	Std	1.302×10^{2}	1.335×10^{2}	2.281×10^{2}	1.130×10^{2}	1.967×10^{2}	1.409×10^{2}	2.289×10^{2}	1.394×10^{2}	1.405×10^{2}
	Best	8.102×10^{2}	1.145×10^{3}	1.290×10^{3}	6.484×10^{2}	1.267×10^{3}	1.049×10^{3}	7.903×10^{2}	8.929×10^{2}	8.240×10^{2}
	Worst	1.282×10^{3}	1.724×10^{3}	2.225×10^{3}	1.326×10^{3}	2.080×10^{3}	1.761×10^{3}	1.600×10^{3}	1.577×10^{3}	1.443×10^{3}
	Rank	2	7	9	1	8	6	3	5	4
22	Mean	2.166×10^{4}	1.800×10^{4}	2.085×10^{4}	$\mathbf{1.476 \times 10^{4}}$	1.969×10^{4}	1.730×10^{4}	2.692×10^{4}	2.551×10^{4}	2.484×10^{4}
	Std	8.214×10^{3}	2.021×10^{3}	3.096×10^{3}	1.398×10^{3}	1.251×10^{3}	1.118×10^{3}	5.593×10^{3}	5.254×10^{3}	8.665×10^{3}
	Best	1.341×10^{4}	1.397×10^{4}	1.660×10^{4}	1.157×10^{4}	1.537×10^{4}	1.492×10^{4}	1.829×10^{4}	1.133×10^{4}	1.276×10^{4}
	Worst	3.406×10^{4}	2.250×10^{4}	2.691×10^{4}	1.859×10^{4}	2.211×10^{4}	2.052×10^{4}	3.415×10^{4}	2.974×10^{4}	3.452×10^{4}
	Rank	6	3	5	1	4	2	9	8	7
23	Mean	1.203×10^{3}	1.789×10^{3}	2.420×10^{3}	$\mathbf{1.100 \times 10^{3}}$	2.264×10^{3}	1.418×10^{3}	1.436×10^{3}	1.301×10^{3}	1.298×10^{3}
	Std	9.380×10^{1}	1.926×10^{2}	2.668×10^{2}	6.199×10^{1}	2.421×10^{2}	1.267×10^{2}	2.203×10^{2}	1.001×10^{2}	1.120×10^{2}
	Best	1.051×10^{3}	1.370×10^{3}	1.855×10^{3}	9.835×10^{2}	1.804×10^{3}	1.134×10^{3}	1.169×10^{3}	1.116×10^{3}	1.046×10^{3}
	Worst	1.376×10^{3}	2.222×10^{3}	3.034×10^{3}	1.235×10^{3}	2.739×10^{3}	1.726×10^{3}	1.949×10^{3}	1.479×10^{3}	1.508×10^{3}
	Rank	2	7	9	1	8	5	6	4	3
24	Mean	1.753×10^{3}	2.492×10^{3}	3.476×10^{3}	$\mathbf{1.530 \times 10^{3}}$	3.168×10^{3}	2.330×10^{3}	1.952×10^{3}	1.927×10^{3}	1.815×10^{3}
	Std	1.663×10^{2}	2.807×10^{2}	4.040×10^{2}	9.677×10^{1}	2.494×10^{2}	1.994×10^{2}	2.455×10^{2}	1.607×10^{2}	1.550×10^{2}
	Best	1.462×10^{3}	1.758×10^{3}	2.537×10^{3}	1.297×10^{3}	2.619×10^{3}	1.899×10^{3}	1.587×10^{3}	1.629×10^{3}	1.484×10^{3}
	Worst	2.172×10^{3}	3.261×10^{3}	4.256×10^{3}	1.706×10^{3}	3.718×10^{3}	2.763×10^{3}	2.540×10^{3}	2.250×10^{3}	2.172×10^{3}
	Rank	2	7	9	1	8	6	5	4	3

Table 4. The simulation results of SSO, DBO, WOA, GWO, HHO, AVOA, GTO, ARBBPSO an ETBBPSO, f25–f29, best results are shwon in bold.

F	Type	SSO	DBO	WOA	GWO	HHO	AVOA	GTO	ARBBPSO	ETBBPSO
25	Mean	7.624×10^{2}	2.310×10^{3}	1.113×10^{3}	2.813×10^{3}	1.021×10^{3}	8.076×10^{2}	1.766×10^{3}	$\mathbf{7.614 \times 10^{2}}$	7.745×10^{2}
	Std	6.221×10^{1}	3.353×10^{3}	5.379×10^{1}	6.758×10^{2}	7.246×10^{1}	6.130×10^{1}	1.783×10^{2}	5.359×10^{1}	6.691×10^{1}
	Best	6.506×10^{2}	7.359×10^{2}	1.010×10^{3}	1.868×10^{3}	8.443×10^{2}	6.832×10^{2}	1.467×10^{3}	6.328×10^{2}	6.448×10^{2}
	Worst	9.385×10^{2}	1.559×10^{4}	1.203×10^{3}	4.875×10^{3}	1.140×10^{3}	9.031×10^{2}	2.123×10^{3}	8.624×10^{2}	8.997×10^{2}
	Rank	2	8	6	9	5	4	7	1	3
26	Mean	1.346×10^{4}	1.788×10^{4}	2.922×10^{4}	$\mathbf{9.924 \times 10^{3}}$	2.127×10^{4}	1.854×10^{4}	1.510×10^{4}	1.433×10^{4}	1.445×10^{4}
	Std	1.774×10^{3}	3.869×10^{3}	3.190×10^{3}	7.893×10^{2}	1.640×10^{3}	2.705×10^{3}	2.739×10^{3}	1.675×10^{3}	2.077×10^{3}
	Best	1.043×10^{4}	1.018×10^{4}	2.117×10^{4}	8.657×10^{3}	1.867×10^{4}	1.417×10^{4}	1.026×10^{4}	1.085×10^{4}	1.021×10^{4}
	Worst	1.875×10^{4}	2.475×10^{4}	3.657×10^{4}	1.159×10^{4}	2.513×10^{4}	2.298×10^{4}	2.191×10^{4}	1.800×10^{4}	1.965×10^{4}
	Rank	2	6	9	1	8	7	5	3	4
27	Mean	$\mathbf{5.000 \times 10^{2}}$	1.206×10^{3}	2.317×10^{3}	1.153×10^{3}	1.396×10^{3}	1.265×10^{3}	1.018×10^{3}	5.000×10^{2}	5.000×10^{2}
	Std	3.785×10^{-4}	2.531×10^{2}	5.599×10^{2}	9.433×10^{1}	2.785×10^{2}	1.821×10^{2}	1.125×10^{2}	5.206×10^{-4}	5.264×10^{-4}
	Best	5.000×10^{2}	7.054×10^{2}	1.422×10^{3}	9.829×10^{2}	1.056×10^{3}	9.671×10^{2}	8.444×10^{2}	5.000×10^{2}	5.000×10^{2}
	Worst	5.000×10^{2}	1.959×10^{3}	3.740×10^{3}	1.351×10^{3}	2.277×10^{3}	1.732×10^{3}	1.268×10^{3}	5.000×10^{2}	5.000×10^{2}
	Rank	1	6	9	5	8	7	4	2	3
28	Mean	$\mathbf{5.000 \times 10^{2}}$	1.200×10^{4}	9.275×10^{2}	4.120×10^{3}	7.670×10^{2}	5.559×10^{2}	1.835×10^{3}	5.000×10^{2}	5.000×10^{2}
	Std	4.422×10^{-4}	7.865×10^{3}	4.799×10^{1}	1.190×10^{3}	4.914×10^{1}	3.666×10^{1}	3.997×10^{2}	4.211×10^{-4}	5.674×10^{-4}
	Best	5.000×10^{2}	6.750×10^{2}	8.462×10^{2}	1.866×10^{3}	6.391×10^{2}	5.000×10^{2}	1.188×10^{3}	5.000×10^{2}	5.000×10^{2}
	Worst	5.000×10^{2}	2.256×10^{4}	1.051×10^{3}	8.132×10^{3}	8.968×10^{2}	6.575×10^{2}	2.845×10^{3}	5.000×10^{2}	5.000×10^{2}
	Rank	1	9	6	8	5	4	7	2	3
29	Mean	$\mathbf{3.553 \times 10^{3}}$	6.307×10^{3}	1.093×10^{4}	4.494×10^{3}	5.608×10^{3}	4.691×10^{3}	6.756×10^{3}	4.502×10^{3}	4.514×10^{3}
	Std	8.641×10^{2}	1.198×10^{3}	1.755×10^{3}	5.604×10^{2}	6.617×10^{2}	5.392×10^{2}	1.300×10^{3}	9.586×10^{2}	1.046×10^{3}
	Best	1.911×10^{3}	3.721×10^{3}	8.172×10^{3}	3.257×10^{3}	4.514×10^{3}	3.146×10^{3}	4.721×10^{3}	2.849×10^{3}	3.025×10^{3}
	Worst	5.768×10^{3}	9.492×10^{3}	1.531×10^{4}	6.276×10^{3}	6.787×10^{3}	5.560×10^{3}	9.779×10^{3}	7.920×10^{3}	8.239×10^{3}
	Rank	1	7	9	2	6	5	8	3	4
Average Rank		2.6207	6.6897	7.2759	4.3793	5.6897	3.5862	5.7931	4.3448	4.6207

Numerical Analysis

In a total of nine algorithms, SSO achieved eight first places, nine second places, five thir places, two fourth places, three fifth places, and two sixth places, with an average ranking c 2.62, ranking first among all algorithms. Specific results comparison are listed below:

1. In f_1, The rank of SSO is 3, the first method is AVOA, the difference between th two methods is 76.174%;
2. In f_2, The rank of SSO is 3, the first method is AVOA, the difference between th two methods is 100%;
3. In f_3, The rank of SSO is 6, the first method is AVOA, the difference between th two methods is 98.343%;
4. In f_4, The rank of SSO is 1, the second method is ARBBPSO, the difference betwee the two methods is 5.42%;
5. In f_5, The rank of SSO is 4, the first method is GWO, the difference between th two methods is 34.647%;
6. In f_6, The rank of SSO is 3, the first method is GWO, the difference between th two methods is 21.938%;
7. In f_7, The rank of SSO is 1, the second method is ARBBPSO, the difference betwee the two methods is 6.178%;
8. In f_8, The rank of SSO is 2, the first method is GWO, the difference between th two methods is 26.383%;
9. In f_9, The rank of SSO is 4, the first method is GWO, the difference between th two methods is 39.515%;

10. In f_{10},The rank of SSO is 5, the first method is GWO, the difference between the two methods is 20.694%;
11. In f_{11}, The rank of SSO is 1, the second method is ARBBPSO, the difference between the two methods is 14.329%;
12. In f_{12}, The rank of SSO is 2, the first method is AVOA, the difference between the two methods is 34.981%;
13. In f_{13}, The rank of SSO is 3, the first method is ETBBPSO, the difference between the two methods is 18.873%;
14. In f_{14}, The rank of SSO is 2, the first method is AVOA, the difference between the two methods is 51.522%;
15. In f_{15}, The rank of SSO is 1, the second method is ETBBPSO, the difference between the two methods is 14.277%;
16. In f_{16}, The rank of SSO is 3, the first method is GWO, the difference between the two methods is 28.221%;
17. In f_{17}, The rank of SSO is 5, the first method is GWO, the difference between the two methods is 42.281%;
18. In f_{18}, The rank of SSO is 5, the first method is AVOA, the difference between the two methods is 86.234%;
19. In f_{19}, The rank of SSO is 1, the second method is ARBBPSO, the difference between the two methods is 27.192%;
20. In f_{20}, The rank of SSO is 2, the first method is GWO, the difference between the two methods is 28.455%;
21. In f_{21}, The rank of SSO is 2, the first method is GWO, the difference between the two methods is 22.436%;
22. In f_{22}, The rank of SSO is 6, the first method is GWO, the difference between the two methods is 31.869%;
23. In f_{23}, The rank of SSO is 2, the first method is GWO, the difference between the two methods is 8.565%;
24. In f_{24}, The rank of SSO is 2, the first method is GWO, the difference between the two methods is 12.753%;
25. In f_{25}, The rank of SSO is 2, the first method is ARBBPSO, the difference between the two methods is 0.131%;
26. In f_{26}, The rank of SSO is 2, the first method is GWO, the difference between the two methods is 26.263%;
27. In f_{27}, The rank of SSO is 1, the second method is ARBBPSO, the difference between the two methods is 0%;
28. In f_{28},The rank of SSO is 1, the second method is ARBBPSO, the difference between the two methods is 0%;
29. In f_{29}, The rank of SSO is 1, the second method is GWO, the difference between the two methods is 20.934%;

In summary, SSO demonstrates outstanding performance, versatility, efficiency, robustness, and broad applicability in addressing high-dimensional optimization problems.

4.2. Numerical Experiments with CEC2022

To further validate the performance of SSO, CEC2022 was utilized for simulation experiments. The experimental results showed that out of a total of 12 test functions, SSO secured eight first-place rankings, one second-place ranking, and three third-place rankings, with an average ranking of 1.58. Numerical and ranked results of CEC2022 are shown in Tables 5 and 6.

Table 5. The simulation results of SSO, DBO, WOA, GWO, HHO, AVOA, and GTO, CEC2022, f1–f best results are shwon in bold.

F	Type	SSO	DBO	WOA	GWO	HHO	AVOA	GTO
1	Mean	$\mathbf{8.450 \times 10^{-14}}$	9.707×10^{0}	1.223×10^{1}	5.130×10^{3}	1.502×10^{0}	2.827×10^{-13}	5.531×10^{4}
	Std	3.934×10^{-14}	2.158×10^{1}	1.619×10^{1}	3.105×10^{3}	6.404×10^{-1}	1.164×10^{-13}	2.432×10^{5}
	Best	5.684×10^{-14}	5.684×10^{-14}	6.837×10^{-1}	8.533×10^{2}	3.373×10^{-1}	1.137×10^{-13}	8.955×10^{2}
	Worst	1.705×10^{-13}	8.868×10^{1}	7.809×10^{1}	1.232×10^{4}	2.876×10^{0}	6.253×10^{-13}	1.487×10^{6}
	Rank	1	4	5	6	3	2	7
2	Mean	$\mathbf{2.039 \times 10^{0}}$	4.625×10^{1}	5.740×10^{1}	6.977×10^{1}	5.700×10^{1}	3.560×10^{1}	5.656×10^{1}
	Std	1.824×10^{0}	2.145×10^{1}	1.793×10^{1}	2.433×10^{1}	2.279×10^{1}	2.364×10^{1}	1.401×10^{1}
	Best	1.986×10^{-1}	6.377×10^{0}	6.135×10^{0}	4.497×10^{1}	4.462×10^{0}	9.160×10^{-4}	9.619×10^{0}
	Worst	6.481×10^{0}	1.207×10^{2}	9.476×10^{1}	1.759×10^{2}	1.454×10^{2}	6.775×10^{1}	7.504×10^{1}
	Rank	1	3	6	7	5	2	4
3	Mean	$\mathbf{7.457 \times 10^{-3}}$	1.299×10^{1}	5.617×10^{1}	1.395×10^{0}	3.816×10^{1}	1.102×10^{1}	1.680×10^{1}
	Std	2.705×10^{-2}	6.720×10^{0}	1.235×10^{1}	1.513×10^{0}	1.011×10^{1}	6.929×10^{0}	9.457×10^{0}
	Best	1.137×10^{-13}	1.712×10^{0}	3.042×10^{1}	3.802×10^{-2}	2.057×10^{1}	6.057×10^{-1}	2.112×10^{0}
	Worst	1.232×10^{-1}	2.645×10^{1}	8.445×10^{1}	6.357×10^{0}	6.284×10^{1}	2.599×10^{1}	5.112×10^{1}
	Rank	1	4	7	2	6	3	5
4	Mean	5.623×10^{1}	8.186×10^{1}	1.172×10^{2}	$\mathbf{3.746 \times 10^{1}}$	8.384×10^{1}	8.659×10^{1}	7.928×10^{1}
	Std	2.091×10^{1}	2.469×10^{1}	3.671×10^{1}	1.430×10^{1}	1.422×10^{1}	2.900×10^{1}	2.119×10^{1}
	Best	2.288×10^{1}	3.285×10^{1}	4.975×10^{1}	1.431×10^{1}	5.699×10^{1}	2.885×10^{1}	4.020×10^{1}
	Worst	1.224×10^{2}	1.304×10^{2}	2.060×10^{2}	9.085×10^{1}	1.219×10^{2}	1.662×10^{2}	1.598×10^{2}
	Rank	2	4	7	1	5	6	3
5	Mean	$\mathbf{3.504 \times 10^{1}}$	3.804×10^{2}	2.158×10^{3}	7.464×10^{1}	1.415×10^{3}	1.552×10^{3}	8.221×10^{2}
	Std	7.383×10^{1}	3.077×10^{2}	1.139×10^{3}	5.327×10^{1}	2.363×10^{2}	4.757×10^{2}	8.107×10^{2}
	Best	1.791×10^{-1}	4.877×10^{0}	6.416×10^{2}	1.225×10^{0}	9.214×10^{2}	2.568×10^{2}	4.134×10^{1}
	Worst	4.034×10^{2}	1.065×10^{3}	5.899×10^{3}	2.777×10^{2}	1.843×10^{3}	2.986×10^{3}	2.835×10^{3}
	Rank	1	3	7	2	5	6	4

Table 6. The simulation results of SSO, DBO, WOA, GWO, HHO, AVOA, and GTO, CEC2022, f8–f1. best results are shwon in bold.

F	Type	SSO	DBO	WOA	GWO	HHO	AVOA	GTO
6	Mean	5.617×10^{3}	1.832×10^{4}	4.887×10^{3}	9.515×10^{5}	7.657×10^{3}	$\mathbf{3.968 \times 10^{3}}$	5.974×10^{3}
	Std	6.412×10^{3}	4.973×10^{4}	5.665×10^{3}	3.539×10^{6}	6.873×10^{3}	4.852×10^{3}	9.473×10^{3}
	Best	6.222×10^{1}	2.311×10^{2}	1.899×10^{2}	2.544×10^{2}	3.994×10^{2}	1.379×10^{2}	1.454×10^{2}
	Worst	2.102×10^{4}	3.079×10^{5}	1.867×10^{4}	1.773×10^{7}	2.848×10^{4}	1.861×10^{4}	5.192×10^{4}
	Rank	3	6	2	7	5	1	4
7	Mean	$\mathbf{3.543 \times 10^{1}}$	8.382×10^{1}	1.520×10^{2}	4.246×10^{1}	1.010×10^{2}	7.208×10^{1}	1.589×10^{2}
	Std	1.194×10^{1}	3.448×10^{1}	5.237×10^{1}	1.361×10^{1}	3.128×10^{1}	3.763×10^{1}	8.407×10^{1}
	Best	2.171×10^{1}	3.203×10^{1}	5.223×10^{1}	2.271×10^{1}	5.024×10^{1}	3.137×10^{1}	6.022×10^{1}
	Worst	6.577×10^{1}	1.829×10^{2}	2.940×10^{2}	8.749×10^{1}	1.558×10^{2}	1.919×10^{2}	4.697×10^{2}
	Rank	1	4	6	2	5	3	7
8	Mean	$\mathbf{2.105 \times 10^{1}}$	4.558×10^{1}	4.910×10^{1}	3.523×10^{1}	3.935×10^{1}	2.643×10^{1}	1.045×10^{2}
	Std	4.800×10^{-1}	3.979×10^{1}	3.129×10^{1}	3.342×10^{1}	2.003×10^{1}	7.248×10^{0}	7.315×10^{1}
	Best	2.003×10^{1}	2.186×10^{1}	2.824×10^{1}	2.164×10^{1}	2.816×10^{1}	2.108×10^{1}	3.293×10^{1}
	Worst	2.244×10^{1}	1.649×10^{2}	1.717×10^{2}	1.471×10^{2}	1.492×10^{2}	4.390×10^{1}	3.282×10^{2}
	Rank	1	5	6	3	4	2	7

Table 6. *Cont.*

F	Type	SSO	DBO	WOA	GWO	HHO	AVOA	GTO
9	Mean	**1.653 × 10^2**	1.809 × 10^2	1.813 × 10^2	1.921 × 10^2	1.813 × 10^2	1.808 × 10^2	1.809 × 10^2
	Std	3.039 × 10^{-13}	5.324 × 10^{-2}	5.481 × 10^{-1}	1.362 × 10^1	3.341 × 10^{-1}	9.752 × 10^{-9}	2.468 × 10^{-1}
	Best	1.653 × 10^2	1.808 × 10^2	1.808 × 10^2	1.808 × 10^2	1.808 × 10^2	1.808 × 10^2	1.808 × 10^2
	Worst	1.653 × 10^2	1.809 × 10^2	1.828 × 10^2	2.329 × 10^2	1.821 × 10^2	1.808 × 10^2	1.823 × 10^2
	Rank	1	4	5	7	6	2	3
10	Mean	1.272 × 10^2	1.206 × 10^2	1.431 × 10^3	4.844 × 10^2	3.147 × 10^2	2.645 × 10^2	**1.016 × 10^2**
	Std	9.533 × 10^1	9.312 × 10^1	1.078 × 10^3	5.006 × 10^2	2.695 × 10^2	2.349 × 10^2	5.654 × 10^{-1}
	Best	1.525 × 10^1	1.003 × 10^2	1.008 × 10^2	1.003 × 10^2	5.298 × 10^1	4.508 × 10^1	1.007 × 10^2
	Worst	4.689 × 10^2	6.317 × 10^2	3.247 × 10^3	1.850 × 10^3	9.701 × 10^2	8.968 × 10^2	1.032 × 10^2
	Rank	3	2	7	6	5	4	1
11	Mean	3.216 × 10^2	3.426 × 10^2	**2.985 × 10^2**	6.291 × 10^2	3.496 × 10^2	3.270 × 10^2	3.152 × 10^2
	Std	4.173 × 10^1	1.483 × 10^2	9.834 × 10^1	2.079 × 10^2	6.459 × 10^1	4.502 × 10^1	1.122 × 10^2
	Best	3.000 × 10^2	4.547 × 10^{-13}	6.195 × 10^{-1}	3.006 × 10^2	2.661 × 10^1	3.000 × 10^2	3.556 × 10^{-3}
	Worst	4.000 × 10^2	7.009 × 10^2	4.001 × 10^2	1.171 × 10^3	4.050 × 10^2	4.000 × 10^2	7.605 × 10^2
	Rank	3	5	1	7	6	4	2
12	Mean	**2.000 × 10^2**	2.754 × 10^2	3.029 × 10^2	2.534 × 10^2	3.098 × 10^2	2.616 × 10^2	2.486 × 10^2
	Std	2.000 × 10^{-4}	2.589 × 10^1	4.893 × 10^1	1.284 × 10^1	5.648 × 10^1	2.173 × 10^1	8.500 × 10^0
	Best	2.000 × 10^2	2.459 × 10^2	2.446 × 10^2	2.351 × 10^2	2.517 × 10^2	2.392 × 10^2	2.406 × 10^2
	Worst	2.000 × 10^2	3.662 × 10^2	4.432 × 10^2	2.884 × 10^2	5.061 × 10^2	3.453 × 10^2	2.864 × 10^2
	Rank	1	5	6	3	7	4	2
Average Rank		1.58	4.08	5.42	4.42	5.17	3.25	4.08

4.3. Optimization Problem Formulation for Deep-Sea Probe Design

In this extended design optimization problem for a deep-sea probe, we aim to minimize the total weight of the probe while accounting for various critical design variables and constraints. The problem is formulated with eight design variables, including wall thickness, radius, length, material density, internal pressure capacity, battery energy storage, and sensor diameter. The objective function reflects the total weight, which is influenced by these variables, while the constraints ensure the probe's structural integrity, energy sufficiency, and appropriate sensor arrangement within the probe's internal volume. In this optimization problem, we aim to minimize the total weight of the deep-sea probe by adjusting eight design variables. The design variables are shown in Equation (5).

$$\mathbf{x} = [x_1, x_2, x_3, x_4, x_5, x_6, x_7, x_8] = [t_s, t_h, r, l, \rho_{\text{material}}, P_{\max}, E_{\text{battery}}, d_{\text{sensor}}] \quad (5)$$

where:

- t_s: Wall thickness of the cylindrical section,
- t_h: Wall thickness of the end caps,
- r: Radius of the probe,
- l: Length of the probe,
- ρ_{material}: Material density,
- $P_{\max}$: Maximum internal pressure capacity,
- E_{battery}: Battery energy storage,
- d_{sensor}: Sensor diameter.

The objective function, which represents the total weight of the probe, is shown in Equation (6).

$$\begin{aligned} \text{Minimize} \quad f(\mathbf{x}) = c_1x_1x_3x_4 + c_2x_2x_3^2 + c_3x_1^2x_4 + c_4x_1^2x_3 + \\ c_5\rho_{\text{material}}x_3^2x_4 + c_6E_{\text{battery}} + c_7d_{\text{sensor}}^2 \end{aligned} \quad (6)$$

where $c_1, c_2, \ldots, c_7$ are constants related to material properties, gravitational acceleration, and design specifics. The problem is subject to the severe constraints which are shown in Equation (7):

$$
\begin{aligned}
g_1(\mathbf{x}) &= -\frac{P_{\text{max}} r}{t_s} + \sigma_{\text{allowable}} \leq 0 \\
g_2(\mathbf{x}) &= -\frac{P_{\text{external}} r}{t_h} + \sigma_{\text{allowable}} \leq 0 \\
g_3(\mathbf{x}) &= W_{\text{total}} - \rho_{\text{water}} g V_{\text{displaced}} \leq 0 \\
g_4(\mathbf{x}) &= -\pi r^2 l - \frac{4\pi r^3}{3} + V_{\text{required}}(d_{\text{sensor}}, E_{\text{battery}}) \leq 0 \\
g_5(\mathbf{x}) &= E_{\text{battery}} - E_{\text{required}} \geq 0 \\
g_6(\mathbf{x}) &= d_{\text{sensor}} - d_{\text{max}} \leq 0
\end{aligned}
$$

The goal is to find the optimal set of variables $\mathbf{x}$ that minimizes the total weight whi satisfying all the constraints. Experimental results are shown in Table 7. Convergenc graph are shown in Figure 8, standard deviation are shown in Figure 9. The experiment results demonstrate that SSO provides high-precision solutions for the Deep-Sea Prob Design Problem.

Table 7. The simulation results of SSO, DBO, WOA, GWO, HHO, AVOA, and GTO, best results a shwon in bold.

Type	SSO	DBO	WOA	GWO	HHO	AVOA	GTO
Mean	$\mathbf{5.265 \times 10^2}$	5.414×10^2	6.669×10^2	5.295×10^2	6.939×10^2	5.396×10^2	5.357×10^2
Std	2.496×10^0	1.613×10^1	5.779×10^1	2.538×10^0	5.889×10^1	1.169×10^1	7.616×10^0
Best	5.248×10^2	5.248×10^2	5.532×10^2	5.251×10^2	5.908×10^2	5.270×10^2	5.264×10^2
Worst	5.327×10^2	5.909×10^2	7.907×10^2	5.324×10^2	8.123×10^2	5.918×10^2	5.563×10^2
Rank	1	5	6	2	7	4	3

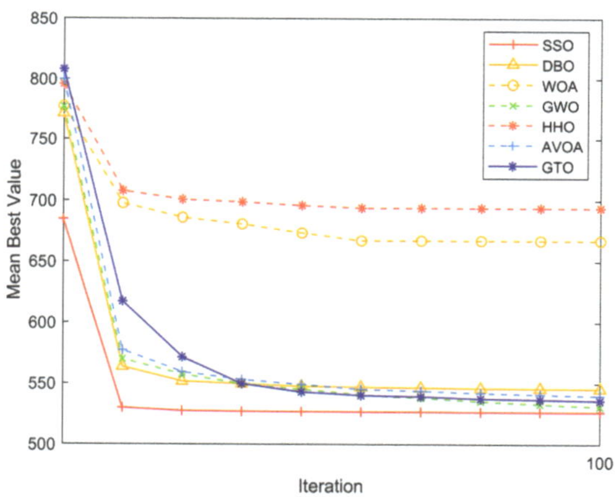

Figure 8. Convergence graph of simulated Deep-Sea Probe Design Problem.

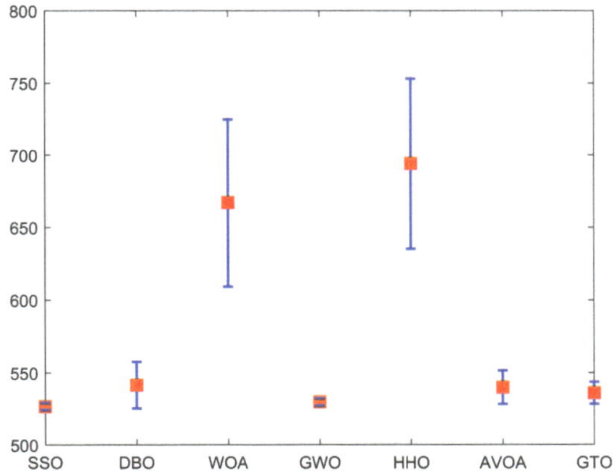

Figure 9. Standard deviation of simulated Deep-Sea Probe Design Problem. The red square stands for the mean value, and the blue line stands for the error.

4.4. Discussion

The Salmon Salar Optimization (SSO) algorithm offers several key advantages in high-dimensional optimization problems, primarily due to its ability to balance exploration and exploitation effectively. By simulating the collective behavior of salmon, the algorithm dynamically adjusts its search strategy, allowing it to explore new areas of the search space while refining known solutions. This enables SSO to avoid local optima, a common issue in high-dimensional problems, and consistently move toward global optimal solutions. The algorithm's adaptability further enhances its performance by allowing it to adjust key parameters based on the nature and complexity of the problem, ensuring robust and efficient optimization without the need for extensive parameter tuning.

However, in each iteration, SSO performs three fitness function evaluations (FFEs), leading to a significant increase in computational time, particularly for high-dimensional and complex optimization problems. This increased computational load can hinder the algorithm's efficiency, making it less suitable for large-scale or time-sensitive applications that demand fast convergence. The need to perform multiple FFEs per iteration poses a challenge, as it directly impacts the algorithm's scalability and practicality in real-world scenarios. One of the primary limitations of SSO lies in this computational cost, which can become a bottleneck in certain applications. Future research should focus on reducing the number of FFEs or optimizing the evaluation process to improve the algorithm's efficiency without compromising solution accuracy. Addressing this limitation is essential for enhancing SSO's competitiveness and applicability in complex engineering and scientific problems.

5. Conclusions

A novel metaheuristic approach, Salmon Salar Optimization (SSO), is proposed to address the complex design challenges associated with deep-sea probes, which are crucial for unconventional oil exploration. SSO distinguishes itself as an innovative solution to the intricate issues of high-dimensional optimization by emulating the social structure and collective behavior of salmon populations.

To rigorously evaluate SSO's effectiveness, a series of simulation experiments were conducted using the challenging CEC2017 benchmark functions. In these benchmarks, SSO was tested against eight historically top-performing algorithms, serving as a stringent control group. The results were remarkable. Out of the nine algorithms considered, SSO achieved an impressive record: eight first-place rankings, nine second-place rankings, five third-place rankings, and several additional top placements. With an average ranking c 2.62, SSO clearly emerged as the leading algorithm among its competitors.

SSO demonstrated its ability to effectively tackle the complex, high-dimensional desig challenges associated with deep-sea probes, which are critical for offshore exploration an extraction of unconventional oil resources. By optimizing both the structural and functiona aspects of these probes under stringent conditions, SSO has made significant advancement in deep-sea exploration technologies, facilitating the efficient and sustainable developmen of unconventional oil resources.

In summary, SSO exhibits outstanding characteristics, including high performanc adaptability, efficiency, robustness, and broad applicability to the multifaceted challenge of high-dimensional optimization problems. These qualities underscore SSO's position as powerful tool for addressing complex optimization issues and pave the way for promisin future research directions.

Future work should focus on further fine-tuning SSO parameters to enhance opt mization performance across diverse problem domains. Additionally, investigating SSO' potential in multi-objective optimization scenarios and extending its application to rea world problem-solving contexts represent exciting research avenues that merit exploratio

Author Contributions: Conceptualization, J.G.; methodology, J.G.; software, J.G.; validation, J.G.; forma analysis, J.G.; investigation, J.G.; resources, J.G.; data curation, Z.Y., Q.Z. and Y.S.; writing—original dra preparation, J.G.; writing—review and editing, Y.S.; visualization, Z.Y., Q.Z. and Y.S.; supervisio Y.S.; project administration, Q.Z.; funding acquisition, J.G., Z.Y. and Y.S. All authors have read an agreed to the published version of the manuscript.

Funding: This research was funded by Natural Science Foundation of Hubei Province (2023AFB00 2023AFB004); Education Department Scientific Research Program Project of Hubei Province of Chin (Q20222208, Q20232206); JSPS KAKENHI Grant Numbers JP22K12185.

Institutional Review Board Statement: Not applicable.

Informed Consent Statement: Not applicable.

Data Availability Statement: Data available on request.

Conflicts of Interest: The authors declare no conflicts of interest. The funders had no role in the desig of the study; in the collection, analyses, or interpretation of data; in the writing of the manuscript; c in the decision to publish the results.

References

1. Petrovic, A.; Damaševičius, R.; Jovanovic, L.; Toskovic, A.; Simic, V.; Bacanin, N.; Zivkovic, M.; Spalević, P. Marine Vessel Class fication and Multivariate Trajectories Forecasting Using Metaheuristics-Optimized eXtreme Gradient Boosting and Recurrer Neural Networks. *Appl. Sci.* **2023**, *13*, 9181 . [CrossRef]
2. Yaseen, Z.M.; Melini Wan Mohtar, W.H.; Homod, R.Z.; Alawi, O.A.; Abba, S.I.; Oudah, A.Y.; Togun, H.; Goliatt, L.; Ul Hassa Kazmi, S.S.; Tao, H. Heavy metals prediction in coastal marine sediments using hybridized machine learning models wit metaheuristic optimization algorithm. *Chemosphere* **2024**, *352*, 141329. [CrossRef] [PubMed]
3. Zhang, Y.H.; Wang, X.J.; Zhang, X.Z.; Saad, M.; Zhao, R.J. Numerical Investigation of the Impacts of Large Particles on th Turbulent Flow and Surface Wear in Series-Connected Bends. *J. Mar. Sci. Eng.* **2024**, *12*, 164. [CrossRef]
4. Nguyen, T.H.H.; Hou, T.H.; Pham, H.A.; Tsai, C.C. Oil Spill Sensitivity Analysis of the Coastal Waters of Taiwan Using a Integrated Modelling Approach. *J. Mar. Sci. Eng.* **2024**, *12*, 155. [CrossRef]
5. Xing, R.; Zhang, Y.; Feng, Y.; Ji, F. Performance Analysis of a WPCN-Based Underwater Acoustic Communication System. *J. Ma Sci. Eng.* **2023**, *12*, 43. [CrossRef]
6. Kennedy, J.; Eberhart, R. Particle swarm optimization. In Proceedings of the ICNN'95—International Conference on Neural Network Perth, Australia, 27 November–1 December 1995; Volume 4, pp. 942–1948. [CrossRef]
7. Fister, I.; Yang, X.S.; Brest, J. A comprehensive review of firefly algorithms. *Swarm Evol. Comput.* **2013**, *13*, 34–46. [CrossRef]
8. Zhou, X.G.; Zhang, G.J. Differential evolution with underestimation-based multimutation strategy. *IEEE Trans. Cybern.* **201** *49*, 1353–1364. [CrossRef] [PubMed]
9. Chen, L.; Liu, Y.; Gao, Y.; Wang, J. Carbon Emission Trading Policy and Carbon Emission Efficiency: An Empirical Analysis c China's Prefecture-Level Cities. *Front. Energy Res.* **2021**, *9*, 793601. [CrossRef]
10. Zhang, X.; Zou, D.; Shen, X. A novel simple particle swarm optimization algorithm for global optimization. *Mathematics* **201** *6*, 287. [CrossRef]

1. Qiao, J.; Zhou, H.; Yang, C. Bare-Bones Multiobjective Particle Swarm Optimization Based on Parallel Cell Balanceable Fitness Estimation. *IEEE Access* **2018**, *6*, 32493–32506. [CrossRef]
2. Zhang, X.; Melbourne, S.; Sarkar, C.; Chiaradia, A.; Webster, C. Effects of green space on walking: Does size, shape and density matter? *Urban Stud.* **2020**, *57*, 3402–3420. [CrossRef]
3. Singh, G.; Singh, A. A hybrid algorithm using particle swarm optimization for solving transportation problem. *Neural Comput. Appl.* **2020**, *32*, 11699–11716. [CrossRef]
4. Meng, X.; Li, J.; Member, S.; Dai, X.; Dou, J. Variable Neighborhood Search for a Colored Traveling Salesman Problem. *IEEE Trans. Intell. Transp. Syst.* **2018**, *19*, 1018–1026. [CrossRef]
5. Meng, X.; Li, J.; Member, S.; Zhou, M.; Dai, X.; Dou, J. Population-Based Incremental Learning Algorithm for a Serial Colored Traveling Salesman Problem. *IEEE Trans. Syst. Man, Cybern. Syst.* **2018**, *48*, 277–288. [CrossRef]
5. Al-Andoli, M.; Tan, S.C.; Cheah, W.P. Parallel stacked autoencoder with particle swarm optimization for community detection in complex networks. *Appl. Intell.* **2022**, *52*, 3366–3386. [CrossRef]
7. Ahandani, M.A.; Abbasfam, J.; Kharrati, H. Parameter identification of permanent magnet synchronous motors using quasi-opposition-based particle swarm optimization and hybrid chaotic particle swarm optimization algorithms. *Appl. Intell.* **2022**, *52*, 13082–13096. [CrossRef]
3. Zhang, J.; Zhao, X.; Jin, S.; Greaves, D. Phase-resolved real-time ocean wave prediction with quantified uncertainty based on variational Bayesian machine learning. *Appl. Energy* **2022**, *324*, 119711. [CrossRef]
9. Hu, P.; Pan, J.S.; Chu, S.C.; Sun, C. Multi-surrogate assisted binary particle swarm optimization algorithm and its application for feature selection. *Appl. Soft Comput.* **2022**, *121*, 108736. [CrossRef]
0. Wang, H.; Zhang, Z. Forecasting Chinese provincial carbon emissions using a novel grey prediction model considering spatial correlation. *Expert Syst. Appl.* **2022**, *209*, 118261. [CrossRef]
1. Lu, B.; Zhou, C. Particle Swarm Algorithm and Its Application in Tourism Route Design and Optimization. *Comput. Intell. Neurosci.* **2022**, *2022*, 6467086. [CrossRef]
2. Pan, J.; Bardhan, R. Evaluating the risk of accessing green spaces in COVID-19 pandemic: A model for public urban green spaces (PUGS) in London. *Urban For. Urban Green.* **2022**, *74*, 127648. [CrossRef] [PubMed]
3. Gao, Y.; Du, W.; Yan, G. Selectively-informed particle swarm optimization. *Sci. Rep.* **2015**, *5*, 9295. [CrossRef] [PubMed]
4. Liang, X.; Li, W.; Zhang, Y.; Zhou, M. An adaptive particle swarm optimization method based on clustering. *Soft Comput.* **2015**, *19*, 431–448. [CrossRef]
5. Li, J.; Zhang, J.; Jiang, C.; Zhou, M. Composite Particle Swarm Optimizer with Historical Memory for Function Optimization. *IEEE Trans. Cybern.* **2015**, *45*, 2350–2363. [CrossRef]
6. Pornsing, C.; Sodhi, M.S.; Lamond, B.F. Novel self-adaptive particle swarm optimization methods. *Soft Comput.* **2016**, *20*, 3579–3593. [CrossRef]
7. Guo, J.; Sato, Y. A pair-wise bare bones particle swarm optimization algorithm. In Proceedings of the 2017 IEEE/ACIS 16th International Conference on Computer and Information Science (ICIS), Wuhan, China, 24–26 May 2017; Number 1, pp. 353–358. [CrossRef]
3. Guo, J.; Sato, Y. A Bare Bones Particle Swarm Optimization Algorithm with Dynamic Local Search. In *Advances in Swarm Intelligence: 8th International Conference, ICSI 2017, Fukuoka, Japan, 27 July–1 August 2017, Proceedings, Part I*; Tan, Y., Takagi, H., Shi, Y., Eds.; Springer International Publishing: Cham, Switzerland, 2017; pp. 158–165. [CrossRef]
9. Guo, J.; Sato, Y. A Hierarchical Bare Bones Particle Swarm Optimization Algorithm. In Proceedings of the 2017 IEEE International Conference on Systems, Man, and Cybernetics (SMC), Banff, AB, Canada, 5–8 October 2017; pp. 1936–1941. [CrossRef]
0. Kennedy, J. Bare bones particle swarms. In Proceedings of the 2003 IEEE Swarm Intelligence Symposium, SIS'03 (Cat. No.03EX706), Indianapolis, IN, USA, 26 April 2003; pp. 80–87. [CrossRef]
1. Xu, X.; Rong, H.; Trovati, M.; Liptrott, M.; Bessis, N. CS-PSO: Chaotic particle swarm optimization algorithm for solving combinatorial optimization problems. *Soft Comput.* **2018**, *22*, 783–795. [CrossRef]
2. Tian, D.; Shi, Z. MPSO: Modified particle swarm optimization and its applications. *Swarm Evol. Comput.* **2018**, *41*, 49–68. [CrossRef]
3. Ghasemi, M.; Akbari, E.; Rahimnejad, A.; Razavi, S.E.; Ghavidel, S.; Li, L. Phasor particle swarm optimization: A simple and efficient variant of PSO. *Soft Comput.* **2019**, *23*, 9701–9718. [CrossRef]
4. Guo, J.; Sato, Y. A fission-fusion hybrid bare bones particle swarm optimization algorithm for single-objective optimization problems. *Appl. Intell.* **2019**, *49*, 3641–3651. [CrossRef]
5. Xu, G.; Cui, Q.; Shi, X.; Ge, H.; Zhan, Z.H.; Lee, H.P.; Liang, Y.; Tai, R.; Wu, C. Particle swarm optimization based on dimensional learning strategy. *Swarm Evol. Comput.* **2019**, *45*, 33–51. [CrossRef]
5. Xu, Y.; Pi, D. A reinforcement learning-based communication topology in particle swarm optimization. *Neural Comput. Appl.* **2020**, *32*, 10007–10032. [CrossRef]
7. Yamanaka, Y.; Yoshida, K. Simple gravitational particle swarm algorithm for multimodal optimization problems. *PLoS ONE* **2021**, *16*, e0248470. [CrossRef] [PubMed]
3. Liu, J.; Jin, B.; Yang, J.; Xu, L. Sea surface temperature prediction using a cubic B-spline interpolation and spatiotemporal attention mechanism. *Remote Sens. Lett.* **2021**, *12*, 478–487. [CrossRef]

39. Wang, Z.J.; Zhan, Z.H.; Kwong, S.; Jin, H.; Zhang, J. Adaptive Granularity Learning Distributed Particle Swarm Optimization fo Large-Scale Optimization. *IEEE Trans. Cybern.* **2021**, *51*, 1175–1188. [CrossRef] [PubMed]
40. Li, X.; Wang, Z.; Ying, Y.; Xiao, F. Multipopulation Particle Swarm Optimization Algorithm with Neighborhood Learnin *Sci. Program.* **2022**, *2022*, 8312450. [CrossRef]
41. Tian, H.; Guo, J.; Xiao, H.; Yan, K.; Sato, Y. An electronic transition-based bare bones particle swarm optimization algorithm fo high dimensional optimization problems. *PLoS ONE* **2022**, *17*, e0271925. [CrossRef]
42. Guo, Q.; Su, Z.; Chiao, C. Carbon emissions trading policy, carbon finance, and carbon emissions reduction: Evidence from a quasi-natural experiment in China. *Econ. Chang. Restruct.* **2022**, *55*, 1445–1480. [CrossRef]
43. Zhou, G.; Guo, J.; Yan, K.; Zhou, G.; Li, B. An Atomic Retrospective Learning Bare Bone Particle Swarm Optimization. In *Advanc in Swarm Intelligence. ICSI 2023*; Lecture Notes in Computer Science; Springer: Cham, Switzerland, 2023; Volume 1396 pp. 168–179. [CrossRef]
44. Xue, J.; Shen, B. Dung beetle optimizer: A new meta-heuristic algorithm for global optimization. *J. Supercomput.* **2023**, *7* 7305–7336. [CrossRef]
45. Mirjalili, S.; Lewis, A. The Whale Optimization Algorithm. *Adv. Eng. Softw.* **2016**, *95*, 51–67. [CrossRef]
46. Mirjalili, S.; Mirjalili, S.M.; Lewis, A. Grey Wolf Optimizer. *Adv. Eng. Softw.* **2014**, *69*, 46–61. [CrossRef]
47. Heidari, A.A.; Mirjalili, S.; Faris, H.; Aljarah, I.; Mafarja, M.; Chen, H. Harris hawks optimization: Algorithm and application *Future Gener. Comput. Syst.* **2019**, *97*, 849–872. [CrossRef]
48. Abdollahzadeh, B.; Gharehchopogh, F.S.; Mirjalili, S. African vultures optimization algorithm: A new nature-inspired metaheuristic algorithm for global optimization problems. *Comput. Ind. Eng.* **2021**, *158*, 107408. [CrossRef]
49. Abdollahzadeh, B.; Soleimanian Gharehchopogh, F.; Mirjalili, S. Artificial gorilla troops optimizer: A new nature-inspire metaheuristic algorithm for global optimization problems. *Int. J. Intell. Syst.* **2021**, *36*, 22535. [CrossRef]

Journal of
Marine Science and Engineering

MDPI

Article

Improved Fracture Permeability Evaluation Model for Granite Reservoirs in Marine Environments: A Case Study from the South China Sea

Jianhong Guo [1,2], Baoxiang Gu [3], Hengyang Lv [1,2], Zuomin Zhu [1,2] and Zhansong Zhang [1,2,*]

1 Key Laboratory of Exploration Technologies for Oil and Gas Resources, Ministry of Education, Yangtze University, Wuhan 430100, China; 2022730024@yangtzeu.edu.cn (J.G.)
2 College of Geophysics and Petroleum Resources, Yangtze University, Wuhan 430100, China
3 CNOOC International Limited, Beijing 100028, China
* Correspondence: zhangzhs@yangtzeu.edu.cn

Abstract: Permeability is a crucial parameter in the exploration and development of oil and gas reservoirs, particularly in unconventional ones, where fractures significantly influence storage capacity and fluid flow. This study investigates the fracture permeability of granite reservoirs in the South China Sea, introducing an enhanced evaluation model for planar fracture permeability based on Darcy's law and Poiseuille's law. The model incorporates factors such as fracture heterogeneity, tortuosity, angle, and aperture to improve permeability assessments. Building on a single-fracture model, this research integrates mass transfer equations and trigonometric functions to assess intersecting fractures' permeability. Numerical simulations explore how tortuosity, angle, and aperture affect individual fracture permeability and the influence of relative positioning in intersecting fractures. The model makes key assumptions, including minimal consideration of horizontal stress and the assumption of unidirectional laminar flow in cross-fractures. Granite outcrop samples were systematically collected, followed by full-diameter core drilling. A range of planar models with varying fracture apertures were designed, and permeability measurements were conducted using the AU-TOSCAN-II multifunctional core scanner with a steady-state gas injection method. The results showed consistency between the improved model and experimental findings regarding the effects of fracture aperture and angle on permeability, confirming the model's accuracy in reflecting the fractures' influence on reservoir flow capacity. For intersecting fractures, a comparative analysis of core X-ray computed tomography (X-CT) scanning results and experimental outcomes highlighted discrepancies between actual permeability measurements and theoretical simulations based on tortuosity and aperture variations. Limitations exist, particularly for cross-fractures, where quantifying complexity is challenging, leading to potential discrepancies between simulation and experimental results. Further comparisons between core experiments and logging responses are necessary for model refinement. In response to the challenges associated with evaluating absolute permeability in fractured reservoirs, this study presents a novel theoretical assessment model that considers both single and intersecting fractures. The model's validity is demonstrated through actual core experiments, confirming the effectiveness of the single-fracture model while highlighting the need for further refinement of the dual-fracture model. The findings provide scientific support for the exploration and development of granite reservoirs in the South China Sea and establish a foundation for permeability predictions in other complex fractured reservoir systems, thereby advancing the field of fracture permeability assessment.

Keywords: fracture permeability; cross-fracture model; mass transfer equation; granite reservoir; plate fracture experiment; X-ray computed tomography; South China Sea

Citation: Guo, J.; Gu, B.; Lv, H.; Zhu, Z.; Zhang, Z. Improved Fracture Permeability Evaluation Model for Granite Reservoirs in Marine Environments: A Case Study from the South China Sea. *J. Mar. Sci. Eng.* 2024, 12, 1868. https://doi.org/10.3390/jmse12101868

Academic Editor: Dejan Brkić

Received: 26 September 2024
Revised: 15 October 2024
Accepted: 16 October 2024
Published: 18 October 2024

1. Introduction

Permeability is a fundamental parameter in the exploration and development of oil an gas reservoirs, making its precise evaluation essential for assessing resource recoverabilit and formulating effective development strategies [1,2]. As exploration and productio efforts evolve, there has been a notable shift toward unconventional reservoirs, such as shal oil, tight oil, coalbed methane, and carbonate formations [3]. Among these, the developmen of fractures is recognized as a critical factor for successful resource extraction [4]. Fracture play a significant role in enhancing both storage capacity and fluid flow within thes unconventional reservoirs, underscoring the importance of accurately assessing fractur permeability for their effective exploration and development [5,6].

Currently, two primary methods are recognized for evaluating permeability. The firs method, laboratory measurement, is the most direct and accurate approach for determinin core permeability using experimental instruments on samples sent to the laboratory [7 However, in fractured reservoirs, the permeability obtained from these measurement primarily reflects matrix permeability, as acquiring plunger samples from fractured section is often challenging [8,9]. Subsequent assessments of permeability conducted on full-siz samples from these fractured sections yield a combined measure of both matrix and fractur permeability [10]. Despite the benefits of laboratory methods, challenges associated wit coring in fracture-prone zones—such as high costs, segment discontinuity, and limite borehole coverage—impede their widespread application [11]. The second approach in volves utilizing geophysical exploration data, particularly focusing on the calculation o permeability through logging data analysis [12,13]. The porosity–permeability equation es tablished by Kozeny [14] elucidates the relationship between pore-throat radius, tortuosit and permeability, employing Archie's formula to characterize the numerical correlation among these factors. This relationship has been further examined across various lithologie However, the permeability evaluation model derived from the pore-permeability relatio ship exhibits limitations in its applicability, especially when addressing unconventiona reservoirs characterized by significant heterogeneity [15]. Such limitations constrain the ac curacy of permeability evaluations in these complex geological settings [16]. Recent studie on rock permeability have revealed a consistent negative correlation between permeabilit and depth; specifically, as depth increases, permeability tends to decrease [17]. This tren is influenced by factors such as the stress state of the reservoir, seismic activity, and th long-term geological history [18]. This negative relationship is evident across various roc types [19]. Currently, the evaluation of permeability in complex unconventional pore-typ reservoirs primarily relies on the petrophysical characteristics of the target reservoir, com bined with core permeability experimental data, to develop various tailored permeabilit evaluation models [20,21]. This approach has been widely adopted and has yielded fa vorable results. With advancements in geophysical logging technology, the use of nuclea magnetic resonance (NMR) logging for permeability evaluation has produced significan outcomes, exemplified by classic models, such as the SDR model [22,23], the Timur-Coate model [24,25], the Prince-Rezaee model [26,27], and the Hossain model [28]. Building o these foundational frameworks, several researchers have successfully developed improve permeability evaluation models, achieving notable breakthroughs [29]. Additionally, th application of array acoustic wave logging data for calculating permeability through Stor ley wave attenuation has emerged as a widely utilized evaluation tool [30]. Among th aforementioned methods, NMR logging data effectively reflect the pore structure of fo mations, leading to permeability evaluation models that utilize NMR parameters (e.g., T geometric mean, T_2 cutoff permeability evaluation model) often demonstrating superio accuracy compared to those based on the porosity–permeability relationship [31]. Howeve these models face challenges in accurately assessing fracture permeability, as the changes i reservoir seepage capacity induced by fractures are difficult to capture in the T_2 distributio spectrum. With the rise of big data technology, permeability prediction methods leveragin artificial intelligence have gained traction. These methods utilize machine-learning an deep-learning algorithms to explore the non-linear relationships between logging curv responses and core permeability, subsequently constructing permeability evaluation models [32]. While these approaches have been successfully implemented across various blocks and demonstrate greater accuracy than traditional techniques, they are primarily applicable to pore-permeability reservoirs with favorable pore-permeability relationships [33]. Consequently, despite their higher accuracy in many scenarios, data-driven methods are not well suited for complex fractured reservoirs, as the difficulty in obtaining core samples that meet experimental criteria, combined with the limited availability of full-size cores, hampers the development of effective permeability prediction models.

Consequently, establishing a model analogous to Archie's formula for calculating fracture permeability holds significant engineering importance, particularly in the exploration and development of fractured reservoirs [34]. For instance, the Qiongdongnan Basin, located in the northern South China Sea, is abundant in subducted reservoirs characterized by diverse basement lithologies, including granite, volcanic rocks, metamorphic rocks, and greywacke, all of which are accompanied by numerous fractures [35,36]. Despite advancements in electric imaging logging technology enabling the identification of fractures as narrow as tens of microns, the existing research on fracture permeability prediction predominantly emphasizes the functional relationship between fracture aperture and permeability, often with a limited scope [37]. Given that the shape of fractures is frequently irregular, and the accuracy of aperture calculations is constrained, the calculated fracture permeability usually fails to align with experimental data and field test results [38]. Therefore, it is imperative to investigate the calculation model of fracture permeability through a more comprehensive response mechanism.

This study employs a plate fracture model as its foundational framework, utilizing Darcy's law and Poisson's law. Unlike traditional modeling approaches, this research incorporates the heterogeneity of fractures in the permeability calculation model, introducing the concept of tortuosity while accounting for the angle and aperture of the fractures. By integrating these factors, we establish a model for fluid flow through a single fracture. For the scenario involving two fractures, we introduce a mass transfer equation to examine the effects of inclination angle, aperture, and fracture intersection on permeability. To explore the influence of various factors on permeability, numerical simulations were conducted based on the derived formulae. In the experimental phase, after collecting granite outcrop samples, core drilling was performed to assemble the planar model, and models with varying fracture apertures were designed to measure fracture permeability. An artificial fracturing method was employed to create multiple-fracture cores, and X-ray computed tomography (X-CT) technology was utilized to extract fracture characteristics while measuring permeability. The results of the numerical simulations were compared and analyzed against experimental data, demonstrating the effectiveness of the single-fracture permeability model. Additionally, the discrepancies between theoretical model results and actual measurements in evaluating the permeability of intersecting fractures were discussed. The findings of this study hold promise for advancing the development of a permeability logging assessment system.

Figure 1 illustrates the research workflow. Chapter Two presents the derivation of formulae and results from numerical simulations. Chapter Three details the core acquisition process, the methods for measuring the permeability of single fractures at different apertures, the artificial fracturing techniques, and the application of X-CT technology. Chapter Four showcases the experimental results from Chapter Three, providing a comprehensive comparison between numerical simulation outcomes and core experimental results. Finally, Chapter Five discusses the observed discrepancies, highlighting the innovations and limitations of this study.

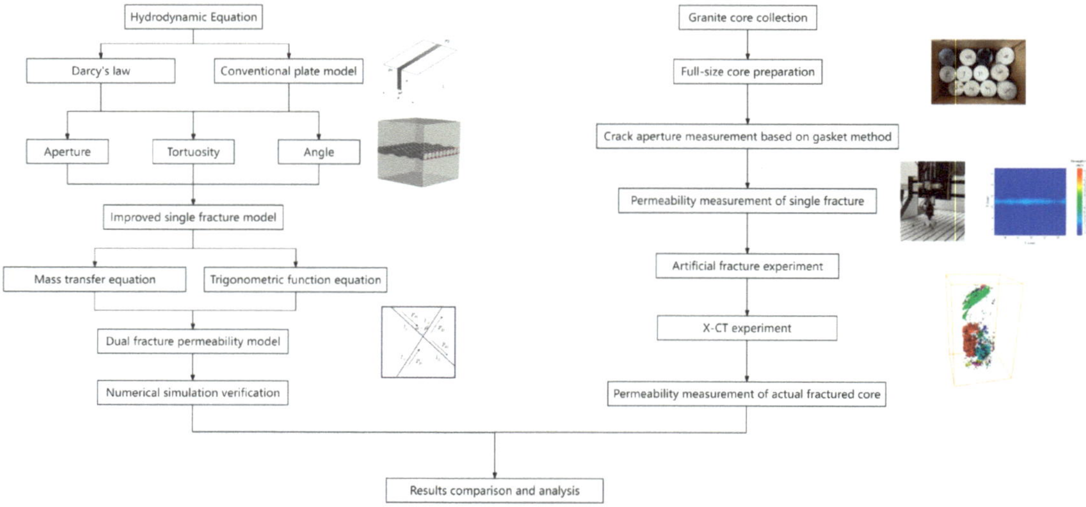

Figure 1. Flow chart of this research.

2. Formula Derivation and Simulation

2.1. Conventional Flat Plate Model for Single Fracture

As a conduit for fluid flow, the flow conductivity of a single fracture can be modele as a finite plane extending from within the rock mass to the well wall [39,40]. In the conte of an ideal rock body containing a fracture, Figure 2 illustrates a rectangular rock structu with a width w and a length l. This structure includes a fracture height of h, a fracture widt of w_f, and experiences a pressure differential ΔP across its sides, defined as $\Delta P = P_1 - F$

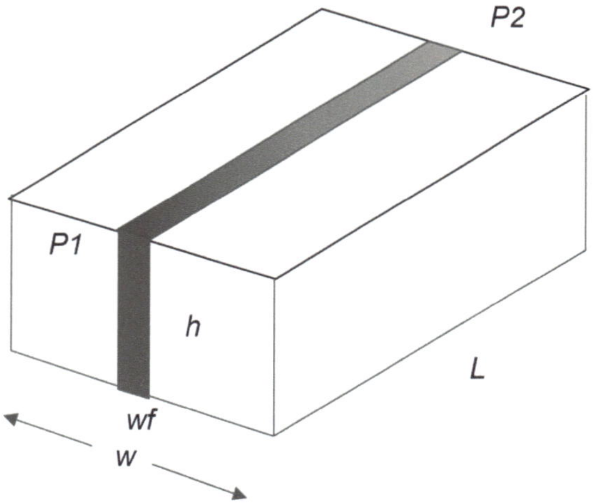

Figure 2. The classic flat plate fracture model.

The classical hydrodynamic equation governing flow [41] through a regularly shape unit width fracture is expressed as follows:

$$Q = \frac{h^3 w_f \Delta P}{12 \mu L} \quad ($$

In Equation (1), Q denotes the flow rate, m^3/s; μ represents the fluid's viscosity, Pa

The flow of fluid within the fracture adheres to Darcy's law [42], articulated as

$$Q = \frac{Ak(P_1 - P_2)}{\mu L} \tag{2}$$

In Equation (2), A signifies the cross-sectional area of flow, m^2; meanwhile, k indicates the permeability of the medium, m^2. By integrating Equations (1) and (2), we derive the expression for permeability:

$$k = \frac{h^3 w_f}{12wL} \tag{3}$$

The volume of the unit length fracture model can be described as

$$V_f = L{\cdot}w_f{\cdot}h \tag{4}$$

Based on this, the porosity of the fracture can be formulated as

$$\varphi = \frac{V_f}{L{\cdot}w{\cdot}h} = \frac{w_f}{w} \tag{5}$$

Combining Equations (3) and (5) yields the relationship between fracture permeability and porosity:

$$k = \frac{h^3 \varphi}{12L} \tag{6}$$

In the plate model, the influence of horizontal stress needs to be explained. Horizontal stress affects the formation, development, and closure of fractures, which in turn alters their characteristics; for instance, increasing horizontal stress can compress the width of fractures, leading to a reduction in permeability. However, in the classic models used for calculating planar permeability, this aspect is not directly considered due to the segmentation of the model.

This study aims to measure absolute air permeability in core samples, assuming unidirectional flow within the fractures. The flat plate model classifies the fluid as air and describes the flow as laminar in accordance with Darcy's law. This characterization indicates a uniform fluid velocity across the fracture cross-section, ensuring stability without transitioning to turbulence. Furthermore, the physical properties of air, including density and viscosity, are treated as constant within the specified ranges of pressure and temperature. The fractures are assumed to be parallel, homogeneous, and straight, with consistent apertures and heights maintained throughout their length.

2.2. Improved Flat Plate Model for Single Fracture

In contrast to the classical model depicted in Figure 2, which fails to account for the non-smooth nature of fractures, the improved model is formulated based on Poiseuille's law [43]. Initially, for the roar channel, the actual configuration is expected to be curved, as illustrated in Figure 3, where L represents the straight-line length of the pipe at both ends, La denotes the actual length of the curved roar channel, and R signifies the radius of the pipe, $\lambda = 2R$, reflecting the degree of aperture.

Figure 3 illustrates the flow within a curved channel. Poiseuille's law [43] delineates the laminar flow of fluid in a pipe, and for the configuration shown in Figure 3, the flow rate is given by

$$Q_v = \frac{\pi(P_1 - P_2)R^4}{8\mu L} \tag{7}$$

The fracture can be subdivided into La/L infinitesimally small straight sections, corresponding to

$$dQ_v = \frac{\pi(d\Delta P)R^4}{8\mu dl} \tag{8}$$

Integrating Equation (8) results in

$$Q_v = \frac{\pi R^4 \int_0^{(P_1-P_2)} (d\Delta P)}{8\mu \int_0^{La} dl} = \frac{\pi (P_1 - P_2) R^4}{8\mu La} \quad (9)$$

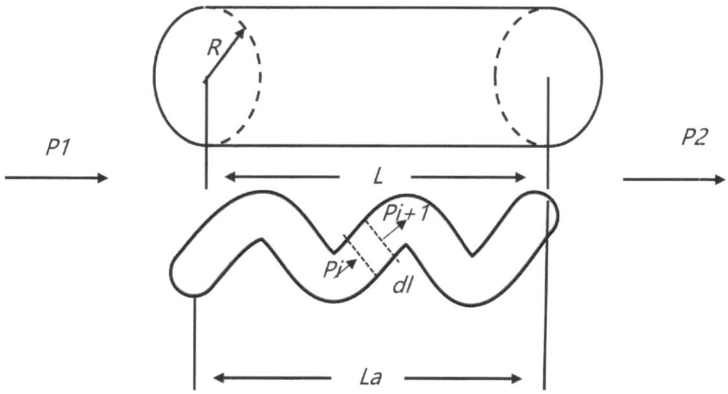

Figure 3. Flow diagram of curved roar.

Combining Equation (9) with Equation (2) leads to

$$k = \frac{\pi \lambda^4 L}{128 A La} \quad (10)$$

By incorporating the tortuosity $\tau = \left(\frac{La}{L}\right)^2$, the permeability equation can be reformulated as [44]

$$k = \frac{\pi \lambda^4 L}{128 A La} \frac{1}{\tau} = \frac{\pi \lambda^4 L}{128 La^3} \quad (11)$$

Equation (11) represents the permeability of the curved interconnected pores, relevant for porous reservoirs, as shown in Figure 4a. For fracture reservoirs, the fractures within the rock can be conceptualized as a plane composed of n interconnected pores, illustrated in Figure 4b.

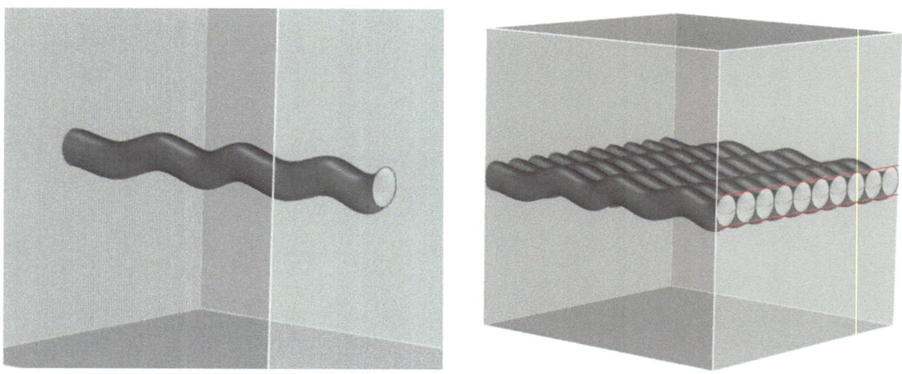

(**a**) (**b**)

Figure 4. A sketch of roars and fractures. (**a**) A diagram of the curved connected pores in the rock mass; (**b**) The fracture in the fractured reservoir is regarded as a plane composed of n connected pores.

Figure 4 depicts the flow in a curved channel within a cylindrical rock mass. For the permeability of fractured reservoirs, based on Equation (11) and considering $n = \frac{L}{\lambda}$, the permeability of horizontal fractures can be expressed as

$$k = n\frac{\pi\lambda^4 L}{128La^3} = \frac{\pi\lambda^3 L^2}{128La^3} \tag{12}$$

To account for the influence of fracture angle on radial permeability, when the inclination angle of the fracture in the model is α (where $\alpha < 90°$),

$$La = \frac{\sqrt{\tau}L}{\cos\alpha} \tag{13}$$

Combining Equation (13) with Equation (12) allows us to express permeability as

$$k = \frac{\pi\lambda^3 \cos\alpha}{128\tau La} \tag{14}$$

Equation (14) is the derived expression for the permeability of the plate fracture model, which integrates the effects of fracture aperture, tortuosity, and angle.

2.3. Derivation of Dual-Fracture Permeability Model

In real reservoirs, the fractures exhibit intricate geometries, particularly when multiple fractures intersect. This section focuses on deriving the permeability formula for two intersecting fractures.

Initially, for two fractures, the flow state of the fluid within the pipeline can be characterized by the Reynolds number [45]:

$$\text{Re} = \frac{\nu\lambda}{\mu} \tag{15}$$

In Equation (15), ν represents the average fluid velocity.

The Reynolds number indicates the flow regime, with the critical Reynolds number marking the transition from laminar to turbulent flow. The subsequent deductions assume that the fluid flow in the pipeline is laminar.

Figure 5 illustrates the laminar flow of fluid in a pipe with constant diameter. The flow state of the fluid is defined by the Reynolds number. The flow remains laminar when the Reynolds number is below a critical value, typically around 2000; exceeding this value may lead to turbulence. In this study, the flow is maintained in a laminar state. As shown, the fluid divides into several layers along the pipe's diameter, each exhibiting varying velocities. A small cylindrical segment aligned with the pipe axis serves as the subject of analysis, where the fluid is in a state of force equilibrium:

$$(p_1 - p_2)\pi r^2 = \Delta p\pi r^2 = F_f \tag{16}$$

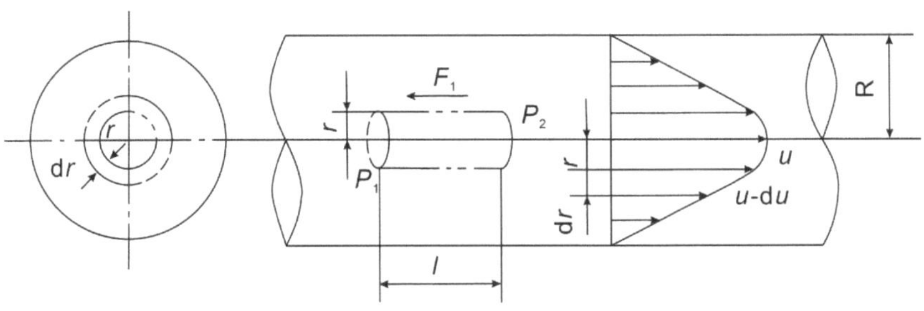

Figure 5. A schematic diagram of laminar flow of liquid in an equal-diameter pipe.

In Equation (16), F_f denotes the internal frictional force of the fluid.
According to Newton's law of viscosity, we have

$$F_f = S \times \mu \times \frac{du}{dr} \tag{17}$$

In Equation (17), S represents the contact area of the friction layer, and $\frac{du}{dr}$ defines th velocity gradient, indicating the rate of change in velocity perpendicular to the flow directio
Thus, Equations (16) and (17) can be combined to yield

$$\Delta p \pi r^2 = F_f = -2\pi r l \mu \frac{du}{dr} \tag{18}$$

Equation (18) can be reformulated as

$$du = -\frac{\Delta p}{2\mu l} r dr \tag{19}$$

Integrating Equation (19) and applying boundary conditions leads to

$$u = \frac{\Delta p}{4\mu l}(R^2 - r^2) \tag{20}$$

It is evident that the velocity distribution in the radial direction follows a paraboli pattern, with maximum velocity at the center of the pipe and decreasing toward the wa where the velocity is zero.

The flow rate of a differential element can be expressed as

$$dq = udA = 2\pi u r dr = 2\pi \frac{\Delta p}{4\mu l}(R^2 - r^2) r dr \tag{21}$$

Integrating Equation (21) leads to a modified form of Poiseuille's equation:

$$q = \frac{\pi \lambda^4}{128\mu l} \Delta p \tag{22}$$

The average velocity is given by

$$v = \frac{q}{A} = \frac{\lambda^2}{32\mu l} \Delta p \tag{23}$$

The pressure loss along the pipe is expressed as

$$\Delta p = \frac{32\mu l v}{\lambda^2} = \frac{32\mu^2 \mathrm{Re}}{\lambda^3} \tag{24}$$

Therefore, under ideal conditions, where other variables remain constant, the flui flow in the fracture exhibits a linear gradient pressure drop with respect to the flow distanc

$$P = P_1 - \left(\frac{32\mu^2 \mathrm{Re}}{\lambda^3}\right) l \tag{25}$$

When calculating the permeability of intersecting fractures, the embedded discret fracture permeability model introduces a mass transfer coefficient T to compute the flo within the grid model of the intersecting fractures:

$$q_{ff} = T_{ff}(p_{fi} - p_{fj}) \tag{26}$$

In Equation (26), T_{ff} is the mass transfer coefficient, while p_{fi} and p_{fj} denote the pressures in fractures i and j, respectively.

The mass transfer coefficient [46] is defined as

$$T_{ff} = \frac{T_{fi}T_{fj}}{T_{fi} + T_{fj}} \tag{27}$$

In Equation (27), $T_{fi} = \frac{k_{fi}\lambda_{fi}}{\mu \hat{d}_i}$, $T_{fj} = \frac{k_{fj}\lambda_{fj}}{\mu \hat{d}_j}$; k_{fi} and k_{ji} are the permeabilities of fractures i and j, derived from Equation (14) in Section 2.2. λ represents the fracture aperture, and $\hat{d}$ is the distance from a single fracture to the intersection point.

Figure 6a illustrates the scenario where fractures i and j intersect at angles α and β, respectively. The intersection point divides fracture i into segments l_{i1} and l_{i2}, and fracture j into segments l_{j1} and l_{j2}; then, $\hat{d}_i = \frac{\int_0^{l_{i1}} ldl + \int_0^{l_{i2}} ldl}{l_{i1} + l_{i2}}$, $\hat{d}_j = \frac{\int_0^{l_{j1}} ldl + \int_0^{l_{j2}} ldl}{l_{j1} + l_{j2}}$. Figure 6b shows the vector diagram of cross-fractures' pressure conduction.

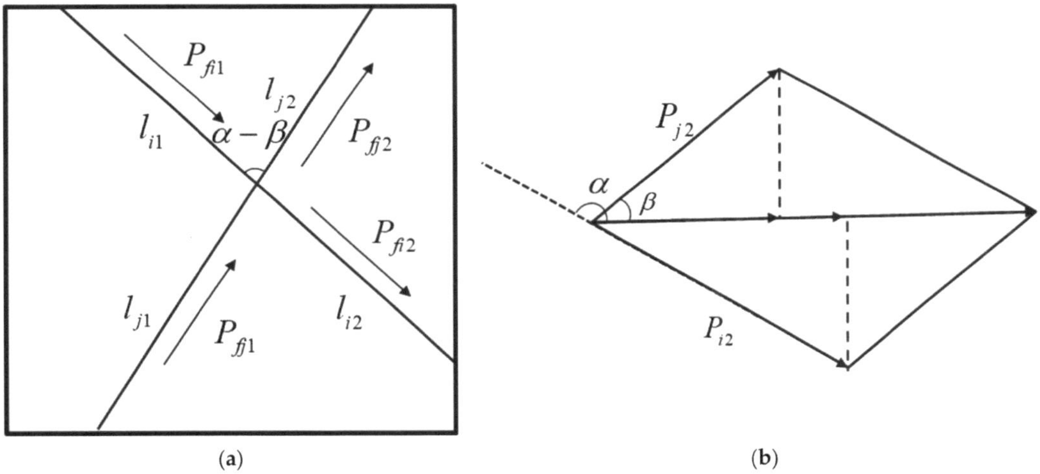

Figure 6. Model diagram of cross-fractures. (**a**) The case where fracture i with an inclination angle of α intersects fracture j with an inclination angle of β; (**b**) Pressure conduction diagram of cross-fractures.

Combining Equations (26) and (27), the flow rate through the intersecting fractures can be expressed as

$$q_{ff} = \frac{2k_{fi}k_{fj}\lambda_{fi}\lambda_{fj}(l_{i1} + l_{i2})(l_{j1} + l_{j2})(p_{fi} - p_{fj})}{\mu\left[k_{fi}\lambda_{fi}(l_{i1} + l_{i2})\left({l_{j1}}^2 + {l_{j2}}^2\right) + k_{fj}\lambda_{fj}\left({l_{i1}}^2 + {l_{i2}}^2\right)(l_{j1} + l_{j2})\right]} \tag{28}$$

In simulating the flow through intersecting fractures using Darcy's law, it is essential to consider the difference in the radial vector sums of the pressures entering and exiting the fractures. The pressures P_1 and P_2 can be expressed as

$$P_1 = -p_{fi1}\cos\alpha + p_{fj1}\cos\beta \tag{29}$$

$$P_2 = -p_{fi2}\cos\alpha + p_{fj2}\cos\beta \tag{30}$$

Given that the pressure loss in the laminar flow state within the fractures is

$$P_2 = -(p_{fi1} - \frac{32\mu^2\text{Re}}{\lambda^3}l_i)\cos\alpha + (p_{fj1} - \frac{32\mu^2\text{Re}}{\lambda^3}l_j)\cos\beta \tag{31}$$

By integrating Equations (29)–(31), the radial pressure difference can be determined a

$$\Delta P = P_1 - P_2 = \frac{32\mu^2 \mathrm{Re}}{\lambda^3}(-l_i \cos\alpha + l_j \cos\beta) \quad (32$$

In Equation (32), $l_j = l_{j1} + l_{j2}$, and $l_i = l_{i1} + l_{i2}$.

The flow rate through the intersecting fractures can be represented using Darcy' law as

$$Q = \frac{Ak_{ff}\left[\frac{32\mu^2\mathrm{Re}}{\lambda^3}(-l_{i1}\cos\alpha + l_{j1}\cos\beta)\right]}{\mu L} \quad (3$$

By combining Equation (33) with the flow rate expression for the embedded discret fracture model (Equation (28)), we can derive the permeability calculation formula for tw intersecting fractures:

$$k_{ff} = \frac{2k_{fi}k_{fj}\lambda_{fi}\lambda_{fj}(l_{i1}+l_{i2})(l_{j1}+l_{j2})(p_{fi}-p_{fj})}{L\left[k_{fi}\lambda_{fi}(l_{i1}+l_{i2})\left({l_{j1}}^2+{l_{j2}}^2\right)+k_{fj}\lambda_{fj}\left({l_{i1}}^2+{l_{i2}}^2\right)(l_{j1}+l_{j2})\right]\left[\frac{32\mu^2\mathrm{Re}}{\lambda^3}(-l_{i1}\cos\alpha+l_{j1}\cos\beta)\right]} \quad (34$$

2.4. *Numerical Simulation Results of the Equation*

Building on the fundamental concepts of fracture mechanics and traditional plate-lik fracture models, a new plate-like fracture model is derived by considering fracture apertur tortuosity, and angle. An improved permeability calculation formula for two intersectin fractures is developed by combining the mass transfer equation and trigonometric function followed by numerical simulations of the derived results.

2.4.1. Improved Single-Fracture Model

Equation (14) delineates the functional relationship between permeability and the de gree of fracture tortuosity, fracture angle, and fracture aperture within the flat model. In thi study, numerical simulations were conducted to explore these factors while maintainin other parameters at constant values. Under conditions where fracture aperture, angle, an tortuosity are uniform—specifically within a homogeneous reservoir—the permeability o the flat model demonstrates an exponential correlation with porosity, as shown in Figure 7 Figure 7b illustrates how flat permeability varies with the tortuosity of the fracture. Whe the fracture angle and aperture remain constant, an increase in the meandering degre leads to a more intricate fluid pathway within the rock, resulting in decreased permeabilit This relationship can be characterized by an exponential function that depicts the declin in permeability with heightened meandering degree. Figure 7c presents the relationshi between flat permeability and fracture angle. It is noteworthy that the fitting employed i polynomial, although it is fundamentally linked to the cosine function; radial permeabi ity diminishes as the fracture angle increases. Figure 7d reveals the correlation betwee flat permeability and fracture aperture. As the fracture aperture expands, permeabilit increases significantly, especially at larger apertures, where the enhancement in permeabi ity becomes particularly pronounced. Combined with the relationship between differen factors and fracture permeability obtained from the above simulation results, the radia permeability is taken as an example. This analysis underscores that the permeability o the rock is governed by a complex interplay of factors, with fracture aperture serving as crucial determinant for permeability enhancement. In contrast, increases in the meanderin degree and angle hinder radial fluid flow, thereby diminishing radial permeability.

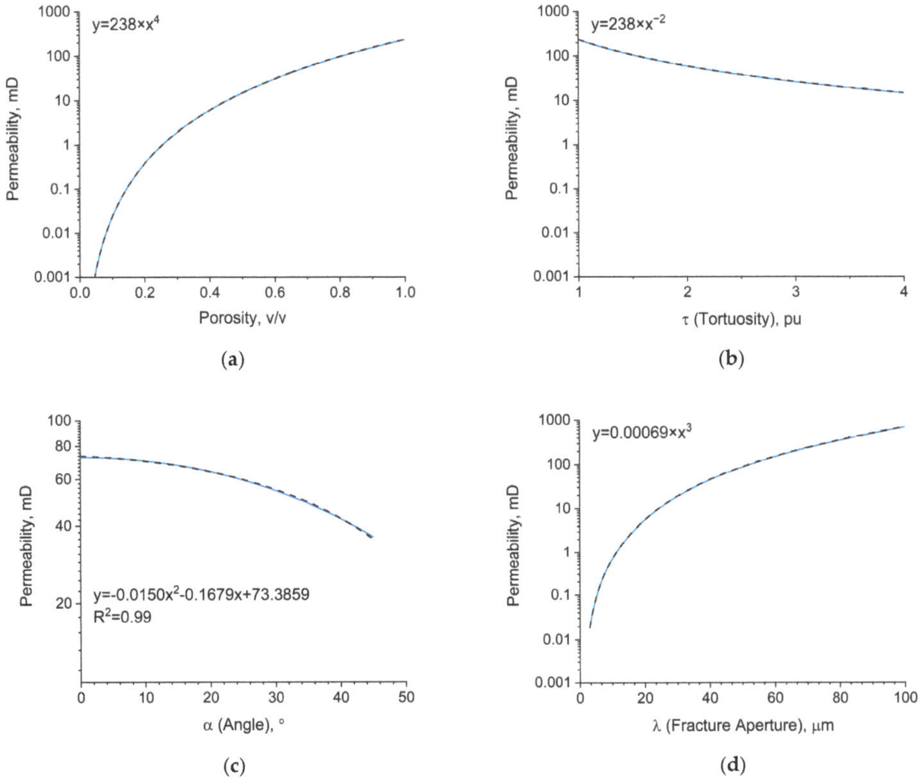

Figure 7. Numerical simulation results of single fracture. (**a**) Fracture porosity vs. permeability in homogeneous reservoir; (**b**) Fracture tortuosity vs. permeability (constant angle and aperture); (**c**) Fracture angle vs. permeability (constant tortuosity and aperture); (**d**) Fracture aperture vs. permeability (constant tortuosity and angle).

2.4.2. Dual-Fracture Permeability Model

Based on Equation (34), cross-fracture permeability is influenced by a variety of factors and their respective variation patterns. This methodology can, in principle, be applied to simulate any cross-fracture scenario. Since the numerical simulation is performed on a per-unit-volume basis for the rock mass, most fixed parameters are dimensionless, allowing for the calculation of only relative permeability values to elucidate the trends and variations.

The calculations presented in this study were conducted using MATLAB 2022b. The model for simulating the intersecting fractures is constructed based on the results obtained from single-fracture simulations. Parameters such as fracture aperture, angle, and tortuosity had already been considered in the single-fracture model, allowing the results to be effectively utilized in the intersecting fracture simulations. For the boundary conditions of the intersecting fractures, both the inlet and outlet were set to fixed fluid pressure to ensure that the fluid flow characteristics within the fractures accurately reflect real-world conditions. Additionally, a constant flow rate was applied to simulate the inflow and outflow of fluids, ensuring a continuity of flow. No-slip boundary conditions were implemented for the walls of the fractures, assuming that the fluid velocity at the wall is zero. This assumption aligns with the actual flow characteristics of fluids near solid surfaces. In the simulation of intersecting fractures, a mass transfer equation was applied to account for the flow characteristics at the intersections. Special attention was given to variations in the angles of the fractures, and scenarios where the angles of the two fractures are identical were not considered. This assumption simplifies the model and better represents the complexity of actual fracture networks. Considering that in the actual reservoir, the development c fractures is more complicated, and the relative error and production of the two fractures i the cross-fracture are also very different, to clarify the influence of the relative position c fractures on the permeability of fractures, the influence of the difference in the angle of th cross-fracture and the change in the position of the intersection point on the permeability simulated under the control variables.

Figure 8 illustrates the numerical simulation of the relative positions of three types c cross-fractures. Figure 8a shows two diagonally crossed fractures within the square grid c the discrete fracture model under ideal conditions, with their inclinations set at $\alpha = 135$ and $\beta = 45°$, and their intersection point located at the center.

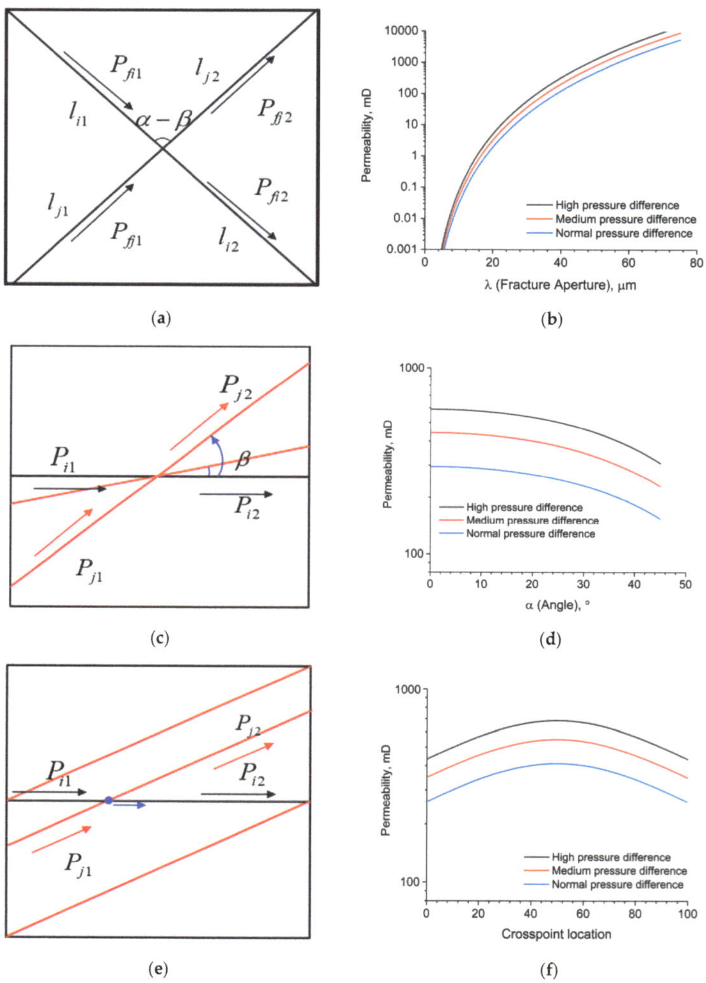

Figure 8. Numerical simulation results of cross-fractures. (**a**) Two diagonally crossed fractures in th discrete fracture model under ideal conditions; (**b**) Relationship between fracture opening and roc mass permeability at fixed cross-fracture angles; (**c**) Model with one fixed fracture and the angle of th other increasing from 1° to 45°; (**d**) Relationship between fracture angle and rock mass permeabili with varying single-fracture angle; (**e**) Model with fixed angles and radial horizontal movemen of one fracture, shifting the intersection point; (**f**) Relationship between the relative position of th intersection and rock mass permeability.

Figure 8b demonstrates the impact of fracture tortuosity on the permeability of the rock mass model, given the relative positions of fractures depicted in Figure 8a, specifically when the cross-fracture angle is held constant. The color gradient of the curve in Figure 8b represents various pressure differential conditions, including high, medium, and normal pressure differentials. The results indicate that permeability increases with larger fracture tortuosity across all pressure differential conditions, with the most pronounced increase observed under high-pressure differentials. Additionally, it is important to note that the tortuosity of the two fractures in the simulation is inherently aligned with the default tortuosity settings.

Figure 8c illustrates a scenario in which one fracture remains fixed, intersecting angle α at a constant 0°. The angle of the other fracture varies from 1° to 45°, with the intersection point of the two fractures centered. It is important to note that the application of the mass transfer equation is not considered in this scenario, as the angle of the two fractures is at 0°, resulting in their recombination into a single crack. The simulation results presented in Figure 8d demonstrate that as the angle of the other fracture increases, permeability exhibits a downward trend across all pressure differential conditions, with a more pronounced decrease under high-pressure differentials.

Figure 8e illustrates a scenario in which the angles of the two fractures remain fixed while one fracture moves horizontally and radially, causing the intersection point to shift from left to right. Figure 8f reveals the impact of intersection location on the permeability model of the rock mass. As the intersection point shifts from the leftmost section of the fracture to the right, its distance is measured relative to the entire length of the rock mass, serving as the horizontal coordinate. The results indicate that permeability initially increases under all pressure differential conditions, reaching a maximum at the midpoint before gradually declining back to the initial permeability. The permeability change curve exhibits axisymmetry.

Unlike the numerical simulation results for a single fracture, a novel permeability calculation for the cross-fracture is derived by incorporating the mass transfer equation and trigonometric functions into the analysis of single fractures. Variations in the angle of the cross-fracture and the relative position of the intersection significantly influence the permeability calculation results of the model.

3. Experimental Materials and Methods

The measurement of fracture permeability is essential for validating the accuracy of the derived permeability equations. Therefore, permeability measurements on actual granite were utilized. Two approaches were used: one was to verify the effect of fracture aperture on core permeability in a single fracture, while the other was to measure permeability after crevassing the core to validate the permeability formula for cross- fracturing.

3.1. Acquisition of Experimental Materials

Considering the extensive development of granite fracture-type reservoirs in the Qiongdongnan Basin of the South China Sea, the outcrop samples were collected (Figure 9b) from granite outcrop rock samples (Figure 9a) at a single location.

The preparation process for drilling full-diameter granite cores consists of five key steps. First, the rock sample is positioned on the drilling platform. Next, it is secured to ensure stability during the drilling operation. Following this, the water supply is activated to cool the drill bit and clean the borehole walls. Subsequently, the power supply is engaged to operate the drilling machine. Finally, the height of the drilling platform is adjusted to achieve optimal contact between the drill bit and the rock sample. The full-diameter drill bit has an inner diameter of 8.5 cm, as depicted in Figure 10a, which illustrates the actual procedure for full-diameter core drilling. Figure 10b presents the core sample obtained, which will be utilized for subsequent measurements of planar fracture permeability. The collected samples are cylindrical, with a length of 50 mm and a diameter of 85 mm. We prepared 12 effective sample groups, each containing multiple samples. The average density measurement of the samples was found to be 2.9 g/cm^3. To accurately depict th fracture features of the South China Sea granite reservoirs, the X-ray diffraction (XRD results of the rock were included in the manuscript. The scanning results indicate tha the primary constituents of the samples are plagioclase (37%), potassium feldspar (20% quartz (20%), clay (11%), with minor amounts of rhodochrosite (4%) and pyrite (1%).

(**a**) (**b**)

Figure 9. Granite sample collection. (**a**) Outcrop rock samples; (**b**) Outcrop sample collection.

(**a**) (**b**)

Figure 10. Full-diameter core drilling schematic diagram. (**a**) Drilling the operating table; (**b**) Cor drilling finished product.

3.2. Flat Plate Fracture Permeability Measurement

For single fractures, the gasket method was used to change the fracture aperture t ensure the reliability of the experiments, especially for the investigation of fracture apertur Following core drilling, the flat plate model was assembled, incorporating five distinc fracture apertures. These apertures were adjusted using shims and grippers, with th designated fracture apertures for the five configurations set at 90 μm, 140 μm, 200 μm 250 μm, and 300 μm, respectively. Figure 11a shows the plate model diagram; Figure 11 is the gasket application diagram (100 μm); and Figure 11c is the gripper diagram.

(a) (b) (c)

Figure 11. Plate model apparatus diagram. (**a**) Flat plate model diagram; (**b**) Gasket schematic diagram; (**c**) Schematic diagram of the gripper.

AUTOSCAN-II Core multifunction scanning instrument is used to measure the permeability of plate fractures, and the permeability of the instrument is measured using a steady-state gas injection method. The standard 4 mm permeability probe is used to measure permeability in the range of 0.1 md~3D during the measurement process. Figure 12 is the schematic diagram of AUTOSCAN-II (NER USA, Inc., White River Junction, VT, USA) Core multifunctional scanner.

Figure 12. AUTOSCAN-II Core multifunctional scanner.

The experiment in this study is carried out by collecting granite outcrop samples, drilling full-diameter cores, based on the plate model, controlling the fracture aperture using gaskets, and measuring the core fracture permeability under different fracture apertures.

3.3. Core Fracturing Methods and X-CT Scanning

Artificial tensile fractures were created based on the Brazilian splitting theory. However, this method may cause fragmentation of the core samples, making them unsuitable for subsequent X-CT scanning. To mitigate this issue, we employed a manual clamping method using a vise during the fracturing process, thereby preventing the core samples from being crushed. Precise control over the initiation time and fracture strength is critica however, the cores are vulnerable to crushing, which limits the number of viable sample that can be collected. Figure 13 presents a schematic representation of the process used t create artificial joints.

Figure 13. Artificial fracture experiment process.

Computed tomography (CT) scanning experiments were performed on core sam ples both before and after the fracture construction process. The resolution achieve for full-diameter samples in these experiments was 50 μm. The micro-pore structure c the reservoir—including the geometry, size, distribution, and connectivity of pores an throats—plays a critical role in influencing reservoir capacity and seepage efficiency. Re cently, X-ray computed tomography (X-CT) has become an essential tool for characterizin the microscopic pore structures in tight sandstone reservoirs. This technique reconstruc the three-dimensional structure of pore throats by using conical X-rays in conjunction wit a 360-degree rotation of the sample (Figure 14).

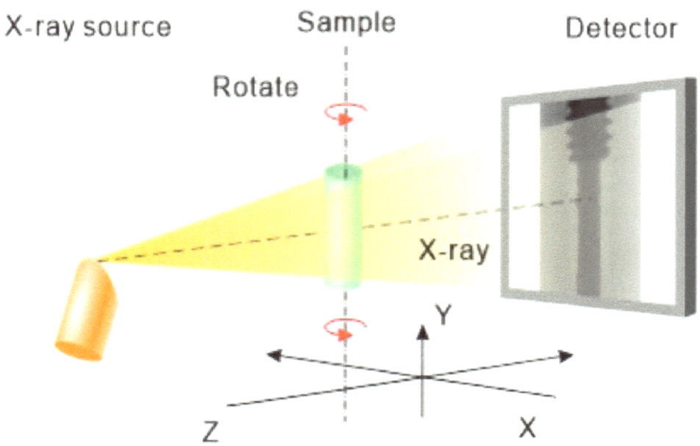

Figure 14. X-CT scan imaging layout diagram.

4. Results

4.1. Core Permeability Test Results

Figure 15 presents the results of five experimental schemes for measuring the permeability of slab fractures, all conducted with a fixed measurement direction and a corresponding fracture angle of 0°. In Figure 15a, a fracture aperture of 90 µm is shown, which corresponds to an average permeability of 30 mD, as illustrated in Figure 15b. Figure 15c displays a fracture aperture of 140 µm, linked to an average permeability of 50 mD, depicted in Figure 15d. In Figure 15e, a fracture aperture of 200 µm results in an increase in average permeability to 230 mD, as shown in Figure 15f. Figure 15g corresponds to a fracture aperture of 250 µm, yielding an average permeability of 350 mD, as demonstrated in Figure 15h. Finally, Figure 15i illustrates a fracture aperture of 300 µm, leading to an impressive average permeability of up to 1300 mD, as presented in Figure 15j. Overall, Figure 15 summarizes the measurement results for the fracture plate models with apertures of 90 µm, 140 µm, 200 µm, 250 µm, and 300 µm.

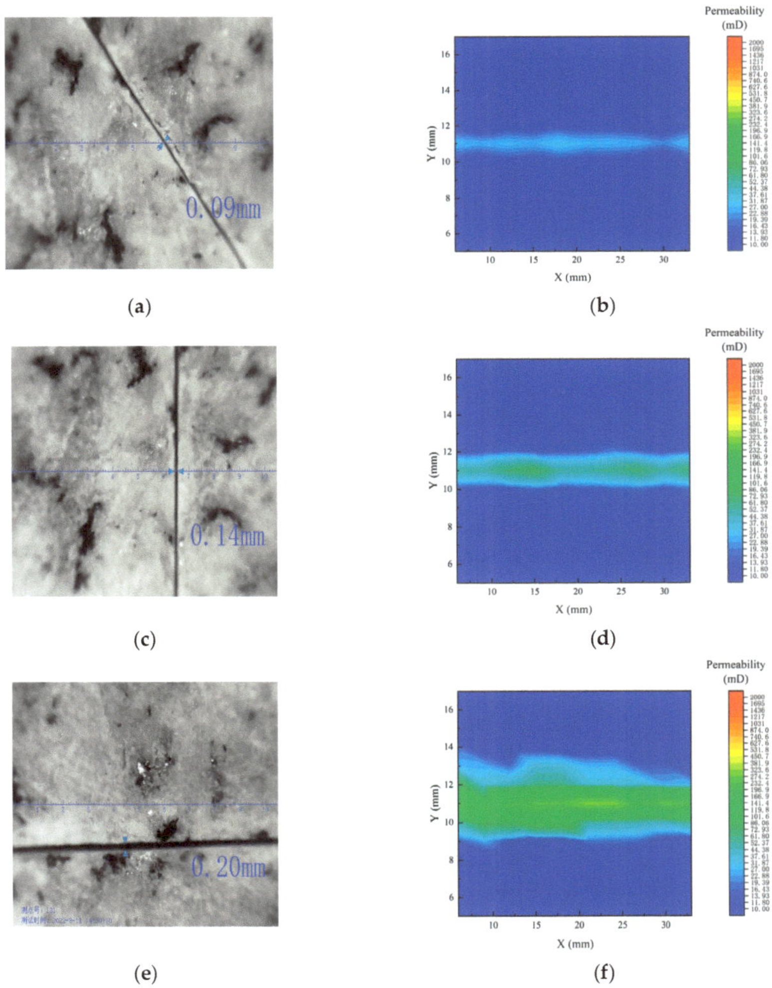

Figure 15. *Cont.*

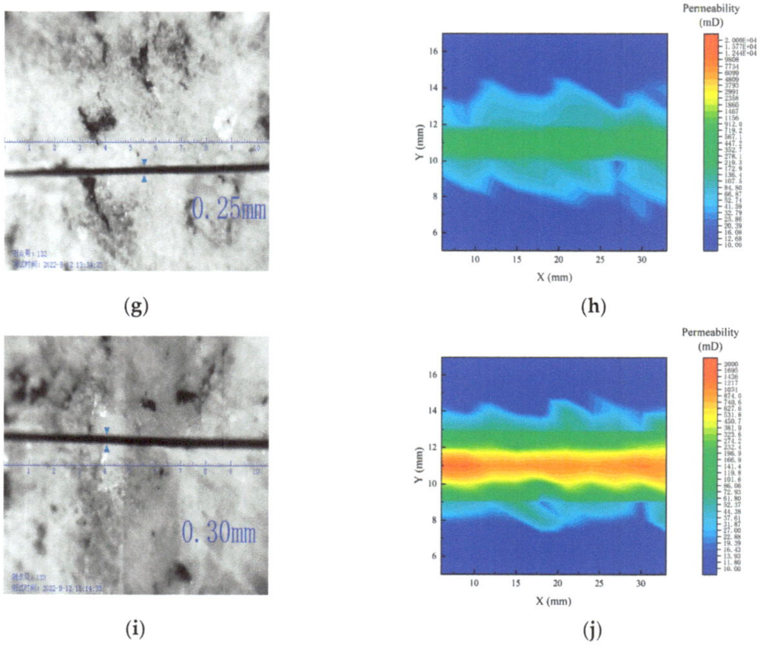

Figure 15. Experimental results of core permeability. (**a**) Measurement results for a core fractur aperture of 90 μm; (**b**) Permeability measurement results for the 90 μm fracture slab model; (**c**) Me surement results for a core fracture aperture of 140 μm; (**d**) Permeability measurement results fo the 140 μm fracture slab model; (**e**) Measurement results for a core fracture aperture of 200 μn (**f**) Permeability measurement results for the 200 μm fracture slab model; (**g**) Measurement results fc a core fracture aperture of 250 μm; (**h**) Permeability measurement results for the 250 μm fracture sla model; (**i**) Measurement results for a core fracture aperture of 300 μm; (**j**) Permeability measuremer results for the 300 μm fracture slab model. The non-English fonts in the lower left corner of Figure (**a**,**c**,**e**,**g**,**i**) are the recorded measurement point numbers and test times.

In this study, the experimentally derived fracture permeability was assessed alongsid the measured fracture apertures. Figure 16 presents the calculated results, where th blue spheres represent the core permeability of a single fracture with varying aperture measured during this experiment under shim action. It is important to note that th permeability measurements were aligned with the fracture direction, which is why th figure does not include information regarding the fracture angle. Due to the technica limitations associated with artificial fractures, the experimental results can only achiev a minimum fracture aperture of 49 μm, making it impossible to investigate the trend c permeability at smaller fracture apertures. The results obtained in this study are consister with the findings of previous research [47,48].

4.2. X-CT Scanning Results and Permeability Test Results of Cores before and after Fracturing

The three-dimensional digital core obtained through CT technology provides a detaile representation of the microstructural features of the rock mass [49]. Figure 17 illustrate the entire application process, with Figure 17a presenting an overall view of the thre dimensional digital core, while Figure 17b and 17c depict the digital reconstruction result of the core samples before and after the fracture construction, respectively. Followin the scanning and three-dimensional reconstruction of the CT images, a series of imag processing steps are required, including filtering and threshold segmentation.

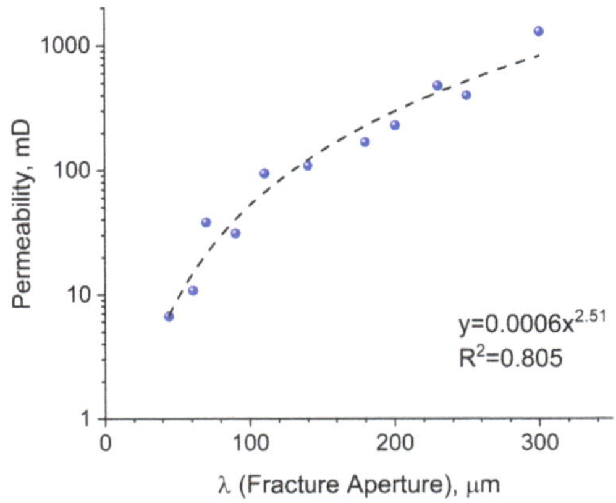

Figure 16. Fitting results of single-fracture aperture and permeability.

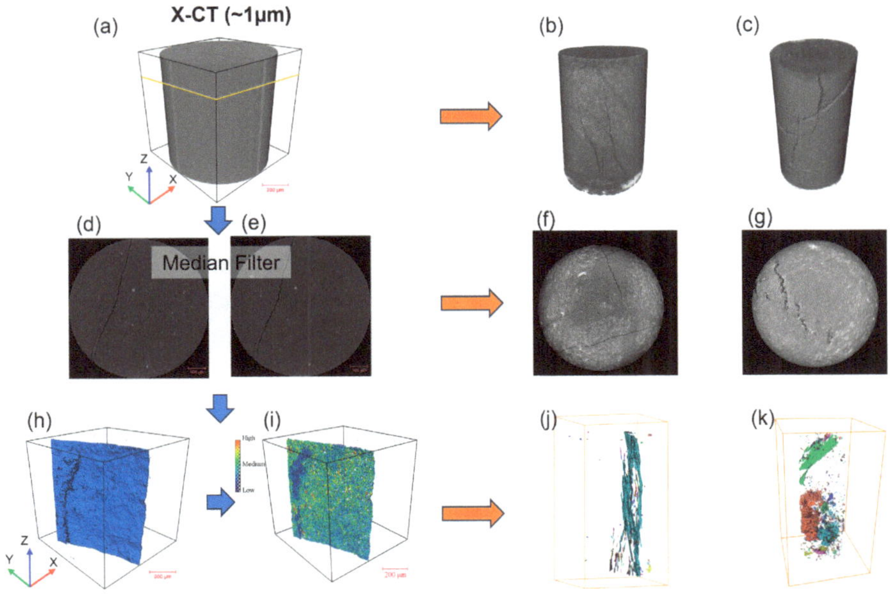

Figure 17. Processing workflow and results of 3D digital core. (**a**) Overall view; (**b**) Reconstruction results of the core before fracturing; (**c**) Reconstruction results of the core after fracturing; (**d**) 2D cross-sectional image before filtering; (**e**) 2D cross-sectional image after filtering; (**f**) Filtered 2D cross-sectional image of the core before fracturing; (**g**) Filtered 2D cross-sectional image of the core after fracturing; (**h**) Preliminary extraction results of fractures; (**i**) Visualization results of the 3D fracture distribution based on threshold segmentation; (**j**) Processing results of the actual full-size core before fracturing; (**k**) Display of fracture morphology of the actual full-size core after fracturing (The different colors in the diagram represent different fracture groups).

In the threshold segmentation process, median filtering is first applied to eliminate noise and enhance image quality, followed by a binarization step to construct the three-dimensional pore network model of the core. Figure 17d,e show the two-dimensional cross-sectional images before and after filtering, while Figure 17f,g correspond to the results obtained before and after the fracture construction, reflecting changes in fracture characteristics. Figure 17h illustrates the fracture extraction process, and Figure 17i presents the visualization of the three-dimensional fracture distribution obtained through threshol segmentation. Finally, Figure 17j,k display the processing results of the full-diamete samples before and after the fracture construction, along with a detailed representation c the fracture morphology.

By comparing the CT scan results of full-size samples before and after fracture form tion, the fracture complexity in the actual rock core is higher, especially the cross-fractur and the aperture and tortuosity of the fracture are significantly different from the ide model. Table 1 measures the porosity and permeability of eight full-diameter cores befor and after fracture formation. After the fractures were formed, only five samples wer complete and did not drop, and valid data were extracted.

Table 1. Porosity and permeability measurement results of eight full-diameter cores before and aft fracture formation.

Core No.	Fracture Information			Before Fracturing				After Fracturing	
	Fracture No.	Angle	Aperture (μm)	Angle	Aperture (μm)	Porosity (%)	Permeability (mD)	Porosity (%)	Permeability (mD)
X1	① ② ③ ④	73° 129° 74° 28°	324 205 172 99	68.26°	34 20.88 15.54 3.13	2.5 8.43 31.75 6.71	0.263	—	—
X2	①	58°	309	—	—	2	0.091	4.3	12.09
X3	① ② ③	84° 98° 9°	324 162 182	99.32°	109	1.5	0.206	4.4	48.841
X4	① ②	91° 98°	359 72	99.348°	80	1.4	0.229	—	—
X5	① ②	84° 83°	354 297	—	—	1.7	0.028	4.7	60.743
X6	①	74°	159	116.68°	79	2.1	0.209	4.2	46.649
X7	①	118°	284	—	—	0.9	0.022	4	20.323
X8	① ② ③	80° 93° 179°	255 76 303	80.57°	121	1.3	0.053	—	—

After excluding certain seamless cores and those affected by falling blocks, we fir compare the relationship between porosity and permeability for the cores before and aft fracturing, as illustrated in Figure 18. Notably, the permeability shows a significant increas following the formation of fractures. Analyzing the permeability variations among the X X3, X5, X6, and X7 samples reveals the emergence of two distinct clusters in the cross-pl with an overall goodness of fit reaching 0.84.

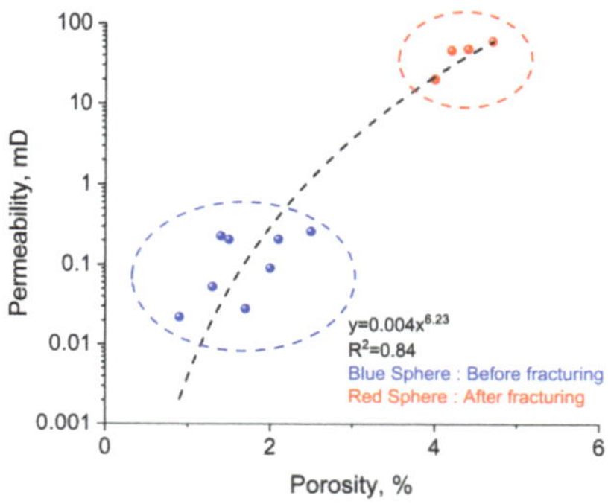

Figure 18. The relationship between porosity and permeability of full-size core before and aft fracture formation.

4.3. Comparative Analysis of Simulation and Experiment Results

To verify the validity of the improved single-fracture flat plate model, the formulae used to fit the permeability of the laboratory core, as shown in Figure 16, were compared with Equation (14). Both equations share the same structure, and the coefficients in the numerical simulation results of the theoretical formulae (Figure 7d) closely align with those from the laboratory results. In Equation (14), permeability is positively proportional to the cubic power of the fracture apertures when only the fracture apertures are considered. However, in the actual results, the exponential coefficient for the fracture apertures is 2.51. This discrepancy arises from the challenges in obtaining the tortuosity of the actual core and the complexity of the fracture surfaces.

To further assess the accuracy of the equation, the model was additionally verified using full-diameter core samples. Permeability was calculated by measuring the fracture aperture and angle within these full-diameter core samples, and the results were subsequently compared with the experimental permeability of the cores. Table 2 presents an analytical summary of the verification results for the plate permeability model. The data indicate that the calculated results of the plate fracture permeability model closely align with the core fracture permeability, exhibiting an average absolute error of 11.47 mD and an average relative error of 25.57%. For the dual-fracture samples after fracture formation, we take X5 as a representative case. Utilizing Equation (34), along with the aperture and angles of the two fractures derived from CT scanning results, we calculate a permeability of 114 mD, which significantly exceeds the laboratory measurements. The comparison between the theoretical derivation and experimental results for a single fracture demonstrates consistency, indicating the effectiveness of the model. The CT scan results align well with the principles established through core experiments and the trends explored in this study. However, for intersecting fractures, the model's calculated permeability is higher than the measured values.

Table 2. Plate fracture permeability model verification table.

Validation Sample Number	Fracture Aperture (μm)	Fracture Angle (°)	Core Fracture Permeability (mD)	Calculated Fracture Permeability (mD)	Absolute Error (mD)	Relative Error (%)
V1	94.9	0	61.58	59.57	2.01	3.26
V2	173.6	0	247.63	268.51	20.88	8.43
V3	109	40	48.94	64.48	15.54	31.75
V4	89	31	46.65	43.52	3.13	6.71
V5	99	58	20.32	36.11	15.79	77.71
	94.9	0	61.58	59.57	2.01	3.26
		Average Error			11.47	25.57

5. Discussion

5.1. Error Analysis of Single-Fracture Model

Building on the classical plate model, this research introduces and validates a permeability calculation model that integrates fracture tortuosity, angle, and aperture, utilizing core experimental data for validation. The identified errors can be summarized as follows.

(1) Although Equation (14) aligns formally with the model depicted in Figure 19, there are discrepancies in the coefficients. The theoretical fracture aperture index in Equation (14) is higher, while the coefficient observed in the numerical simulation (Figure 7d) is lower. Although the tortuosity of fractures is not reflected in the experimental data, the irregularities of fractures within the rock samples result in a non-uniform fracture aperture across the core sample, leading to differences between the final fitting results and the theoretical predictions.

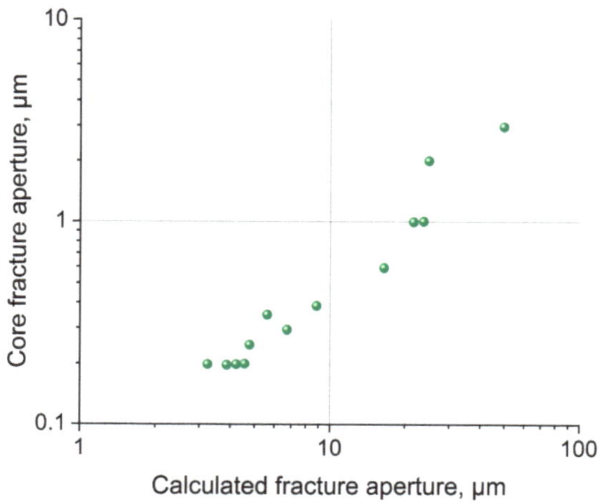

Figure 19. Calculated intersection diagram of fracture aperture and core fracture aperture.

(2) In calculating fracture aperture, electrical imaging logging data may be utilized for calibration with actual logging results. Luthi S.M. et al. [50] developed a formula for determining fracture aperture using the finite element method. This method establishes a correlation between the conductivity anomaly at the fracture and the fracture aperture, with the anomalous conductivity represented by a curve. The area under this curve is influenced by both the fracture aperture and the resistivity of the adjacent intrusion zone near the wellbore. Consequently, a formula is derived for the quantitative assessment of crack opening:

$$W = c \times A_1 \times {R_m}^b \times {R_{xo}}^{1-b} \tag{35}$$

In Equation (35), the parameters b and c represent instrument-specific constants; R_m denotes the resistivity of the mud; R_{xo} indicates the resistivity of the wash zone; and A_1 signifies the area of abnormal current.

Using the ERMI instrument as a case study, the calibrated parameters for the fracture aperture formula are b = 0.7 and c = 3.5, which are applied in the calculation of fracture aperture. Figure 15 illustrates a comparison between the calculated values and the core measurements. Figure 15 reveals a substantial relative error between the fracture aperture obtained using the original parameters and the core-derived fracture aperture, highlighting the necessity for calibration. In the context of calculating actual fracture permeability, inaccuracies in the determination of fracture aperture frequently lead to errors in the computed permeability values.

(3) The computation of tortuosity within the fracture formula presents considerable challenges. In porous reservoirs, researchers such as Li et al. [51] have established numerical relationships among the formation factors, cementation index, and tortuosity through rock electrical experiments. They derived the cementation index by assessing the difference between total porosity and connected porosity, which subsequently allowed for the calculation of tortuosity. However, for fractured reservoirs, establishing a similar numerical relationship for tortuosity calculations remains problematic. Currently, most rock electrical experiments are focused on porous sections, while conducting resistivity experiments on fractured sections is complicated by the inability of core samples to meet experimental standards, preventing the establishment of consistent rules. As a result, the calculation of the cementation index largely remains confined to numerical simulation at this stage [52].

5.2. Limitations of the Dual-Fracture Model

The cross-dual-fracture model is generalized based on the single-fracture model incorporating the mass transfer equation and trigonometric functions. This paper conducts numerical simulations; however, further verification through actual experimental data remains challenging. The study identifies the primary limitations.

5.2.1. Limitations of the Suture Experiment

To create fractures in the core, a clamp is gradually tightened to achieve the fracturing process, resulting in artificial fractures that closely resemble the morphology of natural fractures. However, during the actual operation of core fracturing, there are stringent requirements for controlling the crack initiation time and strength, which makes the core susceptible to crushing [53]. As a result, the number of samples collected is limited. Figure 20 illustrates the outcomes of fracturing in artificial fractures, demonstrating the ability to generate fractures at various angles. However, controlling the opening and angle of these fractures proves to be challenging, and there is a significant incidence of core failure.

Figure 20. Cores with artificial fractures at different angles.

5.2.2. Differences between the Ideal Fracture Model and Actual Core Characteristics

For the cross-fractures observed in the actual core, the model's calculated results are significantly higher than the measured values. This discrepancy arises from the fact that actual data cannot account for the tortuosity, leading to the assumption that the fracture aperture remains constant. This assumption can be identified as one of the primary sources of error in the analysis. At the same time, the X-CT scan results can be used for further demonstration. Figure 21a illustrates the interface of the X5 sample after fracture formation, while Figure 21b depicts the fracture morphology obtained from CT scanning. Analysis reveals that the fractures in the actual core data exhibit greater complexity, particularly regarding the irregularities of the fracture surfaces, which are not smooth and exhibit variable apertures. Additionally, the quantification of tortuosity poses challenges. These factors can occasionally contribute to increased discrepancies in the results.

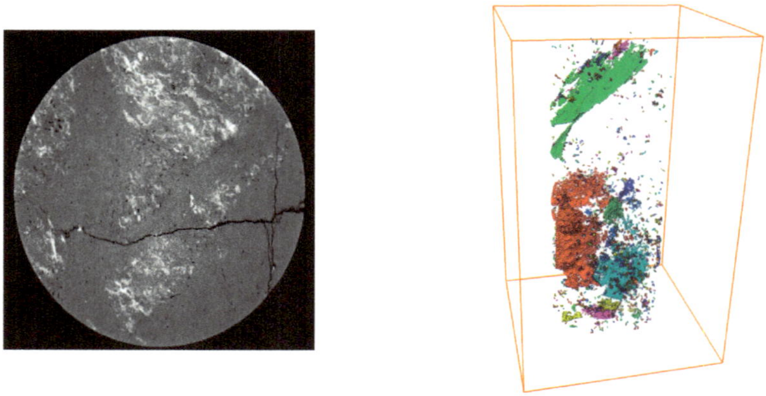

(**a**) (**b**)

Figure 21. CT scan results of X5 sample. (**a**) The cross-section of X5 sample; (**b**) Fracture structur diagram of X5 sample.

5.2.3. Limitations of Logging Methods in Evaluating Cross-Fracture Permeability

Simultaneously, it must be considered that in practical exploration, logging info mation is commonly used to evaluate fracture segments and calculate permeability. Th resistivity of the core was measured both before and after fracturing, revealing a stron correlation with fracture porosity, with a goodness of fit reaching 0.89 (Figure 22a). Ac ditionally, owing to the established relationship between porosity and permeability, th goodness of fit between resistivity and permeability was found to be 0.69 (Figure 22b). Thi finding aligns with the perspective of Zheng et al. [54], who suggest that resistivity data ca be utilized to predict parameters related to fracture occurrence. However, it is importar to note that discrepancies exist between core measurement techniques and logging dat acquisition methods, which introduces several limitations to the direct application of thes findings. The actual resistivity response is influenced by multiple coupled factors, making significantly more complex than laboratory measurements. Additionally, the fluids preser in the reservoir must be considered, as they also affect resistivity. Furthermore, the presenc of high-angle fractures in intersecting fractures can lead to non-unique solutions in th results [54].

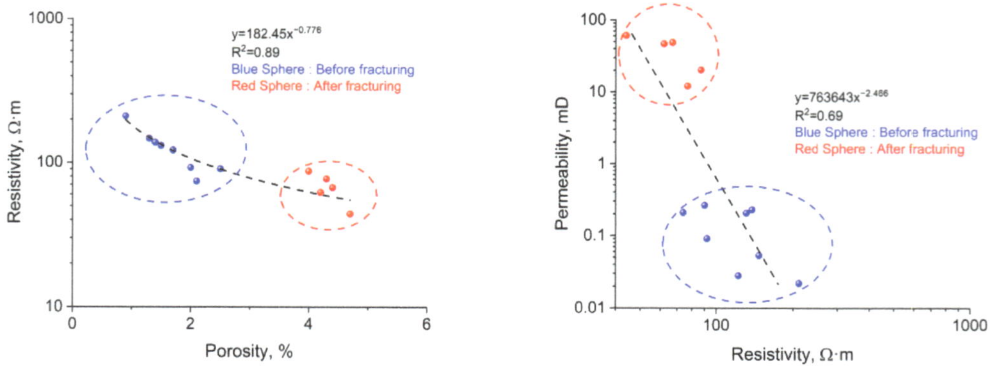

(**a**) (**b**)

Figure 22. The relationship between resistivity and porosity and permeability. (**a**) The relationshi between porosity and resistivity; (**b**) The relationship between permeability and resistivity.

6. Conclusions

This study proposes a novel fracture permeability evaluation model based on the classical planar fracture model, specifically tailored for marine granite reservoirs. The model effectively incorporates the heterogeneity of fractures, accounting for factors such as tortuosity, fracture angle, and aperture, thus enhancing traditional models. Additionally, building upon the single-fracture model, we introduce mass transfer equations and trigonometric functions to develop a dual-crossed fracture model.

By integrating numerical simulations with experiments on full-diameter core samples, we validate the effectiveness of the proposed model. The results indicate that the single-fracture model aligns well with actual experimental data in predicting fracture permeability, demonstrating its practical applicability. Depending on the fracture aperture, when the fracture aperture varies from 90 μm to 130 μm, the permeability values derived from the laboratory tests range from 30 mD to 1300 mD. When the results calculated by the model derived in this paper are compared with the actual measurements, the average absolute error is 11.47 mD, and the relative error is 25.57%, which demonstrates the validity of the model deduced in this paper.

In the case of the dual-crossed fracture model, we elucidate the influence of fracture aperture, relative positioning of the two fractures, and intersection points on permeability. To address the complexities associated with crossed fractures, we employ X-ray computed tomography (X-CT) scanning technology to extract and compute fracture parameters. Although the calculated results from the model exceed the actual measurements, the application of digital core technology confirms that this discrepancy arises from the inherent complexity of fractures in the actual rock samples. Furthermore, we investigate the sensitivity of resistivity in permeability calculations.

The current challenges in this research include the preparation of dual-fracture cores, the computation of tortuosity, and the direct application of simulation results to experimental data.

In summary, the proposed fracture permeability evaluation model offers new insights for theoretical research and provides practical solutions for engineering applications, thus advancing the scientific study and development of fractured reservoirs. Future research could further explore the model's applicability across various geological conditions and optimize parameter extraction methods to enhance the accuracy of permeability assessments.

Author Contributions: Conceptualization, J.G. and Z.Z. (Zhansong Zhang); Methodology, J.G. and H.L.; Software, H.L. and Z.Z.; Validation, J.G., B.G. and H.L.; Formal analysis, J.G.; Investigation, B.G.; Resources, Z.Z. (Zhansong Zhang); Data curation, Z.Z. (Zhansong Zhang); Writing—original draft, J.G.; Writing—review & editing, J.G. and Z.Z. (Zuomin Zhu); Supervision, Z.Z. (Zuomin Zhu); Project administration, Z.Z. (Zhansong Zhang); Funding acquisition, J.G. All authors have read and agreed to the published version of the manuscript.

Funding: This work was financially sponsored by Open Fund of Key Laboratory of Exploration Technologies for Oil and Gas Resources, Ministry of Education (No. K2023–02).

Institutional Review Board Statement: Not applicable.

Informed Consent Statement: Not applicable.

Data Availability Statement: All our data are displayed in the pictures in the paper.

Acknowledgments: The authors would like to express their most sincere gratitude to the field workers in the oil field, Xin Nie, and CNOOC workers.

Conflicts of Interest: Author Baoxiang Gu was employed by the company CNOOC International Limited. The remaining authors declare that the research was conducted in the absence of any commercial or financial relationships that could be construed as a potential conflict of interest.

References

1. Dong, Z.; Aifen, L.; Qiang, S.; Yangqing, F. Measuring oil and water relative permeability in a single fracture and researching it impacting factors. *Petrol. Sci. Technol.* **2013**, *31*, 2191–2201. [CrossRef]
2. Chen, H.; Li, H.; Li, Z.; Li, S.; Wang, Y.; Wang, J.; Li, B. Effects of matrix permeability and fracture on production characteristic and residual oil distribution during flue gas flooding in low permeability/tight reservoirs. *J. Petrol. Sci. Eng.* **2020**, *195*, 10781 [CrossRef]
3. Huang, X.; Kang, Z.; Zhao, J.; Wang, G.; Zhang, H.; Yang, D. Experimental investigation on micro-fracture evolution and fractur permeability of oil shale heated by water vapor. *Energy* **2023**, *277*, 127677. [CrossRef]
4. Zhang, Y.; Hou, S.; Mei, S.; Zhao, T.; Li, D. Experimental study of permeability-hydraulic fracturing characteristics of tight rock *Geomech. Geophys. Geo-Energy Geo-Resour.* **2022**, *8*, 182. [CrossRef]
5. Li, X.; Yang, Z.; Li, S.; Huang, W.; Zhan, J.; Lin, W. Reservoir characteristics and effective development technology in typica low-permeability to ultralow-permeability reservoirs of China National Petroleum Corporation. *Energy Explor. Exploit.* **2021**, *3* 1713–1726. [CrossRef]
6. Gale, J.F.W.; Laubach, S.E.; Olson, J.E.; Eichhubl, P.; Fall, A. Natural fractures in shale: A review and new observations. *AAP Bull.* **2014**, *98*, 2165–2216. [CrossRef]
7. Wang, S.; Elsworth, D.; Liu, J. Permeability evolution in fractured coal: The roles of fracture geometry and water-content. *Int. Coal Geol.* **2011**, *87*, 13–25. [CrossRef]
8. Ghanizadeh, A.; Clarkson, C.R.; Aquino, S.; Ardakani, O.H.; Sanei, H. Petrophysical and geomechanical characteristics c Canadian tight oil and liquid-rich gas reservoirs: II. Geomechanical property estimation. *Fuel* **2015**, *153*, 682–691. [CrossRef]
9. Guo, J.; Zhang, Z.; Xiao, H.; Zhang, C.; Zhu, L.; Wang, C. Quantitative interpretation of coal industrial components using a gra system and geophysical logging data: A case study from the Qinshui Basin, China. *Front. Earth Sci.* **2023**, *10*, 1031218. [CrossRe
10. Zhang, H.; Ait Abderrahmane, H.; Arif, M.; Al Kobaisi, M.; Sassi, M. Influence of heterogeneity on carbonate permeabilit upscaling: A renormalization approach coupled with the pore network model. *Energy Fuels* **2022**, *36*, 3003–3015. [CrossRef]
11. Nabawy, B.S.; Géraud, Y.; Rochette, P.; Bur, N. Pore-throat characterization in highly porous and permeable sandstones. *AAP Bull.* **2009**, *93*, 719–739. [CrossRef]
12. Taghipour, S.A.; Hoseinpour, S.A.; Soltani, B.; Bahadori, A. Accurate estimation of formation permeability by means of petrophy ical logs. *Petrol. Sci. Technol.* **2017**, *35*, 718–725. [CrossRef]
13. Aghli, G.; Moussavi-Harami, R.; Mohammadian, R. Reservoir heterogeneity and fracture parameter determination using electrica image logs and petrophysical data (a case study, carbonate Asmari Formation, Zagros Basin, SW Iran). *Petrol. Sci.* **2020**, *17*, 51–6 [CrossRef]
14. Kozeny, J. Ueber kapillare leitung des wassers im boden. *Sitzungsberichte Akad. Wiss. Wien* **1927**, *136*, 271–306.
15. Wu, F.; Fan, Q.; Huang, D.; Ma, L.; Liang, X.; Sima, L. Predicting gas–water relative permeability using nuclear magnetic resonanc and mercury injection capillary pressure measurements. *J. Nat. Gas Sci. Eng.* **2016**, *32*, 35–47. [CrossRef]
16. Zhu, L.; Zhang, C.; Wei, Y.; Zhang, C. Permeability prediction of the tight sandstone reservoirs using hybrid intelligent algorithr and nuclear magnetic resonance logging data. *Arab. J. Sci. Eng.* **2017**, *42*, 1643–1654. [CrossRef]
17. Piscopo, V.; Baiocchi, A.; Lotti, F.; Ayan, E.A.; Biler, A.R.; Ceyhan, A.H.; Cüylan, M.; Dişli, E.; Kahraman, S.; Taşkın, M. Estimatio of rock mass permeability using variation in hydraulic conductivity with depth: Experiences in hard rocks of western Turke *Bull. Eng. Geol. Environ.* **2018**, *77*, 1663–1671. [CrossRef]
18. Jiang, X.; Wang, X.; Wan, L. Semi-empirical equations for the systematic decrease in permeability with depth in porous an fractured media. *Hydrogeol. J.* **2010**, *18*, 839–850. [CrossRef]
19. Achtziger-Zupančič, P.; Loew, S.; Mariethoz, G. A new global database to improve predictions of permeability distribution i crystalline rocks at site scale. *J. Geophys. Res. Solid Earth* **2017**, *122*, 3513–3539. [CrossRef]
20. Tang, X.; Zhou, W.; Yang, W.; Zhang, C.; Zhang, C. An improved method in petrophysical rock typing based on mercury-injectio capillary pressure data. *Energy Sources Part A* **2020**, *2020*, 1–16. [CrossRef]
21. Chen, L.; Jiang, Z.; Liu, Q.; Jiang, S.; Liu, K.; Tan, J.; Gao, F. Mechanism of shale gas occurrence: Insights from comparative stud on pore structures of marine and lacustrine shales. *Mar. Pet. Geol.* **2019**, *104*, 200–216. [CrossRef]
22. Kenyon, W.E.; Day, P.I.; Straley, C.; Willemsen, J.F. A three-part study of NMR longitudinal relaxation properties of water-saturate sandstones. *SPE Form. Eval.* **1988**, *3*, 622–636. [CrossRef]
23. Morriss, C.E.; Maclnnis, J.; Freedman, R.; Smaardyk, J.; Straley, C.; Kenyon, W.E.; Vinegar, H.S.; Tutunjian, P.N. Field test of a experimental pulsed nuclear magnetism tool. In Proceedings of the SPWLA Annual Logging Symposium, Calgary, AB, Canad 13 June 1993. Paper ID. SPWLA-1993-GGG.
24. Timur, A. An investigation of permeability, porosity, and residual water saturation relationships. In Proceedings of the SPWL Annual Logging Symposium, New Orleans, LA, USA, 23 June 1968. Paper ID. SPWLA-1968-J.
25. Coates, G.R.; Xiao, L.; Prammer, M.G. *NMR Logging. Principles and Interpretation*; Halliburton Energy Service: Huston, TX, USA, 199
26. Prince, C.M.; Steele, D.D.; Devier, C.A. Permeability estimation in tight gas sands and shales using NMR—A new interpretiv methodology. In Proceedings of the 9th AAPG International Conference and Exhibition, Rio de Janeiro, Brazil, 15–18 November 200

7. Rezaee, R.; Saeedi, A.; Clennell, B. Tight gas sands permeability estimation from mercury injection capillary pressure and nuclear magnetic resonance data. *J. Petrol. Sci. Eng.* **2012**, *88*, 92–99. [CrossRef]
8. Hossain, Z.; Grattoni, C.A.; Solymar, M.; Fabricius, I.L. Petrophysical properties of greensand as predicted from NMR measurements. *Petrol. Geosci.* **2011**, *17*, 111–125. [CrossRef]
9. Wu, F.; Li, Y.; Burnham, B.; Zhang, Z.; Yao, C.; Yuan, L.; Zhang, F.; Deng, H.; Xi, Y.; He, J. Fractal-based NMR permeability estimation in tight sandstone: A case study of the Jurassic rocks in the Sichuan Basin, China. *J. Petrol. Sci. Eng.* **2022**, *218*, 110940. [CrossRef]
0. Xia, F.; Su, Y.; Tang, X. Stoneley wave reflection and transmission across permeable formations and fractured zones: Comparison of analytical and numerical modeling results. *Chin. J. Geophys.* **2022**, *65*, 1508–1518.
1. Guo, J.; Zhang, Z.; Nie, X.; Zhao, Q.; Lv, H. Logging Evaluation of Irreducible Water Saturation: Fractal Theory and Data-Driven Approach—Case Study of Complex Porous Carbonate Reservoirs in Mishrif Formation. *Fractal. Fract.* **2024**, *8*, 487. [CrossRef]
2. Tian, Y.; Wang, G.; Li, H.; Huang, Y.; Zhao, F.; Guo, Y.; Gao, J.; Lai, J. A novel deep learning method based on 2-D CNNs and GRUs for permeability prediction of tight sandstone. *Geoenergy Sci. Eng.* **2024**, *238*, 212851. [CrossRef]
3. Li, N.; Wang, K.; Lu, J.; Liu, P.; Xiao, C.; Wu, H.; Guo, Q.; Fan, H.; Men, B.; Feng, Z.; et al. First successful downhole testing of the permeability logging prototype. *J. Geophys. Eng.* **2024**, *21*, 1179–1182. [CrossRef]
4. Moinfar, A.; Varavei, A.; Sepehrnoori, K.; Johns, R.T. Development of an efficient embedded discrete fracture model for 3D compositional reservoir simulation in fractured reservoirs. *SPE J.* **2014**, *19*, 289–303. [CrossRef]
5. Zhang, H.; Hu, Y.; Li, X.; Du, K.; Zeng, T.; Li, C. Application of support vector machines and genetic algorithms to fluid identification in Offshore Granitic subduction hill reservoirs. *Geoenergy Sci. Eng.* **2024**, *240*, 213013. [CrossRef]
6. Li, Z.; Guo, J.; Wu, S. Characteristics and Genetic Mechanism of Granite Weathering Crust of Songnan Low Uplift, Qiongdongnan Basin, South China Sea. *Minerals* **2024**, *14*, 512. [CrossRef]
7. Bisdom, K.; Bertotti, G.; Nick, H.M. The impact of different aperture distribution models and critical stress criteria on equivalent permeability in fractured rocks. *J. Geophys. Res. Solid Earth* **2016**, *121*, 4045–4063. [CrossRef]
8. Zhang, B.; Deng, Z.; Fu, X.; Yu, K.; Zeng, F.B. An experimental study on the effects of acidization on coal permeability: Implications for the enhancement of coalbed methane production. *Energy* **2023**, *280*, 128145. [CrossRef]
9. Nakagawa, S.; Nakashima, S.; Korneev, V.A. Laboratory measurements of guided-wave propagation within a fluid-saturated fracture. *Geophys. Prospect.* **2015**, *64*, 143–156. [CrossRef]
0. Jiao, Y.; Wang, Y.; Feng, D.; Gong, J.; Niu, Y.; Hu, S.; Ren, J. Laboratory study on fluid-induced fracture slip and permeability evolution in marble fractures. *Rock Mech. Rock Eng.* **2023**, *56*, 2497–2513. [CrossRef]
1. Aziz, A. A similarity solution for laminar thermal boundary layer over a flat plate with a convective surface boundary condition. *Commun. Nonlinear Sci.* **2009**, *14*, 1064–1068. [CrossRef]
2. Niessner, J.; Berg, S.; Hassanizadeh, S.M. Comparison of two-phase Darcy's law with a thermodynamically consistent approach. *Transport. Porous. Med.* **2011**, *88*, 133–148. [CrossRef]
3. Pouya, A.; Vu, M.N. Fluid flow and effective permeability of an infinite matrix containing disc-shaped cracks. *Adv. Water Resour.* **2012**, *42*, 37–46. [CrossRef]
4. Su, H.; Zhang, Y.; Xiao, B.; Huang, X.; Yu, B. A fractal-monte carlo approach to model oil and water two-phase seepage in low-permeability reservoirs with rough surfaces. *Fractals* **2021**, *29*, 2150003. [CrossRef]
5. Teruel, F.E. Characterization of a porous medium employing numerical tools: Permeability and pressure-drop from Darcy to turbulence. *Int. J. Heat Mass Tran.* **2009**, *52*, 5878–5888. [CrossRef]
6. Ahmia, A.C.; Idouhar, M.; Wongwailikit, K.; Dietrich, N.; Hébrard, G. Impact of cellulose and surfactants on mass transfer of bubble columns. *Chem. Eng. Technol.* **2019**, *42*, 2465–2475. [CrossRef]
7. Li, X.; Qin, R.; Wei, D.; Cao, J.; Wang, P.; Ye, X.; Yin, Z. Research progress in logging evaluation of offshore buried-hill fractured reservoirs in China. *China Offshore Oil Gas* **2023**, *35*, 69–82.
8. Li, X.; Qin, R.; Cao, J.; Wang, P.; Liu, X.; Ping, H. Method of connected porosity evaluation and quantitative permeability calculation for complex reservoirs. *Oil Geophys. Prospect.* **2022**, *57*, 377–385.
9. Li, B.; Nie, X.; Cai, J.; Zhou, X.; Wang, C.; Han, D. U-Net model for multi-component digital rock modeling of shales based on CT and QEMSCAN images. *J. Petrol. Sci. Eng.* **2022**, *216*, 110734. [CrossRef]
0. Plumb, R.A.; Luthl, S.M. Analysis of borehole images and their application to geologic modeling of an eolian reservoir. *SPE Form. Eval.* **1989**, *4*, 505–514. [CrossRef]
1. Li, X.; Qin, R.; Cao, J.; Wang, P.; Ye, X.; Yin, Z. Dolomite reservoir characteristics and permeability evaluation methods: An example from the Paleogene, Mesopotamian basin, Iraq. *China Offshore Oil Gas* **2024**, *36*, 81–94.
2. Qin, Z.; Wu, J.; Wang, C.; Luo, S.; Yang, H.; Su, K.; Luo, Y.; Liu, L.; Zhu, Y.; Wei, K.; et al. A novel calculation model, characteristics and applications of Archie's cementation exponent in dual porosity reservoirs with intersecting dual fractures. *Geoenergy Sci. Eng.* **2023**, *231*, 212390. [CrossRef]

53. Zhu, Z.; Cui, L.; Dong, Y.; Sheng, Q.; Tian, K.; Guo, Z. A Novel Deformation Analytical Solution and Constitutive Model fo Fractured Rock Masses. *J. Mar. Sci. Eng.* **2023**, *11*, 2351. [CrossRef]
54. Zheng, H.; Zhang, Z.; Guo, J.; Fang, S.; Wang, C. Numerical Simulation Study on the Influence of Cracks in a Full-Size Core o the Resistivity Measurement Response. *Energies* **2024**, *17*, 1386. [CrossRef]

Journal of
Marine Science and Engineering

MDPI

Article

Geological Conditions and Sedimentary Models of Oligocene and Eocene Effective Source Rocks in the Northern Yinggehai Basin

Jianxiang Pei [1,2], Gaowei Hu [2], Zhipeng Huo [3,4,*], Zhihong Chen [2], Yabing Chen [2], Xiaofei Fu [5], Weihong Wang [5], Haiyu Liu [2], Yanan Wang [3,6], Jingshuang Luo [5] and Guofei Chen [5]

1 Institute of Advanced Studies, China University of Geosciences, Wuhan 430074, China; peijx001@163.com
2 Research Institute of Exploration and Development, Hainan Branch of CNOOC (China) Co., Ltd., Haikou 570311, China; hujian1028@126.com.cn (G.H.); chenzhh@cnooc.com.cn (Z.C.); chenyb7@cnooc.com.cn (Y.C.); liuhy83@cnooc.com.cn (H.L.)
3 School of Resources and Materials, Northeastern University at Qinhuangdao, Qinhuangdao 066099, China; 2106210069@email.cugb.edu.cn
4 National Engineering Research Center of Offshore Oil and Gas Exploration, Beijing 100028, China
5 College of Geosciences, Northeast Petroleum University, Daqing 163319, China; fuxiaofei2008@sohu.com (X.F.); wangweihong@nepu.edu.cn (W.W.); 1luojingshuang81@163.com (J.L.); chenguofei@petrochina.com.cn (G.C.)
6 School of Energy Resources, China University of Geosciences, Beijing 100083, China
* Correspondence: huozhipeng521@163.com; Tel.: +86-134-8873-0863

Abstract: The development of the effective source rocks of the Eocene and Oligocene directly determines the oil and gas exploration potential in the northern Yinggehai Basin in China. Based on the analogy with the Hanoi Depression in Vietnam and the Yacheng District in the Qiongdongnan Basin and the comprehensive analysis of self-geological conditions, the development conditions of Eocene and Oligocene source rocks in the northern Yinggehai Basin are examined, focusing on tectonic evolution, sedimentary facies, and the paleoenvironment. Finally, the sedimentary models for the effective source rocks are established. The tectonic activity controlled the formation of the sedimentary deep depression and the migration of the sedimentary trough center, which migrated from east to west and then south from the Eocene to the Oligocene, leading to the sedimentary migration of good muddy source rocks. There are multiple sedimentary facies in favor of source rocks, including lacustrine facies, shallow marine facies, and delta plain swamps. The paleoenvironment indicates that the paleoclimate transitioned from warm and humid to cold and arid, the redox conditions evolved from semi-reducing to oxic, and paleoproductivity increased from the early to late Oligocene. Therefore, the early Oligocene was more conducive to the enrichment of organic matter. It is speculated that the warm and humid paleoclimate, reducing environment, and high paleoproductivity of the Eocene promoted the sedimentation and preservation of more organic matter. The above studies show that the northern Yinggehai Basin, especially the sedimentary period of the Eocene and Oligocene, has favorable geological conditions for the development of effective source rocks. The sedimentary models for Eocene lacustrine mudstones and Oligocene marine mudstones and marine–continental transitional coal-measure source rocks were established. These studies make up for the serious deficiency of previous research and mean that there is great exploration potential for oil and gas in the northern Yinggehai Basin in China.

Keywords: Geological conditions; Eocene lacustrine mudstone; Oligocene marine–continental source rocks; Sedimentary model

Academic Editor: Atilla Incecik

Received: 25 August 2024
Revised: 15 November 2024
Accepted: 22 November 2024
Published: 7 January 2025

Citation: Pei, J.; Hu, G.; Huo, Z.; Chen, Z.; Chen, Y.; Fu, X.; Wang, W.; Liu, H.; Wang, Y.; Luo, J.; et al. Geological Conditions and Sedimentary Models of Oligocene and Eocene Effective Source Rocks in the Northern Yinggehai Basin. *J. Mar. Sci. Eng.* **2025**, *13*, 100. https://doi.org/10.3390/jmse13010100

1. Introduction

Effective source rocks not only produce and discharge hydrocarbons but also signi icantly contribute to the formation of commercial reservoirs [1]. The evaluation criteri for effective source rocks vary depending on the region. In this study, source rocks wit a total organic carbon (TOC) greater than 0.5% are considered effective. Previous studie have identified two main sets of gas source rocks in the Yinggehai Basin: the Oligocen Yacheng Formation and the middle Oligocene Sanya Formation and Meishan Forma tion [2–4], as well as the possibility of another set of source rocks in the Eocene Lingto Formation [5,6]. However, current analyses of source rocks in the Yinggehai Basin ar limited to the marine source rocks of the Miocene Sanya and Meishan Formations, whic are mainly distributed in the Yinggehai Depression with organic matter types dominate by humic types (III-II_2) [4,7–9]. However, few domestic studies have explored the ge ological conditions, distribution, and paleoenvironments of potential Paleogene sourc rocks [5,10,11].

Because of the considerable thickness of the overlying Neogene and Quaternary strat in the Yinggehai Basin, the Paleogene strata are rarely exposed through drilling. Only few exploratory wells have been drilled in the Lingao Bulge in the northwestern sectio and the Yingdong Slope in the northeastern section. However, Paleogene strata have bee encountered in exploratory wells in the Vietnamese portion of the Yinggehai Basin, an the development conditions of Paleogene (mainly Oligocene and Eocene) source rock have been thoroughly studied [12,13]. The Oligocene and Eocene source rocks in th Vietnam region have accumulated more than 3000 $\times$ 10^4 tons of recoverable reserves c hydrocarbons, primarily derived from the Eocene and Oligocene source rocks [14]. Th Oligocene coal-measure source rocks have also been encountered in the Yacheng Distric in the western portion of the Qiongdongdong Basin [15]. This provides a useful referenc for studying the Oligocene and Eocene source rocks in the northern Yinggehai Basin an offers hope for exploration. In summary, the northern Yinggehai Basin has a low level c exploration which has been stopped for 20 years, and the study of the geological condition of oil/gas accumulations is still in its early stages [5]. The oil/gas source conditions ar unclear, particularly whether the deep Eocene and Oligocene have favorable geologica conditions for developing effective source rock. This uncertainty affects the resourc potential evaluation in the northern Yinggehai Basin and has become a prominent problen restricting exploration.

Because the basin genesis and geological conditions of the Hanoi Depression an Yacheng District of the Qiongdongdong Basin are similar to those of the northern Yinggeha Basin, this study discussed the development conditions of the Oligocene and Eocene sourc rock in the northern Yinggehai Basin from the perspectives of tectonics, sedimentary facie and paleo-sedimentary environments by drawing analogies with these conditions an established source rock sedimentary models. These studies will not only guild the oil an gas potential evaluation and exploration in the northern Yinggehai Basin but will also mak up for the serious deficiency of previous research and play a role in encouraging othe marine and petroleum geologists to study the deep Eocene and Oligocene source rocks i the northern Yinggehai Basin or similar basins.

2. Geological Background

The Yinggehai Basin holds significant importance as one of the prominent Cenozoi sedimentary basins located on the northern continental shelf of the South China Sea betwee China's Hainan Province and Vietnam. The basin spans an expansive area of more tha 11 $\times$ 10^4 km^2 and features a rhombic structure with a NNW strike and a length-to-widt ratio of approximately 2.5:1 (Figure 1) [16]. The basin is interconnected with the Beibuwa

Basin to the north, the Kunsong Uplift in the northwest, and the Qiongdongdong Basin in the southeast in a nearly vertical direction across the Yingdong I Fault Belt, which is a strike-slip tensile basin primarily composed of Cenozoic sediments. The maximum thickness of these sediments exceeds 17 km, making them rich sources of hydrocarbons, particularly oil and gas, in the northwestern part of the South China Sea [16,17]. The basin can be classified into three primary tectonic units: the central depression, the Yingdong slope, and the Yingxi slope. The central depression can be further divided into three secondary tectonic units, namely, the Hanoi Depression, Lingao Uplift and the Yinggehai Depression with a NW–SE orientation [5].

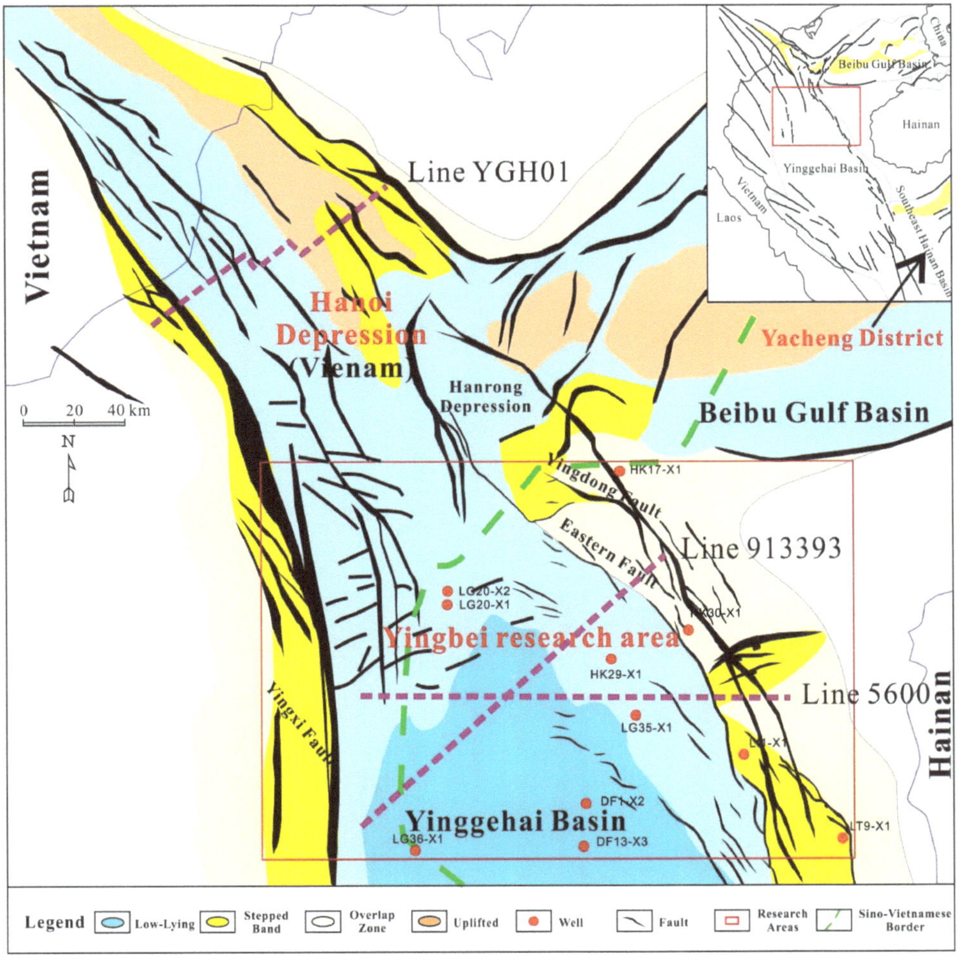

Figure 1. Geological map showing the location and sampling wells of the northern Yinggehai Basin, Hanoi Depression, and Yacheng District.

The Cenozoic tectonic evolution of the basin can be divided into four distinct stages. The early rifting stage characterized by the deposition of the Eocene Lingtou Formation (E_2l), and the late rifting stage characterized by the deposition of the Oligocene Yacheng Formation (E_3y) and the Lingshui Formation (E_3l). E_3y and E_3l may be divided into three sections, namely, the first member (Eyc_1), the second member (Eyc_2), and the third member (Eyc_3) for E_3y, and the first member (Els_1), the second member (Els_2), and the third member (Els_3) for E_3l. During the thermal subsidence zone, the Miocene Sanya Formation (N_1s),

Meishan Formation (N_1m), and Huangliu Formation (N_1h) were deposited, whereas th rapid subsidence zone gave rise to the Pliocene Yinggehai Formation (N_2y) and the Qu ternary Ledong Formation (Ql) (Figure 2). Throughout the Eocene, lacustrine depositio dominated, whereas during the Oligocene, a prevalence of marine—continental intersectio and coastal marine deposition occurred. The Miocene and upper strata are characterize by a predominance of marine deposits [5].

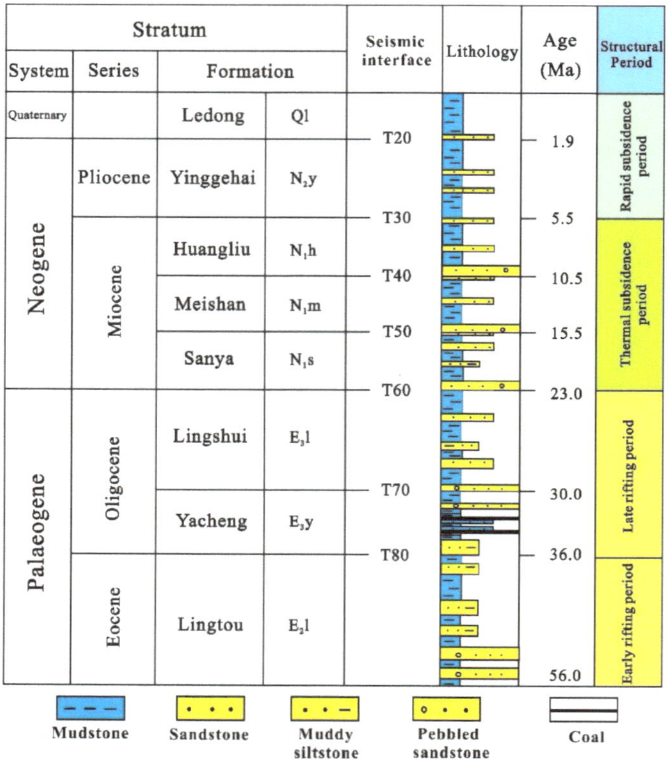

Figure 2. Stratigraphic column of the northern Yinggehai Basin.

In fact, the oil and gas exploration has been stagnant in the past two decades in th northern Yinggehai Basin. There are only about ten exploratory wells (LG20-X1, LG20-X HK17-X1, HK29-X1, HK29-X2, HK30-X1, LG35-X1, and LG36-X1). Only two wells (LG20-X and LG20-X2) drilled into the Oligocene Lingshui Formation, but they did not penetrat into the deeper Oligocene Yacheng Formation and the Eocene. At present, there are no significant oil and gas discoveries in the northern Yinggehai Basin.

3. Samples and Methods

3.1. Samples

In this study, 29 rock debris samples were collected from four wells in the study are These samples were obtained from wells LG20-X2 (6 samples from the Lingshui Formatio northern Yinggehai Basin), YC13-X1 (2 samples from the Yacheng Formation, Yachen District), YC13-X2 (17 samples from the Yacheng Formation, Yacheng District), and YC13-4 (4 samples from the Yacheng Formation, Yacheng District), and the total organic carbo (TOC), major element, and trace element contents were analyzed. Furthermore, the CNOO Hainan Branch provided 28 samples from the Yacheng District in the Qiongdongdon Basin for geological analysis.

3.2. Geological Analogy

The exploration level of the northern Yinggehai Basin was relatively low, and the drilled wells did not encounter the Oligocene Yacheng Formation or the Eocene [18]. Thus, evaluating the geological conditions and models for the effective source rocks of the Oligocene and Eocene is challenging owing to the limited available data. This includes the assessment of both geological development conditions and the hydrocarbon source rock sedimentary model. However, the tectonic and sedimentary conditions of the study area are similar to those of the Yacheng District in the Qiongdongnan Basin and Hanoi Depression in Vietnam [12,19]. This study employed a geological analogy to examine the tectonic evolution, sedimentary types, and paleoenvironmental conditions of each layer of the Oligocene and Eocene in the northern Yinggehai Basin. Although the analogy method cannot fully and accurately reproduce the geological conditions and characteristics of hydrocarbon source rock development in the northern Yinggehai Basin, an analogous study is a feasible and effective method given the lack of deep drilling data.

3.3. Structural Evolution Profile Drawing Methods

In this work, by employing the stratigraphic back-stripping method and adhering to the principle of a "balanced profile", the current evolution profile of the Hanio Depression and the study area is gradually regressed to its original and undeformed state before the deposition of each stratum [10]. This approach allows the reconstruction of the structural development history profile of the area. The tectonic evolution profile encompasses three primary processes: profile selection, application of actual geological data, and profile equilibrium [20]. A profile line perpendicular to the direction of tectonic movement is typically chosen to portray the evolution of underground geological structures reasonably and effectively. In this study area, the prevailing tectonic stress direction is primarily east–west extension and displacement, resulting in the majority of fractures and tectonic developing in a nearly south–north direction. Consequently, several main survey line profiles in the east–west direction were selected to create the evolution profile, and the tectonic period and tectonic style of the study area were analyzed in conjunction with regional geologic features to establish the regional geotectonic framework. In this study, by employing the stratigraphic back-stripping method and adhering to the principle of a balanced profile, the current evolution profile of the Hanio Depression and the study area was gradually regressed to its original and undeformed state before the deposition of each stratum [10]. This approach allows the reconstruction of the structural development history profile of the area.

3.4. TOC Analysis

The total organic carbon (TOC) content was measured via an ELTRR CS-800 Sulfur-Carbon Analyzer (Equipment source: Verder, Shanghai, China) in accordance with the standards outlined in GB/T 19145-2003 [21]. To commence the TOC measurement process, dilute hydrochloric acid was first employed to remove carbonate minerals from the sample under controlled conditions at 68 °C in a water bath. The TOC present in the sample was subsequently completely combusted via a CS-230 analyzer (Equipment source: LECO, St. Joseph, MO, USA) operated under high-temperature conditions. Finally, the TOC content was determined on the basis of the quantity of CO_2 produced, as monitored via an infrared detector.

3.5. Analysis of Major and Trace Element Contents

After the sample was thoroughly dried at 105 °C, it was weighed accurately and placed in a platinum crucible. Next, a mixture of lithium tetraborate, lithium metaborate, and lithium nitrate, which served as the melting agent, was added to ensure homogeneit between the sample and the melting agent. The sample was then melted via a hig precision melting machine at 1050 °C, and the resulting melt was poured into a platinur mold and cooled to form a frit. The quality of the frit was then assessed to determin whether it met the necessary standards (if it did not, it was reweighed and remelted). Onc the frit was deemed acceptable, the flake was weighed and melted again, and the majc element content was determined via a PANalytical PW2424 X-ray fluorescence spectromete (Malvern Panalytical, Malvern City, UK).

To determine the presence of trace elements, perchloric acid, nitric acid, and hydroflu ric acid were added to the samples, which were then evaporated to near dryness by heatin The samples were then dissolved and fixed with dilute hydrochloric acid, and the resultin solution was analyzed via plasma emission spectroscopy and plasma mass spectrometr with an Agilent 7900 (Equipment source: Agilent, Santa Clara, CA, USA) instrument.

4. Results and Discussion

4.1. Tectonic Conditions

The Paleogene Basin underwent initial rifting, during which time the Yinggehai Basi exhibited a characteristic graben–horst structure, with its eastern boundary fault (Yingdon I Fault Belt) experiencing heightened activity. The central part of the Hanoi Depression i double-faulted, whereas the eastern subdepressions are east-faulted and west-superfaulte (Figure 3). However, strong tectonic inversion occurred in the late Oligocene in the Hanc Depression, and obvious truncation was observed at the T60 interface, seeing seismic lin YGH01 (Figures 1 and 3a). To compare the similarities and differences in the evolution of th Hanoi Depression and the northern Yinggehai Basin, the 2D seismic line 913393 (Figure in the northern Yinggehai Basin was selected to construct the evolutionary history sectio (Figure 3b), which is NE-oriented and can clearly reflect the structural characteristics of th basin during the rifting period and fracture activities at the basin boundary. In the earl stage of rift subsidence (Eocene), the northern Yinggehai Basin was characterized by eas breaking and west superposition. In the late stage of rift subsidence (Oligocene), the slidin activity of the main trunk fracture decreased, and the Lingao Inversion Structural Be developed in the central part of the basin. Thus, during the Eocene, the basin experience a rift-based style with east breaking and west superpositioning, and several half-graben developed in the Hanoi Depression and Yingdong slope. The northern Yinggehai Basi is connected to the Hanoi Depression and belongs to the same secondary tectonic uni During the Oligocene, both faults had two dominant strike directions: NW–SE-oriente and E–W-oriented fractures. Most boundary fault zones are NW–SE-oriented and mainl consist of long-term active faults. The E–W-oriented faults, as secondary faults in the basi are characterized by small fracture distances and small extension lengths and are very we developed in the upper plate of the eastern boundary faults. Both the northern Yinggeha Basin and Hanoi Depression experienced multi-deformation processes of cracking an trapping, followed by slip during the rifting period, and were controlled by northeas oriented faults, which resulted in the same stage of overall tectonic evolution and th development of similar source rocks. In summary, the analogy of the tectonic evolutio between the northern Yinggehai Basin and the Hanoi Depression shows that the norther Yinggehai Basin has favorable tectonic conditions for developing effective source rock from the Oligocene and Eocene because it has been proved that the Hanoi Depressio develops good source rocks [12,13,22].

In view of the wide scope of the study area and the lack of directly available drillin data but the rich seismic line resources, this study proposes a well–seismic joint analysi method. This method integrates the analysis of geological background and existing dat and uses impression technology to restore ancient geomorphic features. The tectonic inversion periods, fold development characteristics, and degrees of tectonic evolution differ between the northern Yinggehai Basin and Hanoi Depression. The anticlines formed in the Hanoi Depression region during the late Oligocene–Miocene and Pliocene epochs, reflecting a strong inversion process and a tendency toward weakening in a southerly direction. Conversely, anticlines in the northern Yinggehai Basin emerged in two periods: the end of the Eocene and the late Oligocene–Miocene. The tectonic inversion period in the northern Yinggehai Basin is earlier than the Hanoi Depression, which means that the source rocks of the Eocene and Oligocene are possibly slightly worse than those of the Hanoi Depression.

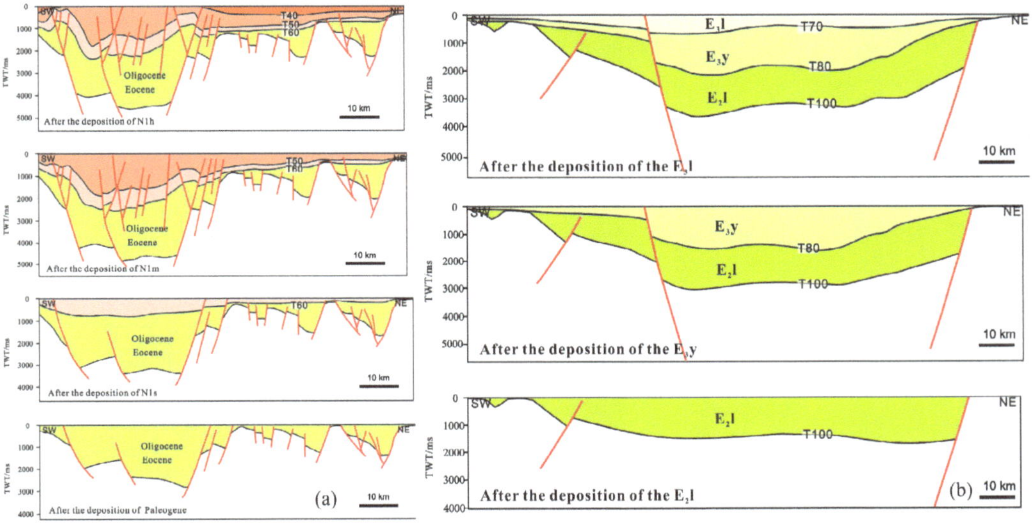

Figure 3. (**a**) Tectonic evolution profile of the Hanoi Depression of seismic line YGH01 [22]. (**b**) Tectonic evolution profile of the northern Yinggehai Basin of the 2D seismic line 913393.

The Hanoi Depression and the northern Yinggehai Basin are both located in the NW-oriented fracture zone, and the controlling effect of the eastern boundary fault zone is stronger; therefore, the deep-lying zones are concentrated in the east. Because the fracture zones controlling the Hanoi Depression and northern Yinggehai Basin are left-ordered rather than through faults, the depression zones are arranged in a bead-like fashion along the eastern boundary fracture zone because of differences in fracture activity. Although there is no complete stratigraphic interpretation of the Cenozoic basement owing to the depth of burial, the distribution of the Eocene sub-sag can be roughly inferred from the study of fracture activity, tectonic evolution, and tectonic deformation. During the depositional period of the Lingtou Formation (Eocene), the Yingdong I Fault Belt was more active, which led to strong subsidence control and the formation of the eastern deep depression zone (Figure 4a). In the late stage of rift sinking in the northern Yinggehai Basin, during the development of E_3y and E_3l, the activity of the Yingxi Fault Belt increased, the sliding activity of the basin's backbone fracture generally decreased, and the Lingao Inversion Structural Belt developed in the central part of the basin. During the development of the Yacheng and Lingshui Formations, the strike-slip movement of the Honghe fracture zone strongly modified the pre-existing tectonics, controlling the migration of the sub-sag, and the center of the sub-sag migrated from east to southwest. During the depositional period of the Yacheng Formation, the controlling effect of the Yingdong I Fault Belt on subsidence decreased, a nose-like structure developed close to the eastern boundary fracture, the water body became shallow, and the center of subsidence migrated to the southwest (Figure 4b). In contrast, the eastern boundary fracture did not control subsidence during the Lingshui Formation, and the subsidence center migrated to the center and the south (Figure 4c). In summary, there is a difference in the activities of the Yingdong I Fault Belt and the Ying Fault Belt, which led to the migration of the center of the sedimentary depression, meaning the migration of good muddy source rocks.

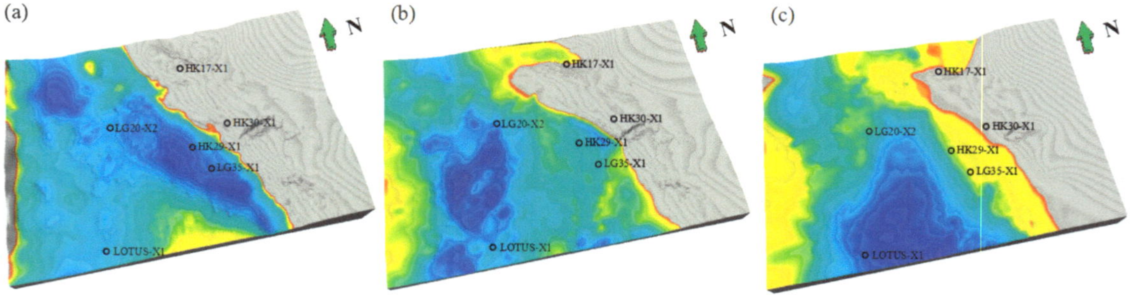

Figure 4. Sedimentary paleotopography of the Eocene–Oligocene in the northern Yinggehai Basin (the water body gradually gets deeper from yellow to blue). (**a**) Sedimentary paleomorphology of the Eocene; (**b**) sedimentary paleomorphology of the Yacheng Formation; (**c**) sedimentary paleomorphology of the Lingshui Formation.

4.2. Depositional Conditions

The northern Yinggehai Basin is characterized by three river systems, including the Red River in the northwest, which follows the direction of the basin, the Ma River in the west, and the Changhua River in the east [23]. The northern Yinggehai Basin and Hanoi Depression are both affected by the Red River source. The Yinggehai Basin is more distant, resulting in greater formation of muddy source rocks. The Ma River is a crucial source of material for the Oligocene and Eocene source rocks on the western slopes of Vietnam and China. In contrast, the Changhua River is a unique source in the eastern part of the northern Yinggehai Basin, providing a continuous supply of material, which is beneficial for the development of mudstone and coal-measure source rocks.

In this study, we relied primarily on the conversion of seismic facies to sedimentary facies to determine the sedimentary characteristics. This conversion allows for multi-resolution analysis because different sedimentary facies types can produce the same seismic facies response. To convert seismic facies to sedimentary facies accurately and reasonably, it is essential to consider the geological meaning of the seismic facies and obtain a thorough understanding of regional sedimentary sequences and features. Typically, taking the seismic line 5600 as an example (Figure 1), the seismic facies in delta plains are wedge-shaped or mat-shaped, parallel or subparallel; in delta fronts, they are wedge-shaped, with large stacked-tile front deposits or diagonal intersections; in seafloor fans, they are lenticular or wedge-shaped, with disorganized reflections; and in shallow seas, they are mat-shaped, with parallel, subparallel, or wave-like patterns (Figure 5).

Considering the sedimentary and seismic facies characteristics of the drilled wells in the Yacheng District of the Qiongdongnan Basin and the sedimentary characteristics of the Hanoi Depression, the sedimentary pattern of the northern Yinggehai Basin is determined by analogy. The northern Yinggehai Basin exhibits seismic facies characteristics similar to those of the coal strata at the edge of the Yacheng District, Hanoi Depression, and other basins, characterized by medium–high continuity, low frequency, and strong amplitude [5,14,24]. Despite the limitations of seismic resolution, the integrated response of multiple thin coal seams was discernible, resulting in the same facies axis displaying low-, medium-, and high-amplitude seismic facies characteristics. The presence of several low-frequency, strong-amplitude reflections and prominent continuous strong-amplitude seismic facies in the northern Yinggehai Basin suggests that lacustrine mudstone, marine mudstone, and coal-measure source rocks were deposited. The stronger the amplitude is, the thicker the coal seams are. The wedge-shaped, subparallel, long-axis, high-frequency, strong-amplitude seismic facies that developed in the northern Yinggehai Basin are indicative of coal strata. The seismic and drilled data both prove that high-organic lacustrine mudstones developed in the Eocene and Lower Oligocene of the Hanoi Depression [5,13,21], and high-quality lacustrine source rocks were also present in the second section of the Eocene Liushagang Formation in the Beibuwan Basin, which is located in the eastern part of the Yinggehai Basin, particularly in deep and semi-deep lakes with thicker layer, parallel, continuous medium and a low frequency and strong amplitude [25–27]. By analogy, it is inferred that deep lake and semi-deep lake facies are also present in the northern Yinggehai Basin, which is favorable for developing effective lacustrine source rocks (Figure 5).

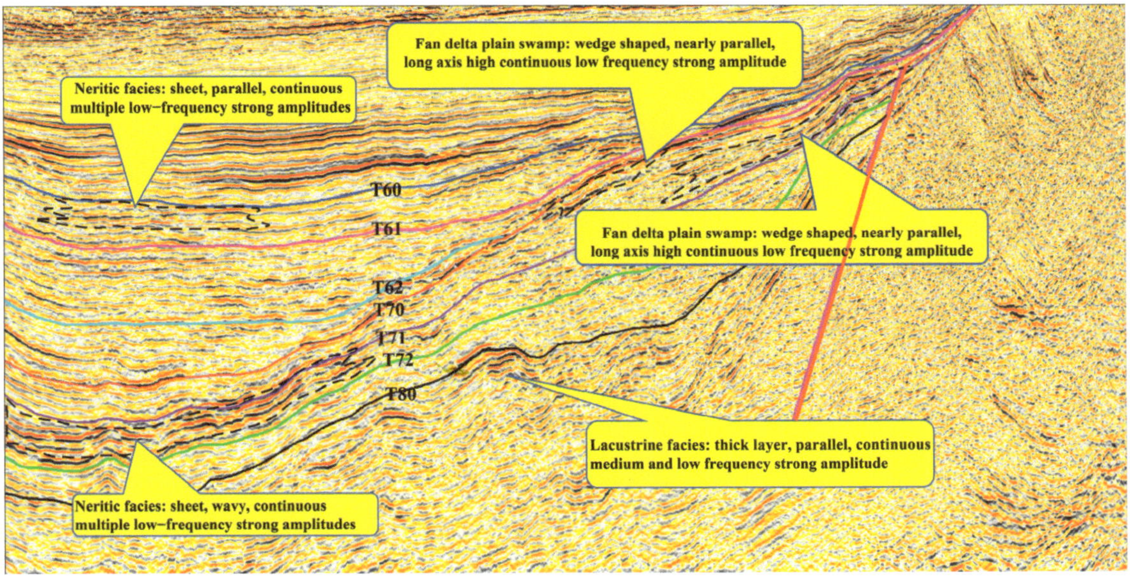

Figure 5. Typical seismic facies and sedimentary facies identification of seismic line 5600 in the northern Yinggehai Basin. T80–T60 represent the seismic (stratigraphic) interface (see Figure 2).

The Yacheng Formation in the northern Yinggehai Basin exhibited specific depositional characteristics. First, three to four fan deltas spanning a length of 35–45 km and a width of 30–35 km were used. Second, two to four fans were developed in the fan delta with a length of 25 km and a width of 20 km. Finally, the coal-measure source rocks are located primarily in coastal plains and the marshy facies of diversion plains, covering an average area of 5.0×10^3 km^2 (Figure 6). The Lingshui Formation also has certain sedimentary characteristics. First, three to four fan deltas, measuring 30–40 km in length and 25–35 km in width, have developed at the front edge of the braided river delta. Second, two to four fans were present in the fan delta, extending 25 km in length and 10–25 km in width. Finally, the coal-measure source rocks are mainly found in the coastal plains and the swamp facies of the diversion plains, covering an average area of 4.5×10^3 km^2 per stratum. Tectonic evolution influences sedimentary development. The Eocene was a fault-bound lake basin in the northern Yinggehai Basin, predominantly characterized by lake facies sedimentation, and a small-scale fan-delta system emerged in the eastern steep-slope area. The Oligocene primarily represented a marine–continental transitional facies with an expanding sea lev and extensive fan-delta and braided river delta formations in the eastern steep-slope zor and northwestern slope. These conditions enabled the deposition of coal-measure sourc rocks in the swampy Delta Huchou plains and the development of marine mudston in the south-central region, which was dominated by shallow coastal marine depositio (Figures 4–6). The distribution of sedimentary facies corresponds to paleogeomorpholog

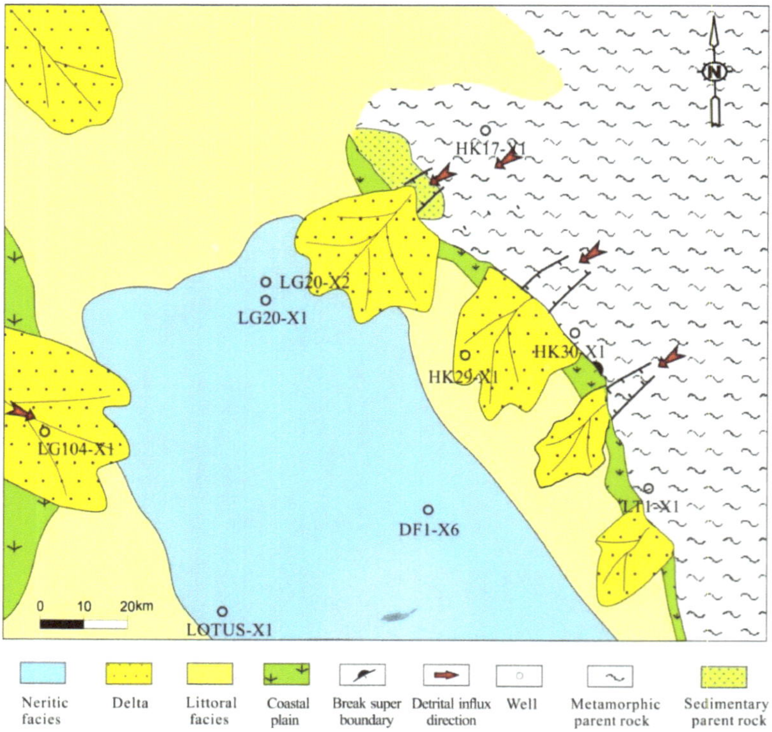

Figure 6. Sedimentary facies of source rocks of the Oligocene Yacheng Formation in the norther Yinggehai Basin, mainly including neritic facies and coastal plain, which are conducive to th sedimentation for mudstone and coal-measure source rocks.

4.3. Paleoenvironmental Conditions

4.3.1. Paleoclimate

The Sr/Cu, Mg/Ca, Al_2O_3/MgO, SiO_2/Al_2O_3, and FeO/MnO ratios are commonl used to distinguish paleoclimatic conditions [28,29]. An arid and hot climate accelerate the evaporation of water, resulting in an increase in water alkalinity. Elements such a Na, Mg, Ca, and Mn in the sedimentary medium precipitate at the bottom of the wate therefore, these elements are more likely to be enriched under arid climatic conditions. Sr i a typical dry element. A high Sr content can reflect arid climatic conditions, whereas a lo Sr content indicates a humid climate [29]. Furthermore, diagenesis may have a profoun impact on the mobility and quantity of these elements. However, in view of the complexit of this field and the large workload needed, we plan to conduct more detailed research i subsequent stages. Therefore, in the present study, diagenesis was not considered in th analysis of the paleo-sedimentary environment of the source rocks with elements.

Because Sr and Cu are very sensitive to changes in climatic conditions, Sr/Cu ratic are commonly used to reflect paleoclimatic conditions, with $w(Sr)/w(Cu) > 5$ indicating cold and dry climate and $w(Sr)/w(Cu) < 5$ reflecting a warm and humid climate [28–3C

Low Mg/Ca ratios indicate a dry and hot climate, whereas high ratios indicate a relatively humid climate [31–33]. Most of the values of Sr/Cu in the early stages of Els_3 and Eyc_3 were less than 5, indicating a warm and humid climate, whereas the values of Sr/Cu in the late stages of Els_2 and Els_1 were greater than 5, indicating a cold and arid environment (Figure 7). In addition, there was an obvious decreasing trend in the Mg/Ca values from the Oligocene Yacheng Formation to the Lingshui Formation (Figure 8), which also indicates that the paleoclimate transitioned from warm and humid to cold and arid from the early to later Oligocene, suggesting that the early Oligocene was more favorable for organic matter enrichment.

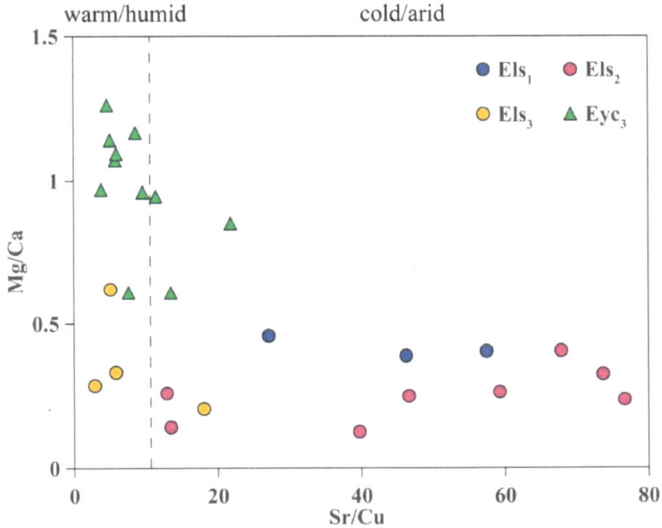

Figure 7. Diagram of the Sr/Cu and Mg/Ca ratios used to discriminate paleoclimatic conditions of the northern Yinggehai Basin and Yacheng District.

4.3.2. Paleoproductivity

The nutrient elements such as phosphorus (P), copper (Cu), iron (Fe), and zinc (Zn) are commonly used as indicators of paleoproductivity, with higher values reflecting greater paleoproductivity [34–36]. To eliminate the influence of terrigenous detrital deposits, P/Ti was used in this study to determine the productivity, and high P/Ti values generally indicate high productivity [37]. The values of P/Ti for the source rocks of Els1 and Els_3 ranged from 0.1 to 0.12, with an average of 0.11, and the values of P/Ti for the source rocks of Els_2 ranged from 0.1 to 0.31, with an average of 0.14. These values are all lower than the P/Ti ratio of the UCC (0.17; [37]), suggesting that the productivity of the late Oligocene stage was moderate (Figure 8). On the other hand, the P/Ti values of the source rocks of Eyc_3 range from 0.06 to 0.14, with an average of 0.08, much lower than the P/Ti of the UCC (0.17; [37]). This may be attributed to the fact that the source rocks of the Yacheng Formation are dominated by marine–continental transitional facies, which have more input of terrestrial organic matter.

In addition, the rate of elemental barium (Ba) accumulation was positively correlated with the organic carbon content and biological productivity. Therefore, Ba enrichment indicates high productivity [38]. There are various sources of sediment Ba, among which only biogenic Ba (Ba_{bio}) accurately reflects the magnitude of primary productivity [38,39], with the following expression:

$$Ba_{bio} = Ba_{total} - Ba_{alusilicate} = Ba_{sample} - Al_{sample} \times (Ba/Al)_{alusilicate} \quad (1)$$

where Al_{sample} and Ba_{sample} are the Al and Ba contents of the measured samples, respectively, and $(Ba/Al)_{alusilicate}$ is a correction factor used to exclude the effect of Ba in terrestrial aluminosilicates [39]. The Ba_{bio} thresholds for low, moderate, and high paleoproductivity were <200 ppm, 200–1000 ppm, and >1000 ppm, respectively [40]. The Ba_{bio} values of Els range from 737.96 ppm to 1057.3 ppm, of which 66.7% of samples have Ba_{bio} values less than 1000, indicating moderate primary productivity, which is consistent with the P/Ti index. The Ba_{bio} values of Eyc_3 samples were all greater than 1000 ppm, indicating high productivity in the early Oligocene (Figure 8). From the early to late Oligocene, the overall paleoproductivity showed a gradual declining trend.

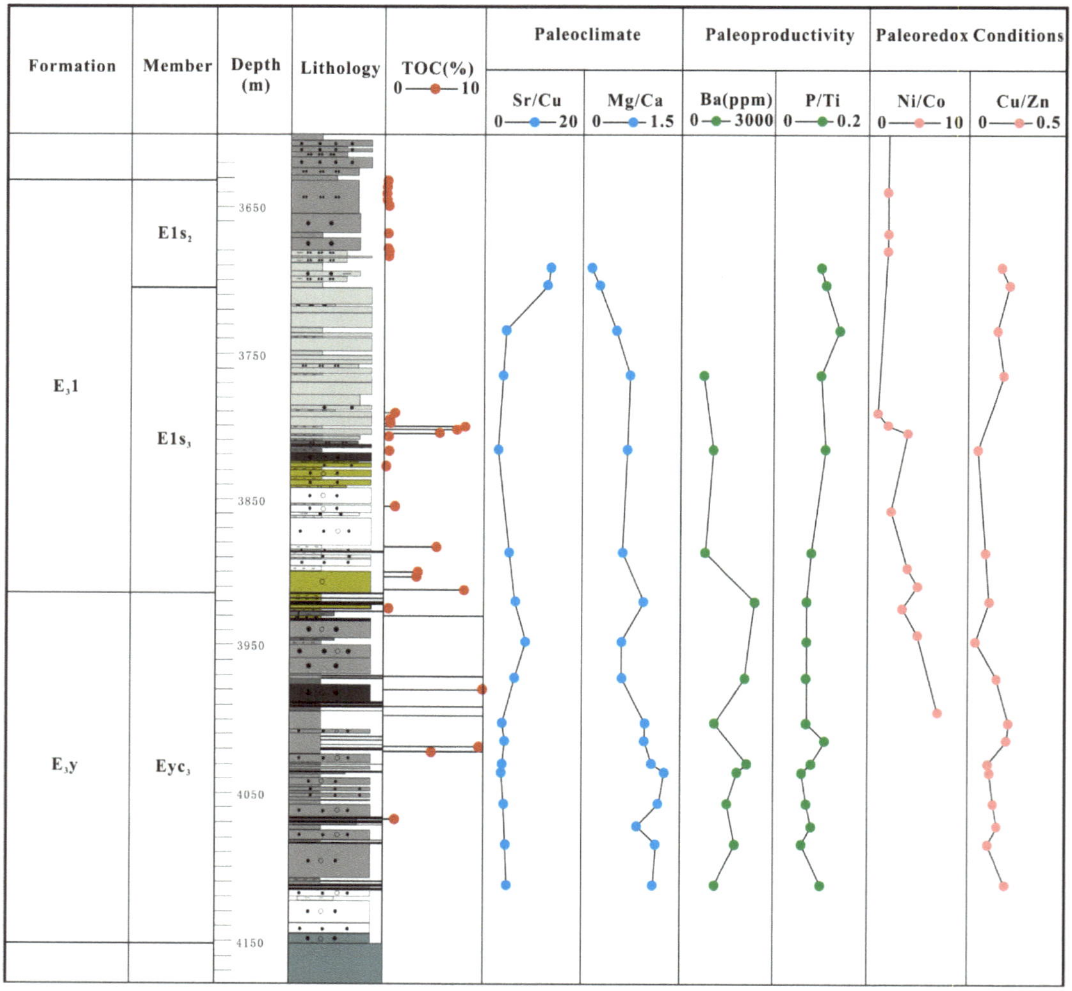

Figure 8. The paleodepositional environment reflected by the major and trace elements in the well YC13-X2 in the Yacheng District. Paleoclimate indicators (Sr/Cu and Mg/Ca), paleoproductivity-related indicators (Babio and P/Ti), and redox indicators (Ni/Co and Cu/Zn).

4.3.3. Redox Conditions

Because trace elements such as vanadium (V), uranium (U), and molybdenum (Mo) are affected by terrigenous components, there is uncertainty in characterizing the redox properties of water bodies only by their absolute content [33,41]. The redox-sensitive indicators can be used to effectively identify the redox status of water bodies. Commonly used elemental ratios include U/Th, Cu/Zn, Ni/Co, and V/(V + Ni); smaller ratios indicate a greater degree of oxidation, whereas larger ratios reflect a greater degree of reduction [42].

A Ni/Co ratio greater than 7 indicates a reducing environment, a Ni/Co ratio ranging from 5 to 7 indicates an anoxic reducing environment, and a Ni/Co ratio less than 5 indicates an oxic sedimentary environment. A Cu/Zn value less than 0.21 indicates a reducing environment, a Cu/Zn value ranging from 0.21 to 0.63 indicates a dysoxic sedimentary environment, and a Cu/Zn value greater than 0.63 indicates a dysoxic environment [42]. The Ni/Co values of the mudstone samples in Els_1 range from 2.68 to 2.8, with an average of 2.74. The Ni/Co values of Els_2 range from 1.8 to 3.04, with an average of 2.47. The Ni/Co values of Els_3 ranged from 1.46 to 1.97, with an average of 1.78. The Ni/Co values of Eyc_3 range from 1.22 to 2.26, with an average of 1.77 (Figure 9). The diagram clearly shows that Ni/Co exhibited a decreasing trend from Eyc_3 to Els_2. The value of Cu/Zn in Els_1 ranges from 0.09 to 0.15, with an average value of 0.11. The Ni/Co values of Els_2 range from 0.03 to 0.21, with an average of 0.08. The Ni/Co values of Els_3 range from 0.05 to 0.18, with an average of 0.12. The Ni/Co values of Els_3 range from 0.04 to 0.21, with an average of 0.13 (Figure 9). The sedimentary environment of E_3y is mainly characterized by suboxic conditions, while the water environment of E_3l becomes more and more oxidized (Figures 8 and 9). From the early to late Oligocene, the redox conditions changed from suboxic (semi-reducing) to an oxidizing environment.

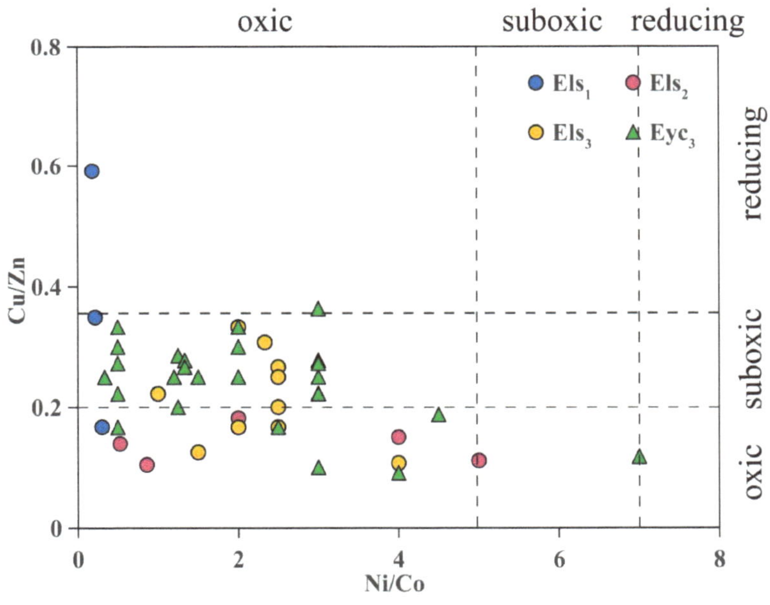

Figure 9. Diagram of Ni/Co versus Cu/Zn ratios of the source rock samples from E_3l and E_3y.

4.4. Sedimentary Model of Source Rock

Overall, the climate in the northern Yinggehai Basin and Yacheng District shifted from a warm and humid environment to a cold and arid environment during the early to late Oligocene. Paleoproductivity tended to decrease during this period, and the redox conditions changed from a semi-reducing to an oxidizing environment. Therefore, it was conducive to the enrichment of organic matter during the early Oligocene. Statistical analysis of the TOC contents of the Oligocene source rocks in the Yacheng District and Hanoi Depression proved this point. From the early to late Oligocene and from deep to shallow depths, the overall TOC content clearly decreased, and the quality of the sourc rocks also decreased [43].

In addition, the early Oligocene source rocks (E3y) were deposited mainly during th marine—continental transitional facies, which were controlled by early deposition an tectonics, and relatively few rocks formed on the northwestern, eastern, and western side of the study area. On the other hand, the upper part of E3l is mainly a shallow coasta marine environment; that is, the water body gradually deepened from the early to lat Oligocene. Based on the study of the paleo-sedimentary environments of E3y and E3l, sedimentary model of the effective source rocks of the Oligocene in the northern Yinggeha Basin was established (Figure 10). The Oligocene Els_3 and Eys_3 were deposited in a warr and humid climate, which was conducive to the growth and prosperity of plants an provided favorable conditions for coal seams. The cold and arid paleoclimate of Els_1 an Els_2 lowered the TOC and was not conducive to the enrichment of organic matter. Th depositional environment of E3y was dominated by oxic—reducing conditions, whic reduced the decomposition of organic matter and favored the preservation of organi matter. In the late Oligocene, the oxidation gradually became stronger, and the wate became deeper, resulting in a reduction in salinity, all of which were unfavorable to th deposition and preservation of organic matter.

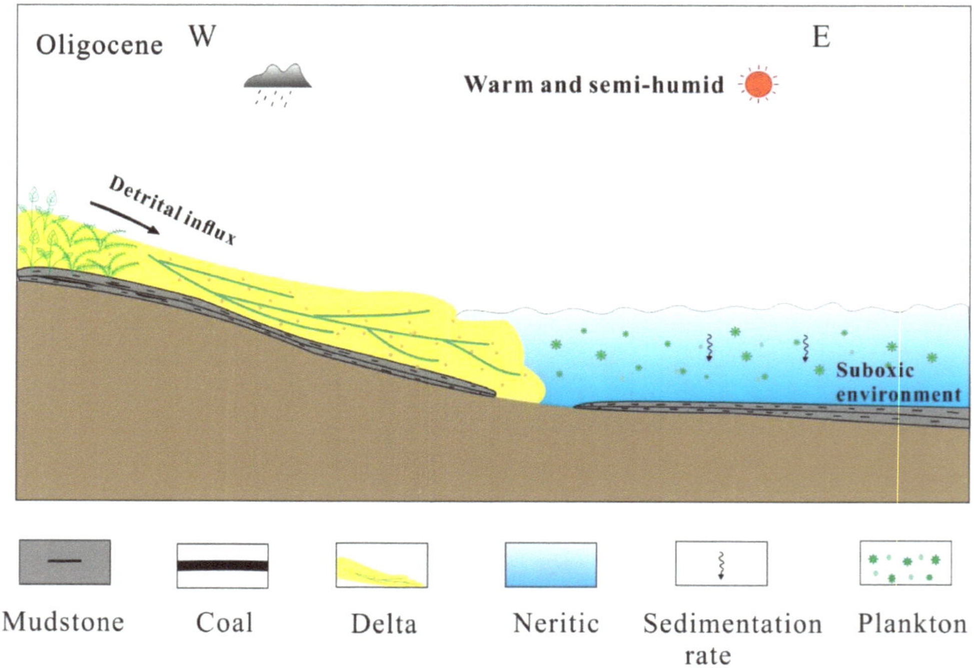

Figure 10. A sedimentary model of source rocks in the marine–continental transitional facies of th Oligocene in the northern Yinggehai Basin.

The Eocene source rocks were deposited in a faulted lake basin with a steep-slop zone in the eastern part and a gently sloping zone in the western part. From the p leodepositional environment of the northern Yinggehai Basin in the Oligocene, it ca be inferred that the Eocene had a warm and humid climate, a reducing environmen and high paleoproductivity, which is consistent with the Hanoi Depression and Beibuwa Basin [5,6,12,13,21,23,25–27]. The enrichment of mudstone organic matter is the result of th combined effects of the paleoclimate, redox conditions, and paleoproductivity (Figure 11

Owing to the influence of the warm and humid paleoclimate, the species and number of paleontological organisms reached unprecedented levels, and large quantities of lake benthic organisms and plankton subsequently appeared and flourished, which promoted an increase in the primary productivity of the paleolake and provided a rich source of material for the enrichment of organic matter. In addition, reducing water decreases the decomposition of organic matter and promotes the preservation of more organic matter. This suboxic–reducing depositional environment provided good preservation conditions for organic matter, which led to the development of lacustrine oil shale and mudstone with more organic matter.

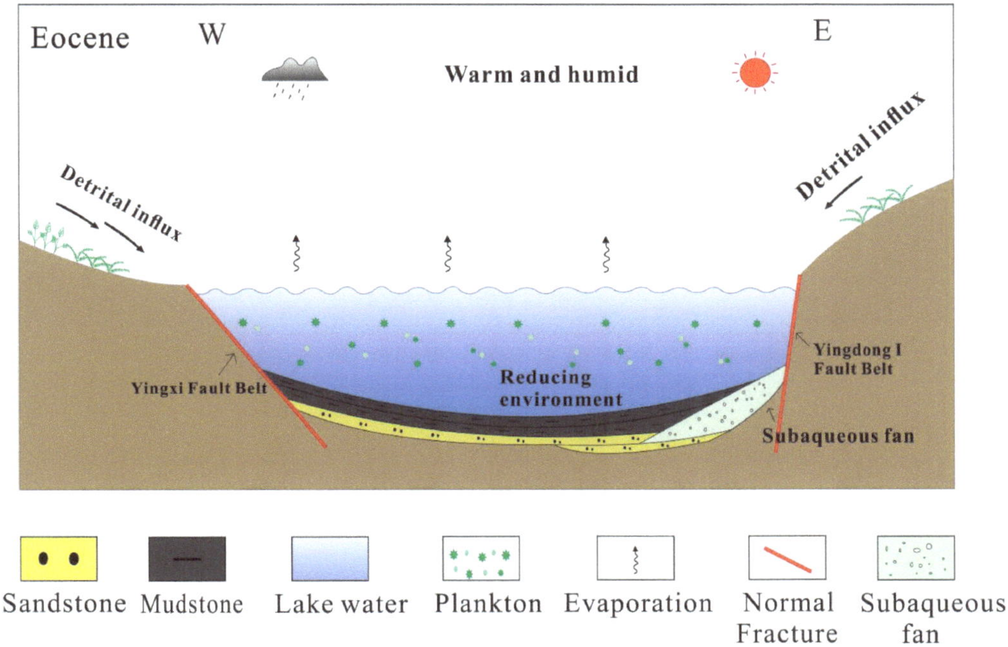

Figure 11. Sedimentary model of the Eocene lacustrine mudstone in the northern Yinggehai Basin.

5. Conclusions

(1) Through the comprehensive analysis of the geological characteristics and the analogy with the Hanoi Depression in Vietnam and Yacheng District in Qiongdongnan Basin, the northern Yinggehai Basin has favorable geological conditions for the development of effective source rocks in terms of tectonic, sedimentary, and paleoenvironmental conditions. Specifically, the tectonic movement not only shaped the sedimentary deep depression but also drove the dynamic migration of the center of sedimentary depression of source rocks. In addition, there is a diversified sedimentary source supply system in the northern Yinggehai Basin, including lacustrine facies, neritic facies, and delta plain swamp facies, which are conducive to the deposition of effective source rocks.

(2) From the early to late Oligocene, the paleoclimate transitioned from warm and humid to cold and arid, and the water depth gradually increased, evolving from a marine–continental transitional facies to a coastal shallow sea. As the water body deepened, the redox conditions shifted from a semi-reducing to an oxidized environment, and the paleoproductivity decreased. From the depositional environment of the Oligocene and the analogy with the Eocene in the Hanoi Depression, it is inferred that the Eocene lacustrine mudstone developed in a warm and humid paleoclimate under a reduced environment

and high paleoproductivity. The paleodepositional environment suggests that the Eocene and early–middle Oligocene periods promoted organic matter enrichment source rocks which means that there is great exploration potential for oil and gas. Some exploration activities are worth pursuing in the future in the northern Yinggehai Basin.

(3) Based on the comprehensive analysis of paleo-sedimentary environments and sedimentary conditions including sedimentary tectonic setting and sedimentary facies the sedimentary model of coal-measure source rocks of marine–continental transitional deltas and mudstone of shallow seas was established in the Oligocene. At the same time the sedimentary model of Eocene lacustrine mudstone deposited in the faulted lake basin has also been established. These studies make up for the serious deficiency of previous studies and play a role in encouraging the study of the deep Eocene and Oligocene source rocks in the northern Yinggehai Basin. As the deep wells are drilled and the seismic data quality is improved, the deep Eocene and Oligocene source rocks in the northern Yinggehai Basin will be studied systematically and meticulously and be recognized more accurately in the future.

Author Contributions: Literature search: J.P., G.H. and Y.C.; Figure: J.P., Z.C., H.L. and G.C.; Study design: J.P. and Z.H.; Data collection: G.H., Z.H., Y.C., W.W., H.L. and G.C.; Data analysis: J.P., Y.C. X.F. and Y.W.; Data interpretation: J.P.; Writing: J.P.; Supervision: G.H., Z.C., W.W., H.L., J.L. and G.C. Writing—Review & Editing: J.P., Z.H. and X.F. All authors have read and agreed to the published version of the manuscript.

Funding: This research was funded by the 14th Five-Year major national science and technology projects, Deep and Ultra-Deep Oil and Gas Accumulation Conditions and Mechanisms in Complex Edge Sea Basins, Yinggehai–Qiongdongnan Basins grant number KJGG2022-0404.

Institutional Review Board Statement: No prior ethical approval was necessary for the study.

Informed Consent Statement: No human subjects were included in the study. Thus, consent was not needed.

Data Availability Statement: The authors have included all relevant data and the sources of freely available data in the manuscript.

Acknowledgments: Four anonymous reviewers are thanked for their insightful comments and suggestions that help to significantly improve the clarity of this manuscript. Gest Editor Mianm Meng and Wenming Ji are also acknowledged for their patient editorial work.

Conflicts of Interest: Jianxiang Pei, Gaowei Hu, Zhihong Chen, Yabing Chen, and Haiyu Liu were employed by the Research Institute of Exploration and Development, Hainan Branch of CNOOC (China) Co., Ltd. The remaining authors declare that the research was conducted in the absence of any commercial or financial relationships that could be construed as a potential conflict of interest.

References

1. Li, J.; Yang, Z.; Wu, S.; Pan, S. Key issues and development direction of petroleum geology research on source rock strata in China. *Adv. Geo-Energy Res.* **2021**, *5*, 121–126. [CrossRef]
2. Huang, B.J. Geneticype Sand Migration Accumulation Dynamics of Natural Cases in the Ying Qiong Basin the South China Sea. Ph.D. Thesis, Guanazhou Institute of Ceochemistry of Chinese Academy of Sciences, Guangzhou, China, 2002.
3. Dong, W.L.; Huang, B.J. Identification mark sand source discrimination of the coal-type gas in Yinggehai Basins of South China Sea. *Nat. Gas Ind.* **2000**, *4*, 36–40.
4. Tong, C.X.; Ma, J.; Pei, J.X.; Xu, X.D.; Liu, P. Geochemical Characteristics and Origin Types of Mid-Deep Natural Gas in Dongfang Area, Yinggehai Basin, Yinggehai Basin. *Xinjiang Pet. Geol.* **2015**, *36*, 258–263.
5. Li, X.T.; Yu, S.Y.; He, J.X.; Ju, Y.W.; Zhang, W. Paleogene hydrocarbon sources and their petroleum geological significance in Yinggehai Basin, Northwestern South China Sea. *Mar. Geol. Front.* **2016**, *32*, 16–25.
6. Guo, X.X.; Xu, X.D.; Xong, X.F.; Hou, J.X.; Liu, H.Y. Gas accumulation characteristics and favorable exploration directions in mid-deep strata of the Yinggehai Basin. *Nat. Gas Geosci.* **2017**, *28*, 1864–1872.

. Wang, Y. Source Composition, Hydrocarbongeneration Potential of Source Rocks and Its Control on Hydrocarbon Accumulation in YinagionaBasin. Ph.D. Thesis, China University of Minina & Technology, Beijing, China, 2018.

. Huang, B.J.; Huang, H.T.; Li, L.; Wang, L.F. Characteristics of marine source rocks and effect of high temperature and overpressure to organic matter maturation in Yinggehai-Qiongdongnan Basins. *Mar. Orig. Pet. Geol.* **2010**, *15*, 11–18.

. Xu, X.D.; Yang, J.H.; Liu, H.; Guo, X.X.; Xiong, X.F. Formafion mechanism of oroanic maer in source rocks under marine environment in Yingehai Basin. *Eat. Sci.* **2019**, *44*, 264–2653.

0. Dahlstom, C.D.A. Balanced cross sections. *Can. J. Earth Sci.* **1969**, *6*, 743–757. [CrossRef]

1. Hu, T.; Wu, G.; Xu, Z.; Pang, X.; Liu, Y.; Yu, S. Potential resources of conventional, tight, and shale oil and gas from Paleogene Wenchang Formation source rocks in the Huizhou Depression. *Adv. Geo-Energy Res.* **2022**, *6*, 402–414. [CrossRef]

2. Petersen, H.I.; Tru, V.Q.; Nielsen, L.H.; Duc, N.A.; Nytoft, H.P. Source rock properties of lacustrine mudstones and coals (Oligogene Dong Ho formation), onshore Sond Hong basin, norththern Vietnam. *J. Pet. Geol.* **2005**, *28*, 19–38. [CrossRef]

3. Andersen, C.; Mathiesen, A.; Nielsen, L.; Tiem, P.V.; Petersen, H.; Dien, P. Distribution of source rocks and maturity modelling in the northern Cenozoic Song Hong basin (gulf of tonkin),Vietnam. *J. Pet. Geol.* **2005**, *28*, 167–184. [CrossRef]

4. Petersen, H.I.; Fyhn, M.; Nytoftt, H.P.; Karen, D.; Lars, H.N. Miocene coals in the Hanoi Trough, onshore northern Vietnam: Depositional environment, vegetation, maturity, and source rock quality. *Int. J. Coal Geol* **2022**, *253*, 103953. [CrossRef]

5. Li, Z.H.; Chen, H.C. Oragnic mechanism and source rocks for natural gas in Qiongdongnan basin, South China sea. *Pet. Geol. Exp.* **2011**, *33*, 639–644.

6. Fan, C.W. Tectonic deformation features ard petroleum geolocical significance in Yinggehai large strike-slip basin, South China Sea. *Pet. Explor. Dev.* **2018**, *45*, 190–199. [CrossRef]

7. He, J.X.; Shi, X.B.; Xia, B.; Liu, H.L.; Yan, B. The satus of the petroleum exploration in the northern South China sea and the resource potential in the deep-water areas. *Adtances Earth Sci.* **2007**, *22*, 261–270.

8. Li, H.; Wang, P.; Xu, H.; Shen, L.; Chen, T.; Lei, G. Comparison of overpressure system and hydrocarbon accumulation between Qaidam basin and Yinggehai basin. *J. Nat. Gas Team* **2012**, *23*, 736–741.

9. Zhang, G.; Deng, Y.; Wu, J.; Li, Y.; Zhao, Z.; Yang, H.; Miao, S.; Chen, Y.; He, Y.; Shen, H.; et al. Coal measure source-rock characteristics and gasexploration directions in Cenozoic superimposedfaulted depressions, offshore China. *China Offshore Oil Gas* **2013**, *25*, 15–25.

0. Wang, Q. The Deformation Mechanism of the Tectonic and the Study of Its Sand-Box Analog Models in Xujiaweizi Fault Depression. Master' s Thesis, Northeast Petroleum University, Daqing, China, 2018.

1. Hoang, B.H.; Fyhn, M.B.W.; Cuong, T.D.; Tuan, N.Q.; Schmidt, W.J.; Boldreel, L.O.; Anh, N.T.K.; Huyen, N.T.; Cuong, T.X. Paleogene structural development of the northern Song Hong Basin and adjacent areas: Implications for the role of extrusion tectonics in basin formation in the Gulf of Tonkin. *Tectonophysics* **2020**, *789*, 228522. [CrossRef]

2. Fyhn, M.B.W.; Thomsen, T.B.; Keulen, N.; Knudsen, C.; Rizzi, M.; Bojesen-Koefoed, J.; Olivarius, M.; Tri, T.V.; Phach, P.V.; Minh, N.Q.; et al. Detrital zircon ages and heavy mineral composition along the Gulf of Tonkin-Implication for sand provenance in the Yinggehai-Song Hong and Qiongdongnan basins. *Mar. Petrol. Geol.* **2019**, *10*, 162–179. [CrossRef]

3. Sahoo, T.R.; Funnell, R.H.; Brennan, S.W.; Sykes, R.; Thrasher, G.P.; Adam, L.; Lawrence, M.J.; Kellett, R.L.; Ma, X. Delineation of coaly source rock distribution and prediction of organic richness from integrated analysis of seismic and well data. *Mar. Petrol. Geol.* **2021**, *125*, 104873. [CrossRef]

4. Huang, B.J.; Tian, H.; Wilkins, R.W.T.; Xiao, X.M.; Li, L. Geochemical characteristics, palaeoenvironment and formation model of Eocene organic-rich shales in the Beibuwan Basin, South China Sea. *Mar. Petrol. Geol.* **2013**, *48*, 77–89. [CrossRef]

5. Li, Y.; Lan, L.; Wang, K.; Yang, Y. Differences in lacustrine source rocks of Liushagang Formation in the Beibuwan Basin. *Pet. Process. Sect.* **2019**, *40*, 1451.

6. Fyhn, M.B.; Hoang, B.H.; Anh, N.T.; Hovikoski, J.; Cuong, T.D.; Dung, B.V.; Olivarius, M.; Tuan, N.Q.; Toan, D.M.; Tung, N.T.; et al. Eocene-Oligocene syn-rift deposition in the northern Gulf of Tonkin, Vietnam. *Mar. Petrol. Geol.* **2020**, *111*, 390–413. [CrossRef]

7. Ren, H.Y.; Ge, Y.H. Late Peman mudstone geochemica features and sedimentary environment sianificance in Zhonchai mine area China Coal field. *Coal Geoloay China* **2016**, *28*, 7–10.

8. Zhang, T.F.; Sun, L.X.; Zhang, Y.; Cheng, Y.H.; Li, Y.F.; Ma, H.L.; Lu, C.; Yang, C.; Guo, G.W. Geochemical characteristics of the Jurassic Yan'an and Zhiluo Formations in the northern margin of Ordos Basin and their paleoenvironmental implications. *Acta Geol. Sin.* **2016**, *90*, 3454–3472.

9. Moradi, A.V.; Saei, A.; Akkaya, P. Geochemistry of the Miocene oil shale (Hanili Formation) in the Ankr-orum Basin, Central Turkey: Implications for Paleoclimate conditions, source-area weathering, provenance and tectonic setting. *Sediment. Geol.* **2016**, *341*, 289–303. [CrossRef]

0. Yang, Z.Y.; Shen, W.Z.; Zheng, L.D. Elements and isotopic geochemistry of Guadalupian-Lopingian boundary profile at the Penglaitan section of Laibin, Guangxi province, and its geological implications. *Acta Geol. Sin.* **2009**, *83*, 1–15.

31. Meng, H.; Ren, Y.; Zhong, D.K.; Gao, C.L.; Gao, Z.; Wang, D.; Jiang, Y.J.F.; Li, J.J. Geochemical characteristic and its paleoen vironmental implication of Cambrian Longwangmiao Formation in eastern Slichuan Basin, China. *Nat. Gas Geosci.* **2016**, 2 1299–1311.
32. Brumsack, H.J. The trace metal content of recent organic carbon-rich sediments:Implications for Cretaceous black shale formatio *Palaeogeogr. Palaeoclimatol. Palaeoecol.* **2006**, *232*, 344–361. [CrossRef]
33. Tribovillard, N.; Algeo, T.J.; Lyons, T.; Riboulleau, A. Trace metals as paleoredox and paleoproductivity proxies: An update. *Chem Geol.* **2006**, *232*, 12–32. [CrossRef]
34. Algeo, T.J.; Ingall, E. Sedimentary Corg: P ratios, paleocean ventilation, and Phanerozoic atmospheric pO2. *Palaeogeog Palaeoclimatol. Palaeoecol.* **2007**, *256*, 130–155. [CrossRef]
35. Ding, J.H.; Zhang, J.C.; Huo, Z.P.; Shen, B.J.; Shi, G.; Yang, Z.H.; Li, X.Q.; Li, C.X. Controlling factors and formation models c organic matter accumulation for the Upper Permian Dalong Formation black shale in the Lower Yangtze region, South Chin Constraints from geochemical evidence. *ACS Omega* **2021**, *6*, 3681–3692. [CrossRef] [PubMed]
36. Taylor, S.R.; McLennan, S.M. The Continental Crust: Its Composition and Evolution. Blackwell Scientific Publication, Oxford. *Geol.* **1985**, *94*, 57–72.
37. Dymond, J.; Suess, E.; Lyle, M. Barium in deep-sea sediment: A geochemical proxy for paleoproductivity. *Paleoceanography* **199** *7*, 163–181. [CrossRef]
38. Shen, J.; Schoepfer, S.D.; Feng, Q.; Zhou, L.; Yu, J.; Song, H.; Wei, H.; Algeo, T.J. Marine productivity changes during th end-Permian crisis and Early Triassic recovery. *Earth Sci. Rev.* **2015**, *149*, 136–162. [CrossRef]
39. Murray, R.W.; Leinen, M. Chemical transport to the seafloor of the equatorial Pacific Ocean across a latitudinal transect at 135 V Tracking sedimentary major, trace, and rare earth element fluxes at the Equator and the Intertropical Convergence Zone. *Geochem Cosmochim. Acta* **1993**, *57*, 4141–4163. [CrossRef]
40. Francois, R. A study on the regulation of the concentrations of some trace metals (Rb, Sr, Zn, Pb, Cu, V, Cr, Ni, Mn and Mo) i Saanich Inlet Sediments, British Columbia, Canada. *Mar. Geol.* **1988**, *83*, 285–308. [CrossRef]
41. Wang, S.F.; Dong, D.Z.; Wang, Y.M.; Li, X.J.; Huang, J.L. Geochemical Characteristics the Sedimentation Environment of th Gas-enriched Shale in the Silurian Longmaxi Formation in the Sichuan Basin. *Bull. Mineral. Petrol. Geochem.* **2015**, *34*, 1203–121
42. Nielsen, L.H.; Mathiesen, A.; Bidstrup, T.; Vejbæk, O.V.; Dien, P.T.; Tiem, P.V. Modelling of hydrocarbon generation in th Cenozoic Song Hong Basin, Vietnam: A highly prospective basin. *J. Asian Earth Sci.* **1999**, *17*, 269–294. [CrossRef]
43. Quan, V.T.; Giao, P.H. Geochemical evaluation of shale formations in the northern Song Hong basin, Vietnam. *J. Petrol. Expl. Pro Technol.* **2019**, *9*, 1839–1853. [CrossRef]

Journal of
Marine Science
and Engineering

MDPI

.rticle

\ Quick Method for Appraising Pore Connectivity and Ultimate mbibed Porosity in Shale Reservoirs

.iqing Hong [1,2], Mianmo Meng [1,2,*], Kong Deng [3], Jingwen Bao [1,2], Qianyou Wang [4] and Xingchen Liu [5]

1 Hubei Key Laboratory of Marine Geological Resources, China University of Geosciences, Wuhan 430074, China; hongziqing1@outlook.com (Z.H.); jingwenbb625@outlook.com (J.B.)
2 College of Marine Science and Technology, China University of Geosciences, Wuhan 430074, China
3 Key Laboratory of Tectonics and Petroleum Resources, China University of Geosciences, Wuhan 430074, China; dengkong@cug.edu.cn
4 Department of Earth, Ocean and Ecological Sciences, University of Liverpool, Liverpool L69 3GP, UK; tsianyou@126.com
5 Cementing Company, Sinopec Zhongyuan Petroleum Engineering Co., Ltd., Puyang 457100, China; liu502147582@163.com
* Correspondence: mengmianmo@outlook.com

Abstract: Pore connectivity and ultimate imbibed porosity are two important parameters used to assess the shale oil reservoir property, the proper appraising of which could facilitate the efficient flow of oil from the matrix and an improvement in recovery efficiency. In previous studies, the uncertainty in sample dimensions and the extra-long stable time during imbibition experiments exploring pore connectivity and ultimate imbibed porosity showed a lack of discussion, which influenced the accuracy and efficiency of the SI experiments. In this study, SI experiments with shale samples of different thicknesses are carried out to acquire the two parameters in a short period of time. As a result, the pore connectivity of sample D86-5 from the Qingshankou Formation (Fm) in the Songliao Basin fluctuates with the increase in thicknesses, with an average of 0.265. The water penetrates sample D86-5 of all thicknesses, so the ultimate imbibed porosity fluctuates around 3.7%, and the stable time increases with thicknesses. The pore connectivity of sample Y172 from the Shahejie Fm in the Bohaiwan Basin fluctuates around an average of 0.026, which is much smaller than that of D86-5. The ultimate imbibed porosity of Y172 decreases with thicknesses because the penetration depth is so small that the pores cannot be fully accessed, and the stable time increases before becoming stable with fluctuations. The method is examined using the samples from the Liushagang Fm in the Beibuwan Basin measuring around 400 μm: the ultimate imbibed porosity of BW1-1 and BW1-3 is 5.8% and 18.1%, respectively, the pore connectivity of BW1-1, BW1-2, and BW1-3 is 0.086, 0.117, and 0.142, respectively, and the results can be obtained within a day. In comparison, the average pore connectivity of the 400 μm samples from Qingshankou, Shahejie, and Liushagang Fms is 0.324, 0.033, and 0.097, respectively, and the average ultimate imbibed porosity of these Fms is 3.7%, 3.1%, and 12.0%, respectively. Based on the above results, a quick method for measuring the two parameters with thin samples by spontaneous imbibition is established, providing a fast solution for the evaluation of the sweet spot.

Keywords: spontaneous imbibition; pore connectivity; ultimate imbibed porosity; capillary force; shale oil/gas

.cademic Editor: Sergei Chernyi

.eceived: 26 December 2024
.evised: 16 January 2025
.ccepted: 17 January 2025
.ublished: 19 January 2025

.itation: Hong, Z.; Meng, M.; Deng, ; Bao, J.; Wang, Q.; Liu, X. A Quick .ethod for Appraising Pore .onnectivity and Ultimate Imbibed .orosity in Shale Reservoirs. *J. Mar. i. Eng.* **2025**, *13*, 174. https:// .i.org/10.3390/jmse13010174

1. Introduction

With the growing demand for fossil fuels and the diminishment of conventional rese voirs, unconventional reservoirs, such as shale oil reservoirs, have grabbed the attention c researchers worldwide [1]. Shale reservoirs hold abundant oil, but they have low porosit and permeability, which impacts the high-efficiency exploitation of their resources [2]. Th pore network is the storage space of shale oil; connected pores contribute to the fluid flo channel, whereas unconnected pores are called dead pores [3,4]. The pore connectivit of connected pores is a key factor influencing the flowing capacity of shale oil becaus good pore connectivity has the potential to realize high yield [5,6]. Some petrophysic methods have been used to investigate the pore connectivity of hydrocarbon reservoir such as mercury intrusion porosimetry (MIP), fluid tracing, wood's metal impregnatio spontaneous imbibition (SI), and so on [7–12]. Imbibition commonly occurs in the proces of exploiting shale oil/gas, referring to the phenomenon of the wetting phase penetratin into the non-wetting phase of shale rocks spontaneously, which is triggered by capillar force [13]. Comprehensively, in the research of pore connectivity, SI is readily implemente compared with other methods, where the slopes of imbibition curves reflect pore conne tivity; however, it requires a considerable amount of time to complete. In some studie imbibition could take more than 600 h before reaching a stable status, and it is hard t determine whether the imbibition fluid has entered all connected pores [14,15]. Therefor it is critical to raise the efficiency of SI tests on shale rocks by reducing the imbibition tim

Apart from pore connectivity, the ultimate imbibed porosity of shale rocks durin imbibition is also a crucial parameter when appraising the potential of a shale reservoi which is reflected by the water volume that can penetrate the rock by the time when th imbibition curve reaches a stable status [16,17]. During the process of imbibition, th wettability of shale rocks plays a significant role in the affinitive ability of the rocks t a certain kind of fluid [18]. When oil-wet pores are predominant in the rock, the roc demonstrates a preference for oil instead of water, and vice versa, thereby influencing th volume of the water entering the rock. Therefore, ultimate imbibed porosity is also affecte by the wettability and the pore structure [19,20]. When the ultimate imbibed porosit is large in the imbibition process, the efficiency of enhanced oil recovery (EOR) can b improved, which is a set of techniques employed to increase the amount of crude oil that ca be extracted from an oil field after the primary and secondary recovery methods have bee exhausted [21]. The main methods of EOR can be summarized into three techniques, i.e solvent, chemical, and thermal, which are mainly involved with the injection of differer substances that are not naturally found in the reservoir, where SI is highly engaged in an contributes to the displacement of oil by water [22–24]. By defining the ultimate imbibe porosity of shale rock, the recovery efficiency of the exploitation can be estimated.

To ascertain pore connectivity and ultimate imbibed porosity, a number of experimen were conducted in the past. Laboratory SI tests were widely conducted as a crucial metho combined with other laboratory methods for the characterization of pore connectivit and ultimate imbibed porosity, for example, contact angle analysis, fluid tracing, nucle magnetic resonance (NMR), mercury intrusion porosimetry (MIP), and so on [25–29 Some studies were conducted extensively on different aspects of the pore structure, por connectivity, and ultimate imbibed porosity of shale samples from a specific source, fc instance, the influence of changing the imbibition liquid on the matrix, the flowing featur of the tracer fluid in the shale matrix, or the comprehensive analysis of the shale in particular location [27,30–33], but the dimensions of the samples selected in the researc were out of focuses. In addition, some researchers observed that the distinct depths at whic tracer fluids permeated into the rock from the solid–liquid boundary during SI reflecte certain differences in the imbibition results [34], which also stressed the importance c defining the thickness of the tested shale rock. Apart from that, by constructing a three-dimensional (3D) pore structure with stacking scanning electron microscopy (SEM) images of an Eagle Ford sample, Davudov et al. discovered that the amount of connected pores decreases with an increase in the thickness of digital samples by modeling, showing the effect of thickness [35]. By setting samples with different thicknesses and origins, the identification of the penetration depth of certain shale rocks during SI could not only improve the scientific research on the shale structure but also provide a possibility for the productive practical exploitation of shale oil/gas [36].

The results of SI tests on ultimate imbibed porosity and pore connectivity are also influenced by mineral components and the amount of organic matter. Mineral components and TOC were also found to remarkably influence the shale pore structure. Clay minerals tend to form more complex pores compared with brittle minerals like quartz, which create larger pores [15,37,38]. When SI happens, the clay in shale swells, and cracks are created, making it more efficient to exploit shale oil/gas [14]. TOC is essential in describing the organic matter in the pore space, which helps in shaping the pores in the early stage of layer formation [39].

Based on the previous studies, this research was designed to explore the following problems: the extremely long time of imbibition tests, the difficulty in the measurement of pore connectivity and ultimate imbibed porosity, and the unknown influence of parameter thickness during SI.

By conducting SI tests with samples of different thicknesses, this research aims to explore how the penetrating process during SI is related to thickness and to further analyze the pore connectivity, ultimate imbibed porosity, and their distributions according to different formations. Samples from the Qingshankou, Shahejie, and Liushagang Fms were chosen and evenly sliced into cylinders in certain thicknesses and approximately the same diameter, covering onshore and offshore shale rocks. In this study, a quick method for evaluating the pore connectivity and ultimate imbibed porosity of the shale matrix using small-thickness samples was established, introducing thickness as a critical parameter and emphasizing the significance of length control in future assessments of pore structure. Utilizing this method, the pore connectivity and ultimate imbibed porosity of all samples were observed to vary with thicknesses.

2. Sample Background, Experimental Methods, and Theory Preparation

2.1. Sample Background and Experimental Methods

Shale samples were chosen from Qingshankou Fm in Songliao Basin, Shahejie Fm in Bohaiwan Basin, and Liushagang Fm in Beibuwan Basin. The Liushagang Fm samples were collected from offshore drilling boreholes, and the samples from other formations were onshore cores. Some experimental methods were conducted in this research, including FEI Helios Nanolab 650 dual beam FIB and SEM, Rigaku TTR III diffractometer, ELTRACsi, and Mettler Toledo ME204, from which the SEM pictures and the basic properties of the samples, as well as the SI test results, were obtained accordingly.

The ion-milled mudstone blocks were tested by Zeiss GeminiSEM450 and FEI Helios Nanolab 650 dual beam FIB-SEM to detect the mineral and organic matter particles and pore spaces with a working distance of 10 mm, following the methods of SEM equipment and sample preparation in prior studies [40,41]. The mineral properties were tested by the Rigaku TTR III diffractometer, while the TOC content was examined by ELTRACsi, presenting the mineral content and the content of organic matter in the examined samples. In the meantime, the samples of each origin were sliced into cylinders, whose thicknesses increased at an interval of approximately 200 μm from about 400 to 2000 μm, which varied with the different conditions of the samples, as can be seen in Figure 1. The minimum thickness of 400 μm represented the finest increment achievable by the cutting machin and was therefore selected as the starting point for the increasing thicknesses. Given tha liquid penetrates from both sides of the samples, it was theoretically expected that wit each 200 μm increase in thickness, there would be a corresponding 100 μm increase in th depth of water penetration. A final thickness of 2000 μm was chosen, but more samples c larger thicknesses are expected in future studies in case the liquid cannot penetrate. Th thicknesses and volumes of the samples are listed in Table 1. The cylindrical samples whos thicknesses were >1400 μm had their flank covered with epoxy adhesive in order to prever the imbibition liquid from injecting via the flank side. The imbibition liquid in this researc was deionized water, reducing the influence of other impurity substances.

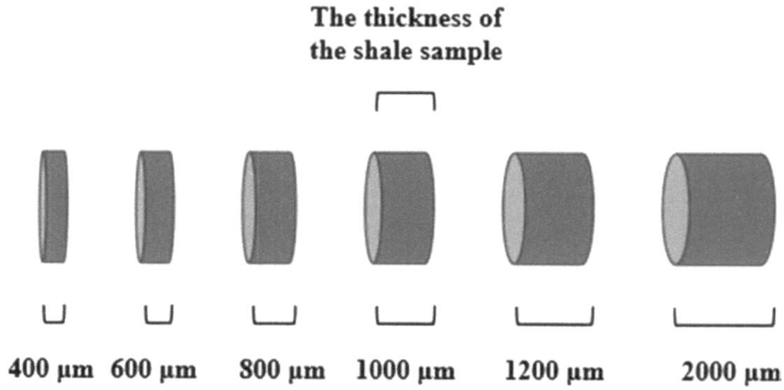

Figure 1. The target thickness of shale samples, from 400 μm to 2000 μm.

Table 1. The thicknesses and volumes of the samples.

	Thickness, mm	Buoyant Sample Volume, cm^3
D86-5	0.45	0.2022
	0.64	0.2965
	0.75	0.3605
	1.1	0.5251
	1.27	0.6353
	1.94	0.9492
Y172	0.46	0.2067
	0.68	0.313
	0.84	0.4171
	1.07	0.5226
	1.14	0.5436
BW1-1	0.44	0.0621
BW1-2	0.40	0.0591
BW1-3	0.41	0.042

The cylindrical samples were placed and examined in the equipment below, whic was improved based on a former research study [16], as shown in Figure 2a, consistin of a data-recording computer and a thermostatic testing part. A balance (Mettler Toled ME204) is attached to a small beaker and connected to the computer so that the data of th content in the small beaker can be recorded every minute (see Figure 2a). The samples ar put into the small beaker that hangs in a bigger beaker filled with imbibition liquid using fishing line. The equipment is connected to the balance so that the recorded data represer the weight of the imbibed water. This method guarantees that even if any fragments of th samples fall, the small beaker makes sure that the data are not affected. Dissimilar to th sample placement of the previous research, which has no requirements for the samples' positions, the samples are placed perpendicularly to the bottom of the small beaker, with supports on both undersides of the cylindrical samples, ensuring that the gaseous phase products generated during the imbibition can be removed in time.

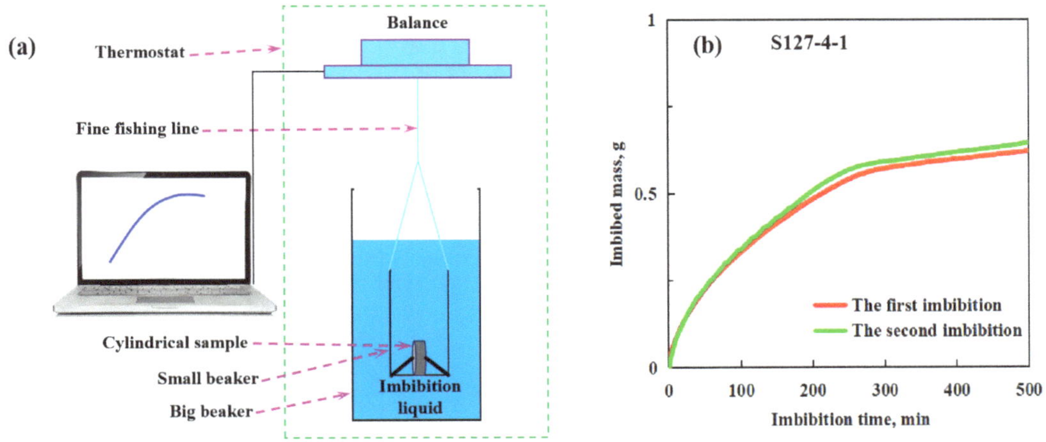

Figure 2. The equipment and repeated group of spontaneous imbibition experiments: (**a**) the equipment; (**b**) the repeated group of sandstone sample S127-4-1.

The imbibition method was tested with repeated groups of a sandstone sample, and the change in the mass of sample S127-4-1 during two SI experiments is exhibited in Figure 2b. It can be seen that the imbibition curves between two SI experiments on the same sample are highly similar, which proves that the method in this research can be repeatable. In comparison, the sliced shale samples tested in repeated groups have more fluctuations than the sandstone samples. This is because the heterogeneity in shales and the expansion of the clay content influence the imbibition curves, making the SI experiments of thin shale rocks less repeatable [42].

2.2. Theory Preparation

Washburn [43] found that the pores of porous materials can be seen as cylindrical capillaries, and at the end of time t (in s), the volume of the liquid (V, cm^3) that enters into n cylindrical capillaries with radii $r_1, r_2, \ldots, r_n$ can be presented as:

$$V = \pi \sum r^2 l = \frac{\pi}{2\eta^{1/2}} t^{1/2} \sum \left(P_E + \frac{2\gamma}{r} \right)^{1/2} r^3 \quad (1)$$

where l is the length of the capillary tube, cm; η is the viscosity of the intruding liquid, mP; P_E is the total external pressure the liquid receives from the back, MPa; and γ is the surface tension of the liquid, Newton per cm. If the pressure of the liquid is constant and small relative to γ/η, this equation can be simplified as:

$$V = k' \left(\frac{\gamma}{\eta} \right)^{1/2} t^{1/2} \quad (2)$$

where k' is a constant that is relevant to the total pore diameter, and the latter also remains unchanged in the same sample during imbibition, so it is not involved with the characteristics of the liquid. In this way, the volume that penetrates a porous body within SI is proportional to the imbibition time t, s.

Furthermore, the Handy equation [44] is widely used to describe the imbibitio process. The Handy equation assumes that the liquid imbibes as pistons in a vertic direction, neglecting the gas pressure gradient in the front of the imbibed liquid. Th equation is presented as follows:

$$\nu_w = \frac{k_w}{\mu_w}\left(\frac{P_c}{x} - \Delta\rho g\right) \quad (3)$$

where ν_w is flow rate, $cm^3/cm^2/sec$; k_w is the effective water permeability, D; μ_w is th water viscosity, centipoises; P_c is the capillary pressure, MPa; $\Delta\rho$ is the density differenc for water and air; g is the acceleration due to gravity; x is the position of front, cm; and P_c i a constant. When there is piston displacement,

$$\nu_w = \phi S_w \frac{\partial x}{\partial t} \quad (4)$$

where ϕ is the fractional porosity; S_w is the fractional water content of pore spaces; and t i the imbibition time, s. Combining Equation (3) and Equation (4), the result would be:

$$x + \frac{P_c}{\Delta\rho g} ln\left(1 - \frac{\Delta\rho g x}{P_c}\right) = -\frac{k_w \Delta\rho g}{\phi S_w \mu_w} \quad (5)$$

When the gravity force is dramatically smaller than the capillary force and $x = \frac{Q_w}{\phi A S_w}$ the equation would become:

$$Q_w^2 = \frac{2P_c \phi k_w A_c^2 S_w}{\mu_w} t \quad (6)$$

where Q_w is the total imbibed liquid, cm^3; P_c is the capillary pressure, MPa; ϕ is fractiona porosity, %; k_w is effective water permeability, D; A_c is the cross-sectional area of the samp cm^2; S_w is the fractional water content of the pore spaces, %; μ_w is the water viscosity, m and t is the imbibition time, s. When the buoyancy effect of the sample volume needs to b removed to obtain the imbibed porosity, this equation can become:

$$Q_w / Q_d = a t^{1/2} \quad (7)$$

where Q_d is the volume of the dry sample, cm^3, and a is the quality of the sample that i not influenced by other conditions but by the nature of the shale. In this equation, Q_w/Q represents the imbibed porosity per unit volume of the samples, %, and a represents the po connectivity of the shale sample. This provided simplified and concrete evidence to investigat spontaneous imbibition, leaving the imbibed porosity and time in a linear relationship.

The slope of the curve showing the changes in imbibed porosity with time (Equation (7 reflects the pore connection of the samples [27,45]. In this research, pore connectivity i normalized to mitigate the impact of thickness, which is a critical parameter:

$$a_n = a\frac{T}{2} \quad (8)$$

where a_n is the pore connectivity, a is the slope in the linear fitting of the imbibition curv which is called the initial imbibition rate, and T is the thicknesses of the samples, cm. Th slope was normalized by multiplying half of the length of the corresponding sample sinc the injection of imbibition liquid occurs from both undersides of the cylinder sample.

3. Results and Discussion

3.1. Basic Properties and Material Components

The measured basic properties and material components of the samples are presente in Table 2, where the percentages of the figures signify the mass percentages of the pro erties versus the whole samples. The samples are from different origins, and their depth varies from 1957.6 m to 3561.9 m. The lithofacies of samples include mixed and clayey shales. Sample D86-5 from the Qingshankou Fm is clayey shale, and sample Y172 from the Shahejie Fm is mixed shale. The samples from the Liushagang Fm have both clayey and mixed shales, such as BW1-1 and BW1-2 belong to mixed shales, and BW1-3 belongs to clayey shale. Quartz is widely seen in the mineral content of all selected samples; sample BW1-2 holds the highest quartz content of 37.1%, and sample BW1-1 holds the lowest content of 18.3%, showing a great divergence because of the heterogeneity in the samples. The feldspar content in all samples is at a lower level compared with the quartz content, which is the highest in sample D86-5 at 13.7%. Sample Y172 and samples BW1-1, BW1-2, and BW1-3 all have lower contents of feldspar (1.9%, 1.3%, 1.7%, and 0.9%, respectively). The calcite content of sample BW1-1 is the highest at 32.6%, which is dramatically distinct from the other two groups of samples that are also from the Liushagang Fm (BW1-2 and BW 1-3), whose content of calcite is much smaller, at 6.8% and 0.2%, respectively. Sample Y172 from the Shahejie Fm contains 27.4% calcite content, which is also relatively high. Samples D86-5 and BW1-3 contain no ankerite, while sample Y172 shows the highest ankerite content in all samples (20.6%). The siderite and pyrite content of the samples are in the range of below 10%, and the divergence between the different samples is not as significant as the other minerals. The clay content of most samples has a relatively large percentage. Sample D86-5 from the Qingshankou Fm and samples BW1-1, BW1-2, and BW1-3 from the Liushagang Fm are all abundant in clay minerals, at 46.8%, and an average of 41.23%, respectively. Sample Y172 from the Shahejie Fm holds a lower clay content of 24.7%. The TOC content of all the samples is in the range of 2.2% to 5.8%. The samples from the Liushagang Fm have a relatively high level of TOC content (2.7%, 5.8%, and 3.9%, respectively) but vary considerably despite the same origin, while the samples from Qingshankou and Shahejie Fm show a lower level of TOC (2.2% and 3.0%, respectively). In general, the quartz, calcite, and clay content of the samples account for a higher percentage of all minerals, and the TOC content is distinct. The samples exhibit a remarkable nature of heterogeneity in the properties, and the SI test is notably influenced by the contents.

Table 2. The basic properties of the samples.

Samples	Quartz, %	Feldspar, %	Calcite, %	Ankerite, %	Siderite, %	Pyrite, %	Clay, %	TOC, %	Formation	Depth, m
D86-5	22.2	13.7	1.1	/	6.5	2.9	46.8	2.2	Qingshankou	1957.6
Y172	21.5	1.9	27.4	20.6	/	3.9	24.7	3.0	Shahejie	3530.0
BW1-1	18.3	1.3	32.6	10.4	3.4	5.6	28.4	2.7	Liushagang	3556.5
BW1-2	37.1	1.7	6.8	5.1	0.8	9.7	38.8	5.8	Liushagang	3557.8
BW1-3	32.5	0.9	0.2	/	3.1	6.8	56.5	3.9	Liushagang	3561.9

3.2. SEM Analysis

SEM images of the four formations are displayed in Figure 3, where the representative area of the samples in each origin is chosen to show the pore structure of the samples. In Figure 3a, the overall view of the examined sample from Qingshankou Fm is shown, where the existence of narrow long cracks extending for more than 30 μm and interparticle pores along with pyrite minerals is revealed. Figure 3c is the enlargement of the yellow square part of Figure 3b, both showing a substantial amount of organic pores, where the porous space difference in the two kinds of pores can be obviously seen, and the organic matter widely distributes among the minerals. Figure 3d is an image of the pore structure of the sample from the Shahejie Fm. Figure 3e,f are the enlargements of its yellow square parts, where large areas of organic matter along with narrow cracks and inorganic pores can be seen, and the organic matter represents a large portion of the shale rock. Figure 3g shows the prevailing existence of pyrite minerals and organic matter in the examined sample fror the Liushagang Fm. Figure 3h,i are the enlargements of Figure 3g, showing that the miner aggregated in the yellow square of Figure 3g is a mixture of pyrite and a mass of organi matter, with abundant micro-cracks alongside, and the mineral mixture contains plentif pores with organic matter. Similar to what the basic properties of the samples exposed, th SEM images in Figure 3 demonstrate the pronounced heterogeneity characteristics of shal which significantly impact pore connectivity and ultimate imbibed porosity [46,47].

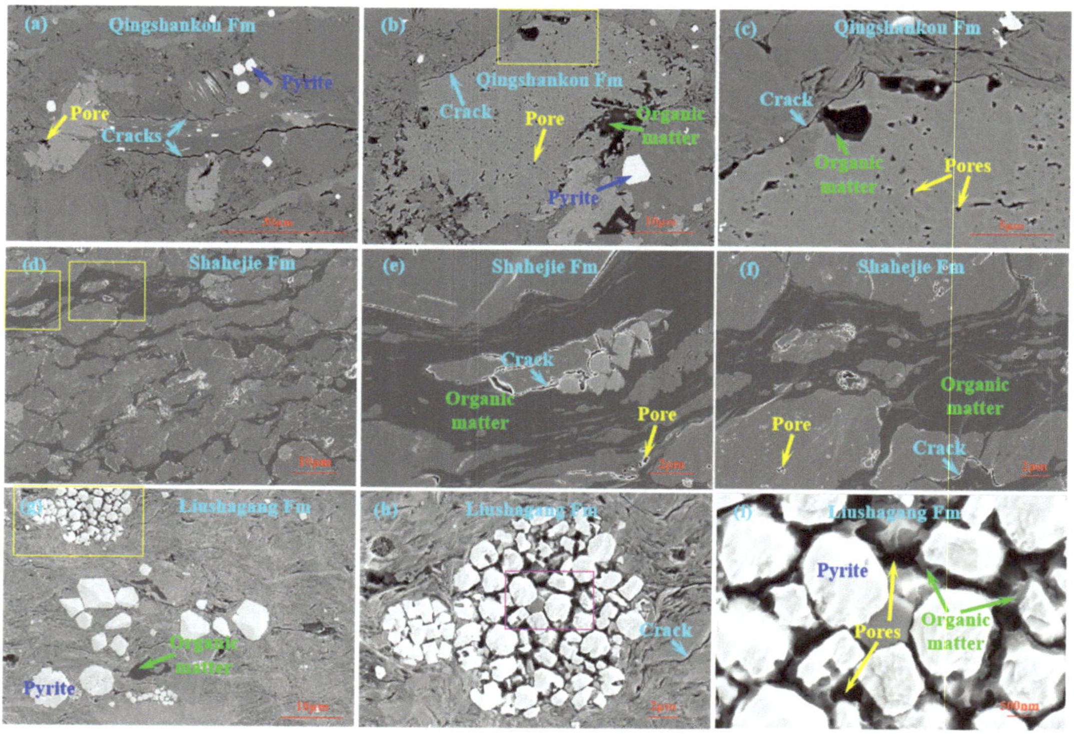

Figure 3. SEM pictures: (**a**–**c**) Qingshankou Fm, (**d**–**f**) Shahejie Fm, and (**g**–**i**) Liushagang Fm. Th square areas in yellow and purple are enlarged, as illustrated in the figures next to them.

3.3. Changes in the Imbibition Curves with Sample Thickness

After the imbibition tests, the imbibition data of various samples were collected an normalized to acquire the ultimate imbibed porosity. The peak of an imbibition curve wa considered to reflect the ultimate imbibed porosity. In some cases, when the rate of wate penetrating into the rocks becomes zero and the imbibition curve reaches a stable status wit time [48], the time that the samples take is recorded as the stable time. The full time of th imbibition is processed into square root, and the linear-fitted slope of the initial imbibitio curve is called the initial imbibition rate; after normalizing the initial imbibition rate, it ca reflect the pore connectivity [34]. In this research, samples with different thicknesses wer used. To compare the pore connectivity from samples of different thicknesses, the por connectivity was normalized by considering the thicknesses of rocks.

Figure 4 illustrates the imbibition curves of sample D86-5 with different thicknesse with the unprocessed time in minutes, with different colors representing the results c different thicknesses. All the curves in Figure 4 show that the imbibed porosity increase rapidly with time, and a stable status is reached in all groups. The stable time of differer samples can be affected by many factors, and in this case, it changes with thicknesses [15,48

In Figure 4a, the stable time of the sample with a thickness of 450 μm is 46 min, which is the smallest in all samples. In Figure 4b–f, the stable time increases with the addition of thicknesses 640 μm, 750 μm, 1100 μm, 1270 μm, and 1940 μm, corresponding to the stable time of 72 min, 102 min, 350 min, 397 min, and 1361 min, respectively.

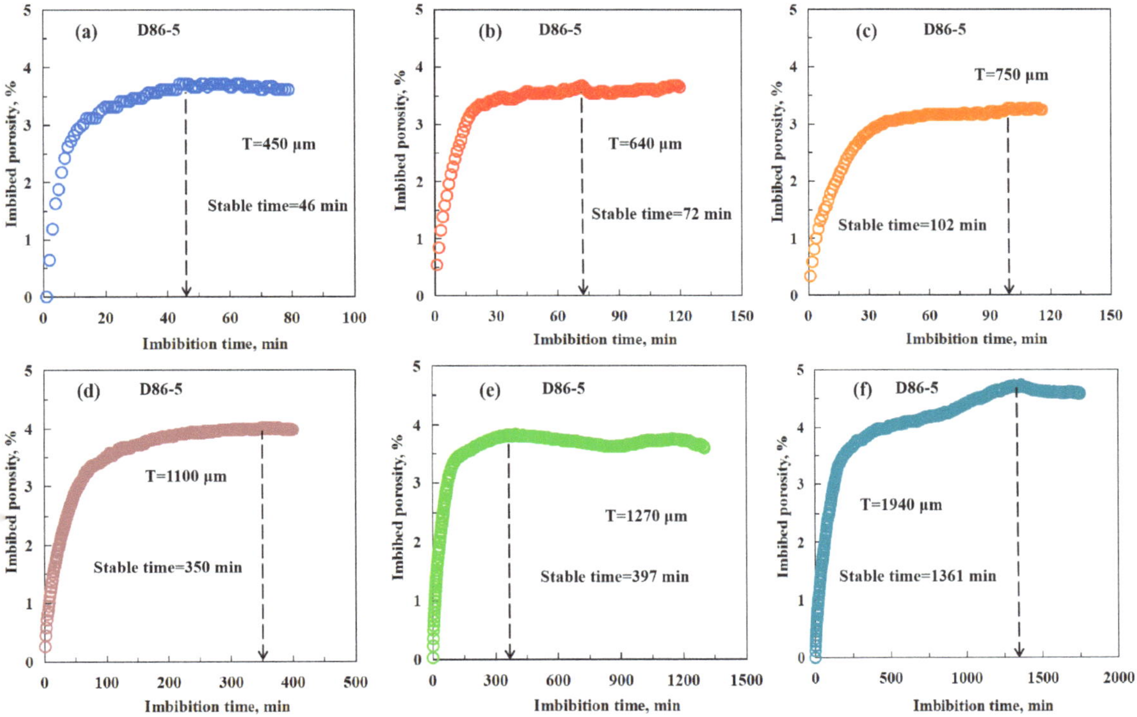

Figure 4. The unprocessed spontaneous imbibition curves of sample D86-5 with different thicknesses: (**a**) 450 μm; (**b**) 640 μm; (**c**) 750 μm; (**d**) 1100 μm; (**e**) 1270 μm; and (**f**) 1940 μm.

The spontaneous imbibition curves of sample D86-5 with square-rooted time are shown in Figure 5, which includes different thicknesses of the sample in different colors. The imbibed porosity increases with imbibition time gradually in all the samples. In Figure 5a, the ultimate imbibed porosity and the initial imbibition rate are 3.8% and 1.441, respectively. In Figure 5b, the ultimate imbibed porosity and the initial imbibition rate are 3.7% and 0.849, respectively. In Figure 5c, the ultimate imbibed porosity and the initial imbibition rate are 3.3% and 0.589, respectively. In Figure 5d, the ultimate imbibed porosity and the initial imbibition rate are 4.0% and 0.441, respectively. In Figure 5e, the ultimate imbibed porosity and the initial imbibition rate are 3.8% and 0.382, respectively. In Figure 5f, the initial imbibition rate is 0.294 and the ultimate imbibed porosity is 4.7%

Figure 6 shows the imbibition curves of sample Y172 with different thicknesses with the unprocessed time in minutes. The characteristics of these curves are different from the previous curves (sample D86-5). In all the groups of sample Y172, the whole periods of imbibition fluctuate. In Figure 6a,c, the samples with thicknesses of 460 μm and 840 μm reach a stable status and continue to remain stable, with stable times of 730 min and 1297 min. In contrast, in Figure 6b,d,e, the samples with thicknesses of 680 μm, 1070 μm, and 1140 μm reach a stable status but decline immediately, with stable times of 1429 min, 1353 min, and 1246 min, respectively.

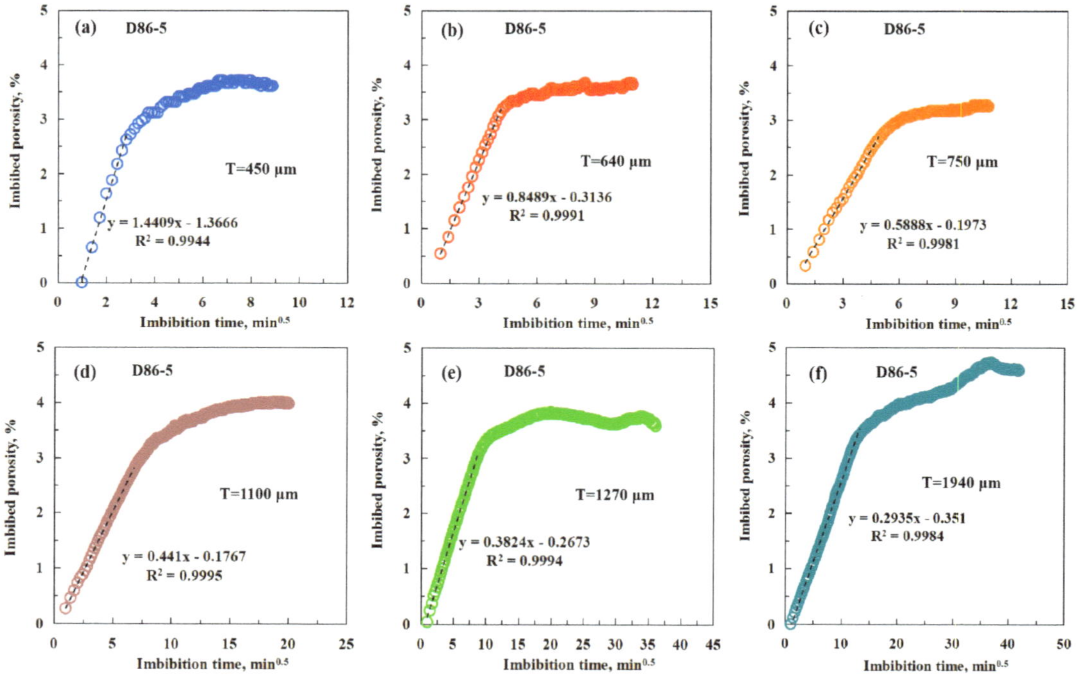

Figure 5. The spontaneous imbibition curves of sample D86-5 with different thicknesses: (**a**) 450 μm; (**b**) 640 μm; (**c**) 750 μm; (**d**) 1100 μm; (**e**) 1270 μm; and (**f**) 1940 μm.

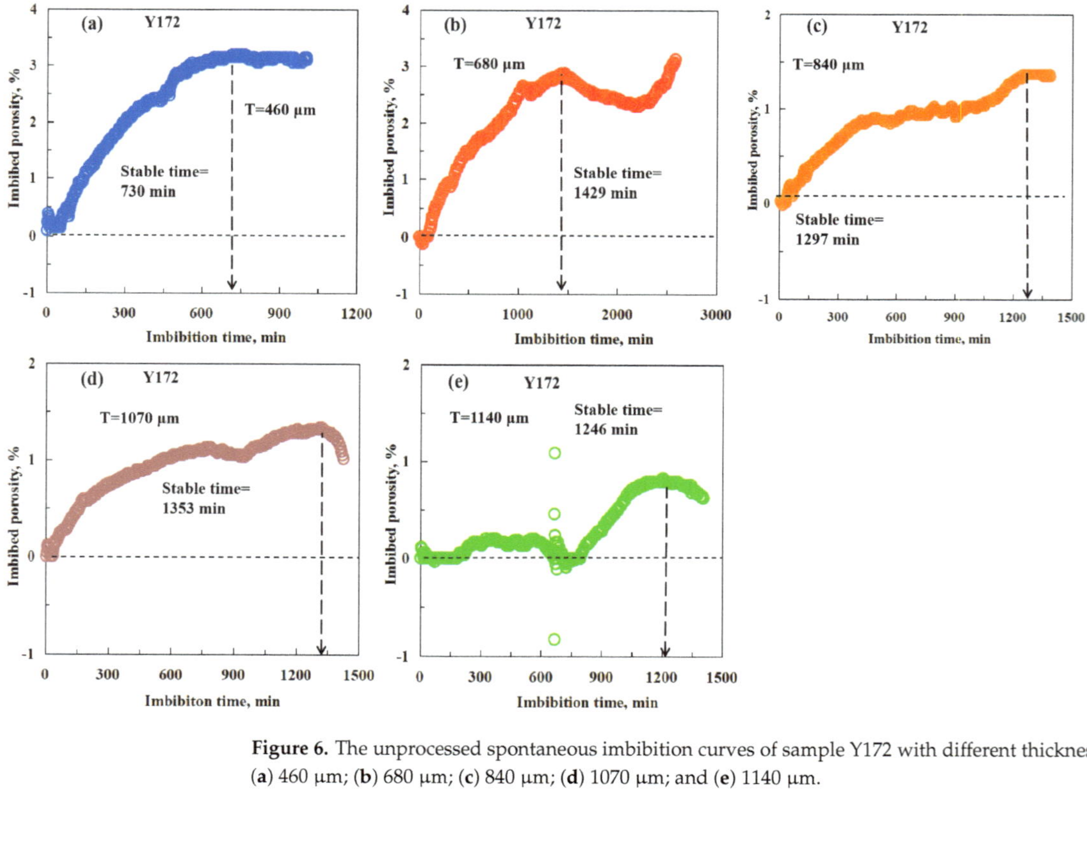

Figure 6. The unprocessed spontaneous imbibition curves of sample Y172 with different thicknesses: (**a**) 460 μm; (**b**) 680 μm; (**c**) 840 μm; (**d**) 1070 μm; and (**e**) 1140 μm.

The imbibition curves of sample Y172 with square-rooted time are shown in Figure 7, which includes samples with different thicknesses. The processed imbibition curves reveal more fluctuation in the initial periods of imbibition. In Figure 7a, hardly any imbibition occurs in the initial period, which is mainly caused by the fact that it requires more interaction time between water and rock before liquid fills into shales. In Figure 7b, the initial imbibed porosity is about zero, which indicates that water has scarcely filled into the rock. Likewise, in Figure 7c, the initial imbibed porosity is also approximately zero and then increases gradually. In Figure 7d, the initial imbibed porosity has some fluctuation but is still near zero. In Figure 7e, the interaction time between the water and rock is long before the fluctuation period ends, and then water fills into the rock quickly. From Figure 7a,e, the ultimate imbibed porosity of sample Y172 is 3.1%, 2.8%, 1.4%, 1.3%, and 0.8%, respectively.

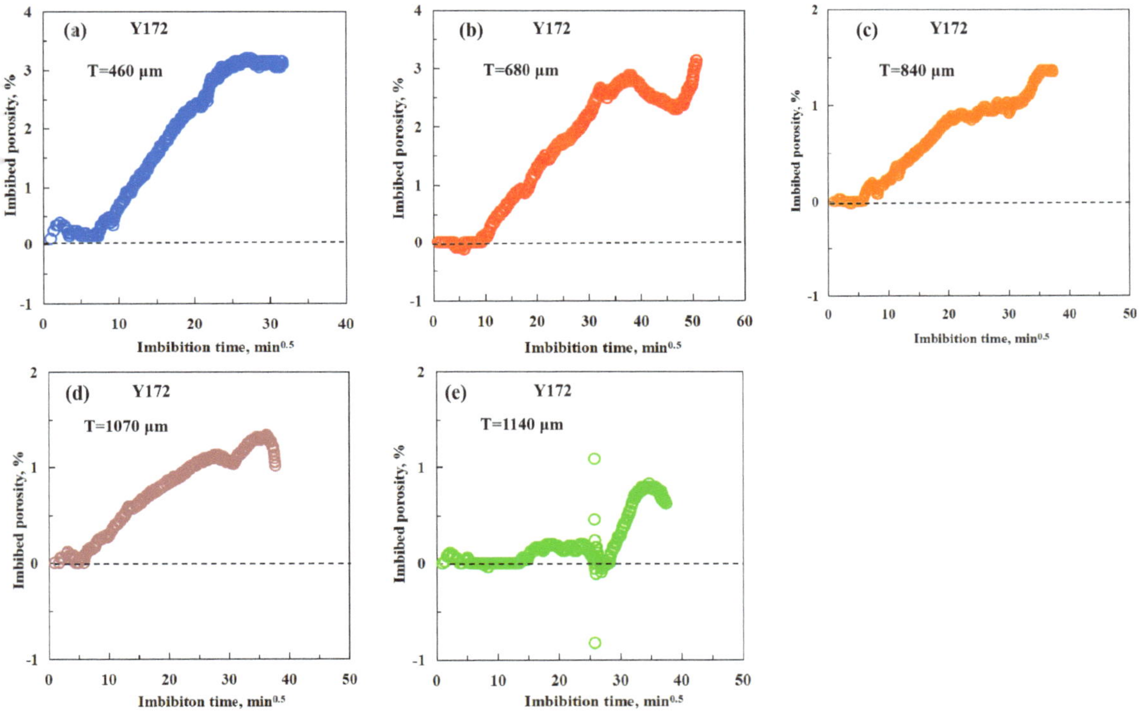

Figure 7. The spontaneous imbibition curves of sample Y172 with different thicknesses: (**a**) 460 μm; (**b**) 680 μm; (**c**) 840 μm; (**d**) 1070 μm; and (**e**) 1140 μm.

At the initial time of the imbibition curves, some special phenomena appear, such as the imbibed rate staying at about zero, the imbibition curve decreasing, and the apparent imbibition porosity dropping to below zero, which influence the selection of the initial curve slopes. Therefore, a turning point aiming to remove the influence of abnormal phenomena during the initial imbibition time is chosen as the starting point for calculating the initial curve slopes. After the turning point, the imbibed porosity increases quickly until reaching stability or the point of declining. The modified imbibition curves of sample Y172 with different thicknesses are shown in Figure 8, whose starting points are adjusted to the turning point. The initial imbibition rates are 0.143, 0.088, 0.050, 0.042, and 0.031, respectively. Upon converting the initial imbibition rate into pore connectivity, the results are 0.033, 0.030, 0.021, 0.023, and 0.018, respectively.

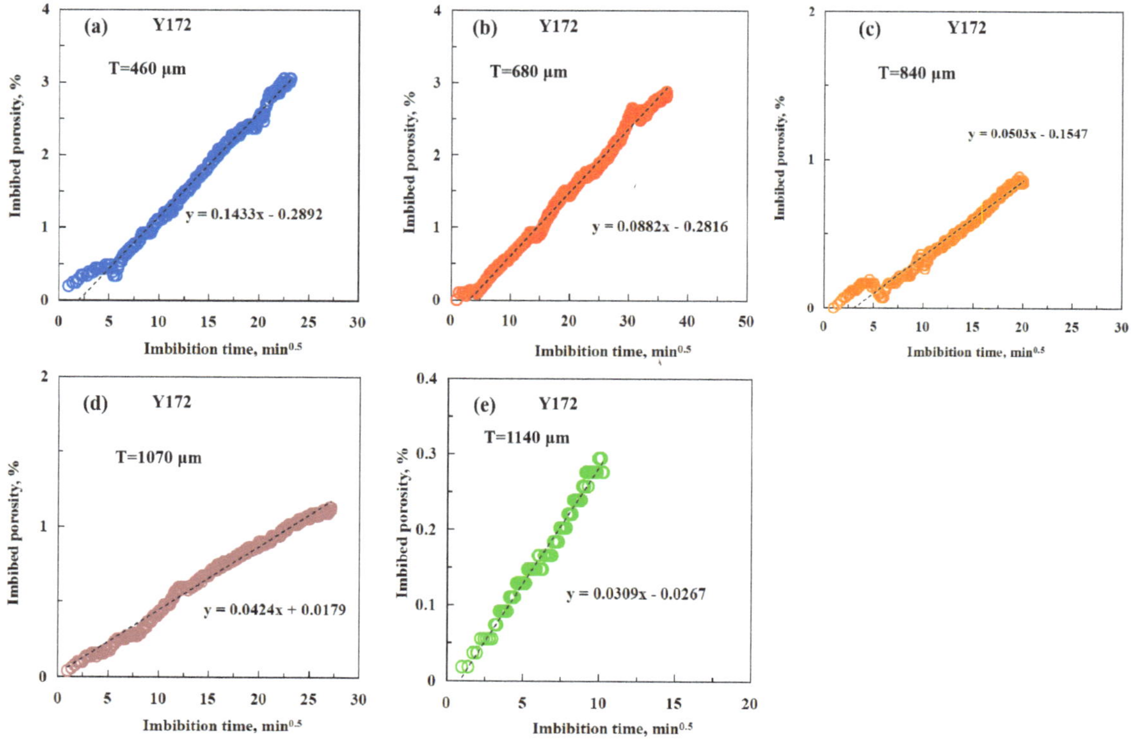

Figure 8. The modified spontaneous imbibition curves of sample Y172 with different thicknesse (**a**) 460 μm; (**b**) 680 μm; (**c**) 840 μm; (**d**) 1070 μm; and (**e**) 1140 μm.

The stable time, ultimate imbibed porosity, and pore connectivity of samples D86- and Y172 were obtained from the imbibition tests, and their variations with the thicknesse of the corresponding samples are shown in Figure 9. Figure 9a–c show the variation in th stable time, ultimate imbibed porosity, and pore connectivity of sample D86-5 with differer thicknesses, while Figure 9d–f show those of sample Y172 with different thicknesses. I Figure 9a, the stable time of D86-5 gradually increases from 46 min to 1361 min with th added thicknesses, with an average of 388 min; the ultimate imbibed porosity varies fror 3.3% to 4.0%, with an average of 3.7%, with the pore connectivity varying from 0.221 t 0.324, with an average of 0.265. In contrast, the stable time of sample Y172 initially increase from 730 min to 1429 min when the thickness increases from 460 μm to 680 μm and the remains at a high level, ranging from 1246 min to 1353 min, with an average of 1211 mi which is much higher than the average stable time of sample D86-5. The ultimate imbibe porosity of sample Y172 decreases dramatically with the increase in thicknesses, rangin from 0.8% to 3.1%, with an average of 1.8%, and the pore connectivity values all remain an extremely low level, changing from 0.018 to 0.033, with an average of 0.026, which ar also much smaller than those of sample D86-5.

The differences between the two groups of samples are due to their different por connectivity values. In Figure 9b,e, the models show the influences caused by differer pore connectivity, with blue representing the water and water-penetrated rock and gre representing the rock that water does not enter. Theoretically, as the natural characterist of shale rocks, the pore connectivity of a shale rock is constant, so the penetration depth also constant with an increase in thickness. In this study, the average pore connectivity c sample D86-5 was 0.265, which is almost ten times larger than that of sample Y172, despi some fluctuations. With better pore connectivity, water can completely penetrate sample D86-5, and the penetration depth is over 970 μm, which is half of the largest thickness of sample D86-5, because water enters from both sides of the sample. As a result, the ultimate imbibed porosity stays around the average value because all the connected pores are accessed. With a small pore connectivity, the penetration depth of sample Y172 is much smaller. In Figure 9d, the stable time increases dramatically from samples at 460 μm to 680 μm, so the penetration depth is between 230 μm and 340 μm. Thus, the ultimate imbibed porosity of sample Y172 declines with thickness because the volume of the sample has increased, but the water cannot enter the volume deeper than 340 μm. The complex pore structure of Y172 influences and declines the efficiency of water entering into the pores with the increase in sample thicknesses, resulting in a long stable time.

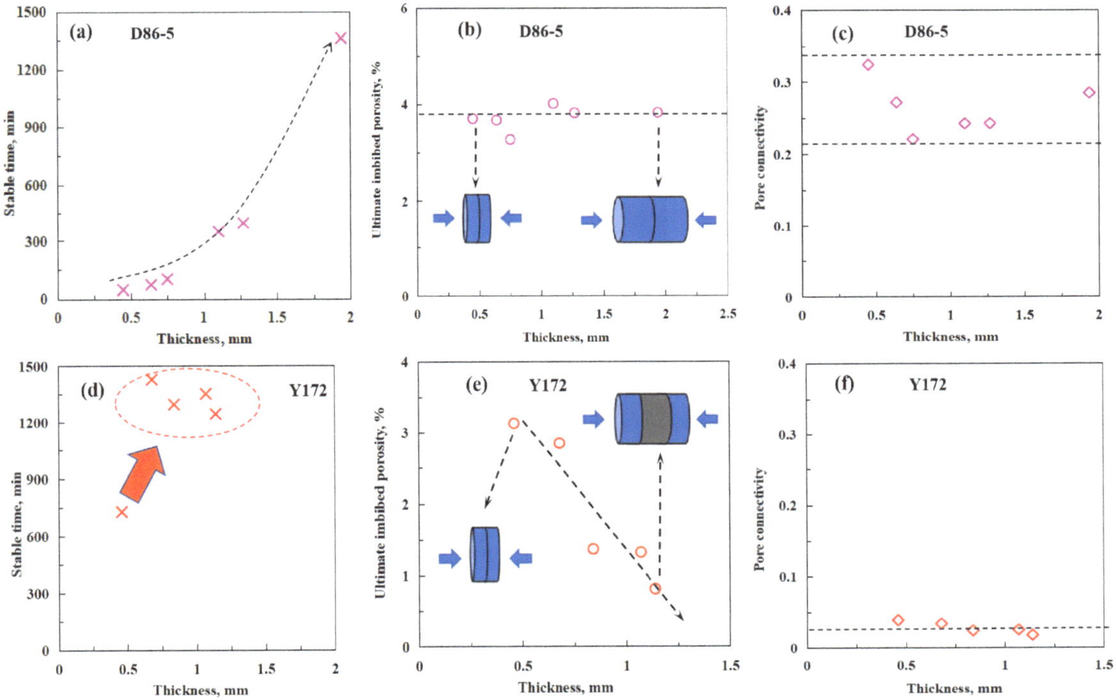

Figure 9. The relationship between stable time, ultimate imbibed porosity, and pore connectivity and thickness: (**a**–**c**) sample D86-5 and (**d**–**f**) sample Y172.

Due to the accurate value of pore connectivity and ultimate imbibed porosity relying on sample thickness, when determining these two parameters, the thickness of the sample should be considered. And when the pore connectivity is small, the ultimate imbibed porosity and stable time are influenced by not only the thickness but also the penetration depth.

3.4. The Imbibition Curves of the Offshore Shale Samples

Based on the results of the quick method of appraising imbibed porosity and pore connectivity, three offshore samples (BW1-1, BW1-2, and BW1-3) were tested to examine the method. Figure 10 exhibits the imbibition curves with the original imbibition time, the imbibition curves with the square-rooted time, and the modified imbibition curves with the square-rooted time, with different colors representing different samples, from which the stable time, ultimate imbibed porosity, and pore connectivity can be acquired. Samples

BW1-1, BW1-2, and BW1-3 have thicknesses of 410 μm, 440 μm, and 400 μm, which ca be seen as approximately the same value and are also the smallest thicknesses of a slice shale rock the experiment machine could produce.

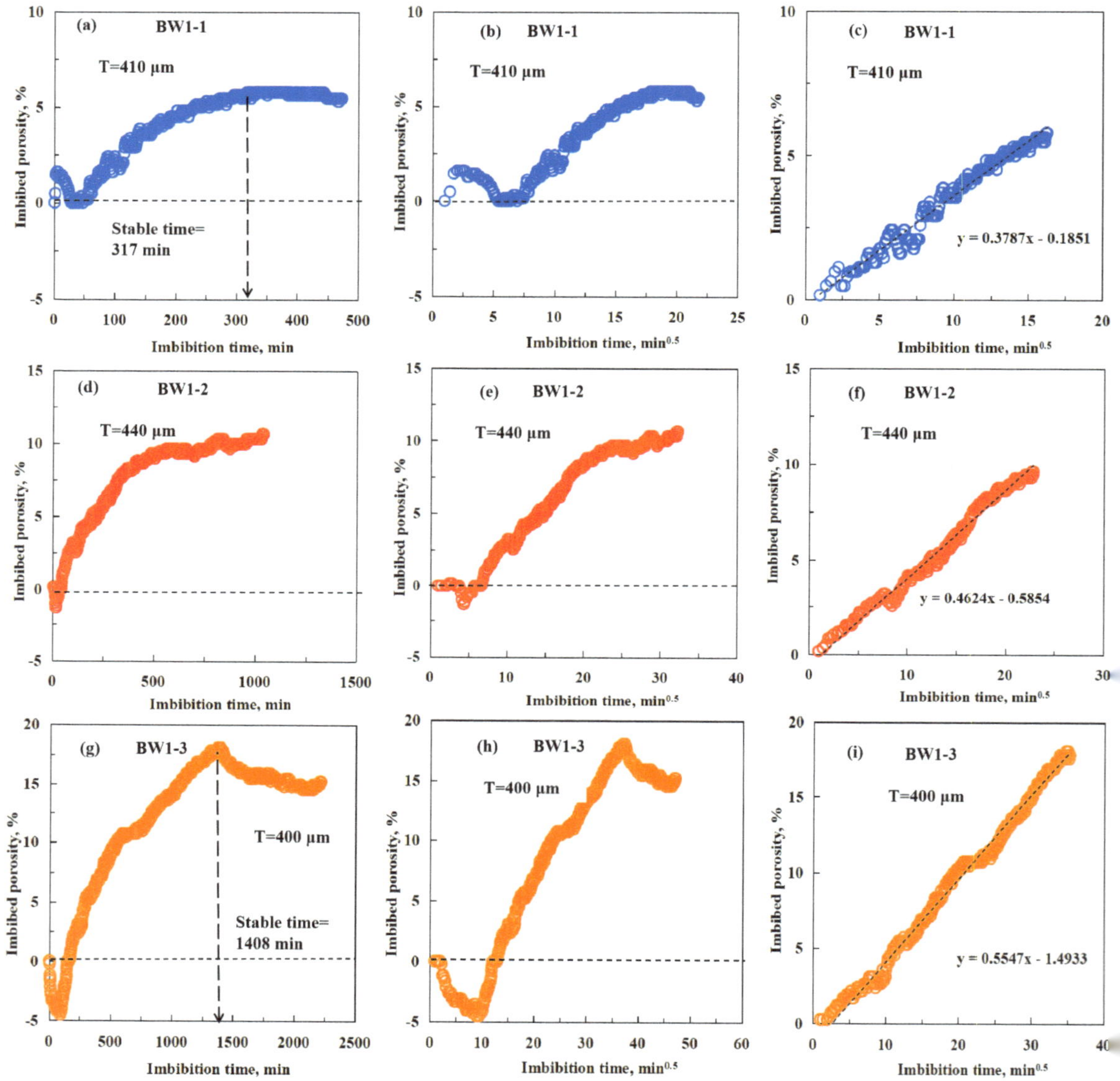

Figure 10. The unprocessed spontaneous imbibition curves, imbibition curves, and modified ir bibition curves of different samples: (**a–c**) BW1-1 with a thickness of 410 μm; (**d–f**) BW1-2 with thickness of 440 μm; and (**g–i**) BW1-3 with a thickness of 400 μm.

In Figure 10a,d,g, the fluctuations are obvious in all three samples during the who imbibition process, and sample BW1-2 does not reach a stable status until the end imbibition. The stable times of BW1-1 and BW1-3 are 317 min and 1408 min, respective which are shorter than a day's time. In Figure 10b, the imbibition curves with the processe time increase sharply and then decrease quickly. This phenomenon is mainly caused b imbibition-induced cracks, which are caused by clay expansion. Induced cracks are main oil-wet and increase the buoyancy of the samples, making the imbibition curves decreas

In Figure 10e, the imbibition curve almost remains stable in the initial period because it requires the interaction time, and then it decreases because of the imbibition-induced cracks. In Figure 10h, the imbibition curve decreases quickly in the initial period, which is also caused by the imbibition-induced cracks. The ultimate imbibed porosity of BW1-1 and BW1-3 is 5.80% and 18.10%, respectively.

The modified imbibition curves with adjustments to the starting point in the shales are shown in Figure 10c,f,i. The initial imbibition rates of samples BW1-1, BW1-2, and BW1-3 are 0.379, 0.462, and 0.555, respectively. Converting the initial imbibition rates to pore connectivity, the pore connectivity values are 0.086, 0.117, and 0.142, respectively. The results of the offshore samples prove the reliability of the quick method, reducing the time of experiments to less than one day and improving the efficiency of evaluating pore connectivity and ultimate imbibed porosity.

3.5. The Distribution of the Three Parameters

Based on the previous analysis, the samples with a thickness of about 400 μm were processed to compare the corresponding parameters. The distribution of pore connectivity and ultimate imbibed porosity of five chosen samples are shown in Figure 11. In Figure 11a, the stable times of samples D86-5 (from Qingshankou Fm), Y172 (from Shahejie Fm), and BW1-1 and BW1-3 (from Liushagang Fm) are 46 min, 730 min, and 863 min, respectively. BW1-2 never reaches a stable status, whereas BW1-3 has the longest stable time, and D86-5 has the smallest. The average ultimate imbibed porosity of the three Fms is 3.7%, 3.1%, and 12.0%, respectively, among which sample BW1-2 has no ultimate imbibed porosity, whereas the sample from Shahejie Fm has the smallest ultimate imbibed porosity, and the samples from Liushagang Fm have the largest. In Figure 11c, the average pore connectivity values of the three Fms are 0.324, 0.033, and 0.097, respectively, where the sample from Qingshankou Fm has the biggest pore connectivity and the sample from Shahejie Fm has the smallest. In Figure 11, the heterogeneity in the shale rocks still has a significant influence on the appraisal of ultimate imbibed porosity and pore connectivity.

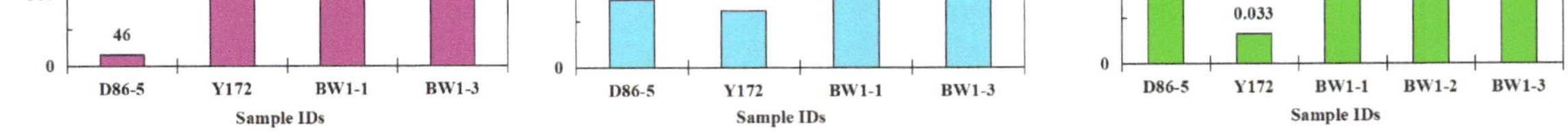

Figure 11. The distribution of three parameters: (**a**) stable time, (**b**) ultimate imbibed porosity, and (**c**) pore connectivity.

4. Conclusions

This study aimed to bring out a quick method to investigate the pore connectivity and ultimate imbibed porosity of shale, leading to the following conclusions:

(1) A quick method was established to measure the pore connectivity and ultimate imbibed porosity of shale by using spontaneous imbibition with thin samples. Within the time of a day, the imbibition results of 400 μm samples can be acquired.

(2) The three parameters change with sample thicknesses. For sample D86-5, the average pore connectivity is 0.265 and the penetration depth is large, so water enters all

connected pores, resulting in a stable ultimate imbibed porosity of around 3.7%. An the stable time increases with the addition of thicknesses. For sample Y172, when th thickness increases, the average pore connectivity is 0.026 and the penetration dept is much smaller with inaccessible pores, so the ultimate imbibed porosity decline with increasing thickness from 3.1% to 0.8%. The stable time of Y172 increases wit the thickness from 460 μm to 680 μm before fluctuating around 1211 min, so th penetration depth of Y172 is between 230 μm and 340 μm.

(3) Comparing the samples around 400 μm, the sample from the Qingshankou Fm ha the smallest stable time and the largest pore connectivity, but the ultimate imbibe porosity is only 3.7%. The sample from the Shahejie Fm has a longer stable time an the smallest ultimate imbibed porosity and pore connectivity. The samples from th Liushagang Fm have the longest stable time and ultimate imbibed porosity and a average pore connectivity of 0.097.

For further research on pore connectivity and ultimate imbibed porosity with ou method, different imbibition fluids and samples are suggested. In this study, we only use deionized water as the imbibition fluid, which can only appraise water-wet pores. Oil-we fluid can investigate oil-wet pores, and amphiphilic fluid can examine mixed-wet pore Simultaneously, fracturing fluid and formation water are also suggested. We only use shale as tested samples, and more samples are suggested, such as tight sandstone, carbonat clastic volcanic rock, and so on. Furthermore, more pore structure characterizations shoul be conducted, such as nitrogen adsorption (NA), mercury intrusion porosimetry (MIP nuclear magnetic resonance (NMR), and so on, which could help to analyze the controllin factors of pore connectivity and ultimate imbibed porosity.

Author Contributions: Conceptualization, Z.H. and M.M.; formal analysis, Z.H., M.M., K.D., J.B Q.W., and X.L.; investigation, Z.H.; resources, M.M.; writing—original draft, Z.H.; writing—revie and editing, M.M., K.D., J.B., Q.W., and X.L. All authors have read and agreed to the publishe version of the manuscript.

Funding: This research was funded by the National Natural Science Foundation of China (Grar Nos. 42372167 and 42102160), the "CUG Scholar" Scientific Research Funds at China University c Geosciences (Wuhan) (Project No.2022077), and the Fundamental Research Funds for the Centra Universities, China University of Geosciences (Wuhan) (No. CUG240632).

Data Availability Statement: The data is contained within the article.

Conflicts of Interest: Author Xingchen Liu was employed by the company Cementing Company Zhongyuan Petroleum Engineering Co., Ltd. under China Petrochemical Company Limited. Th remaining authors declare that the research was conducted in the absence of any commercial c financial relationships that could be construed as a potential conflict of interest.

References

1. Marchionna, M. Fossil energy: From conventional oil and gas to the shale revolution. *EPJ Web Conf.* **2018**, *189*, 00004. [CrossRe
2. National Energy Administration. China's Oil and Gas Industry Analysis and Outlook Series Blue Book. 2024. Available onlin http://www.nea.gov.cn/2024-04/26/c_1310772762.htm (accessed on 26 April 2024).
3. Loucks, R.; Reed, R.; Ruppel, S.; Hammes, U. Spectrum of pore types and networks in mudrocks and a descriptive classificatic for matrix related pores. *AAPG Bull.* **2012**, *96*, 1071–1098. [CrossRef]
4. Chen, Y.; Jiang, C.; Leung, J.Y.; Wojtanowicz, A.K.; Zhang, D. Multiscale characterization of shale pore-fracture system: Geologic controls on gas transport and pore size classification in shale reservoirs. *J. Pet. Sci. Eng.* **2021**, *202*, 108442. [CrossRef]
5. Zhang, P.F.; Lu, S.F.; Li, J.Q.; Chang, X.C.; Zhang, J.J.; Pang, Y.M.; Lin, Z.Z.; Chen, G.; Yin, Y.J.; Liu, Y.Q. Quantitative characteriz tion of shale pore connectivity and controlling factors using spontaneous imbibition combined with nuclear magnetic resonanc T2 and T1-T2. *Pet. Sci.* **2023**, *20*, 1947–1960. [CrossRef]

Ji, W.; Hao, F.; Gong, F.; Zhang, J.; Bai, Y.; Liang, C.; Tian, J. Petroleum migration and accumulation in a shale oil system of the Upper Cretaceous Qingshankou Formation in the Songliao Basin, northeastern China. *AAPG Bull.* **2024**, *108*, 1611–1648. [CrossRef]

Jia, C.; Xiao, B.; Lijun, Y.; Yang, Z.; Yili, K. Experimental study of water imbibition characteristics of the lacustrine shale in Sichuan Basin. *Petroleum* **2023**, *9*, 572–578. [CrossRef]

Wei, J.; Zhang, A.; Jiangtao, L.; Demiao, S.; Xiaofeng, Z. Study on microscale pore structure and bedding fracture characteristics of shale oil reservoir. *Energy* **2023**, *278*, 127829. [CrossRef]

Yuan, Y.; Rezaee, R.; Mei-Fu, Z.; Stefan, I. A comprehensive review on shale studies with emphasis on nuclear magnetic resonance (NMR) technique. *Gas Sci. Eng.* **2023**, *120*, 205163. [CrossRef]

Wang, Y.; Wang, Z.; Zhengchen, Z.; Shanshan, Y.; Hong, Z.; Guoqing, Z.; Feifei, L.; Lele, F.; Kouqi, L.; Liangliang, J. Recent techniques on analyses and characterizations of shale gas and oil reservoir. *Energy Rev.* **2024**, *3*, 100067. [CrossRef]

Zhang, J.; Xiao, X.; Jianguo, W.; Wei, L.; Denglin, H.; Chenchen, W.; Yu, L.; Yan, X.; Xiaochan, Z. Pore structure and fractal characteristics of coal-bearing Cretaceous Nenjiang shales from Songliao Basin, Northeast China. *J. Nat. Gas. Geosci.* **2024**, *9*, 197–208. [CrossRef]

Yasin, Q.; Liu, B.; Mengdi, S.; Ghulam Mohyuddin, S.; Atif, I.; Mariusz, M.; Naser, G.; Yan, M.; Xiaofei, F. Automatic pore structure analysis in organic-rich shale using FIB-SEM and attention U-Net. *Fuel* **2024**, *358*, 130161. [CrossRef]

Norman, R.M.; Geoffrey, M. Recovery of oil by spontaneous imbibition. *Curr. Opin. Colloid. Interface Sci.* **2001**, *6*, 321–337. [CrossRef]

Ding, Y.; Liu, X.; Liang, L.; Xiong, J.; Hou, L. Experimental and model analysis on shale spontaneous imbibition and its influence factors. *J. Nat. Gas Sci. Eng.* **2022**, *99*, 104462. [CrossRef]

Meng, M.; Ge, H.; Yinghao, S.; Qinhong, H.; Longlong, L.; Zhiye, G.; Tonghui, T.; Jing, C. The effect of clay-swelling induced cracks on imbibition behavior of marine shale reservoirs. *J. Nat. Gas. Sci. Eng.* **2020**, *83*, 103525. [CrossRef]

Meng, M.; Ge, H.; Yinghao, S.; Fei, R.; Wenming, J. A novel method for monitoring the imbibition behavior of clay-rich shale. *Energy Rep.* **2020**, *6*, 1811–1818. [CrossRef]

Hu, J.; Zhao, H.; Du, X.; Zhang, Y. An analytical model for shut-in time optimization after hydraulic fracturing in shale oil reservoirs with imbibition experiments. *J. Pet. Sci. Eng.* **2022**, *210*, 110055. [CrossRef]

Lan, Q.; Xu, M.; Binazadeh, M.; Dehghanpour, H.; Wood, J.M. A comparative investigation of shale wettability: The significance of pore connectivity. *J. Nat. Gas Sci. Eng.* **2015**, *27*, 1174–1188. [CrossRef]

Vasheghani Farahani, M.; Mousavi Nezhad, M. On the effect of flow regime and pore structure on the flow signatures in porous media. *Phys. Fluids* **2022**, *34*, 115139. [CrossRef]

Cheng, Z.; Tong, S.; Shang, X.; Yu, J.; Li, X.; Dou, L. Lattice Boltzmann simulation of counter-current imbibition of oil and water in porous media at the equivalent capillarity. *AIP Adv.* **2024**, *14*, 085314. [CrossRef]

Alvarado, V.; Manrique, E. Chapter 2—Enhanced Oil Recovery Concepts. In *Enhanced Oil Recovery*; Alvarado, V., Manrique, E., Eds.; Gulf Professional Publishing: Boston, MA, USA, 2010; pp. 7–16.

Yuan, S.; Han, H.; Wang, H.; Luo, J.; Wang, Q.; Lei, Z.; Xi, C.; Li, J. Research progress and potential of new enhanced oil recovery methods in oilfield development. *Pet. Explor. Dev.* **2024**, *51*, 963–980. [CrossRef]

Valluri, S.; Claremboux, V.; Kawatra, S. Opportunities and challenges in CO2 utilization. *J. Environ. Sci.* **2022**, *113*, 322–344. [CrossRef] [PubMed]

Cheng, Z.; Ning, Z.; Yu, X.; Wang, Q.; Zhang, W. New insights into spontaneous imbibition in tight oil sandstones with NMR. *J. Pet. Sci. Eng.* **2019**, *179*, 455–464. [CrossRef]

Hamid, S.; Syed Mohammad, M.; Reza, R.; Ali, S. Conventional methods for wettability determination of shales: A comprehensive review of challenges, lessons learned, and way forward. *Mar. Pet. Geol.* **2021**, *133*, 105288. [CrossRef]

Lin, Z.; Hu, Q.; Na, Y.; Shengyu, Y.; Huimin, L.; Jing, C. Nanopores-to-microfractures flow mechanism and remaining distribution of shale oil during dynamic water spontaneous imbibition studied by NMR. *Geoenergy Sci. Eng.* **2024**, *241*, 213202. [CrossRef]

Wang, X.; Wang, M.; Li, Y.; Zhang, J.; Li, M.; Li, Z.; Guo, Z.; Li, J. Shale pore connectivity and influencing factors based on spontaneous imbibition combined with a nuclear magnetic resonance experiment. *Mar. Pet. Geol.* **2021**, *132*, 105239. [CrossRef]

Liu, Z.; Bai, B.; Wang, Y.; Qu, H.; Xiao, Q.; Zeng, S. Spontaneous imbibition characteristics of slickwater and its components in Longmaxi shale. *J. Pet. Sci. Eng.* **2021**, *202*, 108599. [CrossRef]

Liu, J.; Sheng, J.J. Experimental investigation of surfactant enhanced spontaneous imbibition in Chinese shale oil reservoirs using NMR tests. *J. Ind. Eng. Chem.* **2019**, *72*, 414–422. [CrossRef]

Sun, M.; Yu, B.; Qinhong, H.; Rui, Y.; Yifan, Z.; Bo, L. Pore connectivity and tracer migration of typical shales in south China. *Fuel* **2017**, *203*, 32–46. [CrossRef]

Lin, M.; Xi, K.; Cao, Y.; Liu, Q.; Zhang, Z.; Li, K. Petrographic features and diagenetic alteration in the shale strata of the Permian Lucaogou Formation, Jimusar sag, Junggar Basin. *J. Pet. Sci. Eng.* **2021**, *203*, 108684. [CrossRef]

32. Wang, W.; Xie, Q.; Li, J.; Sheng, G.; Lun, Z. Fracturing fluid imbibition impact on gas-water two phase flow in shale fracture-matr system. *Nat. Gas Ind. B* **2023**, *10*, 323–332. [CrossRef]
33. Yang, R.; Hu, Q.; Yi, J.; Zhang, B.; He, S.; Guo, X.; Hou, Y.; Dong, T. The effects of mineral composition, TOC content an pore structure on spontaneous imbibition in Lower Jurassic Dongyuemiao shale reservoirs. *Mar. Pet. Geol.* **2019**, *109*, 268–27 [CrossRef]
34. Yang, R.; Hu, Q.; Sheng, H.; Fang, H.; Xusheng, G.; Jizheng, Y.; Xipeng, H. Pore structure, wettability and tracer migration in fo leading shale formations in the Middle Yangtze Platform, China. *Mar. Pet. Geol.* **2018**, *89*, 415–427. [CrossRef]
35. Davudov, D.; Moghanloo, R.G.; Zhang, Y. Interplay between pore connectivity and permeability in shale sample. *Int. J. Coal Ge* **2020**, *220*, 103427. [CrossRef]
36. Yang, J.; Yang, B.; Liang, W.; Junliang, P.; Yunhui, F.; Yun, T. Invasion depths of fracturing fluid imbibition displacement in matr pores of Da'an Zhai shale oil reservoirs in central Sichuan Basin. *Pet. Geol. Recovery Effic.* **2023**, *30*, 84–91.
37. Ye, Y.; Tang, S.; Zhaodong, X.; Dexin, J.; Yang, D. Quartz types in the Wufeng-Longmaxi Formations in southern Chin Implications for porosity evolution and shale brittleness. *Mar. Pet. Geol.* **2022**, *137*, 105479. [CrossRef]
38. Tuzingila, R.M.; Kong, L.; Kasongo, R.K. A review on experimental techniques and their applications in the effects of miner content on geomechanical properties of reservoir shale rock. *Rock Mech. Bull.* **2024**, *3*, 100110. [CrossRef]
39. Xiao, D.; Zheng, L.; Jilin, X.; Min, W.; Rui, W.; Xiaodie, G.; Xueyi, G. Coupling control of organic and inorganic rock componen on porosity and pore structure of lacustrine shale with medium maturity: A case study of the Qingshankou Formation in th southern Songliao Basin. *Mar. Pet. Geol.* **2024**, *164*, 106844. [CrossRef]
40. Liu, B.; Mastalerz, M.; Schieber, J. SEM petrography of dispersed organic matter in black shales: A review. *Earth-Sci. Rev.* **202** *224*, 103874. [CrossRef]
41. Wang, Q.; Li, Y.; Utley, J.E.P.; Gardner, J.; Liu, B.; Hu, J.; Shao, L.; Wang, X.; Gao, F.; Liu, D.; et al. Terrestrial dominance of organ carbon in an Early Cretaceous syn-rift lake and its correlation with depositional sequences and paleoclimate. *Sediment. Geol.* **202** *455*, 106472. [CrossRef]
42. Meng, M.; Ge, H.; Shen, Y.; Ji, W. Evaluation of the Pore Structure Variation During Hydraulic Fracturing in Marine Sha Reservoirs. *J. Energy Resour. Technol.* **2020**, *143*, 083002. [CrossRef]
43. Washburn, E.W. The Dynamics of Capillary Flow. *Phys. Rev.* **1921**, *17*, 273–283. [CrossRef]
44. Handy, L.L. Determination of Effective Capillary Pressures for Porous Media from Imbibition Data. *Trans. AIME* **1960**, *219*, 75–8 [CrossRef]
45. Hu, Q.; Ewing, R.P.; Dultz, S. Low pore connectivity in natural rock. *J. Contam. Hydrol.* **2012**, *133*, 76–83. [CrossRef] [PubMed]
46. Zheng, Y.; Liao, Y.; Wang, J.; Xiong, Y.; Wang, Y.; Peng, P.a. Factors controlling the heterogeneity of shale pore structure and sha gas production of the Wufeng–Longmaxi shales in the Dingshan plunging anticline of the Sichuan Basin, China. *Int. J. Coal Ge* **2024**, *282*, 104434. [CrossRef]
47. Hao, L.; Qing, L.; Yue, D.; Wu, Y.; Gao, J.; Wu, S.; Wang, W.; Li, M.; An, K. Quantitative characterization and formation mechanis of the pore system heterogeneity: Examples from organic-rich laminated and organic-poor layered shales of the upper triass chang 7 member in the southern Ordos Basin, China. *Mar. Pet. Geol.* **2023**, *147*, 105999. [CrossRef]
48. Ge, H.-K.; Yang, L.; Shen, Y.-H.; Ren, K.; Meng, F.-B.; Ji, W.-M.; Wu, S. Experimental investigation of shale imbibition capacity an the factors influencing loss of hydraulic fracturing fluids. *Pet. Sci.* **2015**, *12*, 636–650. [CrossRef]

MDPI AG
Grosspeteranlage 5
4052 Basel
Switzerland
Tel.: +41 61 683 77 34

Journal of Marine Science and Engineering Editorial Office
E-mail: jmse@mdpi.com
www.mdpi.com/journal/jmse

www.ingramcontent.com/pod-product-compliance
Lightning Source LLC
Chambersburg PA
CBHW041417210926
51850CB00013B/3297
9783725836710